# Audio Engineering

# The Newnes Know It All Series

**PIC Microcontrollers: Know It All**
Lucio Di Jasio, Tim Wilmshurst, Dogan Ibrahim, John Morton, Martin Bates, Jack Smith, D.W. Smith, and Chuck Hellebuyck
ISBN: 978-0-7506-8615-0

**Embedded Software: Know It All**
Jean Labrosse, Jack Ganssle, Tammy Noergaard, Robert Oshana, Colin Walls, Keith Curtis, Jason Andrews, David J. Katz, Rick Gentile, Kamal Hyder, and Bob Perrin
ISBN: 978-0-7506-8583-2

**Embedded Hardware: Know It All**
Jack Ganssle, Tammy Noergaard, Fred Eady, Lewin Edwards, David J. Katz, Rick Gentile, Ken Arnold, Kamal Hyder, and Bob Perrin
ISBN: 978-0-7506-8584-9

**Wireless Networking: Know It All**
Praphul Chandra, Daniel M. Dobkin, Alan Bensky, Ron Olexa, David A. Lide, and Farid Dowla
ISBN: 978-0-7506-8582-5

**RF & Wireless Technologies: Know It All**
Bruce Fette, Roberto Aiello, Praphul Chandra, Daniel Dobkin, Alan Bensky, Douglas Miron, David Lide, Farid Dowla, and Ron Olexa
ISBN: 978-0-7506-8581-8

**Electrical Engineering: Know It All**
Clive Maxfield, Alan Bensky, John Bird, W. Bolton, Izzat Darwazeh, Walt Kester, M.A. Laughton, Andrew Leven, Luis Moura, Ron Schmitt, Keith Sueker, Mike Tooley, D.F. Warne, and Tim Williams
ISBN: 978-1-85617-528-9

**Audio Engineering: Know It All**
Douglas Self, Richard Brice, Ben Duncan, John Linsley Hood, Ian Sinclair, Andrew Singmin, Don Davis, Eugene Patronis, and John Watkinson
ISBN: 978-1-85617-526-5

**Circuit Design: Know It All**
Darren Ashby, Bonnie Baker, Stuart Ball, John Crowe, Barrie Hayes-Gill, Ian Grout, Ian Hickman, Walt Kester, Ron Mancini, Robert A. Pease, Mike Tooley, Tim Williams, Peter Wilson, and Bob Zeidman
ISBN: 978-1-85617-527-2

**Test and Measurement: Know It All**
Jon Wilson, Stuart Ball, G.M.S de Silva, Tony Fischer-Cripps, Dogan Ibrahim, Kevin James, Walt Kester, Michael Laughton, Chris Nadovich, Alex Porter, Ed Ramsden, Steve Scheiber, Douglas Warne, and Tim Williams
ISBN: 978-1-85617-530-2

**Wireless Security: Know It All**
Praphul Chandra, Alan Bensky, Tony Bradley, Chris Hurley, Steve Rackley, James Ransome, John Rittinghouse, Timothy Stapko, George Stefanek, Frank Thornton, and Jon Wilson
ISBN: 978-1-85617-529-6

For more information on these and other Newnes titles visit: www.newnespress.com

# *Audio Engineering*

Douglas Self
Richard Brice
Ben Duncan
John Linsley Hood
Ian Sinclair
Andrew Singmin
Don Davis
Eugene Patronis
John Watkinson

ELSEVIER

AMSTERDAM • BOSTON • HEIDELBERG • LONDON • NEW YORK • OXFORD
PARIS • SAN DIEGO • SAN FRANCISCO • SINGAPORE • SYDNEY • TOKYO
Newnes is an imprint of Elsevier

Newnes

Newnes is an imprint of Elsevier
30 Corporate Drive, Suite 400, Burlington, MA 01803, USA
Linacre House, Jordan Hill, Oxford OX2 8DP, UK

 Recognizing the importance of preserving what has been written, Elsevier prints its books on
acid-free paper whenever possible.

**Library of Congress Cataloging-in-Publication Data**
Audio engineering : know it all / by Ian Sinclair … [et al.].
    p. cm.
  Includes bibliographical references and index.
  ISBN 978-1-85617-526-5 (alk. paper)
  1. Sound—Recording and reproducing—Handbooks, manuals, etc.   2. Sound—Recording
and reproducing—Digital techniques—Handbooks, manuals, etc.   I. Sinclair, Ian Robertson.
  TK7881.4.A9235 2008
  621.389′3—dc22
                                                                          2008033305

**British Library Cataloguing-in-Publication Data**
A catalogue record for this book is available from the British Library.

ISBN: 978-1-85617-526-5

For information on all Newnes publications
visit our Web site at www.books.elsevier.com

Typeset by Charon Tec Ltd., A Macmillan Company (www.macmillansolutions.com)

Transferred to Digital Printing, 2010

Printed and bound in the United Kingdom

Working together to grow
libraries in developing countries

www.elsevier.com  |  www.bookaid.org  |  www.sabre.org

ELSEVIER    BOOK AID    Sabre Foundation
            International

# Contents

# *About the Authors*

**Dave Berriman** (Chapter 25) is a contributor to *Audio and Hi-Fi Handbook*.

**Richard Brice** (Chapters 18, 19, 20, 26, 27, and 28) is the author of *Music Engineering*. He has combined a career as composer, music arranger, and producer with a management career in the broadcast television business. He is currently President of Miranda Technologies Asia, based in Hong Kong. He taught Sound Engineering as a Visiting Fellow of Oxford Brookes University and is the author of three books and many articles about television and audio.

**Don Davis** (Chapters 2, 3, and 22) is the co-author of *Sound System Engineering, Third Edition*. Davis is the co-founder of Synergetic Audio Concepts, USA. Don has received a Fellowship Award from the AES for his work in sound system design and audio education.

**Ben Duncan** (Chapters 8 and 24) is the author of *High Performance Audio Power Amplifiers*. Duncan is a *prolific* British polymath audio scientist/researcher, independent electronics engineer; manufacturing trouble-shooter; music technologist; author (900+ articles); electronic and audio product designer (200+), including high-end audio kits; and inventor, inspired by a very wide range of music. As a landowner, Duncan has created organic gardens, a nature reserve, and parkland with 2000 trees. He organized a rock concert in 1974; today, music events are held in the park. Duncan's audio designs are recognized for engineering finesse and exceptional sonic qualities, with equipment he co-designed and also his own bespoke units being known across the diversity of "high-end" hi-fi, recording studios, show production, and by many astute musicians, sound engineers, academics and physicists. As senior engineer at BDResearch, he operates highly-resourced test labs, with hundreds of restored legacy instruments used to make new discoveries. See BDResearch's websites and 1100+ 3rd-party websites and forum mentions.

**Stan Kelley** (Chapter 23) is a contributor to *Audio and Hi-Fi Handbook*.

**John Linsley Hood** (Chapters 5, 6, 7, 9, 11, 16, and 30) is the author of *Audio Electronics* and *Valve and Transistor Audio Amplifiers*. Linsley Hood was head of the electronics research laboratories at British Cellophane, for nearly 25 years. He worked on many instrumentation projects, including width gauges and moisture meters, and made several inventions which were patented under the Cellophane name. Prior to his work at British Cellophane he worked in the electronics laboratory of the Department of Atomic Energy at Sellafield, Cumbria. He studied at Reading University after serving in the military as a radar mechanic. Linsley Hood published more than 30 technical feature articles in *Wireless World* magazine and its later incarnation *Electronics World*. He also contributed to numerous other magazines, including *Electronics Today*.

**Peter Mapp BSc, MSc, CPhys, CEng, FIOA, FASA, FAES, MinstP, FinstSCE, MIEE** (Chapter 29) is a contributor to *Audio and Hi-Fi Handbook*. Mapp is a principal of Peter Mapp Associates, an acoustic consultancy based in Colchester, England, which specializes in the fields of room acoustics, electro-acoustics, and sound system design. Peter holds degrees in applied physics and acoustics and has particular interests in the fields of speech intelligibility of sound systems, small room acoustics, and the interaction between loudspeakers and rooms. He has authored and presented many papers and articles on these subjects both in Europe and the USA. Peter is well known for his research into speech intelligibility and its measurement and developing new measurement techniques in relation to room acoustics.

He is a regular contributor to the audio technical press, having written over 100 articles and technical papers, and is a contributing author to several international audio and acoustics reference books.

**Allen Mornington-West** (Chapter 15) is a contributor to *Audio and Hi-Fi Handbook*.

**Eugene Patronis** (Chapter 2, 3, and 22) is the co-author of *Sound System Engineering, Third Edition*. Patronis is Professor of Physics Emeritus at the Georgia Institute of Technology in Atlanta, Georgia, USA. He has also served as an industrial and governmental consultant in the fields of acoustics and electronics.

**Douglas Self** (Chapters 10, 12, and 13) is the author of *Audio Power Amplifier Design Handbook*. He is a senior designer of high-end audio amplifiers and a contributor to *Electronics World* magazine

**Ian Sinclair** (Chapters 15, 17, 21, 23, 25, and 29), author of *Audio and Hi-Fi Handbook*, was born in 1932 and educated at Madras College, St.Andrews and then at the University of St. Andrews, majoring in chemistry. In 1956 a fascination with the hobby of electronics led him to a post of junior engineer with English Electric Valve Co. (in Essex), where he was researching vacuum electron-optical devices. In 1966 he moved to the position of lecturer in Physics and Electronics at Braintree College, and began writing articles and books on electronics and computing. In 1983 he resigned from college to become a freelance author, as he still is today.

**Andrew Singmin** (Chapter 4) is the author of *Practical Audio Amplifier Circuit Projects*. He currently is a Quality Assurance Manager at Accelerix in Ottawa, Canada, with over 25 years of experience in electronics/semiconductor device technology. Singmin has written for *Popular Electronics* and the *Electronics Handbook*, as well as *Beginning Analog Electronics Through Projects Second Edition*, *Beginning Digital Electronics Through Projects*, *Modern Electronics Soldering Techniques*, *Dictionary of Modern Electronics Technology*, and *Practical Audio Amplifier Circuit Projects*.

**John Watkinson** (Chapters 1, 14, and 17) is the author of *Introduction to Digital Audio, Second Edition* and was a contributor to *Audio and Hi-Fi Handbook*. Watkinson is an international consultant in audio, video, and data recording.

# PART 1

# Fundamentals of Sound

# CHAPTER 1

# Audio Principles

John Watkinson

## 1.1 The Physics of Sound

Sound is simply an airborne version of vibration. The air which carries sound is a mixture of gases. In gases, the molecules contain so much energy that they break free from their neighbors and rush around at high speed. As Figure 1.1(a) shows, the innumerable elastic collisions of these high-speed molecules produce pressure on the walls of any gas container. If left undisturbed in a container at a constant temperature, eventually the pressure throughout would be constant and uniform.

Sound disturbs this simple picture. Figure 1.1(b) shows that a solid object which moves *against* gas pressure increases the velocity of the rebounding molecules, whereas in Figure 1.1(c) one moving *with* gas pressure reduces that velocity. The average velocity and the displacement of all the molecules in a layer of air near a moving body is the same as the velocity and displacement of the body. Movement of the body results in a local increase or decrease in pressure of some kind. Thus sound is both a pressure and a velocity disturbance.

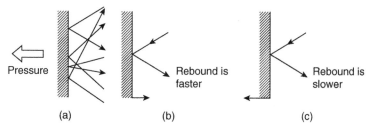

Pressure                    Rebound is              Rebound is
                            faster                  slower

(a)                 (b)                 (c)

**Figure 1.1: (a) The pressure exerted by a gas is due to countless elastic collisions between gas molecules and the walls of the container. (b) If the wall moves against the gas pressure, the rebound velocity increases. (c) Motion with the gas pressure reduces the particle velocity.**

Despite the fact that a gas contains endlessly colliding molecules, a small mass or *particle* of gas can have stable characteristics because the molecules leaving are replaced by new ones with identical statistics. As a result, acoustics seldom need to consider the molecular structure of air and the constant motion can be neglected. Thus when particle velocity and displacement are considered, this refers to the average values of a large number of molecules. In an undisturbed container of gas, the particle velocity and displacement will both be zero everywhere.

When the volume of a fixed mass of gas is reduced, the pressure rises. The gas acts like a spring; it is compliant. However, a gas also has mass. Sound travels through air by an interaction between the mass and the compliance. Imagine pushing a mass via a spring. It would not move immediately because the spring would have to be compressed in order to transmit a force. If a second mass is connected to the first by another spring, it would start to move even later. Thus the speed of a disturbance in a mass/spring system depends on the mass and the stiffness. Sound travels through air without a net movement of the air.

The speed of sound is proportional to the square root of the absolute temperature. On earth, temperature changes with respect to absolute zero ($-273°C$) also amount to around 1% except in extremely inhospitable places. The speed of sound experienced by most of us is about 1000 ft per second or 344 m per second.

## 1.2  Wavelength

Sound can be due to a one-off event known as percussion, or a periodic event such as the sinusoidal vibration of a tuning fork. The sound due to percussion is called transient, whereas a periodic stimulus produces steady-state sound having a frequency $f$.

Because sound travels at a finite speed, the fixed observer at some distance from the source will experience the disturbance at some later time. In the case of a transient sound caused by an impact, the observer will detect a single replica of the original as it passes at the speed of sound. In the case of the tuning fork, a periodic sound source, the pressure peaks and dips follow one another away from the source at the speed of sound. For a given rate of vibration of the source, a given peak will have propagated a constant distance before the next peak occurs. This distance is called the wavelength lambda. Figure 1.2 shows that wavelength is defined as the distance between any two identical points on the whole cycle. If the source vibrates faster, successive peaks get closer together and the wavelength gets shorter. Figure 1.2 also shows that the wavelength

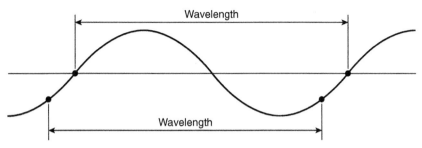

**Figure 1.2: Wavelength is defined as the distance between two points at the same place on adjacent cycles. Wavelength is inversely proportional to frequency.**

is inversely proportional to the frequency. It is easy to remember that the wavelength of 1000 Hz is a foot (about 30 cm).

## 1.3 Periodic and Aperiodic Signals

Sounds can be divided into these two categories and analyzed either in the time domain in which the waveform is considered or in the frequency domain in which the spectrum is considered. The time and frequency domains are linked by transforms of which the best known is the Fourier transform.

Figure 1.3(a) shows that an ideal periodic signal is one which repeats after some constant time has elapsed and goes on indefinitely in the time domain. In the frequency domain such a signal will be described as having a fundamental frequency and a series of harmonics or partials that are at integer multiples of the fundamental. The timbre of an instrument is determined by the harmonic structure. Where there are no harmonics at all, the simplest possible signal results that has only a single frequency in the spectrum. In the time domain this will be an endless sine wave.

Figure 1.3(b) shows an aperiodic signal known as white noise. The spectrum shows that there is an equal level at all frequencies, hence the term "white," which is analogous to the white light containing all wavelengths. Transients or impulses may also be aperiodic. A spectral analysis of a transient [Figure 1.3(c)] will contain a range of frequencies, but these are not harmonics because they are not integer multiples of the lowest frequency. Generally the narrower an event in the time domain, the broader it will be in the frequency domain and vice versa.

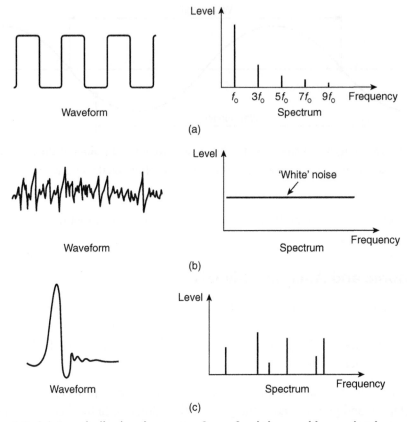

Figure 1.3: (a) A periodic signal repeats after a fixed time and has a simple spectrum consisting of fundamental plus harmonics. (b) An aperiodic signal such as noise does not repeat and has a continuous spectrum. (c) A transient contains an anharmonic spectrum.

## 1.4 Sound and the Ear

Experiments can tell us that the ear only responds to a certain range of frequencies within a certain range of levels. If sound is defined to fall within those ranges, then its reproduction is easier because it is only necessary to reproduce those levels and frequencies that the ear can detect.

Psychoacoustics can describe how our hearing has finite resolution in both time and frequency domains such that what we perceive is an inexact impression. Some aspects of the original disturbance are inaudible to us and are said to be masked. If our goal is the highest quality, we can design our imperfect equipment so that the shortcomings are masked. Conversely, if our goal is economy we can use compression and hope that masking will disguise the inaccuracies it causes.

A study of the finite resolution of the ear shows how some combinations of tones sound pleasurable whereas others are irritating. Music has evolved empirically to emphasize primarily the former. Nevertheless, we are still struggling to explain why we enjoy music and why certain sounds can make us happy whereas others can reduce us to tears. These characteristics must still be present in digitally reproduced sound.

The frequency range of human hearing is extremely wide, covering some 10 octaves (an octave is a doubling of pitch or frequency) without interruption.

By definition, the sound quality of an audio system can only be assessed by human hearing. Many items of audio equipment can only be designed well with a good knowledge of the human hearing mechanism. The acuity of the human ear is finite but astonishing. It can detect tiny amounts of distortion and will accept an enormous dynamic range over a wide number of octaves. If the ear detects a different degree of impairment between two audio systems in properly conducted tests, we can say that one of them is superior.

However, any characteristic of a signal that can be heard can, in principle, also be measured by a suitable instrument, although in general the availability of such instruments lags the requirement. The subjective tests will tell us how sensitive the instrument should be. Then the objective readings from the instrument give an indication of how acceptable a signal is in respect of that characteristic.

The sense we call hearing results from acoustic, mechanical, hydraulic, nervous, and mental processes in the ear/brain combination, leading to the term psychoacoustics. It is only possible to briefly introduce the subject here. The interested reader is referred to Moore[1] for an excellent treatment.

Figure 1.4 shows that the structure of the ear is divided into outer, middle, and inner ears. The outer ear works at low impedance, the inner ear works at high impedance, and the middle ear is an impedance matching device. The visible part of the outer ear is called the pinna, which plays a subtle role in determining the direction of arrival of sound at high frequencies. It is too small to have any effect at low frequencies. Incident sound enters the auditory canal or meatus. The pipe-like meatus causes a small resonance at around

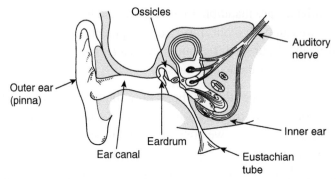

**Figure 1.4: The structure of the human ear. See text for details.**

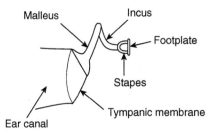

**Figure 1.5: The malleus tensions the tympanic membrane into a conical shape. The ossicles provide an impedance-transforming lever system between the tympanic membrane and the oval window.**

4 kHz. Sound vibrates the eardrum or tympanic membrane, which seals the outer ear from the middle ear. The inner ear or cochlea works by sound traveling though a fluid. Sound enters the cochlea via a membrane called the oval window.

If airborne sound were to be incident on the oval window directly, the serious impedance mismatch would cause most of the sound to be reflected. The middle ear remedies that mismatch by providing a mechanical advantage. The tympanic membrane is linked to the oval window by three bones known as ossicles, which act as a lever system such that a large displacement of the tympanic membrane results in a smaller displacement of the oval window but with greater force. Figure 1.5 shows that the malleus applies a tension to the tympanic membrane, rendering it conical in shape. The malleus and the incus are firmly joined together to form a lever. The incus acts on the stapes through a spherical joint. As the area of the tympanic membrane is greater than that of the oval window, there is further multiplication of the available force. Consequently, small pressures over the large area of the tympanic membrane are converted to high pressures over the small area of the oval window.

The middle ear is normally sealed, but ambient pressure changes will cause static pressure on the tympanic membrane, which is painful. The pressure is relieved by the Eustachian tube, which opens involuntarily while swallowing. The Eustachian tubes open into the cavities of the head and must normally be closed to avoid one's own speech appearing deafeningly loud.

The ossicles are located by minute muscles, which are normally relaxed. However, the middle ear reflex is an involuntary tightening of the *tensor tympani* and *stapedius* muscles, which heavily damp the ability of the tympanic membrane and the stapes to transmit sound by about 12 dB at frequencies below 1 kHz. The main function of this reflex is to reduce the audibility of one's own speech. However, loud sounds will also trigger this reflex, which takes some 60 to 120 ms to occur, too late to protect against transients such as gunfire.

## 1.5 The Cochlea

The cochlea, shown in Figure 1.6(a), is a tapering spiral cavity within bony walls, which is filled with fluid. The widest part, near the oval window, is called the *base* and the distant end is the *apex*. Figure 1.6(b) shows that the cochlea is divided lengthwise into three volumes by Reissner's membrane and the basilar membrane. The *scala vestibuli* and the *scala tympani* are connected by a small aperture at the apex of the cochlea known as the *helicotrema*. Vibrations from the stapes are transferred to the oval window and become fluid pressure variations, which are relieved by the flexing of the round window. Essentially the basilar membrane is in series with the fluid motion and is driven by it except at very low frequencies where the fluid flows through the helicotrema, bypassing the basilar membrane.

The vibration of the basilar membrane is sensed by the organ of Corti, which runs along the center of the cochlea. The organ of Corti is active in that it contains elements that can generate vibration as well as sense it. These are connected in a regenerative fashion so that the $Q$ factor, or frequency selectivity of the ear, is higher than it would otherwise be. The deflection of hair cells in the organ of Corti triggers nerve firings and these signals

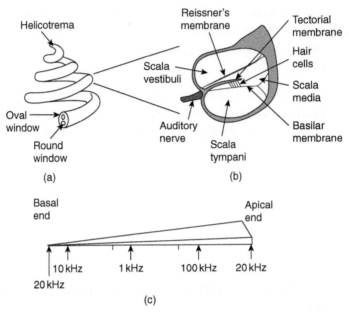

Figure 1.6: (a) The cochlea is a tapering spiral cavity. (b) The cross section of the cavity is divided by Reissner's membrane and the basilar membrane. (c) The basilar membrane tapers so that its resonant frequency changes along its length.

are conducted to the brain by the auditory nerve. Some of these signals reflect the time domain, particularly during the transients with which most real sounds begin and also at low frequencies. During continuous sounds, the basilar membrane is also capable of performing frequency analysis.

Figure 1.6(c) shows that the basilar membrane is not uniform, but tapers in width and varies in thickness in the opposite sense to the taper of the cochlea. The part of the basilar membrane that resonates as a result of an applied sound is a function of the frequency. High frequencies cause resonance near the oval window, whereas low frequencies cause resonances further away. More precisely, the distance from the apex where the maximum resonance occurs is a logarithmic function of the frequency. Consequently, tones spaced apart in octave steps will excite evenly spaced resonances in the basilar membrane. The prediction of resonance at a particular location on the membrane is called *place theory*. Essentially the basilar membrane is a mechanical frequency analyzer.

Nerve firings are not a perfect analog of the basilar membrane motion. On continuous tones, a nerve firing appears to occur at a constant phase relationship to the basilar vibration, a phenomenon called phase locking, but firings do not necessarily occur on every cycle. At higher frequencies firings are intermittent, yet each is in the same phase relationship.

The resonant behavior of the basilar membrane is not observed at the lowest audible frequencies below 50 Hz. The pattern of vibration does not appear to change with frequency and it is possible that the frequency is low enough to be measured directly from the rate of nerve firings.

## 1.6 Mental Processes

The nerve impulses are processed in specific areas of the brain that appear to have evolved at different times to provide different types of information. The time domain response works quickly, primarily aiding the direction-sensing mechanism and is older in evolutionary terms. The frequency domain response works more slowly, aiding the determination of pitch and timbre and evolved later, presumably as speech evolved.

The earliest use of hearing was as a survival mechanism to augment vision. The most important aspect of the hearing mechanism was the ability to determine the location of the sound source. Figure 1.7 shows that the brain can examine several possible differences between the signals reaching the two ears. In Figure 1.7(a), a phase shift is apparent. In Figure 1.7(b), the distant ear is shaded by the head, resulting in a different frequency response compared to the nearer ear. In Figure 1.7(c), a transient sound arrives later at the more distant ear. The interaural phase, delay, and level mechanisms vary in their effectiveness depending on the nature of the sound to be located. At some point a fuzzy logic decision has to be made as to how the information from these different mechanisms will be weighted.

There will be considerable variation with frequency in the phase shift between the ears. At a low frequency such as 30 Hz, the wavelength is around 11.5 m so this mechanism must be quite weak at low frequencies. At high frequencies the ear spacing is many wavelengths, producing a confusing and complex phase relationship. This suggests a frequency limit of around 1500 Hz, which has been confirmed experimentally.

At low and middle frequencies, sound will diffract round the head sufficiently well that there will be no significant difference between the levels at the two ears. Only at high

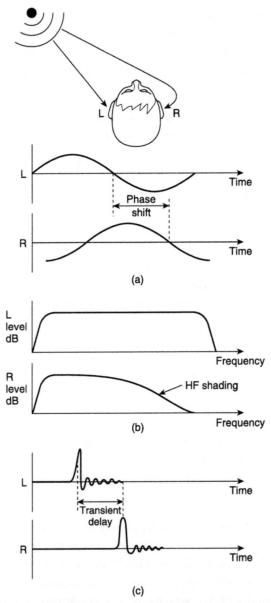

Figure 1.7: Having two spaced ears is cool. (a) Off-center sounds result in a phase difference. (b) The distant ear is shaded by the head, producing a loss of high frequencies. (c) The distant ear detects transient later.

frequencies does sound become directional enough for the head to shade the distant ear, causing what is called interaural intensity difference.

Phase differences are only useful at low frequencies and shading only works at high frequencies. Fortunately, real-world noises and sounds are broadband and often contain transients. Timbral, broadband, and transient sounds differ from tones in that they contain many different frequencies. Pure tones are rare in nature.

A transient has a unique aperiodic waveform, which, as Figure 1.7(c) shows, suffers no ambiguity in the assessment of interaural delay (IAD) between two versions. Note that a one-degree change in sound location causes an IAD of around 10 μs. The smallest detectable IAD is a remarkable 6 μs. This should be the criterion for spatial reproduction accuracy.

Transient noises produce a one-off pressure step whose source is accurately and instinctively located. Figure 1.8 shows an idealized transient pressure waveform following an acoustic event. Only the initial transient pressure change is required for location. The time of arrival of the transient at the two ears will be different and will locate the source laterally within a processing delay of around a millisecond.

Following the event that generated the transient, the air pressure equalizes. The time taken for this equalization varies and allows the listener to establish the likely size of

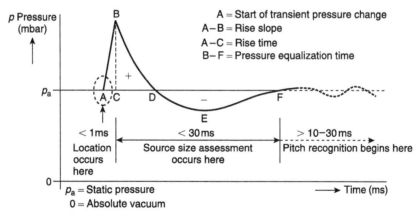

Figure 1.8: A real acoustic event produces a pressure step. The initial step is used for spatial location; equalization time signifies the size of the source. (Courtesy of Manger Schallwandlerbau.)

the sound source. The larger the source, the longer the pressure–equalization time. Only after this does the frequency analysis mechanism tell anything about the pitch and timbre of the sound.

The aforementioned results suggest that anything in a sound reproduction system that impairs the reproduction of a transient pressure change will damage localization and the assessment of the pressure–equalization time. Clearly, in an audio system that claims to offer any degree of precision, every component must be able to reproduce transients accurately and must have at least a minimum phase characteristic if it cannot be phase linear. In this respect, digital audio represents a distinct technical performance advantage, although much of this is later lost in poor transducer design, especially in loudspeakers.

## 1.7 Level and Loudness

At its best, the ear can detect a sound pressure variation of only $2 \times 10^{-5}$ Pascals root mean square (rms) and so this figure is used as the reference against which the sound pressure level (SPL) is measured. The sensation of loudness is a logarithmic function of SPL; consequently, a logarithmic unit, the decibel, was adopted for audio measurement. The decibel is explained in detail in Section 1.12.

The dynamic range of the ear exceeds 130 dB, but at the extremes of this range, the ear either is straining to hear or is in pain. The frequency response of the ear is not at all uniform and it also changes with SPL. The subjective response to level is called loudness and is measured in *phons*. The phon scale is defined to coincide with the SPL scale at 1 kHz, but at other frequencies the phon scale deviates because it displays the actual SPLs judged by a human subject to be equally loud as a given level at 1 kHz. Figure 1.9 shows the so-called equal loudness contours, which were originally measured by Fletcher and Munson and subsequently by Robinson and Dadson. Note the irregularities caused by resonances in the meatus at about 4 and 13 kHz.

Usually, people's ears are at their most sensitive between about 2 and 5 kHz; although some people can detect 20 kHz at high level, there is much evidence to suggest that most listeners cannot tell if the upper frequency limit of sound is 20 or 16 kHz.[2,3] For a long time it was thought that frequencies below about 40 Hz were unimportant, but it is now clear that the reproduction of frequencies down to 20 Hz improves reality and ambience.[4]

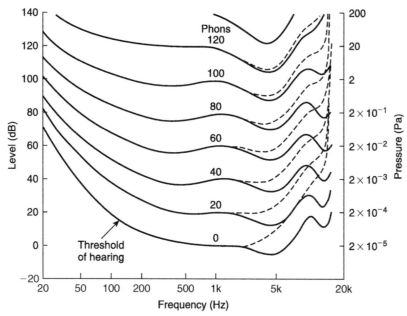

**Figure 1.9: Contours of equal loudness showing that the frequency response of the ear is highly level dependent (solid line, age 20; dashed line, age 60).**

The generally accepted frequency range for high-quality audio is 20 to 20,000 Hz, although an upper limit of 15,000 Hz is often applied for broadcasting.

The most dramatic effect of the curves of Figure 1.9 is that the bass content of reproduced sound is reduced disproportionately as the level is turned down. This would suggest that if a sufficiently powerful yet high-quality reproduction system is available, the correct tonal balance when playing a good recording can be obtained simply by setting the volume control to the correct level. This is indeed the case. A further consideration is that many musical instruments, as well as the human voice, change timbre with the level and there is only one level that sounds correct for the timbre.

Audio systems with a more modest specification would have to resort to the use of tone controls to achieve a better tonal balance at lower SPL. A loudness control is one where the tone controls are automatically invoked as the volume is reduced. Although well meant, loudness controls seldom compensate accurately because they must know the original level at which the material was meant to be reproduced as well as the actual level in use.

A further consequence of level-dependent hearing response is that recordings that are mixed at an excessively high level will appear bass light when played back at a normal level. Such recordings are more a product of self-indulgence than professionalism.

Loudness is a subjective reaction and is almost impossible to measure. In addition to the level-dependent frequency response problem, the listener uses the sound not for its own sake but to draw some conclusion about the source. For example, most people hearing a distant motorcycle will describe it as being loud. Clearly, at the source, it *is* loud, but the listener has compensated for the distance.

The best that can be done is to make some compensation for the level-dependent response using *weighting curves*. Ideally, there should be many, but in practice the A, B, and C weightings were chosen where the A curve is based on the 40-phon response. The measured level after such a filter is in units of dBA. The A curve is almost always used because it most nearly relates to the annoyance factor of distant noise sources.

## 1.8 Frequency Discrimination

Figure 1.10 shows an uncoiled basilar membrane with the apex on the left so that the usual logarithmic frequency scale can be applied. The envelope of displacement of the basilar membrane is shown for a single frequency at Figure 1.10(a). The vibration of the membrane in sympathy with a single frequency cannot be localized to an infinitely small area, and nearby areas are forced to vibrate at the same frequency with an amplitude that decreases with distance. Note that the envelope is asymmetrical because the membrane is tapering and because of frequency-dependent losses in the propagation of vibrational energy down the cochlea. If the frequency is changed, as in Figure 1.10(b), the position of maximum displacement will also change. As the basilar membrane is continuous, the position of maximum displacement is infinitely variable, allowing extremely good pitch discrimination of about one-twelfth of a semitone, which is determined by the spacing of hair cells.

In the presence of a complex spectrum, the finite width of the vibration envelope means that the ear fails to register energy in some bands when there is more energy in a nearby band. Within those areas, other frequencies are mechanically excluded because their amplitude is insufficient to dominate the local vibration of the membrane. Thus the $Q$ factor of the membrane is responsible for the degree of auditory masking, defined as the decreased audibility of one sound in the presence of another. Masking is important because audio compression relies heavily on it.

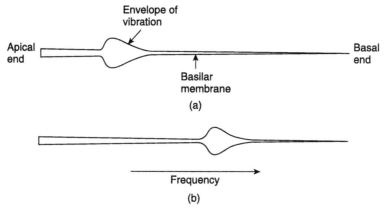

**Figure 1.10: The basilar membrane symbolically uncoiled. (a) Single frequency causes the vibration envelope shown. (b) Changing the frequency moves the peak of the envelope.**

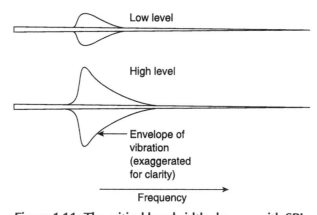

**Figure 1.11: The critical bandwidth changes with SPL.**

The term used in psychoacoustics to describe the finite width of the vibration envelope is *critical bandwidth*. Critical bands were first described by Fletcher.[5] The envelope of basilar vibration is a complicated function. It is clear from the mechanism that the area of the membrane involved will increase as the sound level rises. Figure 1.11 shows the bandwidth as a function of level.

As seen elsewhere, transform theory teaches that the higher the frequency resolution of a transform, the worse the time accuracy. As the basilar membrane has finite frequency resolution measured in the width of a critical band, it follows that it must have finite time resolution. This also follows from the fact that the membrane is resonant, taking time to start and stop vibrating in response to a stimulus. There are many examples of this. Figure 1.12 shows the impulse response. Figure 1.13 shows that the perceived loudness of a tone burst increases with duration up to about 200 ms due to the finite response time.

The ear has evolved to offer intelligibility in reverberant environments, which it does by averaging all received energy over a period of about 30 ms. Reflected sound that arrives within this time is integrated to produce a louder sensation, whereas reflected sound that arrives after that time can be temporally discriminated and perceived as an echo.

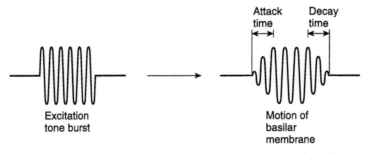

**Figure 1.12: Impulse response of the ear showing slow attack and decay as a consequence of resonant behavior.**

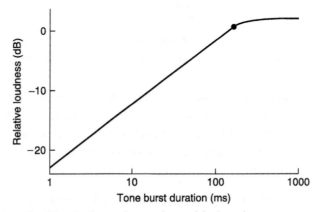

**Figure 1.13: Perceived level of tone burst rises with duration as resonance builds up.**

Microphones have no such ability, which is why acoustic treatment is often needed in areas where microphones are used.

A further example of the finite time discrimination of the ear is the fact that short interruptions to a continuous tone are difficult to detect. Finite time resolution means that masking can take place even when the masking tone begins after and ceases before the masked sound. This is referred to as forward and backward masking.[6]

Figure 1.14(a) shows an electrical signal in which two equal sine waves of nearly the same frequency have been added together linearly. Note that the envelope of the signal varies as the two waves move in and out of phase. Clearly the frequency transform calculated to infinite accuracy is that shown at Figure 1.14(b). The two amplitudes are constant and there is no evidence of envelope modulation. However, such a measurement requires an infinite time. When a shorter time is available, the frequency discrimination of the transform falls and the bands in which energy is detected become broader.

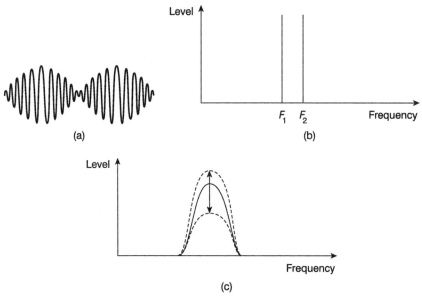

Figure 1.14: (a) Result of adding two sine waves of similar frequency. (b) Spectrum of (a) to infinite accuracy. (c) With finite accuracy, only a single frequency is distinguished whose amplitude changes with the envelope of (a) giving rise to beats.

When the frequency discrimination is too wide to distinguish the two tones as shown in Figure 1.14(c), the result is that they are registered as a single tone. The amplitude of the single tone will change from one measurement to the next because the envelope is being measured. The rate at which the envelope amplitude changes is called *beat* frequency, which is not actually present in the input signal. Beats are an artifact of finite frequency resolution transforms. The fact that human hearing produces beats from pairs of tones proves that it has finite resolution.

## 1.9  Frequency Response and Linearity

It is a goal in high-quality sound reproduction that the timbre of the original sound shall not be changed by the reproduction process. There are two ways in which timbre can inadvertently be changed, as Figure 1.15 shows. In Figure 1.15(a), the spectrum of the original shows a particular relationship between harmonics. This signal is passed through a system [Figure 1.15 (b)] that has an unequal response at different frequencies.

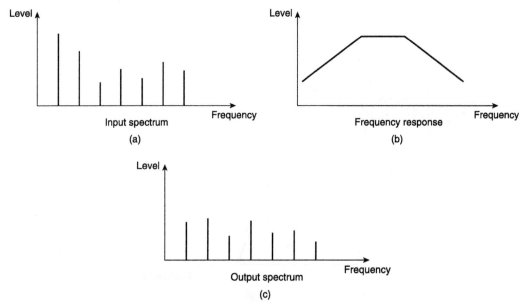

Figure 1.15: Why frequency response matters. (a) Original spectrum determines the timbre of sound. If the original signal is passed through a system with a deficient frequency response (b), the timbre will be changed (c).

The result is that the harmonic structure [Figure 1.15(c)] has changed, and with it the timbre. Clearly a fundamental requirement for quality sound reproduction is that the response to all frequencies should be equal.

Frequency response is easily tested using sine waves of constant amplitude at various frequencies as an input and noting the output level for each frequency.

Figure 1.16 shows that another way in which timbre can be changed is by nonlinearity. All audio equipment has a transfer function between the input and the output, which form the two axes of a graph. Unless the transfer function is exactly straight or *linear*, the output waveform will differ from the input. A nonlinear transfer function will cause distortion, which changes the distribution of harmonics and changes timbre.

At a real microphone placed before an orchestra a multiplicity of sounds may arrive simultaneously. Because the microphone diaphragm can only be in one place at a

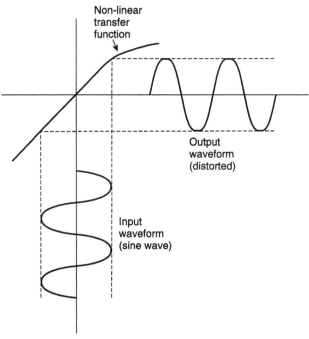

**Figure 1.16: Nonlinearity of the transfer function creates harmonies by distorting the waveform. Linearity is extremely important in audio equipment.**

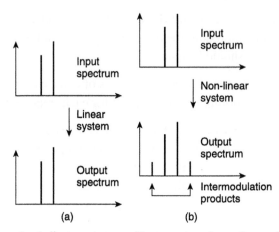

**Figure 1.17: (a) A perfectly linear system will pass a number of superimposed waveforms without interference so that the output spectrum does not change. (b) A nonlinear system causes intermodulation where the output spectrum contains sum and difference frequencies in addition to the originals.**

time, the output waveform must be the sum of all the sounds. An ideal microphone connected by ideal amplification to an ideal loudspeaker will reproduce all of the sounds simultaneously by linear superimposition. However, should there be a lack of linearity anywhere in the system, the sounds will no longer have an independent existence, but will interfere with one another, changing one another's timbre and even creating new sounds that did not previously exist. This is known as *intermodulation*. Figure 1.17 shows that a linear system will pass two sine waves without interference. If there is any nonlinearity, the two sine waves will intermodulate to produce sum and difference frequencies, which are easily observed in the otherwise pure spectrum.

## 1.10 The Sine Wave

As the sine wave is such a useful concept it will be treated here in detail. Figure 1.18 shows a constant speed rotation viewed along the axis so that the motion is circular. Imagine, however, the view from one side in the plane of the rotation. From a distance, only a vertical oscillation will be observed and if the position is plotted against time the resultant waveform will be a sine wave. Geometrically, it is possible to calculate the height or displacement because it is the radius multiplied by the sine of the phase angle.

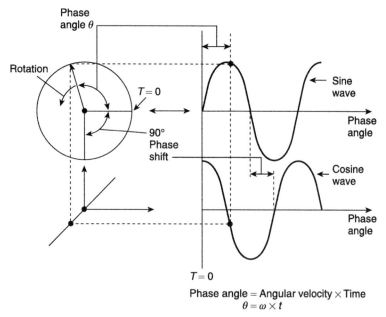

**Figure 1.18:** A sine wave is one component of a rotation. When a rotation is viewed from two places at places at right angles, one will see a sine wave and the other will see a cosine wave. The constant *phase shift* between sine and cosine is 90° and should not be confused with the time variant *phase angle* due to the rotation.

The phase angle is obtained by multiplying the angular velocity ω by the time **t**. Note that the angular velocity is measured in radians per second, whereas frequency **f** is measured in rotations per second or hertz. As a radian is unit distance at unit radius (about 57°), then there are $2\pi$ radians in one rotation. Thus the phase angle at a time **t** is given by sinωt or sin2π**ft**.

A second viewer, who is at right angles to the first viewer, will observe the same waveform but with different timing. The displacement will be given by the radius multiplied by the cosine of the phase angle. When plotted on the same graph, the two waveforms are *phase shifted* with respect to one another. In this case the phase shift is 90° and the two waveforms are said to be *in quadrature*. Incidentally, the motions on each side of a steam locomotive are in quadrature so that it can always get started (the term used is quartering). Note that the *phase angle* of a signal is constantly changing with time, whereas the *phase shift* between two signals can be constant. It is important that these two are not confused.

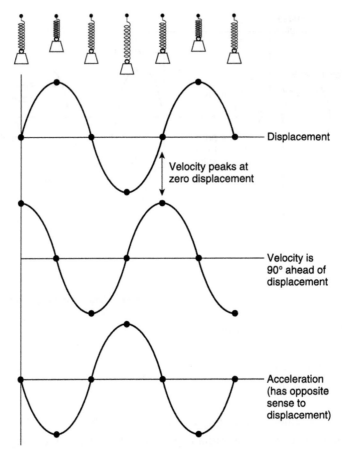

**Figure 1.19: The displacement, velocity, and acceleration of a body executing simple harmonic motion (SHM).**

The velocity of a moving component is often more important in audio than the displacement. The vertical component of velocity is obtained by differentiating the displacement. As the displacement is a sine wave, the velocity will be a cosine wave whose amplitude is proportional to frequency. In other words, the displacement and velocity are in quadrature with the velocity lagging. This is consistent with the velocity reaching a minimum as the displacement reaches a maximum and vice versa. Figure 1.19 shows displacement, velocity, and acceleration waveforms of a body executing simple harmonic motion (SHM). Note that the acceleration and the displacement are always antiphase.

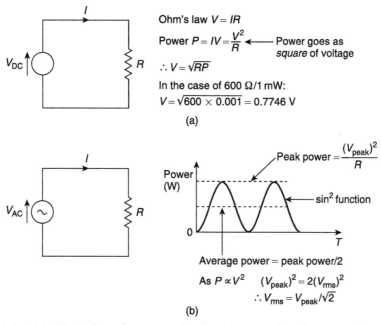

Ohm's law $V = IR$

Power $P = IV = \dfrac{V^2}{R}$ ◄——— Power goes as *square* of voltage

$\therefore V = \sqrt{RP}$

In the case of 600 $\Omega$/1 mW:

$V = \sqrt{600 \times 0.001} = 0.7746$ V

(a)

Peak power $= \dfrac{(V_{peak})^2}{R}$

Power (W)

sin$^2$ function

0 ——————————————— $T$

Average power = peak power/2

As $P \propto V^2$   $(V_{peak})^2 = 2(V_{rms})^2$

$\therefore V_{rms} = V_{peak}/\sqrt{2}$

(b)

Figure 1.20: (a) Ohm's law: the power developed in a resistor is proportional to the square of the voltage. Consequently, 1 mW in 600 $\Omega$ requires 0.775 V. With a sinusoidal alternating input (b), the power is a sine-squared function, which can be averaged over one cycle. A DC voltage that delivers the same power has a value that is the square root of the mean of the square of the sinusoidal input to be measured and the reference. The Bel is too large so the decibel (dB) is used in practice. (b) As the dB is defined as a power ratio, voltage ratios have to be squared. This is conveniently done by doubling the logs so that the ratio is now multiplied by 20.

## 1.11 Root Mean Square Measurements

Figure 1.20(a) shows that, according to Ohm's law, the power dissipated in a resistance is proportional to the square of the applied voltage. This causes no difficulty with direct current (DC), but with alternating signals such as audio it is harder to calculate the power. Consequently, a unit of voltage for alternating signals was devised. Figure 1.20(b) shows that the average power delivered during a cycle must be proportional to the mean of the square of the applied voltage. Since power is proportional to the square of applied

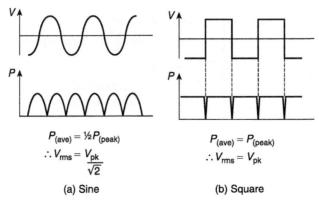

$$P_{(ave)} = \tfrac{1}{2} P_{(peak)}$$
$$\therefore V_{rms} = \frac{V_{pk}}{\sqrt{2}}$$

$$P_{(ave)} = P_{(peak)}$$
$$\therefore V_{rms} = V_{pk}$$

(a) Sine                    (b) Square

**Figure 1.21: (a) For a sine wave the conversion factor from peak to rms is $\frac{1}{\sqrt{2}}$. (b) For a square wave the peak and rms voltage are the same.**

voltage, the same power would be dissipated by a DC voltage whose value was equal to the square root of the mean of the square of the AC voltage. Thus the volt rms was specified. An AC signal of a given number of volts rms will dissipate exactly the same amount of power in a given resistor as the same number of volts DC.

Figure 1.21(a) shows that for a sine wave the rms voltage is obtained by dividing the peak voltage $V_{pk}$ by the square root of 2. However, for a square wave [Figure 1.21(b)] the rms voltage and the peak voltage are the same. Most moving coil AC voltmeters only read correctly on sine waves, whereas many electronic meters incorporate a true rms calculation.

On an oscilloscope it is often easier to measure the peak-to-peak voltage, which is twice the peak voltage. The rms voltage cannot be measured directly on an oscilloscope since it depends on the waveform, although the calculation is simple in the case of a sine wave.

## 1.12  The Decibel

The first audio signals to be transmitted were on telephone lines. Where the wiring is long compared to the electrical wavelength (not to be confused with the acoustic wavelength) of the signal, a transmission line exists in which the distributed series inductance and the parallel capacitance interact to give the line a characteristic impedance. In telephones this turned out to be about $600\,\Omega$. In transmission lines the best power delivery occurs when the source and the load impedance are the same; this is the process of matching.

It was often required to measure the power in a telephone system, and $1\,mW$ was chosen as a suitable unit. Thus the reference against which signals could be compared was the dissipation of $1\,mW$ in $600\,\Omega$. Figure 1.20(a) shows that the dissipation of $1\,mW$ in $600\,\Omega$ will be due to an applied voltage of $0.775\,V$ rms. This voltage became the reference against which all audio levels are compared.

The decibel is a logarithmic measuring system and has its origins in telephony[7] where the loss in a cable is a logarithmic function of the length. Human hearing also has a logarithmic response with respect to sound pressure level. In order to relate to the subjective response, audio signal level measurements also have to be logarithmic and so the decibel was adopted for audio.

Figure 1.22 shows the principle of the logarithm. To give an example, if it is clear that $10^2$ is 100 and $10^3$ is 1000, then there must be a power between 2 and 3 to which 10 can be raised to give any value between 100 and 1000. That power is the logarithm to base 10 of the value, for example, $\log_{10} 300 = 2.5$ approximately. Note that $10^0$ is 1.

Logarithms were developed by mathematicians before the availability of calculators or computers to ease calculations such as multiplication, squaring, division, and extracting roots. The advantage is that, armed with a set of log tables, multiplication can be performed by adding and division by subtracting. Figure 1.22 shows some examples. It will be clear that squaring a number is performed by adding two identical logs and the same result will be obtained by multiplying the log by 2.

The slide rule is an early calculator, which consists of two logarithmically engraved scales in which the length along the scale is proportional to the log of the engraved number. By sliding the moving scale, two lengths can be added or subtracted easily and, as a result, multiplication and division are readily obtained.

The logarithmic unit of measurement in telephones was called the Bel after Alexander Graham Bell, the inventor. Figure 1.23(a) shows that the Bel was defined as the log of the *power* ratio between the power to be measured and some reference power. Clearly the reference power must have a level of 0 Bels, as $\log_{10} 1$ is 0.

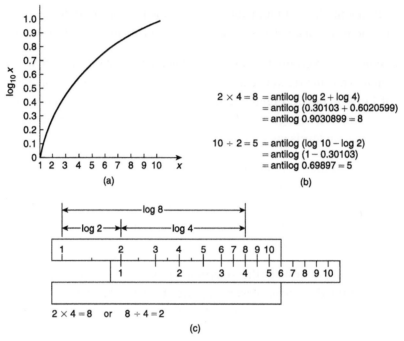

1.0
0.9
0.8
0.7
0.6
0.5
0.4
0.3
0.2
0.1
0
$\log_{10} x$
1 2 3 4 5 6 7 8 9 10 x
(a)
$2 \times 4 = 8$ = antilog (log 2 + log 4)
= antilog (0.30103 + 0.6020599)
= antilog 0.9030899 = 8
$10 \div 2 = 5$ = antilog (log 10 − log 2)
= antilog (1 − 0.30103)
= antilog 0.69897 = 5
(b)
log 8
log 2
log 4
1 2 3 4 5 6 7 8 9 10
1 2 3 4 5 6 7 8 9 10
$2 \times 4 = 8$ or $8 \div 4 = 2$
(c)


**Figure 1.22: (a) The logarithm of a number is the power to which the base (in this case 10) must be raised to obtain the number. (b) Multiplication is obtained by adding logs, division by subtracting. (c) The slide rule has two logarithmic scales whose length can be added or subtracted easily.**

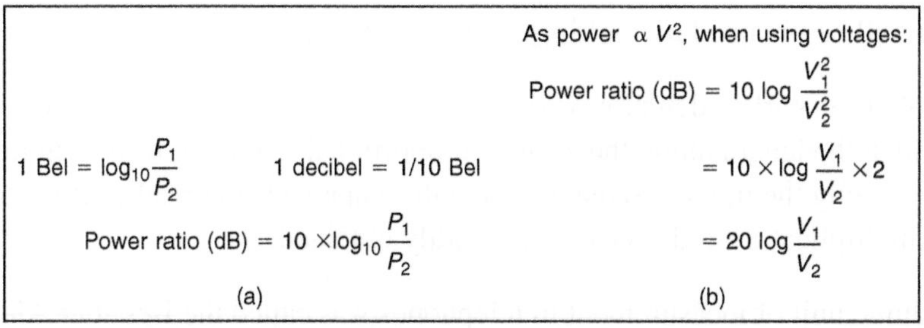

$1 \text{ Bel} = \log_{10}\frac{P_1}{P_2}$
1 decibel = 1/10 Bel
$\text{Power ratio (dB)} = 10 \times \log_{10}\frac{P_1}{P_2}$
(a)
As power $\alpha\ V^2$, when using voltages:
$\text{Power ratio (dB)} = 10 \log \frac{V_1^2}{V_2^2}$
$= 10 \times \log \frac{V_1}{V_2} \times 2$
$= 20 \log \frac{V_1}{V_2}$
(b)


**Figure 1.23: (a) The Bel is the log of the ratio between two powers, that between two powers, that to be measured, and the reference. The Bel is too large so the decibel is used in practice. (b) As the decibel is defined as a power ratio, voltage ratios have to be squared. This is done conveniently by doubling the logs so that the ratio is now multiplied by 20.**

The Bel was found to be an excessively large unit for practical purposes and so it was divided into 10 decibels, abbreviated dB with a small d and a large B and pronounced deebee. Consequently, the number of dBs is 10 times the log of the power ratio. A device such as an amplifier can have a fixed power gain that is independent of signal level and this can be measured in dBs. However, when measuring the power of a signal, it must be appreciated that the dB is a ratio and to quote the number of dBs without stating the reference is about as senseless as describing the height of a mountain as 2000 without specifying whether this is feet or meters. To show that the reference is 1 mW into 600 Ω the units will be dB(m). In radio engineering, the dB(W) will be found, which is power relative to 1 W.

Although the dB(m) is defined as a power ratio, level measurements in audio are often done by measuring the signal voltage using 0.775 V as a reference in a circuit whose impedance is not necessarily 600 Ω. Figure 1.23(b) shows that as the power is proportional to the square of the voltage, the power ratio will be obtained by squaring the voltage ratio. As squaring in logs is performed by doubling, the squared term of the voltages can be replaced by multiplying the log by a factor of two. To give a result in dBs, the log of the voltage ratio now has to be multiplied by 20.

While 600 Ω matched impedance working is essential for the long distances encountered with telephones, it is quite inappropriate for analog audio wiring in a studio. The wavelength of audio in wires at 20 kHz is 15 km. Studios are built on a smaller scale than this and clearly analog audio cables are *not* transmission lines and their characteristic impedance is not relevant.

In professional analog audio systems, impedance matching is not only unnecessary but also undesirable. Figure 1.24(a) shows that when impedance matching is required, the output impedance of a signal source must be raised artificially so that a potential divider is formed with the load. The actual drive voltage must be twice that needed on the cable as the potential divider effect wastes 6 dB of signal level and requires unnecessarily high power supply rail voltages in equipment. A further problem is that cable capacitance can cause an undesirable HF roll-off in conjunction with the high source impedance.

In modern professional analog audio equipment, shown in Figure 1.24(b), the source has the lowest output impedance practicable. This means that any ambient interference

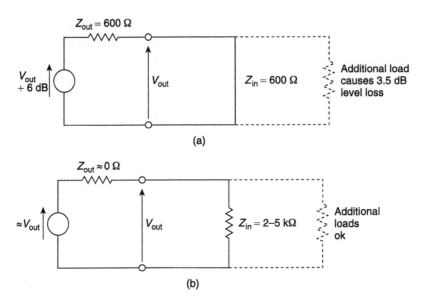

**Figure 1.24: (a) Traditional impedance matched source wastes half the signal voltage in the potential divider due to the source impedance and the cable. (b) Modern practice is to use low-output impedance sources with high-impedance loads.**

is attempting to drive what amounts to a short circuit and can only develop very small voltages. Furthermore, shunt capacitance in the cable has very little effect. The destination has a somewhat higher impedance (generally a few kΩ to avoid excessive currents flowing and to allow several loads to be placed across one driver).

In the absence of fixed impedance, it is meaningless to consider power. Consequently, only signal voltages are measured. The reference remains at 0.775 V, but power and impedance are irrelevant. Voltages measured in this way are expressed in dB(u), the most common unit of level in modern analog systems. Most installations boost the signals on interface cables by 4 dB. As the gain of receiving devices is reduced by 4 dB, the result is a useful noise advantage without risking distortion due to the drivers having to produce high voltages.

## 1.13  Audio Level Metering

There are two main reasons for having level meters in audio equipment: to line up or adjust the gain of equipment and to assess the amplitude of the program material.

Line up is often done using a 1-kHz sine wave generated at an agreed level such as 0 dB(u). If a receiving device does not display the same level, then its input sensitivity must be adjusted. Tape recorders and other devices that pass signals through are usually lined up so that their input and output levels are identical, that is, their insertion loss is 0 dB. Line up is important in large systems because it ensures that inadvertent level changes do not occur.

In measuring the level of a sine wave for the purposes of line up, the dynamics of the meter are of no consequence, whereas on program material the dynamics matter a great deal. The simplest (and least expensive) level meter is essentially an AC voltmeter with a logarithmic response. As the ear is logarithmic, the deflection of the meter is roughly proportional to the perceived volume, hence the term volume unit (VU) meter.

In audio, one of the worst sins is to overmodulate a subsequent stage by supplying a signal of excessive amplitude. The next stage may be an analog tape recorder, a radio transmitter, or an ADC, none of which respond favorably to such treatment. Real audio signals are rich in short transients, which pass before the sluggish VU meter responds. Consequently, the VU meter is also called the virtually useless meter in professional circles.

Broadcasters developed the peak program meter (PPM), which is also logarithmic, but which is designed to respond to peaks as quickly as the ear responds to distortion. Consequently, the attack time of the PPM is carefully specified. If a peak is so short that the PPM fails to indicate its true level, the resulting overload will also be so brief that the ear will not hear it. A further feature of the PPM is that the decay time of the meter is very slow so that any peaks are visible for much longer and the meter is easier to read because the meter movement is less violent. The original PPM as developed by the British Broadcasting Corporation was sparsely calibrated, but other users have adopted the same dynamics and added dB scales.

In broadcasting, the use of level metering and line-up procedures ensures that the level experienced by the listener does not change significantly from program to program. Consequently, in a transmission suite, the goal would be to broadcast recordings at a level identical to that which was determined during production. However, when making a recording prior to any production process, the goal would be to modulate the recording as fully as possible without clipping as this would then give the best signal-to-noise ratio. The level could then be reduced if necessary in the production process.

# References

1. Moore, B. C. J., 'An introduction to the psychology of hearing', London: Academic Press, 1989.

2. Muraoka, T., Iwahara, M., and Yamada, Y., 'Examination of audio bandwidth requirements for optimum sound signal transmission', *J. Audio Eng. Soc.,* 2–9, **29**, 1982.

3. Muraoka, T., Yamada, Y., and Yamazaki, M., 'Sampling frequency considerations in digital audio', *J. Audio Eng. Soc.,* 252–256, **26**, 1978.

4. Fincham, L. R., The subjective importance of uniform group delay at low frequencies. Presented at the 74th Audio Engineering Society Convention (New York, 1983), Preprint 2056(H-1).

5. Fletcher, H., 'Auditory patterns', *Rev. Modern Phys.,* 47–65, **12**, 1940.

6. Carterette, E. C. and Friedman, M. P., 'Handbook of perception', 305–319, New York: Academic Press, 1978.

7. Martin, W. H., 'Decibel—The new name for the transmission unit', *Bell System Tech. J.,* January 1929.

# CHAPTER 2

# Measurement

Don Davis and Eugene Patronis

## 2.1 Concepts Underlying the Decibel and its Use in Sound Systems

Most system measurements of level start with a voltage amplitude. Relative level changes at a given point can be observed on a voltmeter scale when it is realized that

$$10 \log \frac{E_1^2}{E_2^2} = 10 \log \frac{P_1}{P_2} \tag{2.1}$$

which is only true if both values are measured at an identical point in their circuit. A common usage has been to remove the exponent from the ratio and apply it to the multiplier.

$$2 \times 10 \log \frac{E_1}{E_2} = 20 \log \frac{E_1}{E_2} \tag{2.2}$$

Bear in mind that the decibel is always and only based on a power ratio. Any other kind of ratio (i.e., voltage, current, or sound pressure) must first be turned into a power ratio by squaring and then converted into a power level in decibels.

### 2.1.1 Converting Voltage Ratios to Power Ratios

Many audio technicians are confused by the fact that doubling the voltage results in a 6-dB increase while doubling the power only results in a 3-dB increase. Figure 2.1 demonstrates what happens if we simultaneously check both the voltage and the power in a circuit where we double the voltage. Note that for a doubling of the voltage, the power increases four times.

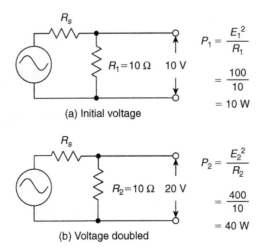

(a) Initial voltage

$$P_1 = \frac{E_1^2}{R_1}$$
$$= \frac{100}{10}$$
$$= 10 \text{ W}$$

(b) Voltage doubled

$$P_2 = \frac{E_2^2}{R_2}$$
$$= \frac{400}{10}$$
$$= 40 \text{ W}$$

**Figure 2.1: Voltage and power relationships in a circuit.**

$$10 \log \frac{P_1}{P_2} = 10 \log \frac{40 \text{ W}}{10 \text{ W}}$$
$$= 6.02 \text{ dB} \tag{2.3}$$

$$10 \log \frac{E_1^2}{E_2^2} = 20 \log \frac{20 \text{ V}}{10 \text{ V}}$$
$$= 6.02 \text{ dB} \tag{2.4}$$

### 2.1.2 The dBV

One of the most common errors when using the decibel is to regard it as a voltage ratio (i.e., so many decibels above or below a reference voltage). To compound the error, the result is referred to as a "level." The word "level" is reserved for power; an increase in the voltage magnitude is properly referred to as "amplification."

However, the decibel can be legitimately used with a voltage reference. The reference is 1.0 V. When voltage magnitudes are referenced logarithmically, they are called dBV (i.e., dB above or below 1.0 V). This use is legitimate because *all* such measurements are made open circuit and can easily be converted into power levels at any impedance interface.

The following definition is from the *IEEE Standard Dictionary of Electrical and Electronics Terms, Second Edition*:

### 2.1.2.1 244.62

Voltage Amplification (1) (general). An increase in signal voltage magnitude in transmission from one point to another or the process thereof. See also: amplifier. 210 (2) (transducer). The scalar ratio of the signal output voltage to the signal input voltage. Warning: By incorrect extension of the term decibel, this ratio is sometimes expressed in decibels by multiplying its common logarithm by 20. It may be currently expressed in decilogs. Note: If the input and/or output power consist of more than one component, such as multifrequency signal or noise, then the particular components used and their weighting must be specified. See also: Transducer.

### 2.1.2.2 239.210

Decilog (dg). A division of the logarithmic scale used for measuring the logarithm of the ratio of two values of any quantity. Note: Its value is such that the number of decilogs is equal to 10 times the logarithm to the base 10 of the ratio. One decilog therefore corresponds to a ratio of $10^{0.1}$ (that is 1.25829+).

### 2.1.3 The Decibel as a Power Ratio

Note that 20 W/10 W and 200 W/100 W both equal 3.01 dB, which means that a 2 to 1 (2:1) power ratio exists but reveals nothing about the actual powers. The human ear hears the same small difference between 1 and 2 W as it does between 100 and 200 W.

Changing decibels back to a power ratio (exponential form) is the same as for any logarithm with the addition of a multiplier (Figure 2.2). The arrows in Figure 2.2 indicate the transposition of quantities. Table 2.1 shows the number of decibels corresponding to various power ratios.

### 2.1.4 Finding Other Multipliers

Occasionally in acoustics, we may need multipliers other than 10 or 20. Once the $\Delta$dB (the number of dB for a 2:1 ratio change) is known, calculate the multiplier by

$$log\,multiplier = \frac{\log{(Base)} \times \Delta\text{dB}}{\log{(Ratio)}} \tag{2.5}$$

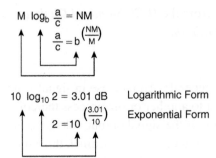

Figure 2.2: Conversion of dB from logarithmic form to exponential form.

Table 2.1: Power Ratios in Decibels

| Power ratio | Decibels (dB) |
| --- | --- |
| 2 | 3.01030 |
| 3 | 4.77121 |
| 4 | 6.02060 |
| 5 | 6.98970 |
| 6 | 7.78151 |
| 7 | 8.45098 |
| 8 | 9.03090 |
| 9 | 9.54243 |
| 10 | 10.00000 |
| 100 | 20.00000 |
| 1000 | 30.00000 |
| 10,000 | 40.00000 |
| 100,000 | 50.00000 |
| 1,000,000 | 60.00000 |

For example, if a 2:1 change is equivalent to 3.01 dB, then

$$log\ multiplier = \frac{\log{(Base)} \times 3.01}{\log 2} = 10 \qquad (2.6)$$

or

$$10 \log 2 = 3.01.$$

If a 2:1 change is equivalent to 6.02 dB, then

$$log \, multiplier = \frac{log \, (Base) \times 6.02}{log \, 2} = 20$$

or

$$20 \, log \, 2 = 6.02.$$

Finally, if a 2:1 change is equivalent to 8 dB, then

$$log \, multiplier = \frac{log \, (Base) \times 8}{log \, 2} = 26.58$$

or

$$26.58 \, 2 \, log = 8.$$

For any $\Delta$dB corresponding to a 2:1 ratio change involving logarithms to the base 10, this may be reduced to

$$log \, multiplier = 3.332 \times \Delta dB. \tag{2.7}$$

### 2.1.5 The Decibel as a Power Quantity

We have seen that a number of decibels by themselves are only ratios. Given any reference (such as 50 W), we can use decibels to find absolute values. A standard reference for power in audio work is $10^{-3}$ W (0.001 W) or $x$ V across $Z\Omega$. Note that when a level is expressed as a wattage, it is not necessary to state an impedance, but when it is stated as a voltage, an impedance is mandatory. This power is called 0 dBm. The small "m" stands for milliwatt (0.001 W) or one-thousandth of a watt.

### 2.1.6 Example

The power in watts corresponding to +30 dBm is calculated as follows:

$$10 \, log \, \frac{x}{0.001} = 30$$

$$x = 0.001 \times 10^{\frac{30}{10}} = 1 \text{ W}.$$

For a power of $-12\,$dBm:

$$0.001 \times 10^{\frac{-12}{10}} = 0.00006309\,\text{W}.$$

The voltage across $600\,\Omega$ is

$$
\begin{aligned}
E &= \sqrt{WR} \\
&= \sqrt{0.00006309 \times 600} \\
&= 0.195\,\text{V}.
\end{aligned}
$$

Note that this $-12$-dBm power level can appear across any impedance and will always be the same power level. Voltages will vary to maintain this power level. In constant-voltage systems the power level varies as the impedance is changed. In constant-current systems the voltage changes as the impedance varies (i.e., $-12\,$dBm across $8\,\Omega = \sqrt{0.00006309 \times 8} = 0.022\,\text{V}$).

## 2.2  Measuring Electrical Power

$$W = EI \cos \theta \qquad\qquad (2.8)$$

$$W = I^2 Z \cos \theta \qquad\qquad (2.9)$$

$$W = \frac{E^2}{Z} \cos \theta$$

where $W$ is the power in watts, $E$ is the electromotive force in rms volts, $I$ is the current in rms amperes, $Z$ is the magnitude of the impedance in ohms [in audio (AC) circuits $Z$ (impedance) is used in place of $R$ (AC resistance)], and $\theta$ is the phase difference between $E$ and $I$ in degrees.

These equations are only valid for single frequency rms sine wave voltages and currents.

### 2.2.1  Most Common Technique

1.  Measure $Z$ and $\theta$.

2.  Measure $E$ across the actual load $Z$ so that

$$W = \frac{E^2}{Z} \cos \theta.$$

## 2.3 Expressing Power as an Audio Level

The reference power is 0.001 W (1 mW). When expressed as a level, this power is called 0 dBm (0 dB referenced to 1 mW).

Thus to express a power level we need two powers—first the measured power $W_1$ and second the reference power $W_2$. This can be written as a power change in dB:

$$
\frac{W_1}{W_2} = \left( \frac{\dfrac{E_1^2}{1}}{\dfrac{E_2^2}{1}} \right) \left( \frac{\dfrac{1}{R_1}}{\dfrac{1}{R_2}} \right)
$$

$$
= \left( \frac{E_1^2}{E_2^2} \right) \left( \frac{R_2}{R_1} \right). \tag{2.10}
$$

This can be written as a power level:

$$
10 \log \left[ \left( \frac{E_1^2}{E_2^2} \right) \left( \frac{R_2}{R_1} \right) \right] = power\ change\ in\ dB \tag{2.11}
$$

or

$$
20 \log \frac{E_1}{E_2} + 10 \log \frac{R_2}{R_1} = power\ change\ in\ dB. \tag{2.12}
$$

### 2.3.1 Special Circumstance

When $R_1 = R_2$ and *only* then:

$$
Power\ level\ in\ dB = 20 \log \frac{E_1}{E_2} \tag{2.13}
$$

where $E_2$ is the voltage associated with the reference power.

## 2.4 Conventional Practice

When calculating power level in dBm, we commonly make $E_2 = 0.775$ V and $R_2 = 600\,\Omega$. Note that $E_2$ may be any voltage and $R_2$ any resistance so long as together they represent 0.001 W.

### 2.4.1 Levels in dB

1. The term "level" is always used for a power expressed in decibels.

2. $10 \log \dfrac{E_1^2}{E_2^2} = 10 \log \dfrac{W_1}{W_2}$

   when $R_1 = R_2$

   $$2 \times 10 \log \frac{E_1}{E_2} = 20 \log \frac{E_1}{E_2}$$

   $$= 10 \log \frac{W_1}{W_2}.$$

3. Power definitions:

   Apparent power $= E \times I$ or $E^2/Z$,
   The average real or absorbed power is $(E^2/Z)\cos\theta$,
   The reactive power is $(E^2/Z)\sin\theta$,
   Power factor $= \cos\theta$.

4. The term "gain" or "loss" always means the power gain or power loss *at the system's output* due to the device under test.

### 2.4.2 Practical Variations of the dBm Equations

When the reference is the audio standard, that is, 0.77459 V and $600\,\Omega$, then

$$dB\ level\ to\ a\ reference\ =\ 10 \log \left[ \left( \frac{E_1^2}{E_2^2} \right) \left( \frac{R_2}{R_1} \right) \right] \qquad (2.14)$$

where $E_2 = 0.77459 \ldots$ V, $R_2 = 600\,\Omega$. Then

$$\frac{R_2}{R_1} = 1000$$

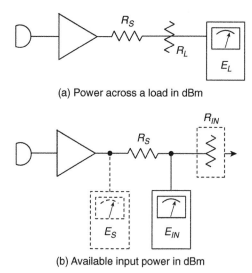

(a) Power across a load in dBm

(b) Available input power in dBm

**Figure 2.3: Power in dB across a load versus available input power.**

and $1/1000 = 0.001$. Note that any $E_2$ and $R_2$ that result in a power of 0.001 W may be used. We can then write:

$$Level\ (in\ dBm) = 10 \log \frac{E_1^2}{0.001 R_1} \tag{2.15}$$

and

$$E_1 = \sqrt{0.001 R_1 \left( 10^{\frac{dBm}{10}} \right)}$$

$$R_1 = \frac{E^2}{0.001 \left( 10^{\frac{dBm}{10}} \right)}. \tag{2.16}$$

See Figure 2.3.

For all of the values in Table 2.2 the only thing known is the voltage. The indication is not a level. The apparent level can only be true across the actual reference impedance. Finally, the presence or absence of an attenuator or other sensitivity control is not known. See Section 2.20 for an explanation of VU.

**Table 2.2: Root Mean Square Voltages Used as Nonstandard References**

| Voltage (V) | Meter indication | Apparent level (VU) | User |
|:---:|:---:|:---:|:---|
| 1.950 | 0 | +8 | Broadcast |
| 1.230 | 0 | +4 | Recording |
| 0.245 | 0 | −10 | Home recording |
| 0.138 | 0 | −15 | Musical instruments |

The power output of Boulder Dam is said to be approximately 3,160,000,000 W. Expressed in dBm, this output would be

$$10 \log \frac{3.16 \times 10^9}{10^{-3}} = 125 \, \text{dBm}.$$

## 2.5  The Decibel in Acoustics—$L_P$, $L_W$, and $L_I$

In acoustics, the ratios encountered most commonly are changes in pressure levels. First, there must be a reference. The older level was 0.0002 dyn/cm$^2$, but this has recently been changed to 0.00002 N/m$^2$ (20 μN/m$^2$). Note that 0.0002 dyn/cm$^2$ is exactly the same sound pressure as 0.00002 N/m$^2$. Even more recently the standards group has named this same pressure pascals (Pa) and arranged this new unit so that

$$20 \, \mu\text{Pa} = \frac{0.0002 \, \text{dyn}}{\text{cm}^2}. \tag{2.17}$$

This means that if the pressure is measured in pascals,

$$L_P = 20 \log \frac{xPa}{20 \, \mu\text{Pa}}. \tag{2.18}$$

If the pressure is measured in dynes per square centimeter (dyn/cm$^2$), then

$$L_P = 20 \log \frac{x \, \text{dyn/cm}^2}{0.0002 \, \text{dyn/cm}^2}. \tag{2.19}$$

The root mean square (rms) sound pressure $P$ can be found by

$$P_{rms} = 2\pi fA\rho c, \tag{2.20}$$

where $P_{rms}$ is in pascals, $f$ is the frequency in Hertz (Hz), $A$ is particle displacement in meters (rms value), $\rho$ is the density of air in kilograms per cubic meter (kg/m³), $c$ is the velocity of sound in meters per second (m/s), $\rho c = 406$ RAYLS and is called the characteristic acoustic resistance (this value can vary), or

$$L_P = 20\log\frac{P_{rms}}{20\,\mu Pa}. \tag{2.21}$$

These are identical sound pressure levels bearing different labels. Sound pressure levels were identified as dB-SPL, and sound power levels were identified as dB-PWL. Currently, $L_P$ is preferred for sound pressure level and $L_W$ for sound power level. Sound intensity level is $L_I$:

$$L_I = 10\log\frac{x\,\text{W/m}^2}{10^{-12}\,\text{W/m}^2}. \tag{2.22}$$

At sea level, atmospheric pressure is equal to 2116 1 b/ft². Remember the old physics laboratory stunt of partially filling an oil can with water, boiling the water, and then quickly sealing the can and putting it under the cold water faucet to condense the steam so that the atmospheric pressure would crush the can as the steam condensed, leaving a partial vacuum?

1Atm = 101,300 Pa

Therefore

$$L_P = 20\log\frac{101,300}{0.00002}$$
$$= 194\,\text{dB}.$$

This represents the complete modulation of atmospheric pressure and would be the largest possible sinusoid. Note that the sound pressure ($SP$) is analogous to voltage. An $L_P$ of 200 dB is the pressure generated by 50 lb of TNT at 10 ft. Table 2.3 shows the equivalents of sound pressure levels.

For additional insights into these basic relationships, the *Handbook of Noise Measurement* by Peterson and Gross is thorough, accurate, and readable.

**Table 2.3: Equivalents of Pressure Levels**

$$L_p = 20 \log \frac{1 \, \text{N/m}^2}{0.00002 \, \text{N/m}^2}$$

$$= 93.98 \, \text{dB}$$

Older values of a similar nature are:

1 microbar $\cong$ 1/1,000,000 of atmospheric pressure

$\cong 74 \, \text{dB}$

therefore

1 Pa $= 10 \, \text{dyn/cm}^2$

Other interesting figures:

Atmospheric pressure fully modulated $L_p \cong 194 \, \text{dB}$

1 lb/ft$^2$ $L_p = 127.6 \, \text{dB}$

1 lb/in$^2$ $L_p = 170.8 \, \text{dB}$

50 lb of TNT measured at 10 ft $L_p = 200 \, \text{dB}$

12-inch cannon, 12 ft in front of and below muzzle $L_p$

$= 220+ \, \text{dB}$

Courtesy of GenRad Handbook.

## 2.6 Acoustic Intensity Level ($L_I$), Acoustic Power Level ($L_W$), and Acoustic Pressure Level ($L_P$)

### 2.6.1 Acoustic Intensity Level, $L_I$

The acoustic intensity $I_a$ (the acoustic power per unit of area—usually in W/m$^2$ or W/cm$^2$) is found by

$$L_I = 10 \log \frac{x \text{W/m}^2}{10^{-12} \, \text{W/m}^2}$$

$$L_I = 10 \log \frac{1.0 \, \text{W/m}^2}{10^{-12} \, \text{W/m}^2}$$

$$= 120 \, \text{dB}. \tag{2.23}$$

### 2.6.2 Acoustic Power Level, $L_W$

The total acoustic power can also be expressed as a level ($L_W$):

$$L_W = 10 \log \frac{Total\ acoustic\ watts}{10^{-12}\,W}. \qquad (2.24)$$

### 2.6.3 Acoustic Pressure Level, $L_P$

To identify each of these parameters more clearly, consider a sphere with a radius of 0.282 m. (Since the surface area of a sphere equals $4\pi r^2$, this yields a sphere with a surface area of 1 m$^2$.) An omnidirectional point source radiating one acoustic watt is placed into the center of this sphere. Thus we have, by definition, an acoustic intensity at the surface of the sphere of 1 W/m$^2$. From this we can calculate the $P_{rms}$:

$$P_{rms} = \sqrt{10 W_a \times \rho c} \qquad (2.25)$$

where $W_a$ is the total acoustic power in watts and $\rho c$ equals 406 RAYLS and is called the characteristic acoustic resistance.

Knowing the acoustic watts, $P_{rms}$ is easy to find:

$$\begin{aligned} P_{rms} &= \sqrt{10 W_a \times 406} \\ &= 20.15\,Pa. \end{aligned}$$

Thus the $L_p$ must be

$$\begin{aligned} L_p &= 20 \log \frac{20.15\,Pa}{20\,\mu Pa} \\ &= 120\,dB. \end{aligned}$$

and the acoustic power level in $L_W$ must be

$$\begin{aligned} L_W &= 10 \log \frac{1\,W}{10^{-12}\,W} \\ &= 120\,dB. \end{aligned}$$

Thus the $L_P$, $L_I$, and $L_W$ at 0.282 m are the same numerical value if the source is omnidirectional (see Figure 2.4).

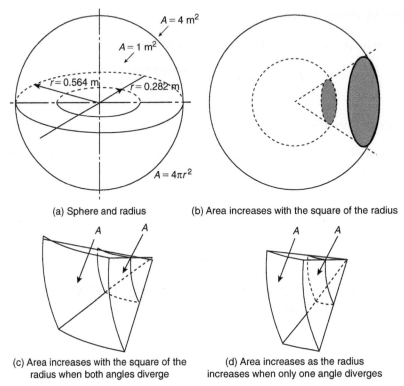

(a) Sphere and radius     (b) Area increases with the square of the radius

(c) Area increases with the square of the     (d) Area increases as the radius
radius when both angles diverge              increases when only one angle diverges

**Figure 2.4: Relationship of spherical surface area to radius.**

## 2.7 Inverse Square Law

If we double the radius of the sphere to 0.564 m, the surface area of the sphere quadruples because the radius is squared in the area equation ($A = 4\pi r^2$). Thus our intensity will drop to one-fourth its former value. (Note, however, that the total acoustic power is still 1 W so the $L_W$ still is 120 dB.) Now an intensity change from 1 W to 0.25 W/m$^2$ can be written as a decibel change. The acoustic intensity (i.e., the power per unit of area) has dropped 6 dB in any given area:

$$L_I = 10 \log \frac{0.25\,(\text{W/m}^2)(\text{new measurement})}{(1\,\text{W/m}^2)\left(\begin{array}{c}\text{original reference}\\\text{at the shorter radius}\end{array}\right)}$$

$$= -6.02\,\text{dB}.$$

Therefore our $L_P$ had to also drop 6 dB and would now be approximately 114 dB.

This effect is commonly called the inverse square law change in level. Gravity, light, and many other physical effects exhibit this rate of change with varying distance from a source. Obviously, if you halve the radius, the levels all rise by 6 dB.

## 2.8 Directivity Factor

Finally, make the point source radiating one acoustic watt a hemispherical radiator instead of an omnidirectional one. Thus at 0.282 m the surface area is now half of what our sphere had or 0.5 m². Therefore our intensity is now 1 W/0.5 m² or the equivalent 2 W/m²:

$$10 \log \frac{2 \text{ W/m}^2}{1 \text{ W/m}^2} = 3.01 \text{ dB}.$$

Therefore our $L_P$ is 123.01 dB. $L_w$ remains 120 dB. This 3.01-dB change represents a 2:1 change in the power per unit area; thus, a hemispherical radiator is said to have twice the directivity factor a spherical radiator has. The directivity factor is identified by a number of symbols—$D_F$, $Q$, $R\theta$, $\lambda$, $M$, etc. $Q$ is the most widely used in the United States so we have chosen it for this text. Directivity can also be expressed as a solid angle in steradians or $sr = 4\pi/Q$.

## 2.9 Ohm's Law

Recall that the use of the term "decibel" always implies a power ratio. Power itself is rarely measured as such. The most common quantity measured is voltage. If in measuring the voltage of a sine wave signal (oscillators are the most reliable and common of the test-signal sources) you obtain the rms voltage, you can calculate the average power developed by using Ohm's law. Figure 2.5 is a reminder of its many basic forms and uses the following definitions:

$W$ is the average electrical power in watts (W).

$I$ is the rms electrical current in amperes (A).

$R$ is the electrical resistance in ohms (Ω).

$E$ is the electromotive force in rms volts (V).

$PF$ is the power factor (cos θ).

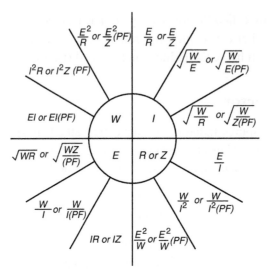

Figure 2.5: Ohm's law nomograph for AC or DC.

## 2.10 A Decibel is a Decibel is a Decibel

The decibel is always a power ratio; therefore, when dealing with quantities that are not power ratios, that is, voltage, use the multiplier 20 in place of 10. As we encounter each reference for the dB, we will indicate the correct multiplier. Table 2.4 lists all the standard references, and Tables 2.5 through 2.8 contain additional information regarding reference labels and quantities. The decibel is not a unit of measurement like an inch, a watt, a liter, or a gram. It is the logarithm of a nondimensional ratio of two power-like quantities.

For $L_P = 20 \log (x\,\text{Pa}/0.00002\,\text{Pa})$, use Eq. (2-29).

$$L_P = (20 \log x\,\text{Pa} + 94)\,\text{dB} \tag{2.26}$$

## 2.11 Older References

Much earlier, but valuable, literature used $10^{-13}\,\text{W}$ as a reference. In that case, the $L_P$ value approximately equals the $L_W$ value at 0.282 ft from an omnidirectional radiator in a free field (i.e., the number values are the same but, of course, different quantities are

**Table 2.4: Common Decibel Notations and References**

| Quantity | Standard reference | Symbol | Log multiplier |
|---|---|---|---|
| Sound pressure | Water: 1 dyn/cm$^2$ | SPL or | 20 |
| | Air: 0.0002 dyn/cm$^2$ | $L_P$ | |
| | or 0.00002 N/m$^2$ | | |
| Sound intensity | 10$^{-16}$ W/cm$^2$ | | 10 |
| | 10$^{-12}$ W/m$^2$ | | |
| Sound power | 10$^{-12}$ W (new) | PWL | 10 |
| | 10$^{-13}$ W (old) | or $L_w$ | |
| Audio power | 10$^{-3}$ W | dBm | 10 |
| EMF | 1 V | dBV | 20 |
| Amperes | 1 mA | | 20 |
| Acceleration | 1 gRMS | | 20 |
| Acceleration | 1 g$^2$/Hz | | 10 |
| Spectral density | | | |
| Volume units | 10$^{-3}$ W | VU | 10 |
| Distance | 1 ft or 1 m | $\Delta D_x$ | 20 |
| Noise-ref | −90 dBm at 1 kHz | dBm | 10 |

$$dB = Logarithm\ Multiplier \times \log \frac{Quantity}{Standard\ Reference}$$

being measured). For 1 W using $10^{-12}$ W at 0.283 m, $L_W \cong L_P = 120$ dB. For 1 W using $10^{-13}$ W at 0.282 ft, $L_W \cong L_P = 130$ dB as found with the equation:

$$L_P = L_W - 10 \log(4\pi r^2) \tag{2.27}$$

where $L_W$ is 10 log the wattage divided by the reference power $10^{-13}$ and $r$ is the distance in meters from the center of the sound source.

Figure 2.6 requires that you either know the distance from the source or assumes you are in the steady reverberant sound field of an enclosed space. $L_P$ readings without one of these is meaningless.

Figure 2.7 shows typical power and $L_W$ values for various acoustic sources.

### Table 2.5: Preferred Reference Labels for Acoustic

| Name | Definition |
|------|-----------|
| Sound pressure squared level | $L_P = 20 \log (p/p_o)$ dB |
| Vibratory acceleration level | $L_a = 20 \log (a/a_o)$ dB |
| Vibratory velocity level | $L_V = 20 \log (v/v_o)$ dB |
| Vibratory force level | $L_F = 20 \log (F/F_o)$ dB |
| Power level | $L_W = 10 \log (P/P_o)$ dB |
| Intensity level | $L_I = 10 \log (I/I_o)$ dB |
| Energy density level | $L_E = 10 \log (E/E_o)$ dB |

### Table 2.6: A-Weighted Recommended Descriptor List

| Term | Symbol |
|------|--------|
| A-weighted sound level | $L_A$ |
| A-weighted sound power level | $L_{WA}$ |
| Maximum A-weighted sound level | $L_{max}$ |
| Peak A-weighted sound level | $L_{pk}$ |
| Level exceeded $\times$ % of the time | $L_x$ |
| Equivalent sound level | $L_{eq}$ |
| Equivalent sound level over time ($T$) | $L_{eq(T)}$ |
| Day sound level | $L_d$ |
| Night sound level | $L_n$ |
| Day–night sound level | $L_{dn}$ |
| Yearly day–night sound level | $L_{dn(Y)}$ |
| Sound exposure level | $L_{SE}$ |

### Table 2.7: Associated Standard Reference Values

| |
|---|
| 1 atm = 1.013 bar = 1.033 kpa/cm$^2$ = 14.70 lb/in$^2$ = 760 mm Hg = 29.92 in Hg |
| Acceleration of gravity: g = 980.665 cm/s$^2$ = 32.174 ft/s$^2$ (standard or accepted value) |
| Sound level: The common reference level is the audibility threshold at 1000 Hz, i.e., 0.0002 dyn/cm$^2$, $2 \times 10^{-4}$ μbar, $2 \times 10^{-5}$ N/m$^2$, $10^{-16}$ W/cm$^2$ |

**Table 2.8: Recommended Descriptor List**

| Term | A weighting | Alternative[a] A weighting | Other weighting[b] | Unweighted |
|---|---|---|---|---|
| Sound (pressure) level[c] | $L_A$ | $L_{pA}$ | $L_B$, $L_{pB}$ | $L_p$ |
| Sound power level | $L_{WA}$ | | $L_{WB}$ | $L_W$ |
| Maximum sound level | $L_{max}$ | $L_{Amax}$ | $L_{Bmax}$ | $L_{pmax}$ |
| Peak sound (pressure) level | $L_{Apk}$ | | $L_{Bpk}$ | $L_{pk}$ |
| Level exceeded $x\%$ of the time | $L_x$ | $L_{Ax}$ | $L_{Bx}$ | $L_{Px}$ |
| Equivalent sound level | $L_{eq}$ | $L_{Aeq}$ | $L_{Beq}$ | $L_{peq}$ |
| Equivalent sound level over time $(T)^d$ | $L_{eq(T)}$ | $L_{Aeq(T)}$ | $L_{Beq(T)}$ | $L_{peq(T)}$ |
| Day sound level | $L_d$ | $L_{Ad}$ | $L_{Bd}$ | $L_{pd}$ |
| Night sound level | $L_n$ | $L_{An}$ | $L_{Bn}$ | $L_{pn}$ |
| Day–night sound level | $L_{dn}$ | $L_{Adn}$ | $L_{Bdn}$ | $L_{pdn}$ |
| Yearly day–night sound level | $L_{dn(Y)}$ | $L_{Adn(Y)}$ | $L_{Bdn(Y)}$ | $L_{pdn(Y)}$ |
| Sound exposure level | $L_S$ | $L_{SA}$ | $L_{SB}$ | $L_{Sp}$ |
| Energy average value over (nontime domain) set of observations | $L_{eq(e)}$ | $L_{Aeq(e)}$ | $L_{Beq(e)}$ | $L_{peq(e)}$ |
| Level exceeded $x\%$ of the total set of (nontime domain) observations | $L_{x(e)}$ | $L_{Ax(e)}$ | $L_{Bx(e)}$ | $L_{px(e)}$ |
| Average $L_x$ value | $L_x$ | $L_{Ax}$ | $L_{Bx}$ | $L_{px}$ |

[a]"Alternative" symbols may be used to assure clarity or consistency.

[b]Only B weighting is shown. Applies also to C, D, and E weighting.

[c]The term "pressure" is used only for the unweighted level.

[d]Unless otherwise specified, time is in hours [e.g., the hourly equivalent level is $L_{eq(1)}$]. Time may be specified in nonquantitative terms [e.g., could be specified as $L_{eq(WASH)}$ to mean the washing cycle noise for a washing machine].

## 2.12 The Equivalent Level ($L_{EQ}$) in Noise Measurements

Increasingly, acoustical workers in the noise control field are erecting an interesting edifice of measurement systems. A number of these measurement systems are based on the concept of average energy. Suppose, for example, that we have some means of collecting all of the A-weighted sound energy that arrives at a particular location over a

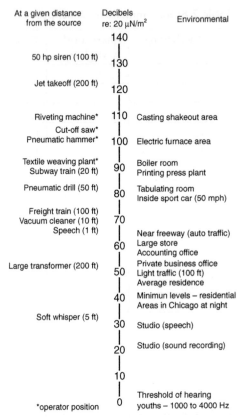

**Figure 2.6: Typical A-weighted sound levels as measured with a sound level meter. (Courtesy of GenRad.)**

certain period of time such as 90 dBA for 3.6 s (this could be a series of levels that lasted seconds, hours, or even days). We can then calculate the decibel level of steady noise for, say, 1 h that would be the equivalent level of the dBA for 3.6 s. That is, we wish to find the energy equivalent level for 1 h:

$$L_{EQ} = 10 \log \left( \frac{1}{3600 \text{ s}} \int_0^{3.6\text{s}} \frac{P_A^2}{P_o^2} \, dt \right) \text{ in decibels} \tag{2.28}$$

where $P_A$ is the acoustic pressure, $P_o$ is the reference acoustic pressure, and 3600 s is the averaging time interval.

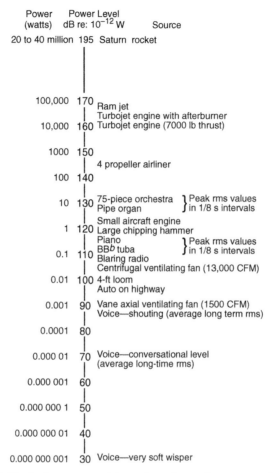

**Figure 2.7: Typical power and $L_W$ values for various acoustic sources.**

This integration reduces to

$$L_{EQ} = 10 \log \left( \frac{10^{\frac{90}{10}} \times 3.6\,\text{s}}{3600\,\text{s}} \right).$$

Thus 1.0 hour of noise energy at 60 dBA is the equivalent energy exposure of 90 dBA for 3.6 s.

$L_{DN}$ (day–night level), *CNEL* (community noise level), and so on all follow similar schemes with variation in weightings for differing times of day, etc.

It is of interest that shooting a 0.458 magnum 174.7 $L_P$ (peak) for 2.5 ms translates into

$$L_{EQ} = 10 \log \left( \frac{10^{\frac{174.7}{10}} \times 0.0025\,\text{s}}{3600\,\text{s}} \right)$$

$$= 113.12\,\text{dB}$$

of steady sound for 1 h. OSHA allows only 15 min of exposure to levels of 110–115 dBA. As Howard Ruark's African guide, Harry Selby, remarked after Ruark had accidentally set off both barrels at once of a 0.470 express rifle while being charged by a Cape buffalo, "One of you ought to get up."

## 2.13  Combining Decibels

### 2.13.1  Adding Decibel Levels

The sum of two or more levels expressed in dB may be found as follows:

$$L_T = 10 \log \left( 10^{\frac{L_1}{10}} + 10^{\frac{L_2}{10}} + K + 10^{\frac{L_N}{10}} \right). \tag{2.29}$$

If, for example, we have a noisy piece of machinery with an $L_P = 90$ dB and wish to turn on a second machine with an $L_P = 90$ dB, we need to know the combined $L_P$. Because both measured levels are the result of the power being applied to the machine, with some percentage being converted into acoustic power, we can determine $L_T$ by using Eq. (2-33). Therefore

$$L_T = 10 \log \left( 10^{\frac{90}{10}} + 10^{\frac{90}{10}} \right)$$

$$= 10 \log \left( 10^9 + 10^9 \right)$$

$$= 10 \log \left( 2 \times 10^9 \right)$$

$$= 93\ \text{dB}.$$

Doubling the acoustic power results in a 3 dB increase.

An alternative dB addition technique is given through the courtesy of Gary Berner.

$$L_T = 10 \log \left( 10^{\frac{-(diff\ in\ dB)}{10}} + 1 \right) + smallest\ number \tag{2.30}$$

### Example

If we wish to add 90 dB to 96 dB, using Eq. (2-33), take the difference in dB (6 dB) and put it in the equation:

$$L_T = 10 \log \left( 10^{\frac{6}{10}} + 1 \right) + 90$$
$$= 96.97 \text{ dB.}$$

Input signals to a mixing network also combine in this same manner, but the insertion loss of the network must be subtracted. Two exactly phase-coherent sine wave signals of equal amplitude will combine to give a level 6 dB higher than either sine wave.

The general case equation for adding sound pressure, voltages, or currents is

$$\text{Combined } L_P = 20 \log \sqrt{\left( 10^{\frac{E_1}{20}} \right)^2 + \left( 10^{\frac{E2}{20}} \right) + 2 \left( 10^{\frac{E_1}{20}} \right) \left( 10^{\frac{E_2}{20}} \right) (\cos[a_1 - a_2])}. \tag{2.31}$$

Table 2.9 shows the effects of adding two equal amplitude signals with different phases together using Eq. (2-36).

### 2.13.2 Subtracting Decibels

The difference of two levels expressed in dB may be found as follows:

$$L_{diff} = 10 \log \left( 10^{\frac{Total\ Level}{10}} - \frac{Level\ with\ one\ source\ off}{10} \right). \tag{2.32}$$

### 2.13.3 Combining Levels of Uncorrelated Noise Signals

When the sound level of a source is measured in the presence of noise, it is necessary to subtract out the effect of the noise on the reading. First, take a reading of the source

**Table 2.9: Combining Pure Tones of the Same Frequency but Differing Phase Angles**

| Signal 1 amplitude, $L_P$ (dB) | Signal 1 phase, in degrees | Signal 2 amplitude, $L_P$ (dB) | Signal 2 phase, in degrees | Combined signal amplitude, $L_P$ (dB) |
|---|---|---|---|---|
| 90 | 0 | +90 | 0 | 96.02 |
| 90 | 0 | +90 | 10 | 95.99 |
| 90 | 0 | +90 | 20 | 95.89 |
| 90 | 0 | +90 | 30 | 95.72 |
| 90 | 0 | +90 | 40 | 95.48 |
| 90 | 0 | +90 | 50 | 95.17 |
| 90 | 0 | +90 | 60 | 94.77 |
| 90 | 0 | +90 | 70 | 94.29 |
| 90 | 0 | +90 | 80 | 93.71 |
| 90 | 0 | +90 | 90 | 93.01 |
| 90 | 0 | +90 | 100 | 92.18 |
| 90 | 0 | +90 | 110 | 91.19 |
| 90 | 0 | +90 | 120 | 90.00 |
| 90 | 0 | +90 | 130 | 88.54 |
| 90 | 0 | +90 | 140 | 86.70 |
| 90 | 0 | +90 | 150 | 84.28 |
| 90 | 0 | +90 | 160 | 80.81 |
| 90 | 0 | +90 | 170 | 74.83 |
| 90 | 0 | +90 | 180 | $-\infty$ |

and the noise combined ($L_{S+N}$). Then take another reading of the noise alone (the source having been shut off). The second reading is designated $L_N$. Then

$$L_S = 10 \log \left( 10^{\frac{L_{S+N}}{10}} - 10^{\frac{L_N}{10}} \right). \tag{2.33}$$

To combine the levels of uncorrelated noise signals we can also use the chart in Figure 2.8.

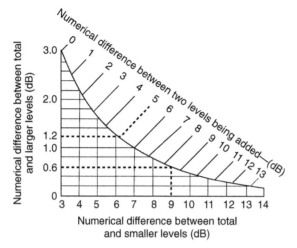

**Figure 2.8: Chart used for determining the combined level of uncorrelated noise signals.**

### 2.13.4 To Add Levels

Enter the chart with the numerical difference between the two levels being added (top of chart). Follow the line corresponding to this value to its intersection with the curved line and then move left to read the numerical difference between the total and larger levels. Add this value to the larger level to determine the total.

### Example
To add 75 dB to 80 dB, subtract 75 dB from 80 dB; the difference is 5 dB. In Figure 2.8, the 5-dB line intersects the curved line at 1.2 dB on the vertical scale. Thus the total value is 80 dB + 1.2 dB, or 81.2 dB.

### 2.13.5 To Subtract Levels

Enter the chart in Figure 2.8 with the numerical difference between the total and larger levels if this value is less than 3 dB. Enter the chart with the numerical difference between the total and smaller levels if this value is between 3 and 14 dB. Follow the line corresponding to this value to its intersection with the curved line and then either left or down to read the numerical difference between total and larger (smaller) levels. Subtract this value from the total level to determine the unknown level.

**Example**

Subtract 81 dB from 90 dB; the difference is 9 dB. The 9-dB vertical line intersects the curved line at 0.6 dB on the vertical scale. Thus the unknown level is 90 dB – 0.6 dB, or 89.4 dB.

## 2.14 Combining Voltage

To combine voltages, use the following equation:

$$E_T = \sqrt{E_1^2 + E_2^2 + 2E_1E_2[\cos(a_1 - a_2)]} \tag{2.34}$$

where $E_T$ is the total sound pressure, current, or voltage; $E_1$ is the sound pressure, current, or voltage of the first signal; $E_2$ is the sound pressure, current, or voltage of the second signal; $a_1$ is the phase angle of signal one; and $a_2$ is the phase angle of signal two.

## 2.15 Using the Log Charts

### 2.15.1 The 10 Log x Chart

There are two scales on the top of the 10 $\log_{10} x$ chart in Figure 2.9. One is in dB above and below a 1-W reference level and the other is in dBm (reference 0.001 W). Power ratios may be read directly from the 1-W dB scale.

**Example**
How many decibels is a 25:1 power ratio?

1. Look up 25 on the power–watts scale.

2. Read 14 dB directly above the 25.

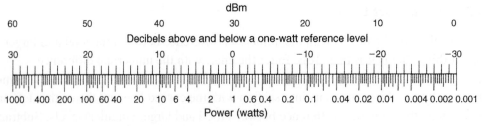

**Figure 2.9: The 10log₁₀ x chart.**

*Example*

We have a 100 W amplifier but plan to use a 12-dB margin for "head room." How many watts will our program level be?

1.  Above 100 W find +50 dBm.

2.  Subtract 12 dB from 50 dBm to obtain +38 dBm. Just below +38 dBm find approximately 6 W.

*Example*

A 100-W amplifier has 64 dB of gain. What input level in dBm will drive it to full power?

1.  Above 100 W read +50 dBm.

2.  +50 dBm − 64-dB gain = −14 dBm.

*Example*

A loudspeaker has a sensitivity of $L_P$ = 99 dB at 4 ft with a 1-W input. How many watts are needed to have an $L_P$ of 115 at 4 ft?

1.  115 $L_P$ − 99 $L_P$ = +16 dB.

2.  At +16 on the 1-W scale read 39.8 W.

### 2.15.2 The 20 Log x Chart

Refer to the chart in Figure 2.10. A 2:1 voltage, distance, or sound pressure change is found by locating 2 on the ratio or D scale and looking directly above to 6 dB.

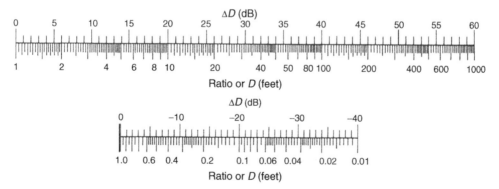

**Figure 2.10: The 20log$_{10}$ x chart.**

*Example*

A loudspeaker has a sensitivity of $L_P = 99\,\text{dB}$ at 4 ft with 1 W of input power. What will the level be at 100 ft?

1. Find the relative dB for 4 ft (relative dB = 12 dB).

2. Find the relative dB for 100 ft (relative dB = 40 dB).

3. Calculate the absolute dB (40 dB – 12 dB = 28 dB).

4. $L_P = 99\,\text{dB} - 28\,\text{dB} = 71\,\text{dB}$.

*Example*

If we raise the voltage from 2 to 10 V, how many decibels would we increase the power?

1. Find the relative dB for a ratio of 2 (relative dB = 6 dB).

2. Find the relative dB for a ratio of 10 (relative dB = 20 dB).

3. Absolute dB change = 20 dB – 6 dB = 14 dB.

4. Because a dB is a dB, the power also changed by 14 dB.

## 2.16 Finding the Logarithm of a Number to Any Base

In communication theory, the base 2 is used. Occasionally, other bases are chosen. To find the logarithm of a number to any possible given base, write

$$x = b^n \tag{2.35}$$

where $x$ is the number for which a logarithm is to be found, $b$ is the base, and $n$ is the logarithm.

Then write

$$\log x = n \log b \tag{2.36}$$

and

$$\frac{\log x}{\log b} = n. \tag{2.37}$$

Suppose we want to find the natural logarithm of 2 (written ln 2). The base of natural logarithms is $e = 2.7188281828$. Then

$$\frac{\log 2}{\log e} = \frac{0.30103}{0.43425}$$

$$= 0.69315$$

To verify this result,

$$e^{0.69315} = 2.$$

To find log 2 of 26,

$$\frac{\log 26}{\log 2} = \frac{1.41497}{0.30103}$$

$$= 4.70044$$

The general case is

$$\frac{\log_{10} \ of \ the \ number}{\log_{10} \ of \ the \ base} = \log_{\text{base}} \ of \ the \ number. \qquad (2.38)$$

## 2.17 Semitone Intervals

Suppose that we need $\sqrt[12]{2}$ (the semitone interval in music). We could write

$$\frac{\log 2}{12} = \log \sqrt[12]{2}. \qquad (2.39)$$

Therefore

$$10^{\frac{\log 2}{12}} = 10^{\frac{0.30}{12}}$$

$$= 10^{0.02508}$$

$$= 1.05946$$

$$= \sqrt[12]{2}.$$

This is the same as multiplying 1.05946 by itself 12 times to obtain 2.

$10^{0.02508}$ is called the antilog of 0.02508. The antilog is also written as $\log^{-1}$, antilog 10, or 10 exp. All these terms mean exactly the same thing.

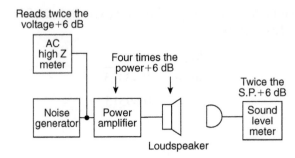

Figure 2.11: Voltage, electrical power, $P_w$, and sound pressure compared.

## 2.18 System Gain Changes

Imagine a noise generator driving a power amplifier and a loudspeaker (Figure 2.11). If the voltage out of the noise generator is raised by 6 dB, what happens?

| Voltage | Electrical power | $L_P$ | $L_W$ |
|---|---|---|---|
| Doubled | Quadrupled | Doubled | Quadrupled |
| +6 dB | +6 dB | +6 dB | +6 dB |

This means that, in a linear system, a level change ahead of any components results in a level change for that same signal in all subsequent components, although it might be measured as quite different voltages or wattages at differing points. The change in level at any point would be the same. We will work with this concept a little later when we plot the gains and losses through a total system.

## 2.19 The VU and the Volume Indicator Instrument

Volts, amperes, and watts can be measured by inserting an appropriate meter into the circuit. If all audio signals were sine waves, we could insert a dBm meter into the circuit and get a reading that would correlate with both electrical and acoustical variations. Unfortunately, audio signals are complex waveforms and their rms value is not 0.707 times peak but can range from as small as 0.04 times peak to as high as 0.99 times peak (Figure 2.12). To solve this problem, broadcasting and telephone engineers got together in 1939 and designed a special instrument for measuring speech and music in communication circuits. They calibrated this new type of instrument in units called

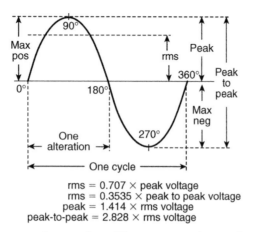

rms = 0.707 × peak voltage
rms = 0.3535 × peak to peak voltage
peak = 1.414 × rms voltage
peak-to-peak = 2.828 × rms voltage

**Figure 2.12: Sine wave voltage values. The average voltage of a sine wave is zero.**

VU. The dBm and the VU are almost identical; the only difference is their usage. The instrument used to measure VU is called the volume indicator (VI) instrument. (Some users ignore this and incorrectly call it a VU meter.) Both dBm meters and volume indicator instruments are specially calibrated voltmeters. Consequently, the VU and dBm scales on these meters give correct readings only when the measurement is being made across the impedance for which they are calibrated (usually 150 or 600 $\Omega$). Readings taken across the design impedance are referred to as true levels, whereas readings taken across other impedances are called apparent levels.

Apparent levels can be useful for relative frequency response measurements, for example. When the impedance is not 600 $\Omega$, the correction factor of 10 log (600/*new impedance*) can be added to the formula containing the reference level as in the following equation:

$$True\ VU\ =\ Apparent\ VU\ +\ 10\log\frac{600}{Z\ measured}.\tag{2.40}$$

The result is the true level.

### 2.19.1 The VU Impedance Correction

When a VI instrument is connected across 600 $\Omega$ and is indicating 0 VU on a sine wave signal, the true level is 4 dB higher, or +4 dBm, instead of 0 dBm or zero level. The reason this is so is shown in Figure 2.13. The VI instrument uses a 50-$\mu$A D'Arsonval

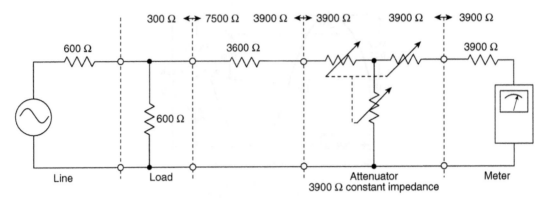

**Figure 2.13: Volume indicator instrument circuit.**

movement in conjunction with a copper-oxide bridge-type rectifier. The impedance of the instrument and rectifier is $3900\,\Omega$. To minimize its effect when placed across a $600$-$\Omega$ line, it is "built out" an additional $3600\,\Omega$ to a total value of $7500\,\Omega$. The addition of this build-out resistance causes a 4-dB loss between the circuit being measured and the instrument. Therefore when a properly installed VI instrument is fed with 0 dBm across a 600 line, the meter would actually read $-4$ VU on its scale. (When the attenuator setting is added, the total reading is indeed 0 VU.)

Presently, no major U.S. manufacturer offers for sale a standard volume indicator that complies with the applicable standard (C16.5). The standard requires that an attenuator be supplied with the instrument and none of the manufacturers do so. What they are doing requires some attention. The instruments (usually high-impedance bridge types) are calibrated so as to act as if the attenuator were present. When the meter reads 0 VU (on a sine wave for calibration purposes), the true level is $+4$ dBm. This means a voltage of 1.23 V across $600\,\Omega$ will cause the instrument to read an apparent 0 VU. Note that when reading sine wave levels, the label used is "dBm." When measuring program levels, the label used is "VU." The VU value is always the instrument indication plus the attenuator value.

Two different types of scales are available for VI meters (Figure 2.14). Scale A is a VU scale (recording studio use), and scale B is a modulation scale (broadcast use). On complex waveforms (speech and music), the readings observed and the peak levels present are about 10 dB apart. This means that with a mixer amplifier having a sine wave output capability of

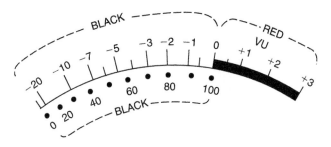

(a) Recording and test equipment

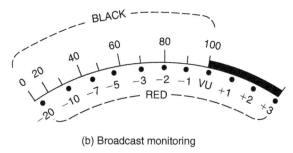

(b) Broadcast monitoring

**Figure 2.14: Volume indicator instrument scales.**

+18 dBm, you are in danger of distortion with any signal indicating more than +8 VU on the VI instrument (+18 dBm – [+10 dB] peaking factor or meter lag equals +8 VU).

Figure 2.15 shows an example of commercially available VI instrument panels used in the past that included the VI instrument and 3900-Ω attenuator, which also contains the 3600-Ω build-out resistor.

### 2.19.2 How to Read the VU Level on a VI Instrument

A VI instrument is used to measure the level of a signal in VU. In calibration: 0 VU = 0 dBm and a 1.0-VU increment is identical to a 1.0-dB increment. The true level reading in VU is found by

$$True\ VU\ level\ =\ Apparent\ level\ +\ Impedance\ correction \tag{2.41}$$

**Figure 2.15: Examples of commercial-type VI instrument panels.**

or

$$\text{True VU level} = \text{Instrument indication} + 10\log\left(\frac{600}{Z_{act}}\right)$$

where, *apparent level = instrument indication + attenuator or sensitivity indicator.*

Thus we can have the following.

1. A direct reading from the face of the instrument (zero preferred).

2. The reading from the face of the instrument plus the reading from the attenuator or other sensitivity adjustment—normally a minimum of +4 dB or higher. When the instrument indicates zero, the apparent level is the attenuator setting.

3. The correction factor for impedance other than the reference impedance. 600 $\Omega$ is the normal impedance chosen for a reference, but any value can be used so long as the voltage across it results in 0.001 W (Figure 2.16).

*Example*

We have an indication on the instrument of $-4$ VU. The sensitivity control is at $+4$. We are across 50 $\Omega$ (a 100-W amplifier with a 70.7-V output). Using Figure 2.16, our true VU would be $-4$ VU $+ (+4$ VU$) + 10.8$ correction factor $= 10.8$ VU.

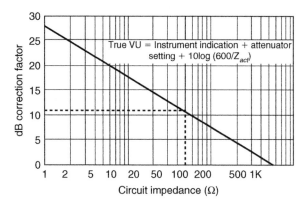

**Figure 2.16: Relationship between circuit impedance and dB correction value.**

### 2.19.3  Calibrating a VI Instrument

The instrument should be calibrated to read a true level of zero VU when an input of a 1000-Hz steady-state sine wave signal of 0 dBm (0.001 W) is connected to it. For example, typical calibration is when the instrument indicates −4, the attenuator value is +4, and it is connected across a 600 circuit. Levels read on a VI instrument when the source is the aforementioned sine wave signal should be stated as dBm levels.

#### 2.19.3.1  Reading a VI Instrument on Program Material

Because of the ballistic properties of VI instruments, they exhibit what has been called "instrument lag." On short-duration peak levels, they will "lag" by approximately 10 dB. Stated another way, if we read a true VU level of +8 VU on a speech signal, then the level in dBm becomes +18 dBm. This means that the associated amplification equipment, when fed a true VU level of +8 VU, must have a steady-state sine wave capability of +18 dBm to avoid overload.

#### 2.19.3.2  Rule

Levels stated in VU are assumed to be program material, and levels stated in dBm are assumed to be steady-state sine wave.

### 2.19.4 Reading Apparent VU Levels

Volume indicator instruments can be used to read apparent or relative levels. If, for example, you know that overload occurs at some apparent level, you can use that reading as a satisfactory guide to the system's operation, even though you do not know the true level. When adjusting levels using the instrument to read the relative change in level, such as turning the system down 6 dB, you do not need to do so in true level readings. Instrument indication serves effectively in such cases.

When being given a level, be sure to ascertain whether it is:

1. An instrument indication.
2. An apparent level.
3. A true level.
4. A relative level.
5. A calibration level.
6. A program level.
7. None of the above but simply an arbitrary meter reading.

**Special Note:** Well-designed mixers have instruments that indicate the available input power level to the device connected to its output. Such levels are true levels.

## 2.20 Calculating the Number of Decades in a Frequency Span

To find the relationship of the number of decades between the lowest and the highest frequencies, use the following equations:

$$\frac{H.F.}{L.F.} = 10^1 = 1\,\text{decade} \tag{2.42}$$

therefore

$$\frac{H.F.}{L.F.} = 10^{x\,\text{decade}} \tag{2.43}$$

$$\frac{\ln H.F. - \ln L.F.}{\ln 10} = x\,\text{decades}$$

or

$$\text{In } H.F. - \text{In } L.F. = \text{In } 10 \times (x \text{ decades}). \tag{2.44}$$

Further,

$$H.F. = e^{(x \text{ decades}) \times (\text{In}10) + \text{In } L.F.} \tag{2.45}$$

and

$$L.F. = e^{[\text{In } H.F. - (x \text{ decades} \times \text{In } 10)]}. \tag{2.46}$$

**Example**

How many decades does the bandpass 500 to 12,500 Hz contain? Using Eq. (2-48),

$$\frac{\text{In } 12,500 - \text{In } 500}{\text{In } 10} = 1.39794 \text{ decades}.$$

If we had 12,500 Hz as a *H.F.* limit and wished to know the low frequency that would give us 1.4 decades, we would calculate:

$$L.F. = e^{[\text{In } 12,500 - (1.4 \text{ decades} \times \text{In } 10)]}$$
$$= 497.63 \text{ Hz}.$$

If we had the L.F. limit and wished to know the H.F., then

$$H.F. = e^{(1.4 \text{ decades} \times \text{In } 10) + \text{In } 497.63}$$
$$= (12,500 \text{ Hz}).$$

## 2.21 Deflection of the Eardrum at Various Sound Levels

If we make the assumption that the eardrum displacement is the same as that of the air striking it, we can write

$$D_{in} = 3 \times 10^{-7} \left( \frac{10^{\frac{L_P}{20}}}{f} \right) \tag{2.47}$$

or

$$D_{cm} = 39 \times 10^{-3} \left( \frac{0.0002 \times 10^{\frac{L_p}{10}}}{f} \right) \quad\quad (2.48)$$

where $D_{in}$ is the displacement in inches (the rms amplitude) of the air, $D_{cm}$ is the displacement in centimeters, $f$ is the frequency in hertz, and $L_p$ is the sound level in decibels referred to 0.00002 N/m$^2$.

**Example**
What is the displacement of the eardrum in inches for a tone at 1000 Hz at a level of 74 dB? Using Eq. (2-51),

$$D_{in} = 3 \times 10^{-7} \left( \frac{10^{\frac{74}{20}}}{1000} \right)$$

$$= 0.0000015 \text{ in}$$

which is a displacement of approximately one-one-millionth of an inch (0.000001 in).

## 2.22 The Phon

Figure 2.17 shows free-field equal-loudness contours for pure tones (observer facing source), determined by Robinson and Dadson at the National Physical Laboratory, Teddington, England, in 1956 (ISO/R226-1961). The phon scale is of equal-loudness level contours. At 1000 Hz every decibel is the equivalent loudness of a phon unit.

For two different sounds within a critical band (for most practical purposes, using ⅓ octave bands suffices) they are added in the same manner as decibel readings.

$$P_T = 10 \log \left( 10^{\frac{L_{p1}}{10}} + 10^{\frac{L_{p2}}{10}} \right)$$

$$= \text{phons} \quad\quad (2.49)$$

where $L_{P1}$ and $L_{P2}$ are the individual sound levels in dB.

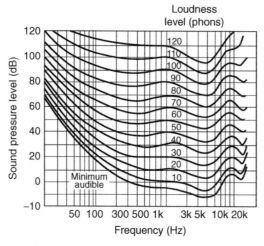

**Figure 2.17: Equal loudness contours.**

For example, suppose that within the same critical band we have two tones each at 70 phons. Using Eq. (2-53),

$$P_T = 10_{\log}\left(10^{\frac{70}{10}} + 10^{\frac{70}{10}}\right)$$
$$= 73 \text{ phons.}$$

An interesting experiment in this regard is to start with two equal level signals 10 Hz apart at 1000 Hz and gradually separate them in frequency while maintaining their phon level.

They will increase in apparent loudness as they separate. This is one of the reasons a distorted system sounds louder than an undistorted system at equal power levels. One final factor worthy of storage in your own mental "read-only memory" is that in the 1000-Hz region most listeners judge a change in level of 10 dB as twice or half the loudness of the original tone.

Figure 2.18 is a chart of frequency and dynamic range for various musical instruments and the upper and lower frequency range of the average young adult.

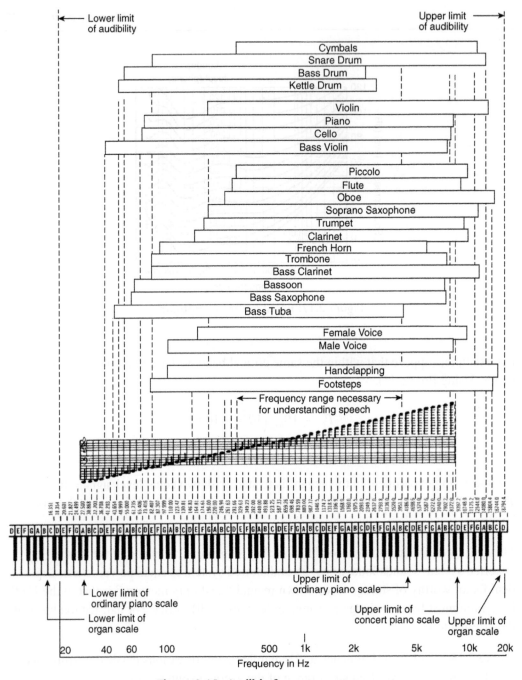

**Figure 2.18: Audible frequency range.**

**Table 2.10: Tempered Scale**

| Note | Frequency ratio | Frequency Hz |
|------|-----------------|--------------|
| C | 1.000 | 262 |
| C#, Db | 1.059 | 277 |
| D | 1.122 | 294 |
| D#, Eb | 1.189 | 311 |
| E | 1.260 | 330 |
| F | 1.335 | 349 |
| F#, Gb | 1.414 | 370 |
| G | 1.498 | 392 |
| G#, Ab | 1.587 | 415 |
| A | 1.682 | 440 |
| A#, Bb | 1.782 | 466 |
| B | 1.888 | 494 |
| C | 2.000 | 523 |

## 2.23 The Tempered Scale

The equal tempered musical scale is composed of 12 equally spaced intervals separated by a factor of $\sqrt[12]{2}$. All notes on the musical scale (excluding sharps and flats), however, are not equally spaced. This is because there are two one-half step intervals on the scale: that between E and F and that between B and C. The 12 tones, therefore, go as follow: C, C#, D, D#, E, F, F#, G, G#, A, A#, B, C (see Table 2.10).

## 2.24 Measuring Distortion

Figure 2.19 illustrates one of the ways of measuring harmonic distortion. Two main methods are employed. One uses a band rejection filter of narrow bandwidth having a rejection capability of at least 80 dB in the center of the notch. This deep notch "rejects" the fundamental of the test signal (usually a known-quality sine wave from a test audio oscillator) and permits reading the noise voltage of everything remaining in the rest of the bandpass. Unfortunately, this also includes the hum and noise, as well as the harmonic content of the equipment being tested (see Figure 2.20).

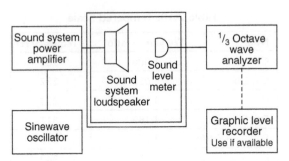

**Figure 2.19: Measurement of harmonic distortion.**

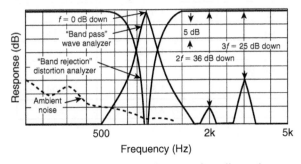

**Figure 2.20: Methods of measuring distortion.**

The second method is more useful. It uses a tunable wave analyzer. This instrument allows measurement of the amplitudes of the fundamental and each harmonic, as well as identifying the hum, the amplitude, and the noise spectrum shape (Figure 2.20). Such analyzers come in many different bandwidths, with a 1/10 octave unit allowing readings down to 1% of the fundamental (it is $-45\,dB$ at $2f$). By looking at Figure 2.20, it is easy to see that harmonic distortion appears as a spurious noise. Today, tracking filter wave analysis allows nonlinear distortion behavior to be "tracked" or measured.

## 2.25 The Acoustical Meaning of Harmonic Distortion

The availability of extremely wide-band amplifiers with distortions approaching the infinitesimal and the gradual engineering of a limited number of loudspeakers with distortions just under 1% at usable levels (90 dB *SPL*–100 dB *SPL* at 10–12 ft) brings up an interesting question: "How low a distortion is really needed?"

### 2.25.1 Calculating the Maximum Allowable Total Harmonic Distortion in an Arena Sound System

The most difficult parameter to achieve in the typical arena sound system is a sufficient signal-to-noise ratio (*SNR*) to ensure acceptable articulation losses for consonants in speech. It must be at least 25 dB. In that case, the total harmonic distortion should be at least 10 dB below the 25-dB *SNR* to avoid the addition of the two signals. If both signals were at the same level, a 3-dB increase in level would occur. Therefore $(-25\,\text{dB}) + (-10\,\text{dB})$ means that the total harmonic distortion (*THD*) should not exceed $-35\,\text{dB}$.

$$Percentage = 100 \times 10^{\frac{+dB}{20}} \qquad (2.50)$$

Therefore we could calculate

$$100 \times 10^{\frac{-35}{20}} = 1.78\%.$$

This is why carefully thought-out designs for use in heavy-duty commercial sound work have a *THD* of 0.8 to 0.9%:

$$20\log\frac{100 \pm x\%}{100} = \text{dB change.}$$

Since the 0.8% already represents $(100 - 99.2)$, we can write

$$20\log\frac{0.8}{100} = -42 \text{ dB.}$$

Now, suppose an amplifier has 0.001% distortion. What sort of dynamic range does this represent?

$$20\log\frac{0.001}{100} = -100$$

That is a power ratio of

$$10^{\frac{100}{10}} = 10,000,000,000.$$

We can conclude that if such a figure were achievable, it would nevertheless not be useful in arena systems.

## 2.26 Playback Systems in Studios

Assume that a monitor loudspeaker can develop $L_P = 110\,dB$ at the mixer's ears and that in an exceptionally quiet studio we reach $L_P = 18\,dB$ at $2000\,Hz$ (NC-20). We then have

$$L_{P_{Diff}} = L_{P_{Total}} - L_{P_{Noise}} \tag{2.51}$$

which is equal to 92 dB. Adding 10 dB to avoid the inadvertent addition of levels gives 102 dB. The distortion now becomes

$$100 \times 10^{\frac{-102}{20}} = 0.00078\%.$$

In this case, extraordinary as it is, the previously esoteric figure becomes a useful parameter.

### 2.26.1 Choosing an Amplifier

As pointed out earlier, the loudspeaker will establish equilibrium around 1% with its acoustic distortion. To the builder of systems, this means that extremely low distortion figures cannot be used within the system as a whole. Therefore systems-oriented amplifier designers have not attempted to extend the bandpass to extreme limits. They know that they must balance bandpass, distortion, noise, and hum against stability with all types of loads, extensions of mean time-before-failure characteristics. Most high-quality sound reinforcement amplifiers incorporate an output transformer, giving us 70 and 25 V and 4, 8, and 16 Ω outputs. In fact, connecting across the 4 and 8 Ω taps yields a 0.69-Ω output.

### Example

Let the rms speech value be $L_P = 65\,dB$ at 2 ft in the 1000- to 2000-Hz octave band (Figure 2.21). Let the ambient noise level be $L_P = 32\,dB$ with the air conditioning on and 16 dB with the air conditioning off in the same octave band (Figure 2.22). With the air conditioning on the signal to noise ratio (*SNR*) is

$$SNR = 65\ dB - 32\ dB \tag{2.52}$$
$$= 33\ dB$$

and with the air conditioning off

$$SNR = 65\ dB - 16\ dB$$
$$= 49\ dB.$$

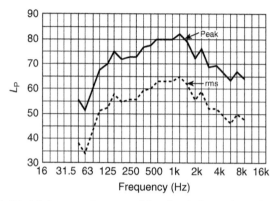

Figure 2.21: Male speech, normal level 2 ft from the microphone.

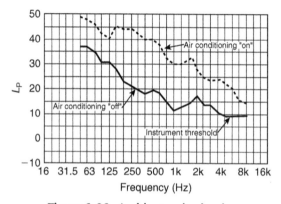

Figure 2.22: Ambient noise levels.

For a harmonic to be equal to $-33\,\text{dB}$, its percentage would be

$$100 \times 10^{\frac{-33}{20}} = 2.24\%.$$

For a harmonic to be equal to $-49\,\text{dB}$, its percentage would be

$$100 \times 10^{\frac{-49}{20}} = 0.355\%.$$

## 2.27 Decibels and Percentages

The comparison of data in decibels often needs to be expressed as percentages. The measurement of *THD* compares the harmonics with the fundamental. After finding out how many dB down each harmonic is compared to the fundamental, sum up all the harmonics and then compare their sum to the fundamental value. The difference is expressed as a percentage. The efficiency of a loudspeaker in converting electrical energy to acoustic energy is also expressed as a percentage. We know that

$20 \log_{10} = 20 \, db$

$20 \log_{100} = 40 \, db$

$20 \log_{1000} = 60 \, db.$

Therefore a signal of $-20 \, dB$ is 1/10 of the fundamental, or $100 \times 1/10 = 10\%$. A signal of $-40 \, dB$ is 1100 of the fundamental, or $100 \times 1/100 = 1\%$. A signal of $-60 \, dB$ is 11,000 of the fundamental, or $100 \times 1/1000 = 0.1\%$. We can now turn this into an equation for finding the percentage when the level difference in decibels is known. For such ratios as voltage, *SPL*, and distance:

$$Percentage = 100 \times 10^{\frac{\pm dB}{20}}. \tag{2.53}$$

For power ratios:

$$Percentage = 100 \times 10^{\frac{\pm dB}{10}}. \tag{2.54}$$

Occasionally, we are presented with two percentages and need the decibel difference between them. For example, two loudspeakers of otherwise identical specifications have differing efficiencies: one is 0.1% efficient and the other is 25% efficient. If the same wattage is fed to both loudspeakers, what will be the difference in level between them in dB?

Since we are now talking about efficiency, we are talking about power ratios, not voltage ratios. We know that

$10 \log_{10} = 20 \, db$

$10 \log_{100} = 40 \, db$

$10 \log_{1000} = 60 \, db$

and so forth.

A 0.1% efficiency is a power ratio of 1000 to 1, or $-30 \, dB$. We also know that $-3 \, dB$ is 50% of a signal, so $-6 \, dB$ would be 25%; $(-6) - (-30) = 24 \, dB$. In other words, there would be a 24-dB difference in level between these two loudspeakers when fed by the same signal. Some consumer market loudspeakers vary this much in efficiency.

## 2.28 Summary

The decibel is the product of the greatest engineering minds in communications early in the last century. When it is combined with the work of Oliver Heaviside and others on impedance at the turn of the 20th century, we are equipped to handle audio levels. The concepts of dB, Z, and dBm are the tools of the professional as well as their language.

## Further Reading

Albers, V. M., 'The world of sound', New York: Barnes, 1970.

Jay, F. (Ed.), 'IEEE standard dictionary of electrical and electronics terms', 2nd ed., New York: The Institute of Electrical and Electronics Engineers, 1977.

Keast, D. N., 'Measurement in mechanical dynamics', New York: McGraw-Hill, 1967.

Read, O., 'The recording and reproduction of sound', Indianapolis, IN: Howard W. Sams, 1952.

Research Council of the Academy of Motion Picture Arts and Sciences, 'Motion picture sound engineering', New York: Van Nostrand, 1938.

Wood, A., 'The physics of music', New York: Dover, 1966.

# CHAPTER 3

# *Acoustic Environment*

Don Davis and Eugene Patronis

## 3.1 The Acoustic Environment

We are concerned about the effect the acoustic environment has on sound. We need to know the effect of a particular acoustic environment on the unaided talker or musician, on the sound system, if installed, and on unwanted sounds (noise) that may be present in the same environment.

An outdoor environment can often be a "free field." "A sound field is said to be a free field if it is uniform, free from boundaries, and is undisturbed by other sources of sound. In practice, it is a field where the effects of the boundaries are negligible over the region of interest." (From the GenRad instruction manual for their precision microphones.)

"Free from boundaries" is the catch phrase here. Anyone who has designed a sound system into a football stadium, a replica of a Greek theater, or a major motor racing course knows first-hand the primary influence of a boundary.

We must also consider:

1. Inverse-square-law level change.
2. Excess attenuation by frequency because of humidity and related factors.

Other factors that can materially affect sound outdoors include:

3. Reflection by and diffraction around solid-objects.
4. Refraction and shadow formation by wind and temperature and wind variations.
5. Reflection and absorption by the ground surface itself.

Research in recent years has advanced the knowledge of atmospheric absorption significantly from the original base laid by Kneser, Knudsen, followed later by Harris, and, more recently, by the work of Sutherland, Piercy, Bass, and Evans (see Figure 3.1). This prediction graph is felt to be reliable within +5% for the temperature indicated (20°C) and 10% over a range of 0 to 40°C.

The June 1977 *Journal of the Acoustical Society of America* had an exceptional tutorial paper entitled "Review of Noise Propagation in the Atmosphere," pages 1403–1418, and included a 96 reference bibliography.

## 3.2 Inverse Square Law

The geometrical spreading of sound from a coherent source (inverse square law rate of level change), which is a change in level of 6 dB for each doubling of distance for a spherical expansion from a point source, is well known to most sound technicians.

$$L_P \text{ at measurement point } = \text{ Ref distance } L_P + 20\log\frac{D_r}{D_m} \qquad (3.1)$$

where $D_r$ is the reference distance and $D_m$ is the measured distance.

Not as well recognized is the change in level of 3 dB per doubling of distance for cylindrical expansion from an infinite line source. The ambient noise from a motor race track with the field of cars evenly spread during the early stages of a race can come very close to being effectively an infinite line source.

$$L_P \text{ at measurement point } = \text{ Ref distance } L_P + 10\log\frac{D_r}{D_m} \qquad (3.2)$$

Finally, there is the case of the parallel "loss free" propagation from an infinite area source—the crowd noise viewed from the center of the audience.

Descriptions of the spreading out of sound for coherent sources remain true for incoherent sources as well. The size of the near field may be more restricted and the propagation less directional but the general rate of level change remains the same. Note that this "spreading out" of sound does not constitute absorption or other loss but merely the reduction of power per unit of area as the distance is increased. Unfortunately, other processes also are going on.

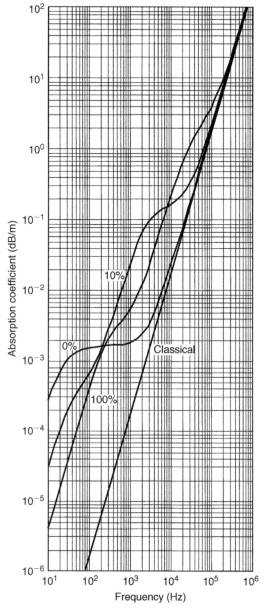

Figure 3.1: Predicted atmospheric absorption in dB/100 m for a pressure of 1 atm, temperature of 20°C, and various values of relative humidity.

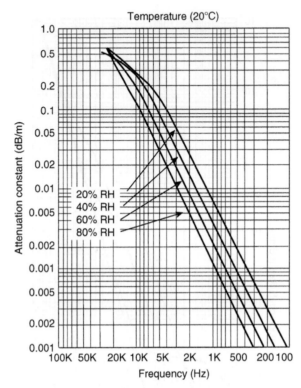

**Figure 3.2: Absorption of sound for different frequencies and values of relative humidity.**

## 3.3 Atmospheric Absorption

These other processes represent actual dissipation of sound energy. Energy is lost due to the combined action of the viscosity and heat conduction of the air and relaxation of behavior in the rotational energy states of the molecules of the air. These losses are independent of the humidity of the air. Additional losses are due to a relaxation of behavior in the vibrational states of the oxygen molecules in the air, as this behavior is strongly dependent on the presence of water molecules in the air (absolute humidity). Both of these energy loss effects cause increased attenuation with increased frequency (Figure 3.2).

This frequency-discriminative attenuation is referred to as excess attenuation and must be added to the level change due to divergence of the sound wave. Total level change is

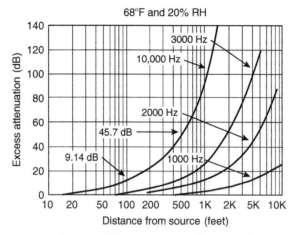

**Figure 3.3: Excess attenuation for different frequencies and distances from the source.**

the sum of inverse-square-law level change and excess attenuation. Figure 3.3 shows the excess attenuation difference between 1000 and 10,000 Hz at various distances.

## 3.4 Velocity of Sound

For a given frequency, the relation of the wavelength to the velocity of sound in the medium is

$$\lambda = \frac{c}{f}$$
$$c = \lambda f \tag{3.3}$$
$$f = \frac{c}{\lambda}$$

where $\lambda$ is the wavelength in feet or meters, $c$ is the velocity of sound in ft/s or m/s, and $f$ is the frequency in Hz.

In dealing with many acoustic interactions, the wavelength involved is significant and the ability to calculate it is important. Therefore we need to be able to both calculate and measure the velocity of sound quickly and accurately.

The velocity of sound varies with temperature to a degree sufficient to require our alertness to it. A knowledge of the exact velocity of sound when using signal-delayed signal analysis allows very precise distance measurements to be made by observing the frequency interval between comb filters from two sources and then converting from frequency to time and finally to distance.

The velocity of sound under conditions likely to be encountered in connection with architectural acoustic considerations is dependent on three fundamental factors. These are:

1. $\gamma$ is the ratio of specific heats and is 1.402 for diatomic molecules (air molecules).
2. $P_S$ is the equilibrium gas pressure in Newtons per square meter ($1.013 \times 10^5$ N/m$^2$).
3. $\rho$ is the density of air in kilograms per cubic meter (kg/m$^3$).

$$c = \sqrt{\frac{\gamma P_s}{\rho}} \tag{3.4}$$

where $c$ is the velocity of sound in m/s.

The density of air varies with temperature, and an examination of the basic equations reveals that, indeed, temperature variations are the predominant influence on the velocity of sound in air.

The equation for calculating the density of air is

$$Density\ of\ air = \left[ \frac{1.293H}{[1 + 0.00367(°C)](76)} \right] \tag{3.5}$$

where *density of air* is in kg/m$^3$; $H$ is the barometric pressure in centimeters of mercury, Hg; °C is the temperature in degrees Celsius; 9/5 (°C) + 32 = °F; and 5/9 (°F) − 32 = °C. Hg in inches times 2.54 equals Hg in centimeters.

### 3.4.1 Example

If we were to measure a temperature of 72°F and a barometric pressure of 29.92 in cm Hg, we would first calculate the density of the air according to data gathered:

$$\frac{5}{9}(72 - 32) \qquad = 22.22°C$$

$$29.92 \text{ in Hg} \times 2.54 = 76 \text{ cm Hg}$$

$$Density = \frac{1.293(76)}{[1 + 0.00367(22.22)](76)}$$

$$= 1.1955 \text{ kg/m}^3.$$

**Table 3.1: Typical Sound Velocities in Various Media (at Approximately 15°C)**

| Media | Velocity | |
|---|---|---|
| | m/s | ft/s |
| Air | 341 | 1119 |
| Water (pure) | 1440 | 4724 |
| Water (sea) | 1500 | 4921 |
| Oxygen | 317 | 1040 |
| Ice | 3200 | 10,499 |
| Marble | 3800 | 12,467 |
| Glass (soft) | 5000 | 16,404 |
| Glass (hard) | 6000 | 19,685 |
| Cast iron | 3400 | 11,155 |
| Steel | 5050 | 16,568 |
| Lead | 1200 | 3937 |
| Copper | 3500 | 11,483 |
| Beryllium | 8400 | 27,559 |
| Aluminum | 5200 | 17,060 |

Having made the metric conversions and obtained the density figure, we can then use the basic equation for velocity

$$c = \sqrt{\frac{1.402(1.013 \times 10^5)}{1.1955}} \qquad (3.6)$$
$$= 344.67 \text{ m/s}$$

Since we started with the dimensions commonly used here in the United States, we then convert back to them by

$$\frac{344.67 \text{ m}}{1 \text{ s}} \times \frac{100 \text{ cm}}{1 \text{ m}} \times \frac{1.0 \text{ in}}{2.54 \text{ cm}} \times \frac{1 \text{ ft}}{12 \text{ in}} = \frac{1130.81 \text{ ft}}{\text{s}}$$

Typical velocities in other media are shown in Table 3.1.

## 3.5 Temperature-Dependent Velocity

The velocity of sound is temperature dependent. The approximate formula for calculating velocity is

$$c = 49\sqrt{459.4°F} \qquad (3.7)$$

where $c$ is the velocity in feet per second (ft/s) and °F is the temperature in degrees Fahrenheit.

For Celsius temperatures:

$$c = 20.6\sqrt{273 + °C} \qquad (3.8)$$

where $c$ is the velocity in meters per second (m/s) and °C is the temperature in degrees Celsius.

Therefore at a normal room temperature of 72.5°F, we can calculate:

$$49\sqrt{459.4 + 72.5} = 1130\,\text{ft/s}.$$

## 3.6 The Effect of Altitude on the Velocity of Sound in Air

The theoretical expression for the speed of sound, $c$, in an ideal gas (air, for example) is

$$c = \sqrt{\frac{\gamma P}{\rho}} \qquad (3.9)$$

where $c$ is the velocity in m/s, $P$ is the ambient pressure, $\rho$ is the gas density, and $\gamma$ is the ratio of the specific heat of the gas at a constant pressure to its heat at constant volume.

Consider the equation

$$PV = RT \qquad (3.10)$$

where $P$ is the ambient pressure, $V$ is the volume, $R$ is the gas constant, and $T$ is the absolute temperature.

Considering the definition of density ($\rho$), our first equation can be rewritten as

$$c = \sqrt{\frac{\gamma RT}{M}} \qquad (3.11)$$

where $M$ is the molecular weight of the gas.

It can be seen that the velocity is dependent only on the type of gas and the temperature and is independent of changes in pressure. This is true because both $P$ and $\rho$ decrease with increasing altitude and the net effect is that atmospheric pressure has only a very slight effect on sound velocity. Therefore the speed of sound at the top of a mountain would be the same as at the bottom of the mountain if the temperature is the same at both locations.

## 3.7 Typical Wavelengths

Some typical wavelengths for midfrequency octave centers are shown in Table 3.2.

Now suppose the temperature increases 20°F to 92.5°F.

$$49\sqrt{459 + 92.5} = 1151 \text{ ft/s}$$

The table of frequencies and wavelengths is shown in Table 3.3.

### Table 3.2: Typical Wavelengths for Midfrequency Octave Centers

| Frequency (Hz) | Wavelength (ft) |
|----------------|------------------|
| 250 | 4.52 |
| 500 | 2.26 |
| 1000 | 1.13 |
| 2000 | 0.57 |
| 4000 | 0.28 |
| 8000 | 0.14 |
| 16,000 | 0.07 |

### Table 3.3: Frequencies and Wavelengths

| Frequency (Hz) | Wavelength (ft) |
|----------------|------------------|
| 250 | 4.60 |
| 500 | 2.30 |
| 1000 | 1.15 |
| 2000 | 0.58 |
| 4000 | 0.29 |
| 8000 | 0.14 |
| 16,000 | 0.07 |

Suppose we had "tuned" to the peak of a 1000-Hz standing wave in a room first at 72.5°F and then later at 92.5°F. The apparent frequency shift would be

$$\frac{1151}{1.13} - 1000 = 18.58 \text{ Hz}$$

where 1151 is the velocity (ft/s) at the temperature of measurement and 1.13 is the wavelength at the original temperature.

## 3.8 Doppler Effect

We have all experienced the Doppler effect—hearing the pitch change from a higher frequency to a lower frequency as a train whistle or a car horn comes toward a stationary listener and then recedes into the distance. The frequency heard by the listener due to the velocity of the source, the listener, or some combination of both is found by

$$F_L = \left[ \frac{c \pm V_L}{c \pm V_S} \right] F_S \tag{3.12}$$

where $F_L$ is the frequency heard by the listener (observer in Hz), $F_S$ is the frequency of the sound source in Hz, $c$ is the velocity of sound in ft/s, $V_L$ is the velocity of the listener in ft/s, and $V_S$ is the velocity of the sound source in ft/s.

Use minus (−) if $V_S$ in the denominator is coming toward the listener. If the listener, $V_L$, in the numerator is moving away from the source, use minus (−), and for the listener moving toward the source, use plus (+).

*Example*
Assume $c = 1130$ ft/s, $V_L = 0$, $V_S = 60$ mi/h (approaching listener), and $F_S = 1000$ Hz

$$\frac{60 \text{ mi}}{1 \text{ h}} \times \frac{1 \text{ h}}{3600 \text{ h}} \times \frac{5280 \text{ ft}}{1 \text{ mi}} = \frac{88 \text{ ft}}{s}$$

$$F = \left[ \frac{1130 - 0}{1130 - 88} \right] 1000$$
$$= 1084 \text{ Hz.}$$

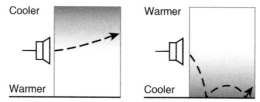

**Figure 3.4: Effect of temperature differences between the ground and the air on the propagation of sound.**

As the sound source passes the listener and recedes, the pitch swings from 1084 Hz to

$$F = \left[ \frac{1130 - 0}{1130 - 88} \right] 1000$$
$$= 928 \text{ Hz.}$$

This rapid sweep of 156 Hz is called the Doppler effect. A very large excursion low-frequency driver can exhibit Doppler distortion of its signal. Moving vanes in reverberation chambers can produce Doppler effects in the reflected signals that can cause unexpected difficulties in modern spectrum analyzers.

## 3.9 Reflection and Refraction

Sound can be reflected by hitting an object larger than one-quarter wavelength of the sound. When the object is one-quarter wavelength or slightly smaller, it also causes diffraction of the sound (bending around the object). Refraction occurs when the sound passes from one medium to another (from air to glass to air, for example, or when it passes through layers of air having different temperatures). The velocity of sound increases with increasing temperature. Therefore sound emitted from a source located on the frozen surface of a large lake on a sunny day will encounter warmer temperatures as the wave diverges upward, causing the upper part of the wave to travel faster than the part of the wave near the surface. This causes a lens-like action to occur, which bends the sound back down toward the surface of the lake (Figure 3.4).

Sound will travel great distances over frozen surfaces on a quiet day. Wind blowing against a sound source causes temperature gradients near the ground surface that result in the sound being refracted upward. Wind blowing in the same direction as the sound produces temperature gradients along the ground surface that tend to refract the sound downward. We hear it said, "The wind blew the sound away." That is not so; it refracted away. Even a 50-mph wind (and that's a strong wind) cannot blow away something traveling 1130 ft/s:

$$\frac{1130\,\text{ft}}{1\,\text{s}} \times \frac{3600\,\text{s}}{1\,\text{h}} \times \frac{1\,\text{m}}{5280\,\text{ft}} = 770.455\ \text{mi/h}$$

770.45 mi/h is the velocity of sound at sea level at 72.5°F.

Wind velocities that vary with elevation can also cause "bending" of the sound velocity plus or minus the wind velocity at each elevation.

Reflections from large boundaries, when delayed in time relative to the direct sound, can be highly destructive of speech intelligibility. It is important to remember, however, that a reflection within a nondestructive time interval can be extremely useful. Reflections that are at or near (within 10 dB) equal amplitude and that are delayed more than 50 ms require careful attention on the part of a sound system designer. Figure 3.5 shows how to calculate probable levels from a reflection. Figure 3.6 shows other influences. Calculation of the time interval is found by:

$$\frac{1000}{c}(D_R - D_D) = \textit{Time interval} \text{ (in ms)} \tag{3.13}$$

where $c$ is the velocity of sound in ft/s or m/s, $D_R$ is the distance in feet or meters traveled by the reflection, and $D_D$ is the distance the direct sound traveled in feet or meters.

A large motor speedway used to make very effective use of ground reflections on the coverage of the grandstands behind the pit area. The very high temperature gradients encountered warp the sound upward during the hot part of the day and in the cool of the morning, the ground reflection helps with the coverage of the near seating area. The directional devices are aimed straight ahead along the ground rather than up at an angle, and when the temperature gradient "bends" the sound upward, it's still covering the audience area effectively (Figure 3.4).

One caution about using ground reflections in northern climes is that a heavy snowfall can provide unbelievable attenuation, as the authors can attest after trying to demonstrate, years ago, a high-level sound system the day after a blizzard in Minnesota.

## 3.10 Effect of a Space Heater on Flutter Echo

The velocity of sound increases with an increase in temperature; therefore, the effect of an increase in temperature with an increase in height is a downward bending of the sound

**Case No. 1**

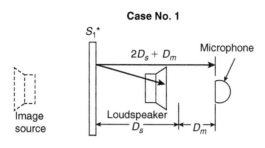

Influence of surface $S_1$ on measured signal at microphone equals:
Reflected signals relative level = $20 \log \left[ \dfrac{D_m}{2D_s + D_m} \right]$

**Case No. 2**

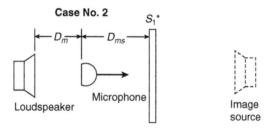

Influence of surface $S_1$ on measured signal at microphone equals:
Reflected signals relative level = $20 \log \left[ \dfrac{D_m}{D_m + 2D_{ms}} \right]$

Where $S_1$ is absorptive then the equation becomes:

Reflected signals relative level =

$$20 \log \left[ \dfrac{D_m}{D_m + 2D_{ms}} \right] + 10 \log (1 - \alpha)$$

In the case of substantial transmission loss then these losses can be added as required.
T.L. = $20 \log fw$ - 47 dB

*Assuming $S_1$ is nonabsorptive, nondiffuse, and nonfocusing.

**Figure 3.5: Calculating relative levels of reflections.**

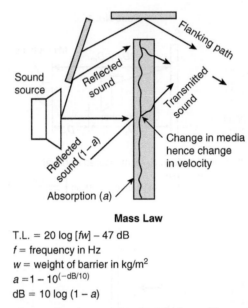

**Mass Law**

T.L. = 20 log [$fw$] – 47 dB

$f$ = frequency in Hz

$w$ = weight of barrier in kg/m$^2$

$a = 1 - 10^{(-dB/10)}$

dB = 10 log (1 – $a$)

**Figure 3.6: Absorption, reflection, and transmission of boundary surface areas.**

path. This illustrates why feedback modes change as air conditioners, heating, or crowds dramatically change the temperature of a room (Figure 3.7).

## 3.11 Absorption

Absorption is the inverse of reflection. When sound strikes a large surface, part of it is reflected and part of it is absorbed. For a given material, the absorption coefficient ($a$) is

$$a = \frac{E_A}{E_I} \tag{3.14}$$

where $E_A$ is the absorbed acoustic energy, $E_I$ is the total incident acoustic energy (i.e., the total sound), and (1 – $a$) is the reflected sound.

This theoretically makes the absorption coefficient some value between 0 and 1. For $a = 0$, no sound is absorbed; it is all reflected. If a material has an $a$ of 0.25, it will absorb 25% of all sound energy having the same frequency as the absorption coefficient rating, and it will reflect 75% of the sound energy having that frequency.

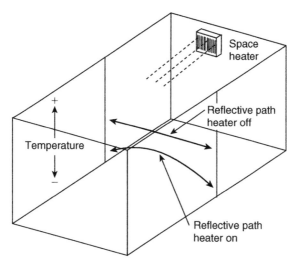

**Figure 3.7: Effect of thermal gradients in a room.**

*Example*

An anechoic room absorbs 99% of the energy received from the sound source. What percentage of the $L_P$ from the source is reflected? Assume 10 W of total energy output from the source. Then the chamber absorbs 9.9 W of it.

---

**Box 3.1 Definitions in Acoustics**

*Sound Energy Density*—is the sound per unit volume measured in joules per cubic meter.

*Sound Energy Flux*—is the average rate of flow of sound energy through any specified area. The unit is joules per second (joules per second are called watts).

*The Sound Intensity* (or sound energy flux density)—in a specified direction at a point is the sound energy transmitted per second in the specified direction through unit area normal to this direction at the point. The unit is watts per square meter.

*Sound Pressure*—is exerted by sound waves on any surface area. It is measured in Newtons per square meter (now called pascals). The sound pressure is proportional to the square root of the sound density.

(Continued)

---

**Box 3.1 (Continued)**

*The Sound Pressure Level* (in decibels of a sound)—20 times the logarithm to the base 10 of the ratio of the pressure of this sound to the reference pressure. Unless otherwise specified, the reference pressure is understood to be 0.00002 N/m$^2$ (20 micropascals or 20 µPa).

*The Velocity Level* (in decibels of a sound)—20 times the logarithm to the base 10 of the ratio of the particle velocity of the sound to the reference particle velocity. Unless otherwise specified, the reference particle velocity is understood to be 50 × 10$^{-9}$ meters per second (m/s).

*The Intensity Level* (in decibels of a sound)—10 times the logarithm to the base 10 of the ratio of the intensity of this sound to the reference intensity. Unless otherwise specified, the reference intensity is 10$^{-12}$ watts per square meter (W/m$^2$).

$$10 \log \frac{10 \text{ W}}{0.1 \text{ W}} = 20 \text{ dB}$$

Therefore the $L_P$ drops by 20 dB also

$$100 \times 10^{-\text{dB}/20} = 10\% \text{ reflected } L_P.$$

In other words, 10% of the $L_P$ returns as a reflection. If the sound source had directed an $L_P$ of a 100-dB signal at the wall of the chamber, a signal of 80 dB would be reflected back. Remembering how dB are combined, we can see that this reflection will not change the 100-dB reading of the direct sound by a discernible amount on any normal sound level meter.

The desirability of a reflective surface can be seen when it is realized that the direct sound and the reflected sound from a single surface can combine to be as much as 3 dB higher than the direct sound alone. If the loudspeakers are directed to reflect off the ground during the cool early morning hours, then when the refraction effect of the sun on the hard surfaces causes the sound to bend upward during the hot part of the day, the sound bends up into the grandstand area. Most of the time, the reflected sound is assisting the direct sound, thereby saving audio power.

# 3.12 Classifying Sound Fields

## 3.12.1 Free Fields

A sound field is said to be a free field if it is uniform, free of boundaries, and is undisturbed by other sources of sound. In practice, it is a field in which the effects of the boundaries are negligible over the region of interest. The flow of sound energy is in one direction only. Anechoic chambers and well-above-the-ground outdoors are free fields. The direct sound level from a sound source in a free field is labeled $L_D$.

## 3.12.2 Diffuse (Reverberant) Fields

A diffuse or reverberant sound field is one in which the time average of the mean square sound pressure is the same everywhere and the flow of energy in all directions is equally probable. This requires an enclosed space with essentially no acoustic absorption. The reverberant sound level is labeled $L_R$.

## 3.12.3 Semireverberant Fields

A semireverberant field is one in which sound energy is both reflected and absorbed. The flow of energy is in more than one direction. Much of the energy is truly from a diffused field; however, there are components of the field that have a definable direction of propagation from the noise source. The semireverberant field is the one encountered in the majority of architectural acoustic environments. The early reflections, that is, under 50 ms after $L_D$, are labeled $L_{RE}$.

## 3.12.4 Pressure Fields

A pressure field is one in which the instantaneous pressure is uniform everywhere. There is no direction of propagation. The pressure field exists primarily in cavities, commonly called couplers, where the maximum dimension of the cavity is less than one-sixth of the wavelength of the sound. Because of ease of repeatability, this type of measurement is used by the National Bureau of Standards when they calibrate microphones. At low frequencies the pressure field can be large, that is, big enough for a listener to sit in.

## 3.12.5 Ambient Noise Field

The ambient noise field is composed of those sound sources not contributing to the desired $L_D$ (i.e., active sources). The ambient noise level is labeled $L_N$.

### 3.12.6  Outdoor Acoustics

If, for example, the ambient noise level measured 70 dBA (not an unreasonable reading outdoors) and the most *SPL* you could generate at 4 ft was 110 dB $L_P$, how far could you reach before your signal was submerged in noise?

$$110\,L_P - 70L_P = 40\,\text{dB}$$

$$20\log\frac{x}{4} = 40\,\text{dB}$$

$$x = 4 \times 10^{40/20}$$

$$= 400\ \text{ft.}$$

The problem actually is more complicated than this outdoors, but this serves as an illustration of how to begin.

We have now touched on the most important basics of the acoustics environment outdoors. Before going indoors, let us apply some of this knowledge to a series of ancient outdoor problems. A simple rule of thumb dictates that when a change of +10 dB occurs, the higher level will be subjectively judged as approximately twice as loud as the level 10 dB below it. While the computation of loudness is more complex than this, the rule is useful for midrange sounds. Using such a rule, we could examine a sound source radiating hemispherically due to the presence of the surface of the earth. Figure 3.8 shows sound in

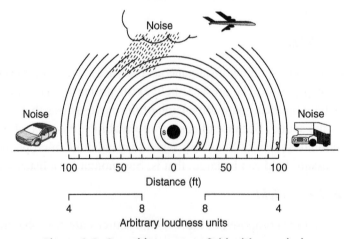

**Figure 3.8: Sound in an open field with no wind.**

an open field with no wind. The sound at 100 ft is one-half as loud as that at 30 ft, although the amplitude of the vibration of the air particles is roughly one-third. Similarly, the sound at 30 ft is one-half as loud as the sound at 10 ft. Because the sound is outdoors, atmospheric effects, ambient noise, and so on cause difficulty for the talker and listener. The ancients learned to place a back wall behind the talker, and many Native American council sites were at the foot of a stone cliff so that the talker could address more of the tribe at one time. Figure 3.9 illustrates how a reflecting structure can double the loudness as compared to totally open space. The weather and some noise still interfere with listening.

Figure 3.10 illustrates the absorptive effect of an audience on the sound traveling to the farthest listener. Figure 3.11 shows the right way and the wrong way to arrange a sound source on a hill. In Figure 3.11(a), the loudness of the sound at the rear of the audience is enhanced by sloping the seating upward. In addition, the noise from sources on the ground is reduced. Figure 3.11(b) is a poor way to listen outdoors.

While the Bible doesn't say which way Jesus addressed the multitudes, we can deduce from the acoustical clues present in the Bible text that the multitude arranged themselves above him because:

1. He addressed groups as large as 5000. This required a very favorable position relative to the audience and a very low ambient noise level.

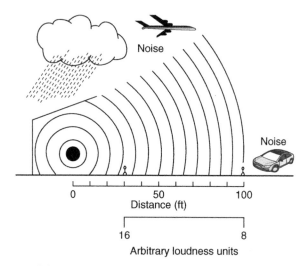

**Figure 3.9: Sound from an orchestra enclosure in an open field with no wind.**

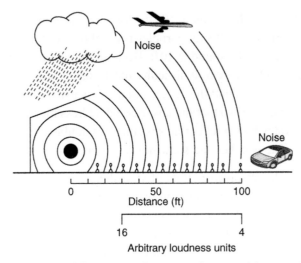

**Figure 3.10: Sound from an orchestra enclosure with an audience.**

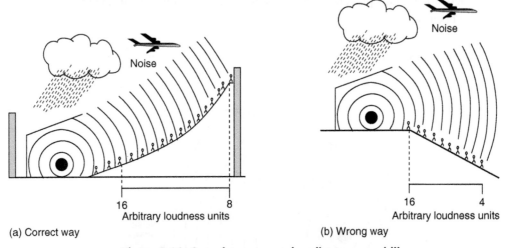

**Figure 3.11: Sound sources and audiences on a hill.**

2. Upon departing from such sessions, he could often step into a boat in the lake, suggesting that he was at the bottom of a hill or mountain.

We can further surmise that the reason Jesus led these multitudes into the countryside was to avoid the higher noise levels present even in small country villages.

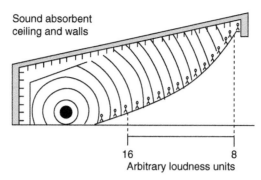

**Figure 3.12: Means of eliminating noise and weather while preserving outdoor conditions.**

The Greeks built their amphitheaters to take advantage of these acoustical facts:

1. They provided a back reflector for the performer.
2. They increased the talker's acoustic output by building megaphones into the special face masks they held in front of their faces to portray various emotions.
3. They sloped the audiences upward and around the talker at an included angle of approximately 120°, realizing, as many modern designers do not seem to, that humans do not talk out of the back of their heads.
4. They defocused the reflective "slapback" by changing the radius at the edges of the seating area.

Because there were no aircraft, cars, motorcycles, air conditioners, and so on, the ambient noise levels were relatively low, and large audiences were able to enjoy the performances. They had discovered absorption and used jars partially filled with ashes (as tuned Helmholtz resonators) to reduce the return echo of the curved stepped seats back to the performers. It remained only for some unnamed innovative genius to provide walls and a roof to have the first auditorium, "a place to hear" (Figure 3.12). No enhancement of sound is provided in Figure 3.12 because there is no reverberation in a room whose walls are highly sound absorbent.

Sometimes acoustic progress was backward. For example, the Romans, when adopting Christianity, took over the ancient echo-ridden pagan temples and had to convert the spoken service into a chanted or sung service pitched to the predominant room modes of these large, hard structures. Today, churches still often have serious acoustical shortcomings and require a very carefully designed sound system in order to allow the normally spoken word to be understood.

It is also of real interest to note that in large halls and arenas the correct place for the loudspeaker system is most often where the roof should have gone if the building had been designed specifically for hearing. A loudspeaker is therefore usually an electroacoustic replacement for a natural reflecting surface that has not been provided.

## 3.13 The Acoustic Environment Indoors

The moment we enclose the sound source, we greatly complicate the transmission of its output. We have considered one extreme when we put the sound source in a well-elevated position and observed the sound being totally absorbed by the "space" around it. Now, let us go to the opposite extreme and imagine an enclosed space that is completely reflective. The sound source would put out sound energy, and none of it would be absorbed. If we continued to put energy into the enclosure long enough, we could theoretically arrive at a pressure that would be explosive. Human speech power is quite small. It has been stated by Harvey Fletcher in his book *Speech and Hearing in Communication* that it would take "…500 people talking continuously for one year to produce enough energy to heat a cup of tea." Measured at 39.37 in (3.28 ft), a typical male talker generates 67.2 dB-SPL, or 34 $\mu$W of power, and a typical female talker generates 64.2 dB-SPL, or 18 $\mu$W. From a shout at this distance (3.28 ft) to a whisper, the dB $L_P$ ranges from 86 to 26 dB, or a dynamic range of about 60 dB. Not only does the produced sound energy tend to remain in the enclosure (dying out slowly), but it tends to travel about in the process.

Let us now examine the essential parameters of a typical room to see what does happen. First, an enclosed space has an internal volume ($V$), usually measured in cubic feet. Second, it has a total boundary surface area ($S$), measured in square feet (floor, ceiling, two side walls, and two end walls). Next, each of the many individual surface areas has an absorption coefficient. The average absorption coefficient ($a$) for all the surfaces together is found by

$$\bar{a} = \frac{s_1 a_1 + s_2 a_2 + \cdots + s_n a_n}{s} \tag{3.15}$$

where $s_{1,2,\ldots n}$ are the individual boundary surface areas in square feet, $\bar{a}_{1,2,\ldots n}$ are the individual absorption coefficients of the individual boundary surface areas, and $S$ is the total boundary surface area in square feet.

The reflected energy is $1 - \bar{a}$.

Table 3.4 gives typical absorption coefficients for common materials. These coefficients are used to calculate the absorption of boundary surfaces (walls, floors, ceilings, etc.).

**Table 3.4: Sound Absorption Coefficients of General Building Materials and Furnishings**

| Materials | Coefficient | | | | | |
|---|---|---|---|---|---|---|
| | 125 Hz | 250 Hz | 500 Hz | 1 kHz | 2 kHz | 4 kHz |
| Acoustical plaster ("Zonolite") | | | | | | |
| ½-in.-thick trowel application | 0.31 | 0.32 | 0.52 | 0.81 | 0.88 | 0.84 |
| 1-in.-thick trowel application | 0.25 | 0.45 | 0.78 | 0.92 | 0.89 | 0.87 |
| Acoustile, surface glazed and perforated structural clay tile, perforate surface backed with 4-in. glass fiber blanket of 1 lb/ft² density | 0.26 | 0.57 | 0.63 | 0.96 | 0.44 | 0.56 |
| Air (Sabins per 1000 ft³) | | | | | 2.3 | 7.2 |
| Brick, unglazed | 0.03 | 0.03 | 0.03 | 0.04 | 0.05 | 0.07 |
| Brick, unglazed, painted | 0.01 | 0.01 | 0.02 | 0.02 | 0.02 | 0.03 |
| Carpet, heavy | | | | | | |
| On concrete | 0.02 | 0.06 | 0.14 | 0.37 | 0.60 | 0.65 |
| On 40-oz hairfelt or foam rubber with impermeable latex backing On 40-oz hairfelt or foam rubber | 0.08 | 0.24 | 0.57 | 0.69 | 0.71 | 0.73 |
| 40-oz hairfelt or foam rubber | 0.08 | 0.27 | 0.39 | 0.34 | 0.48 | 0.63 |
| Concrete block | | | | | | |
| Coarse | 0.36 | 0.44 | 0.31 | 0.29 | 0.39 | 0.25 |
| Painted | 0.10 | 0.05 | 0.06 | 0.07 | 0.09 | 0.08 |
| Fabrics | | | | | | |
| Light velour, 10 oz/yd², hung straight in contact with wall | 0.03 | 0.04 | 0.11 | 0.17 | 0.24 | 0.35 |
| Medium velour, 10 oz/yd², draped to half area | 0.07 | 0.31 | 0.49 | 0.75 | 0.70 | 0.60 |
| Heavy velour, 18 oz/s yd² draped to half area | 0.14 | 0.35 | 0.55 | 0.72 | 0.70 | 0.65 |

(Continued)

Table 3.4: Continued

| Materials | Coefficient | | | | | |
|---|---|---|---|---|---|---|
| | 125 Hz | 250 Hz | 500 Hz | 1 kHz | 2 kHz | 4 kHz |
| Fiberboards, ½-in. normal soft, mounted against solid backing | | | | | | |
|     Unpainted | 0.05 | 0.10 | 0.15 | 0.25 | 0.30 | 0.3 |
|     Some painted | 0.05 | 0.10 | 0.10 | 0.10 | 0.10 | 0.15 |
| Fiberboards, ½-in. normal soft, mounted over 1-in. air space | | | | | | |
|     Unpainted | 0.30 | | 0.15 | | 0.10 | |
|     Some painted | 0.30 | | 0.15 | | 0.10 | |
| Fiberglass insulation blankets | | | | | | |
|     AF100, 1 in., mounting #4 | 0.07 | 0.23 | 0.42 | 0.77 | 0.73 | 0.70 |
|     AF100, 2 in., mounting #4 | 0.19 | 0.51 | 0.79 | 0.92 | 0.82 | 0.78 |
|     AF530, 1 in., mounting #4 | 0.09 | 0.25 | 0.60 | 0.81 | 0.75 | 0.74 |
|     AF530, 2 in., mounting #4 | 0.20 | 0.56 | 0.89 | 0.93 | 0.84 | 0.80 |
|     AF530, 4 in., mounting #4 | 0.39 | 0.91 | 0.99 | 0.98 | 0.93 | 0.88 |
| Flexboard, 3/16-in. unperforated cement asbestos board mounted over 2-in. air space | 0.18 | 0.11 | 0.09 | 0.07 | 0.03 | 0.03 |
| Floors | | | | | | |
|     Concrete or terrazzo | 0.01 | 0.01 | 0.015 | 0.02 | 0.02 | 0.02 |
|     Linoleum, asphalt, rubber, or cork tile on concrete | 0.02 | 0.03 | 0.03 | 0.03 | 0.03 | 0.02 |
|     Wood | 0.15 | 0.11 | 0.10 | 0.07 | 0.06 | 0.07 |
|     Wood parquet in asphalt on concrete | 0.04 | 0.04 | 0.07 | 0.06 | 0.06 | 0.07 |
| Geoacoustic, 13 1/2 in. × 13 1/2 in., 2-in.-thick cellular glass tile installed | 0.13 | 0.74 | 2.35 | 2.53 | 2.03 | 1.73 |
| Glass | | | | | | |
|     Large panes of heavy plate glass | 0.18 | 0.06 | 0.04 | 0.03 | 0.02 | 0.02 |
|     Ordinary window glass | 0.35 | 0.25 | 0.18 | 0.12 | 0.07 | 0.04 |
| Gypsum board, 1/2 in. nailed to 2 in. × 4 in., 16 in. o.c. | 0.29 | 0.10 | 0.05 | 0.04 | 0.07 | 0.09 |
| Hardboard panel, 1/8 in., 1 lb/ft$^2$ with bituminous roofing felt stuck to back, mounted over 2-in. air space | 0.90 | 0.45 | 0.25 | 0.15 | 0.10 | 0.10 |
| Marble or glazed tile | 0.01 | 0.01 | 0.01 | 0.01 | 0.02 | 0.02 |

(Continued)

Table 3.4: Continued

| Materials | Coefficient | | | | | |
|---|---|---|---|---|---|---|
| | 125 Hz | 250 Hz | 500 Hz | 1 kHz | 2 kHz | 4 kHz |
| Masonite, 1/2 in., mounted over 1-in. air space | 0.12 | 0.28 | 0.19 | 0.18 | 0.19 | 0.15 |
| Mineral or glass wool blanket, 1 in., 5–15 lb/ft² density mounted against solid backing | 0.15 | 0.35 | 0.70 | 0.85 | 0.90 | 0.90 |
|   Covered with 5% perforated hardboard | 0.10 | 0.35 | 0.85 | 0.85 | 0.35 | 0.15 |
|   Covered with 10% perforated or 20% slotted hardboard | 0.15 | 0.30 | 0.75 | 0.85 | 0.75 | 0.40 |
| Mineral or glass wool blanket, 2 in., 5–15 lb/ft² density mounted over 1-in. air space | | | | | | |
|   Covered with open weave fabric | 0.35 | 0.70 | 0.90 | 0.90 | 0.95 | 0.90 |
|   Covered with 10% perforated or 20% slotted hardboard | 0.40 | 0.80 | 0.90 | 0.85 | 0.75 | |
| Openings | | | | | | |
|   Stage, depending on furnishings | | | 0.25–0.75 | | | |
|   Deep balcony, upholstered seats | | | 0.50–1.00 | | | |
|   Grills, ventilating | | | 0.15–0.50 | | | |
| Plaster, gypsum or lime | | | | | | |
|   Smooth finish, on tile or brick | 0.013 | 0.015 | 0.02 | 0.03 | 0.04 | 0.05 |
|   Rough finish on lath | 0.02 | 0.03 | 0.04 | 0.05 | 0.04 | 0.03 |
|   Smooth finish on lath | 0.02 | 0.02 | 0.03 | 0.04 | 0.04 | 0.03 |
| Plywood panels | | | | | | |
|   2 in., glued to 2 ½ -in. thick plaster wall on metal lath | 0.05 | | 0.05 | | 0.02 | |
|   1/4 in., mounted over 3-in. air space, with 1-in. glassfiber batts right behind the panel | 0.60 | 0.30 | 0.10 | 0.09 | 0.09 | 0.09 |
|   3/8 in. | 0.28 | 0.22 | 0.17 | 0.09 | 0.10 | 0.11 |
| Rockwool blanket, 2-in. thick batt (Semi-Thik) | | | | | | |
|   Mounted against solid backing | 0.34 | 0.52 | 0.94 | 0.83 | 0.81 | 0.69 |
|   Mounted over 1-in. air space | 0.36 | 0.62 | 0.99 | 0.92 | 0.92 | 0.86 |
|   Mounted over 2-in. air space | 0.31 | 0.70 | 0.99 | 0.98 | 0.92 | 0.84 |

(Continued)

Table 3.4: Continued

| Materials | Coefficient | | | | | |
|---|---|---|---|---|---|---|
| | 125 Hz | 250 Hz | 500 Hz | 1 kHz | 2 kHz | 4 kHz |
| Rockwool blanket, 2-in.-thick batt (Semi-Thik), covered with 3/16 in.-thick perforated cement-asbestos board (Transite), 11% open area | | | | | | |
|   Mounted against solid backing | 0.23 | 0.53 | 0.99 | 0.91 | 0.62 | 0.84 |
|   Mounted over 1-in. air space | 0.39 | 0.77 | 0.99 | 0.83 | 0.58 | 0.50 |
|   Mounted over 2-in. air space | 0.39 | 0.67 | 0.99 | 0.92 | 0.58 | 0.48 |
| Rockwall blanket, 4-in.-thick batt (Full-Thik) | | | | | | |
|   Mounted against solid backing | 0.28 | 0.59 | 0.88 | 0.88 | 0.88 | 0.72 |
|   Mounted over 1-in. air space | 0.41 | 0.81 | 0.99 | 0.99 | 0.92 | 0.83 |
|   Mounted over 2-in. air space | 0.52 | 0.89 | 0.99 | 0.98 | 0.94 | 0.86 |
| Rockwool blanket, 4-in.-thick batt (Full-Thik), covered with 3⁄16-in.-thick perforated cement–asbestos board (Transite), 11% open area | | | | | | |
|   Mounted against solid backing | 0.50 | 0.88 | 0.99 | 0.75 | 0.56 | 0.45 |
|   Mounted over 1-in. air space | 0.44 | 0.88 | 0.99 | 0.88 | 0.70 | 0.30 |
|   Mounted over 2-in. air space | 0.62 | 0.89 | 0.99 | 0.92 | 0.70 | 0.58 |
| Roofing felt, bituminous, two layers, 0.8 lb/ft$^2$, mounted over 10-in. air space | 0.50 | 0.30 | 0.20 | 0.10 | 0.10 | 0.10 |
| Spincoustic blanket | | | | | | |
|   1 in., mounted against solid backing | 0.13 | 0.38 | 0.79 | 0.92 | 0.83 | 0.76 |
|   2 in., mounted against solid backing | 0.45 | 0.77 | 0.99 | 0.99 | 0.91 | 0.78 |
| Spincoustic blanket, 2 in., covered with 3⁄16-in. perforated cement–asbestos board (Transite), 11% open area | 0.25 | 0.80 | 0.99 | 0.93 | 0.72 | 0.58 |
| Sprayed "Limpet" asbestos | | | | | | |
|   3/4 in., 1 coat, unpainted on solid backing | 0.08 | 0.19 | 0.70 | 0.89 | 0.95 | 0.85 |
|   1 in., 1 coat, unpainted on solid backing | 0.30 | 0.42 | 0.74 | 0.96 | 0.95 | 0.96 |
|   3/4 in., 1 coat, unpainted on metal lath | 0.41 | 0.88 | 0.90 | 0.88 | 0.91 | 0.81 |

(Continued)

**Table 3.4: Continued**

| Materials | Coefficient | | | | | |
|---|---|---|---|---|---|---|
| | 125 Hz | 250 Hz | 500 Hz | 1 kHz | 2 kHz | 4 kHz |
| Transite, 3/16-in. perforated, cement–asbestos board, 11% open area | | | | | | |
| Mounted against solid backing | 0.01 | 0.02 | 0.02 | 0.05 | 0.03 | 0.08 |
| Mounted over 1 in. air space | 0.02 | 0.05 | 0.06 | 0.16 | 0.19 | 0.12 |
| Mounted over 2 in. air space | 0.02 | 0.03 | 0.12 | 0.27 | 0.06 | 0.09 |
| Mounted over 4 in. air space | 0.02 | 0.05 | 0.17 | 0.17 | 0.11 | 0.17 |
| Paper-backed board, mounted over 4-in. air space | 0.34 | 0.57 | 0.77 | 0.79 | 0.43 | 0.45 |
| Water surface, as in a swimming pool | 0.008 | 0.008 | 0.013 | 0.015 | 0.02 | 0.025 |
| Wood paneling, 3/8 in. to 1/2 in. thick, mounted over 2-in. to 4-in. air space | 0.30 | 0.25 | 0.20 | 0.17 | 0.15 | 0.10 |

Table 3.5 gives typical absorption units in sabins rather than percentage figures. Sabins are either in per-unit figures or in units per length.

Finally, the room will possess a reverberation time, $RT_{60}$. This is the time in seconds that it will take a steady-state sound, once its input power is terminated, to attenuate 60 dB. For the sake of illustration, assume a room with the following characteristics:

$$V = 500,000\,\text{ft}^3,$$

$$s = 42,500\,\text{ft}^2,$$
$$\bar{a} = 0.128.$$

Therefore the $RT_{60}$ is

$$RT_{60} = \frac{0.049V}{Sa}$$
$$= 4.5\ \text{s}.$$

**Table 3.5: Absorption of Seats and Audience**[a]

| Materials | 125 Hz | 250 Hz | 500 Hz | 1 kHz | 2 kHz | 4 kHz |
|---|---|---|---|---|---|---|
| Audience, seated, depending on spacing and upholstery of seats | 2.5–4.0 | 3.5–5.0 | 4.0–5.5 | 4.5–6.5 | 5.0–7.0 | 4.5–7.0 |
| Seats | | | | | | |
| Heavily upholstered with fabric | 1.5–3.5 | 3.5–4.5 | 4.0–5.0 | 4.0–5.5 | 3.5–5.5 | 3.5–4.5 |
| Heavily upholstered with leather, plastic, etc. | 2.5–3.5 | 3.0–4.5 | 3.0–4.0 | 2.0–4.0 | 1.5–4.0 | 1.0–3.0 |
| Lightly upholstered with leather, plastic, etc. | | | 1.5–2.0 | | | |
| Wood veneer, no upholstery | 0.15 | 0.20 | 0.25 | 0.30 | 0.50 | 0.50 |
| Wood pews | | | | | | |
| No cushions, per 18-in. length | | | 0.40 | | | |
| Cushioned, per 18-in. length | | | 1.8–2.3 | | | |

[a] Values given are in sabins per person or unit of seating.

### 3.13.1 The Mean Free Path (MFP)

The mean free path is the average distance between reflections in a space. For our sample space:

$$MFP = 4\frac{V}{S}$$
$$= 4\left(\frac{500,000}{42,500}\right)$$
$$= 47 \text{ ft.}$$

If a sound is generated in the sample space, part of it will travel directly to a listener and undergo inverse-square-law level change on its way. Some more of it will arrive after having traveled first to some reflecting surface, and still more will finally arrive having undergone several successive reflections (each 47 ft apart on the average). Each of these signals will have had more attenuation at some frequencies than at others because of divergence, absorption, reflection, refraction, diffraction, etc.

We can look at this situation in a different manner. Each sound made will have traveled 4.5 s × 1130 ft/s, or 5085 ft. Since the mean free path is 47 ft, then we can assume each

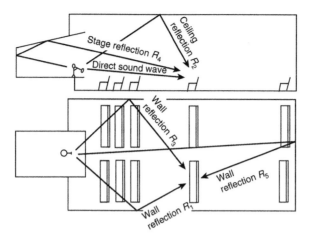

**Figure 3.13: Sound paths in a concert hall.**

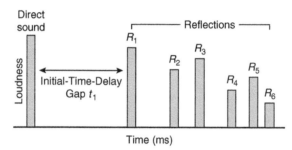

**Figure 3.14: Time relationship of direct and reflected sounds.**

sound underwent approximately 108 reflections in this sample space before becoming inaudible. The result is a lot different than hearing the sound just once.

### 3.13.2 Build-Up of the Reverberant Sound Field

Figure 3.13 shows the paths of direct sound and several reflected sound waves in a concert hall. Reflections also occur from balcony faces, rear wall, niches, and any other reflecting surfaces. We can obtain a chart such as that shown in Figure 3.14 if we plot the amplitude of a short-duration signal vertically and the time interval horizontally. This diagram shows that at listener's ears, the sound that travels directly from the performer arrives first, and after a gap, reflections from the walls, ceiling, stage enclosure, and other reflecting surfaces arrive in rapid succession. The height of a bar suggests the loudness of the sound. This kind of diagram is called a reflection pattern. The initial-signal-delay gap can be measured from it.

Figure 3.14 illustrates the decay of the reverberant field. Here the direct sound enters at the left of the diagram. The initial-signal-delay gap is followed by a succession of sound reflections. The reverberation time of the room is defined as the length of time required for the reverberant sound to decay 60 dB.

We will encounter the effects of delay versus attenuation again when we approach the calculation of articulation losses of consonants in speech.

Figure 3.15 shows measurements from an analyzer made in both large and small rooms. Figure 3.16 shows that the sound arriving at the listener has at least three distinct divisions:

1. The direct sound level $L_D$.
2. The early reflections level $L_{RE}$.
3. The reverberant sound level $L_R$.

The direct sound, by definition, undergoes no reflections and follows inverse-square-law level change. The reverberant sound tends to remain at a constant level if the sound

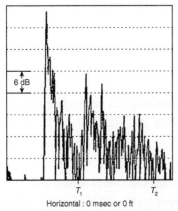

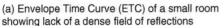

(a) Envelope Time Curve (ETC) of a small room showing lack of a dense field of reflections

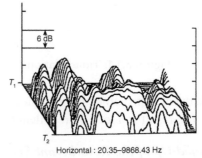

Horizontal : 20.35–9868.43 Hz

(b) Small room without reverberant sound field but with room modes

**Figure 3.15: Vivid proof that there is a fundamental difference between a small reverberant space and a large reverberant hall.**

(Continued)

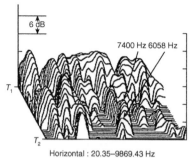

Horizontal : 20.35–9869.43 Hz

(c) Small room without reverberant sound field showing decay side of room modes

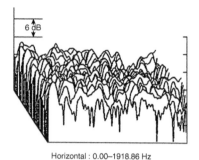

Horizontal : 0.00–1918.86 Hz

(d) Large room with reverberant sound field

**Figure 3.15: (Continued).**

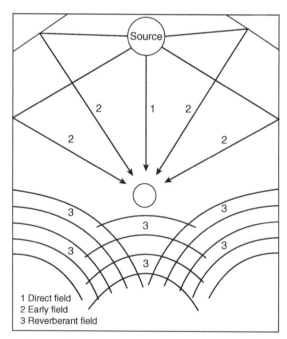

**Figure 3.16: Comparison of direct, early, and reverberant sound fields in an auditorium (reflection adjusted for purposes of illustration).**

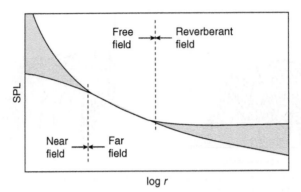

**Figure 3.17: Graphic representation of near field, free field, and reverberant field.**

source continues to put energy into the room at a reasonably regular rate. This gives rise
to a number of basic sound fields (Figure 3.17):

1. The near field.
2. The far free field.
3. The far reverberant field.

The near field does not behave predictably in terms of $L_P$ versus distance because
the particle velocity is not necessarily in the direction of travel of the wave, and
an appreciable tangential velocity component may exist at any point. This is why
measurements are usually not made closer than twice the largest dimension of the
sound source. In the far free field, the inverse-square-law level change prevails. In
the far reverberant field, or diffuse field, the sound-energy density is very nearly
uniform. Measuring low-frequency loudspeakers is an exception to the rule, and such
measurements are often made in the pressure response zone of the device.

## 3.14 Conclusion

The study of acoustics for sound system engineers divides into outdoors and indoors with
indoor acoustics again divided into large room acoustics and small room acoustics. Classical
Sabinian acoustics are rapidly being refined where applicable, discarded where misapplied,
and reexamined where the "fine structure of reverberation" is the meaningful parameter.
The digital computer has fueled basic research into the mathematics of enclosed spaces, and
modern analyzers have served to verify or deny the validity of the theories put forward.

## Further Reading

Acoustical Materials Assoc., The use of architectural materials—theory and practice.

Davis, D. and Davis, C., 'What reverberation is and what it is not', *Syn-Aud-Con Tech Topic,* **12**(13): 1985.

Kinsler, L. E. and Frey, A. R., 'Fundamentals of acoustics', 2nd ed., New York: Wiley, 1962.

Knudsen, V. O. and Harris, C. M., 'Acoustical designing in architecture', New York: Wiley, 1950.

Kuttruff, H., 'Room acoustics', New York: Halstead Press, 1973.

Lindsay, B. R., 'Acoustics—historical and philosophical development', Stroudsburg, PA: Dowden, Hutchinson & Ross, 1973.

MacKenzie, R. (Ed.), 'Auditorium acoustics', London: Applied Science Publishers, 1975.

Olson, H. F., 'Music, physics, and engineering', New York: Dover, 1966.

Pierce, A. D., 'Acoustics: An introduction to its physical principles and applications', New York: McGraw-Hill, 1981.

Pierce, J. R., 'The science of musical sound', New York: Scientific American Books, 1983.

Rossing, T. D., 'The science of sound', Reading, MA: Addison-Wesley, 1982.

Sabine, P. E., 'Acoustics and architecture', New York: McGraw-Hill, 1932.

Sabine, W. C., 'Collected papers on acoustics', Cambridge, MA: Harvard Univ. Press, 1922.

Sivian, L. J., Dunn, H. K., and White, S. D., 'Absolute amplitudes and spectra of certain musical instruments and orchestras', *IRE Trans. on Audio,* 47–75, May–June, 1959.

Strutt, J. W. and Rayleigh, B., 'The theory of sound vols. I and II', 2nd ed., New York: Dover, 1945.

# PART 2

# Audio Electronics

# CHAPTER 4

# *Components*

Andrew Singmin

## 4.1 Building Block Components

### *4.1.1 Resistors*

The humble resistor is by far the most prolific component in use, so it makes a good starting point. A resistor, as the name implies, serves to provide some form of resistance, which is measured in ohms. Even the very name *resistor* already presents an inkling of what it does. In its very simplest form, as a stand-alone component, a resistor presents a resistance to the current flow that would normally take place when voltage is applied to a circuit. A high resistance presents more of an "obstacle," so the resulting current flow is relatively small. However, a low resistance allows more current to flow. If a resistor were connected in series with a current source it would be acting as a current limiter. With resistors you can carry out a lot of simple experiments that are easy to understand and explain. For instance, put a resistance in series with a voltage source and a light bulb: as the resistance goes up, the light dims, and as the resistance goes down, the light brightens. What could be easier to understand?

If limiting current flow through a circuit were all there is to a resistor's function, then we wouldn't have much of a range of circuits to play with. But human ingenuity being what it is, we (electronics designers) have a lot more uses for the resistor. What can we do with two resistors? As you will soon see, the ingenuity or cleverness of the application is tied into the situation in which the resistor is being put to use.

The sole function of humble light-emitting diodes (LEDs) is usually no more than to produce light and to serve as a solid-state indicator lamp. Driven from a low-voltage source, the LED nevertheless has to have a current-limiting resistor inserted in series

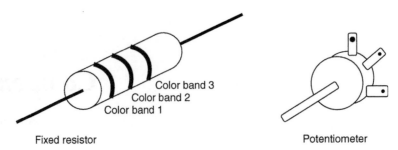

Color band 3
Color band 2
Color band 1

Fixed resistor

Potentiometer

**Figure 4.1: Fixed resistor and potentiometer.**

with the voltage source. Where else do we find the innocent resistor lurking? Operational amplifiers, or op-amps, have a devastating amount of power packed into a tiny eight-pin dual-in-line (DIL) package. Gain setting, the most common feature for an op-amp, is determined by two resistors. Regardless of the sophistication and variety of op-amps (and there are many), they all have to depend on the lowly resistor to function. A resistor is like the mortar holding the bricks together that ultimately form a house. Mortar's not much to look at or get excited about, but where would bricks be without it?

Split bias voltages are found everywhere in op-amp circuits running off a single battery. The positive noninverting pin must be biased in order to halve the supply voltage. Two resistors of equal value placed across the supply voltage and ground nicely provide the required split voltage.

In a slightly different form, but nevertheless still a resistor, there is the *potentiometer*, which is nothing more than a variable resistor. Figure 4.1 shows the two basic resistor types. All radio receivers, stereo amplifiers, cassette recorders, and other such devices have volume controls for obvious reasons. Resistors come in a variety of practically infinite values, from the typically used values of a few ohms to a few megohms. The LED example uses a current-limiting resistor that can vary from a few hundred ohms to a few thousand ohms depending on the supply voltage and the LED brightness required. Gain-setting resistors can range anywhere between a few kohms to a Mohm. Resistors for the split bias supply typically are 100 kohms in value. Resistors are usually associated with DC circuits, as we've seen, and provide a number of useful functions, but most commonly they control current.

Other than limiting current, one of the next most common functions of the resistor is to act as a potential divider circuit. In the simplest case, two equal resistors are placed across a simple voltage source (e.g., a 9-volt battery). The resistor midpoint is half the source voltage, that is, 4.5 volts. This can be checked with a multimeter set to measure DC volts. Ohm's law tells us that we can find the current flow in a resistor by dividing the applied voltage by the resistance. The voltage source for the projects in this book is always a 9-volt battery. For the ease of the arithmetic I just round this up to 10 volts. So if we've got a 10-ohm resistor, the current is just under 1 mA, actually, it is 0.9 mA, as the current is the ratio of the voltage to the resistance. That quick calculation gives us an idea of what to expect for our meter reading.

The same multimeter set to the ohms or resistance range can be used to check out resistor values. There are two precautions if you're going to do this now. The first is to: keep your fingers away from the resistor terminals because your body has a finite resistance, more if your hands are sweaty and less if they're dry. What you're doing when you touch the resistor terminals is adding your body resistance to that of the resistor you're trying to measure. The other precaution is to zero the resistance meter first. Do this by shorting the meter terminals and adjusting the "zero knob" until the meter reads zero. You need only do this with the analog type of multimeter.

The value of a particular resistance is marked on the component body, typically with a three-color band code. A fourth band represents the tolerance, but for the sake of simplicity you may ignore this if you just want to read off the resistor value (which is generally the case). As you almost certainly will want to be able to read resistor color codes, here they are:

| Color band | Equivalent number code |
|------------|------------------------|
| Black      | 0                      |
| Brown      | 1                      |
| Red        | 2                      |
| Orange     | 3                      |
| Yellow     | 4                      |
| Green      | 5                      |
| Blue       | 6                      |
| Violet     | 7                      |
| Gray       | 8                      |
| White      | 9                      |

Often people will make up their own jingle to remember the color codes—you know, something that has meaning for you, such as **B**ye **B**ye **R**eba **O**ff **Y**ou **G**o **B**e **V**aliant **G**o **W**ell. You get the idea.

The next most common format for resistors, and one that you'll come across very often in the circuit projects, is the variable resistor or, as it is more usually called, the *potentiometer*. Relatively speaking, the potentiometer is a much larger device than the resistor; it is more mechanical as opposed to electrical, and it is a three-terminal device. A rotating shaft coupled internally to a movable wiper track follows an arc-shaped path over a track of resistive material. The movable wiper terminal is brought out to a fixed electrical connection point. Further, two fixed terminals are connected electrically to the other two ends of the resistive track. As you can probably tell, the resistance measured across the wiper terminal and either of the other ends will vary continuously as the shaft is rotated. The maximum resistance value will be the value marked on the device; typically, values of 1, 10, and 100 kohms are used.

Resistor values will typically run from 1 ohm to 1 Mohm. I find that with most circuit applications you can get away with using just a few "good" resistor values. My own personal preference is 10 ohms, 100 ohms, 470 ohms, 1 kohm, 2.7 kohms, 4.7 kohms, 10 kohms, 27 kohms, 47 kohms, 100 kohms, 470 kohms, and 1 Mohm. If I had to choose the four most useful values, these values can be further distilled down to 100 ohms, 1 kohm, 10 kohms, and 100 kohms. Look at the circuits later in the book and see how often these values turn up. Intermediate values can be built up by juggling a handful of basic values and learning a bit of "resistor math." Two resistors of equal value connected in parallel produce half the resistor value. So two 1-kohm resistors produce 500 ohms, and two 10-kohm resistors give you 5 kohms. So if a circuit called for a 5.5-kohm resistor and it's late at night and you desperately need that last component to finish, join two 1-kohm resistors connected in parallel to two 10-kohm resistors connected in parallel, and you've got what you need. A useful trick indeed.

The more general rule to follow when the resistors are not equal in value is that for two resistors of unequal value connected in parallel, the total value is the product divided by the sum of the two values. For example, a 1- and a 10-kohm resistor connected in parallel will yield the product $10 \times 1 = 1$, divided by the sum of the resistor values, $10 + 1 = 11$, yields $10 \times 11 = 0.9$ kohm. Another useful trick to remember when connecting two resistors in parallel is that the total is always less than the smaller of the two values. In the example given earlier, 0.9 kohm is less then 1 kohm (the smaller). For more than two resistors connected in parallel (you can use as many resistors as you want), the rule is

$$1/\text{total resistance} = 1/\text{resistor1} + 1/\text{resistor2} + 1/\text{resistor3}.$$

Here's another example. A 1-, 2-, and 3-ohm resistor are connected in parallel. The result is

$$1/\text{total resistance} = 1/1 + 1/2 + 1/3 = 1 + 0.5 + 0.33 = 1.833 \text{ ohms.}$$

To check our math, since 1/total resistance is 1.833, the total resistance is 1/1.833 = 0.545 ohm, and this value is less than the smallest value (1 ohm). However, adding two or more resistors in series (end to end) merely gives you the sum of all the individual resistor values. A 1-kohm resistor and a 100-kohm resistor connected in series thus yield 101 kohms. So by combining resistors in series and parallel you could make up almost any value you want. Figure 4.2 shows the series, parallel combination. However, it's much easier to go out and buy a resistor with the value you want (and that one resistor will take up less space).

Apart from the actual resistance value, there is a second parameter associated with resistors, the tolerance rating, and it is designated by an extra color band. The most commonly specified tolerance is 5% (a gold band), followed by 10% tolerance (indicated with a silver band). In case you encounter them, there are also resistors with no color band that are equal to 20% tolerance, but it is inadvisable to use them because they tend not to be accurate. The tolerance percentage refers to the spread of values on either side

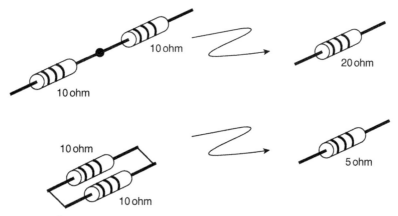

**Figure 4.2: Resistors in series and resistors in parallel.**

of the nominally marked value (the three color bands) that the resistor is allowed to read and still remain within specification. This tolerance designation gives the resistor manufacturer greater latitude in offering resistors with a nominal value than would be otherwise possible. From the user's point of view (you and me), this means a 100-kohm resistor might not exactly read that value when measured, but it is perfectly acceptable from the manufacturer's point of view. For example, if you have a 5% 100-kohm resistor and you measure the actual resistance, it could lie anywhere between

$$100 \text{ kohms} + 5 \text{ percent} = 100 \text{ kohms} + 5 \text{ kohms} = 105 \text{ kohms, or}$$
$$100 \text{ kohms} - 5 \text{ percent} = 100 \text{ kohms} - 5 \text{ kohms} = 95 \text{ kohms.}$$

If this were a 20% 100-kohm resistor, then the limits would run from 120 to 80 kohms, which is an extraordinarily wide variation. All the projects described later in the book use 5% tolerance resistors.

The third parameter associated with resistors is their power rating. The value typically used is 1/4 watt, which is also the wattage specified for the project circuits in this book. The power rating of a resistor refers to its ability to dissipate power, which in turn translates to its ability to dissipate heat. The more current you pass through a resistor, the hotter it gets, and the resistor power rating must be sufficient to stand up to the dissipated power. Larger resistors go up to 1/2 W and more. It's a waste to use these for the projects in this book because these resistors take up more space, cost more, and are unnecessary. However, for the sake of demonstrating the calculations involved, I'll describe what happens to the power rating when we join resistors in series or parallel. In the simple case of two 100-ohm 1/4 watt resistors joined in series, the total resistance is 200 ohms, and the power rating is still 1/4 watt. However, when these resistors are joined in parallel, the resistance drops to 50 ohms, and the power rating increases to 1/2 watt—a nice technique to remember if you want to increase your power rating. Let's say you wanted a 10-ohm 1-watt resistor and the shops are closed. This is quite a large beast. You've got a bunch of common 100-ohm 1/4-watt resistors. Take 10 of these 100-ohm resistors and connect them in parallel. The total resistance is now 10 ohms (one-tenth of the individual values), and the power is increased to 10 1/4 = 1.25 watt. This is another good trick to remember.

### 4.1.2 Capacitors

Capacitors, like resistors, are two-terminal devices and are distinctive in terms of their ability to block DC signals and pass AC signals. For example, a DC signal, that is, voltage from a battery, cannot be passed through a capacitor, but an AC signal, say it's coming from transistor radio's earpiece socket, will pass through a capacitor. A resistor, by comparison, will pass both AC and DC signals by the same amount.

In practical circuit situations there are many instances in which the AC signal has to be passed but the DC component needs to be blocked. One such instance is when a power amplifier's signal is fed to a speaker. You'll always see a capacitor feeding the signal to the speaker. Another area in which you'll always notice the presence of capacitors is at the input and output of AC amplifiers. Capacitors are measured in units of farads, but because these are very large units, the much smaller units of pico-, nano-, and microfarads are most often used. A picofarad is $10^{-12}$ farads, a nanofarad is $10^{-9}$ farads, and a microfarad is $10^{-6}$ farads. The conversion between the units is such that 1 pF equals $10^{-6}$ $\mu$F.

Remember the simple LED circuit we discussed, the one with the resistor acting as the current limiting device? If the resistor were replaced by a capacitor, the LED would not function because no DC current would be allowed to pass through. Capacitors have a property that is equivalent to DC resistance; they have AC resistance or reactance. The capacitor's reactance is calculated in ohms (like that of the resistor), and it is a function of the frequency of the signal under consideration. The capacitive reactance is inversely proportional to frequency; that is, as the frequency increases, the reactance decreases.

Capacitors can be broken into two basic categories based on their physical structure: the simple nonpolarized type, which is also small in size and small in electrical value (i.e., capacitance), and the larger polarized type, with higher associated capacitance values. Figure 4.3 shows the two basic types. Figure 4.4 shows the series, parallel combinations. Figure 4.5 shows the axial, radial types.

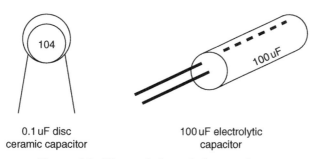

0.1 uF disc
ceramic capacitor

100 uF electrolytic
capacitor

**Figure 4.3: Disc and electrolytic capacitors.**

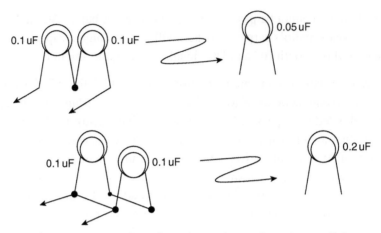

**Figure 4.4: Capacitors in series and capacitors in parallel.**

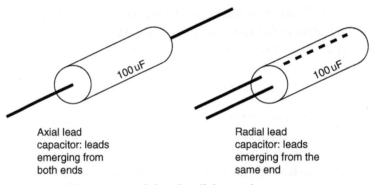

Axial lead
capacitor: leads
emerging from
both ends

Radial lead
capacitor: leads
emerging from the
same end

**Figure 4.5: Axial and radial capacitor types.**

Capacitors such as the electrolytic capacitor are polarity sensitive, which means that they have to be connected in a certain way in the circuit. The electrolytic capacitor is a polarized component, and markings on the body of this capacitor indicate the appropriate negative and positive terminals. As a general rule, capacitors above and including $1\,\mu F$ in value are usually polarized. Capacitance values for the components with larger values are marked on the component's body, as there is sufficient space to print out the value in full; that is, $1\,\mu F$ will actually be printed on the body of the capacitor. The values of capacitors with smaller values are represented with a unique numbering code. The system is similar to the color coding used for resistors, except numbers are used instead of colors. There are three numbers to represent capacitance. It's much easier to understand the system by way of an example. Let's look at the code 104. This is a capacitance value expressed in picofarads. The first and second numbers relate to the actual first two digits of capacitance. The final number indicates the number of zeros following. So 104 is 100,000 picofarads. Because this number is a bit unwieldy, multiply it by $10^{-6}$ to convert to μF, which works out to 0.1 μF, a much more convenient number to work with. This is a very common capacitor value.

Variable capacitors do exist, but they are used less frequently than variable resistors. But variable capacitors are still two-terminal devices. Why? Variable capacitors operate on the principle of varying the overlap between two metal plates, separated by either air or an insulator—the greater the overlap, the greater the capacitance. So you see, just two terminals are needed. There are no variable capacitors used in the projects in this book.

Radial lead capacitors have leads emerging from one side of the body, and if you don't have any height restrictions in your project case, this is the type I recommend you use. Axial lead capacitors, however, have leads emerging one from each end of the body of the component. They take up an awful lot of board space and are used only when the assembly board profile has to be as low as possible, but this is hardly a requirement for simple single-IC hobby projects. (An example of a requirement where you would need a very low profile would be for a pager. Pagers are thin as we know and therefore need an assembly board with a low profile.)

Like resistors, capacitors can also be connected in series and in parallel to form different values. However, the rules are different from those for resistors. To increase a capacitor value, we connect two together in parallel. So two 0.1-μF capacitors connected in parallel give us 0.2 μF. Three capacitors of 0.1 μF value each connected in parallel give us 0.3 μF, and so on. If the capacitors were to be connected in series, then

$$1/\text{total capacitance} = 1/\text{capacitance } 1 + 1/\text{capacitance } 2, \text{ and so on.}$$

For example, two 0.1-μF capacitors connected in series result in a 0.05-μF capacitor, since

$$1/\text{total capacitance} = 1/0.1 \,\mu\text{F} + 1/0.1 \,\mu\text{F} = 10 + 10 = 20.$$

Hence the capacitance is $1/20 = 0.05 \,\mu\text{F}$. Sometimes for timing applications in an oscillator circuit, you might want to change the output frequency a little, and this is one way of obtaining a 0.05- or 0.2-μF capacitor if you don't have one handy (and it's too late to run out to your local component store).

For AC applications, an approximate counterpart to the resistor is the capacitor. Again, a seemingly innocent two-terminal device, the capacitor appears lowly in form, but it is critically needed, like the resistor. Consider any amplifier circuit as an example.

Returning to the AC amplifier example, there is always a capacitor coupling the signal in and coupling the signal out. That's the way to recognize an AC amplifier by the presence of the capacitor at the input and the output. For simple preamplifiers, the coupling capacitors, as they're called, could be around 0.1 μF in value. If we assume the signal to be in the audio frequency range, say 10 kHz, then the capacitive reactance works out to be 159 ohms. This is very low and practically a short circuit. As the capacitive reactance scales inversely with the capacitance, doubling the capacitor to 0.2 μF will reduce the capacitive reactance by half to 79.5 ohms. In our example of the split supply with the resistor we saw that two resistors of equal value gave us the split voltage. Generally in an actual working circuit, you will see a capacitor placed across the lower resistor, that is, the one connected to ground. This is typically a capacity with a large value (100 μF), which is really a short circuit at the audio frequencies we are working with. Another very common way of connecting a capacitor is directly across the supply line, that is, between the plus and the minus voltage rail. With a battery supply this is not so critical, but if you're using a low-voltage line adapter, using a large value smoothing capacitor (several 1000 μF in value) will aid in producing a smoother supply source.

### 4.1.3 Diodes

Diodes are two-terminal devices that have a feature that is totally distinct from the features of resistors or capacitors. They are distinctly polarity sensitive. When DC voltage is applied to a diode, a high current will flow in one direction, but reversing the voltage will, to all intents and purposes, cause no current to flow. Put another way, when the diode is configured in what is called the forward-biased mode, the diode will conduct current. Reverse the bias to the reverse-biased mode and no current will flow. This is defined as a rectifying action. AC voltage, say originating from the line voltage, can be immediately converted into a DC voltage of sorts by feeding it through a diode. The diode essentially passes on only half of the positive and negative going waveform. Electronic circuits are invariably powered with the positive voltage supplying the power rail (Figure 4.6).

To test this out, connect up a resistor, say 100 kohms, across a 9-volt battery with a current meter inserted between the positive battery terminal and one resistor terminal.

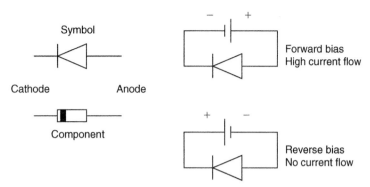

**Figure 4.6: Diode symbol and bias conditions.**

Make sure the current meter's positive terminal goes to the battery's positive terminal. The current will be just under a tenth of a milliamp. The actual current value doesn't matter. If the meter's needle kicks against the end stop, reverse the meter polarity (assuming you've got an analog multimeter); a digital multimeter will automatically compensate whatever polarity is present. When using a digital multimeter to measure DC voltage, there is no need to worry if you get the test leads reversed. The multimeter will still show the correct voltage; there's just a negative sign in front of the numeral. That tells you that the multimeter red test lead, for example, has been connected to the negative voltage potential. There is no damage done to the digital multimeter. If you now take the feed of the positive battery terminal via a diode (it doesn't matter at this stage which way round it goes), one of two things will happen. Either the current will be the same as before or the current will be zero. Whatever it is, take note of it. Then reverse the diode polarity; just reverse the diode's connection in the circuit. An effect opposite to the one you first observed will now take place. You're seeing the rectifying action of the diode.

One really useful function for the diode is as a protective device. Electronic circuits are invariably powered with the positive voltage supplying the power rail. If the voltage is inadvertently reversed, there is a high probability that the components will suffer some damage. Placing a diode (this would be a power type called a rectifier) in series with the positive supply voltage would do the trick. When the polarity is correct, insert the diode in such a way that current starts to flow (trial and error is the quickest way to learn which way to attach the diode if you're not sure about the markings). Now if the voltage polarity should be reversed, no current will flow, thus providing the protection. Try it and see.

### 4.1.4 Transistors

Transistors are totally different from resistors, capacitors, and diodes. The latter are what are termed passive components, performing a singular function as we've seen, useful certainly, but not active in the electronic sense. A transistor is a truly active device. It can take a signal and amplify it. A number of support components are needed to make the transistor into a working amplifier—you guessed it, using a few resistors and capacitors again. Depending on the designer's talent, transistors can be configured into an endless string of circuits, amplifiers, oscillators, filters, alarms, receivers, transmitters, and so on. The versatility of transistors knows no bounds.

Although I do not include transistor-based circuits in this book—the reason being that integrated circuit projects are so much more well behaved and therefore simpler to design—I do provide a brief overview on transistors, as integrated circuits are really just a huge collection of transistor-based circuits. Transistors are three-terminal devices; the terminals are known as the emitter, the base, and the collector. Figure 4.7 shows transistor details. Transistors come in two "flavors" so to speak: the more common NPN type operates with a positive supply voltage, and hence, it is very compatible with integrated circuits, which almost always run on a positive supply. The less common transistor type is the PNP device, which, as you might have guessed, requires a negative supply voltage (not so commonly found in circuits).

Transistors are defined as active devices because they have the capability, given the appropriate support components, to perform useful functions; the most common of these is amplification, but the other is oscillation. A simple, common emitter amplifier can

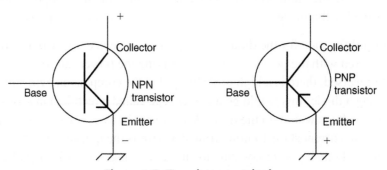

**Figure 4.7: Transistor terminals.**

be designed around four resistors and a capacitor as well as the usual input and output coupling capacitors. However, there are two main reasons to use the integrated circuit (IC). The amplifier's performance is influenced by the transistor's parameters, not so with the IC. Coupling the transistor amplifier into a following stage requires careful consideration of the loading effect. An IC-based amplifier just gets coupled into the next. The IC amplifier is such an effortless pleasure to use. The input, output, and gain are so nicely controlled. You would have had to have labored through the transistor's design quirks to really appreciate how much more controlled the IC is.

Transistors come in a huge variety of types, from general-purpose, small signal (the most common) to large power devices. The frequency range of operation can extend from DC to audio all the way up into the microwave range. Transistors are not as easy to evaluate as ICs. Put together a few resistors and capacitors around an IC and you'll soon know if the circuit is working (and it usually is), as you don't have to even wonder if the IC itself is working. However, try the same with a transistor, and you'll find that determining whether or not the circuit is working is a lot harder. Was the transistor the right type? Was the bias network correct? Is the circuit design right? If the transistor circuit doesn't work, you'll always wonder whether the transistor itself is okay for the application. Isn't it great to know that in the majority of cases, you need only ask for one IC (the LM 741 as it turns out) when working with ICs. Enough said about transistors. They have their uses in specific applications, but you've got to be a bit more circuit smart.

### 4.1.5 Other Components

#### 4.1.5.1 Integrated Circuits

The integrated circuit is an amazingly robust bullet-proof device, by which I mean that you can put practically any design around the IC and know that it is going to behave itself—okay, behave itself within reason, but ICs are brilliantly transparent compared to transistors. A small handful of resistors and capacitors and, hey, presto, we've got a well-behaved amplifier. The transistor could never match that! I know the comparison is a little unfair, especially because the IC itself is composed of a very carefully designed collection of transistor-based circuits, but we're talking user-friendliness here. I recall the difficulty I experienced way back in the mid-1960s getting a simple half-watt transistor power amplifier to function properly. The component count was high, special parts were difficult to come by, setup was tricky, and current consumption was high. Now we've got the LM 386 audio power amplifier on a chip! It's actually been around for a considerable number of years, but it is still very widely used. One IC and two capacitors and you're in business—wow! The current consumption is pretty good, too. The LM 386 IC is an example of a special function IC that is designed to deliver (which it does admirably) just one unique function. The unique one function for the LM 386 IC is as an audio power amplifier. It's hard to believe that a small eight-pin plastic part, a little bigger than one of the buttons on your TV remote, packs such a technological punch. This particular IC runs nicely off a regular 9-volt battery—there are no weird dual supplies to worry about. Many of other higher power ICs require dual supplies, or voltages of 12 volts and higher (a 12-volt battery that you can't buy off the shelf and that would fit in a project case), and consume masses of current.

Integrated circuits fall into two broad categories: analog and digital. They are very easily recognized in terms of their functionality and also in terms of the way they're depicted in circuit schematics. Analog ICs process mostly AC signals, but they also process DC signals. The absence of a coupling capacitor at the input would signify that this is a DC amplifier we're looking at. A DC amplifier has to be capable of amplifying DC signals as well as AC signals. Analog signals, such as audio signals, require coupling capacitors at the input and output because only AC signals are allowed to be coupled through the amplifier. The presence of coupling capacitors removes the DC components. The schematic is also drawn in the form of a sideways triangle representing the IC. Input goes into the wide end on the left and exits as an output from the pointed end on the right. In essence, all analog IC blocks resemble this basic form. Typical examples of analog ICs are the LM 741 general-purpose op-amp in an 8-pin DIL package and the LM 324 quad op-amp package in a 14-pin DIL package. When space is at a premium, the LM 324 is a superb device; it is especially suited for audio applications and occupies far less board space than do four separate LM 741s. Analog ICs, incidentally, are also called linear ICs. Digital ICs only use two voltage states, a logic high (1) and a logic low (0). There are no capacitors in the signal coupling lines, and the schematics are generally drawn in the shape of rectangles or squares. Typical examples can be found in the 7400 series of digital TTL ICs. There are no digital ICs used in this book, but it's worthwhile to make a quick mention of them here because they're such a major portion of the IC family.

The third group of ICs covered in this book are special function ICs, that is, devices falling into neither the analog nor the digital category. Analog or digital ICs don't really do anything by themselves, so to speak. To turn an LM 741 into an amplifier (which is

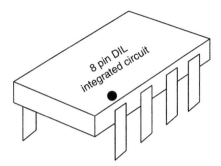

**Figure 4.8: Integrated circuit package outline.**

usually the case), you have to adjust the rest of the circuitry. Alternately the 741 could be designated as an oscillator, and again it is changed accordingly. Digital ICs operate on the principle of responding to just two voltage levels, a low level (also called a '0') and a high level (also called a '1'), and hence, are also called logic devices. Digital ICs can be thought of as a series of logic gates that are configured to perform a certain logic function. Special-function ICs are complete in themselves. The LM 386 audio power amplifier that we'll be focusing on heavily later in the book is just that; it is a self-contained unit that is designed to perform just one task (and it does so admirably at that). Another much-used special function IC is the LM 555, a timer IC, so commonly used to provide a train of square wave pulses. Figure 4.8 shows the basic IC outline.

### 4.1.5.2 Switches

Switches occur in so many places despite their somewhat mundane nature. After all, a switch is just an on/off device. There are actually many different configurations for switches, and it's a good idea to get to know the variations. First of all, there's a terminology specific to switches: *poles* and *throws*. The simplest type of switch, like the type you'd find in a lamp switch, is called a *single pole, single throw,* or *SPST* switch. The simple SPST switch has two terminals, one of these goes to the source (this being typically the positive voltage supply from a battery) and the other goes to the output (typically this would be the circuit that is to receive the power from the battery); hence, the output can only be connected to one terminal. It also has a toggle that flips back and forth. The light flips on one way and off the other. Switches always have to be described with respect to an input signal and an output signal. The pole refers to the number of terminals the input can be connected to. With the SPST switch there is just one. The throw refers to the number of terminals the output can be connected to. In the SPST switch there is just one. What if we had two terminals to which the output could be connected? Because there are now two throws, this kind of switch is called a *single pole, double throw switch*. In this switch there are actually three terminals arranged in a row. The input attaches to the center terminal, and the other two terminals go to the two outputs. The SPST switch, as we've seen, is the type used to switch an appliance on and off. The SPDT can be used to switch either one of two lights on. This kind of switch is not too useful in real life, as there is a chance you may want both lights off. But it illustrates the point. Incidentally, there is a less common type of enhanced version of the SPDT switch with a center off position. The toggle is biased mechanically so it can be positioned in between the two extreme positions. That switch will turn off either light (in our example). In the aforementioned example we have had the switch connected just in the positive supply line (where it is usually connected). The other terminal, that is, the negative terminal, if we were considering, say, a battery being hooked up to a light, would be permanently connected into the circuit. In situations where both sides of the battery need to be switched, we use a switch that is essentially a dual version of the SPST switch. This switch has two sets of terminals, each set identical to the other in function. As you might have guessed, this is a *double pole, single pole*, or *DPST* switch, where a pair of inputs can be switched to a pair of outputs. This switch type is useful because it makes possible more than just the basic on/off function. An even more versatile switch is the *double pole, double throw*, or *DPDT* switch, where two separate inputs can be switched to two separate pairs of outputs.

Table 4.1 illustrates the use of the different switch types. Figures 4.9 and 4.10 depict the switch types very clearly.

The dotted line for the DPST and DPDT switches indicates that these switches have ganged contacts, that is, they are switched together with each mechanical toggle. For a seemingly simple mechanical device, there's certainly more to the humble switch than you

**Table 4.1: Uses of Different Switch Types**

| Switch type | Purpose |
| --- | --- |
| SPST | Used to switch a single monoamplifier speaker on or off |
| SPDT | Used to switch a monoamplifier between two speakers |
| DPST | Used to switch a single pair of stereo amplifier speakers on or off |
| DPDT | Used to switch a stereo amplifier between two pairs of stereo speakers |

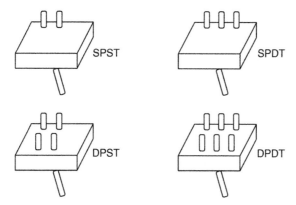

**Figure 4.9: Switch terminals.**

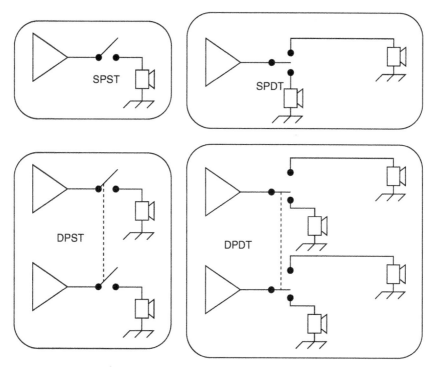

**Figure 4.10: Different switch applications.**

first thought. Apart from the switching differences, switches also come in different current ratings; the higher the current capacity, the larger the physical switch. For the circuit projects shown in this book, choose switches with the smallest current ratings available. Here's a very useful tip that I only found through experience: some small switches (the toggle type) require a huge amount of force to toggle between positions. What this means is that if you've got a very light plastic project case with this type of switch mounted on the front panel, you will most likely tip over the case when you try to flip the switch. I found this out the hard way! So choose small switches that have a very light toggle action. A slight flick of your finger should flip the switch to the other position. Switches are quite costly, and you can save yourself a bundle by not buying the wrong type.

Rotary switches are like super versions of the regular switch and are defined by *poles* and *ways*. For example, a simple, one-pole, four-way switch will switch one input signal to one of four outputs. Let's say we had a two-pole, four-way switch. This switch has two sets of independent contacts that can be coupled to one of four positions. Let's say one pole was used to switch a radio output to one of four speakers. To know which speaker was being powered, the second set of contacts could be wired to four LED indicators, marked as 1 to 4. Each LED would then light up, corresponding to its matching speaker. This setup is shown in Figure 4.11.

### 4.1.5.3 Jack Plugs and Sockets

Audio connections are made much neater and easier with the use of miniature 1/8-inch jack plug/jack socket combinations. If you're using a jack plug, you're going to need a

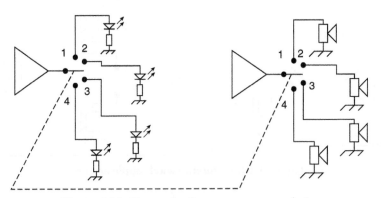

**Figure 4.11: One-pole, four-way rotary switch.**

jack socket. This size of jack plug is almost always found with the headphones provided for portable radios and cassette players. Now you know the size we're talking about. These aren't the huge jack plugs used with electric guitars. The jack plug has a screw-on barrel, often plastic but sometimes metal. Once you remove the cover, and if it's a mono plug, you'll see two connections. There's a short connection to the center pin and a longer connection that goes to the ground terminal. You can recognize a mono jack plug by the single insulator strip near the end of the jack plug tip. The stereo jack plug has two such insulator strips. Jack sockets come in the normally closed and normally open types. In the normally closed type of jack socket, there are two contacts that are in mechanical and electrical contact, that is, it's normally closed. The action of inserting the jack plug causes the two contacts to mechanically spring apart, so the electrical connection is broken. When you remove the jack plug, the electrical connection is made again. The normally open type of jack socket has two close-by terminals that are not electrically connected to each other. When a jack plug is inserted, these two contacts are mechanically brought together and as long as the jack plug remains inserted, the electrical connection is maintained between the two terminals. Figure 4.12 shows the differences for one particular type of popular socket. For the basic application, such as connecting a speaker to an amplifier output, it makes no difference which type is used. But the normally closed type of socket has a special use; it is used where an amplifier is connected normally to an internal speaker, and when an external speaker is plugged in, the internal speaker is disconnected by the action of this jack socket. Portable radios have the same arrangement, whereby plugging in the external headphones disconnects the internal speaker. This

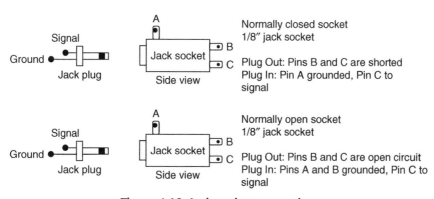

**Figure 4.12: Jack socket conventions.**

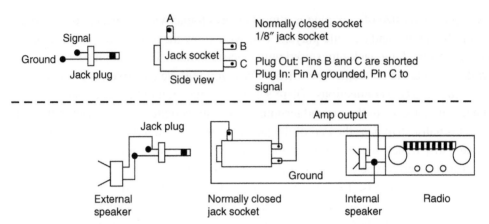

**Figure 4.13: An example of a normally closed jack socket.**

example is seen in Figure 4.13. Like switches, jack plugs and sockets are more complex than they might at first seem.

### 4.1.5.4 Light-Emitting Diodes

The LED is today's solid-state marvel, the equivalent of the filament indicator lamp of years gone by. When I started in hobby electronics, especially in the building of amplifiers, I always had to use filament indicator lamps as power on/off indicators. They took up a lot more space than LEDs, but, more critically, the current they drew was enormous. Fortunately, with the advent of the integrated circuit era came also the solid-state electronics age, with the LED soon becoming the universal indicator device. Small, light, extremely robust, and drawing an economical amount of current, the LED is a natural for panel indicators. In absolute terms, the current drawn is not insignificant, however, but as the rest of the electronics technology speeds ahead to devices that use much less power, the indicator remains locked (at least for the time being) with the LED. Fundamentally, if the LED is to be used as a relatively long-range viewing device, current has to be supplied to produce the visible light energy. Typically, current through the device is limited with a resistor to just a few milliamps for acceptable viewing. LEDs come in a limited range of colors—red, green, yellow—but red is by far the most common and useful color. They come in different shapes (cylindrical and rectangular) and sizes, from pin-head tiny to jumbo sized, the most commonly used size being something like the size of a TV remote button. There are some special LEDs with very high brightness levels, but they draw more current than the plain vanilla variety, so unless you really need extra high brightness, be careful when you choose your LEDs. The LED package is sometimes marked with the brightness and current values, depending on where you buy your components. Of course, you can always increase the brightness level a good deal by increasing the current up to its maximum limit, but your battery life will be shortened. There's always a compromise, isn't there? Who has ever heard of a Corvette that is also economical to run.

# CHAPTER 5

# Power Supply Design

John Linsley Hood

Active systems such as audio amplifiers operate by drawing current from some voltage source—ideally with a fixed and unvarying output—and transforming this into a variable voltage output that can be made to perform some useful function, such as driving a loudspeaker, or some further active or passive circuit arrangement. For most active systems, the ideal supply voltage would be one having similar characteristics to a large lead-acid battery: a constant voltage, zero voltage ripple, and a virtually unlimited ability to supply current on demand. In reality, considerations of weight, bulk, and cost would rule out any such Utopian solution and the power supply arrangements will be chosen, with cost in mind, to match the requirements of the system they are intended to feed. However, because the characteristics of the power supply used with an audio amplifier have a considerable influence on the performance of the amplifier, this aspect of the system is one that cannot be ignored.

## 5.1 High Power Systems

In the early days of valve-operated audio systems, virtually all of the mains-powered DC power supply arrangements were of the form shown in Figure 5.1(a), and the only real choice open to the designers was whether they used a directly heated rectifier, such as a 5U4, or an indirectly heated one, such as a 5V4 or a 5Z4. The indirectly heated valve offered the practical advantage that the cathode of the rectifier would heat up at roughly the same rate as that of the other valves in the amplifier so there would not be an immediate switch-on no-load voltage surge of 1.4, the normal HT supply output voltage. With a directly heated rectifier, this voltage surge would always appear in the interval between the rectifier reaching its operating temperature, which might take only a few seconds, and the 30 s or so that the rest of the valves in the system would need to come

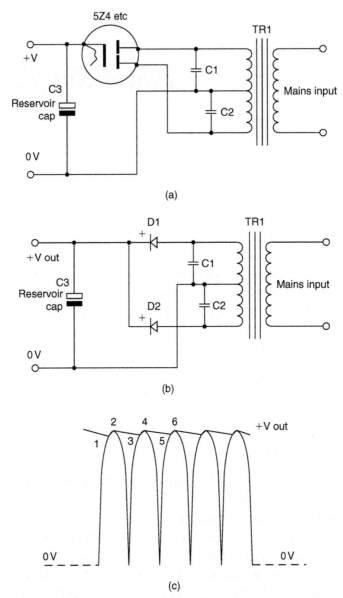

Figure 5.1: Full-wave rectifier systems.

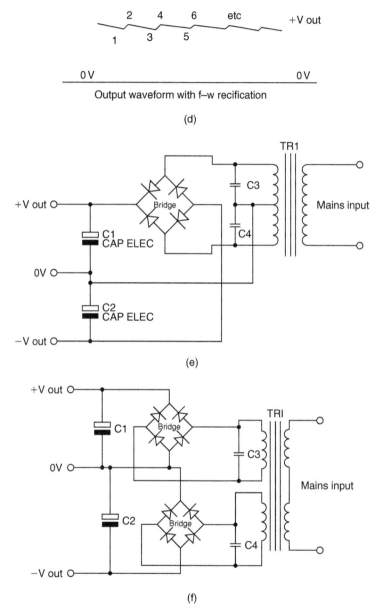

Figure 5.1: (Continued).

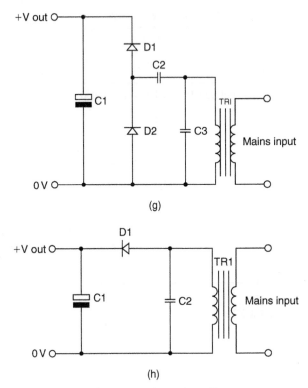

(g)

(h)

**Figure 5.1: (Continued).**

into operation and start drawing current. Using an indirectly heated rectifier would avoid
this voltage surge and would allow lower working voltage components to be used with
safety in the rest of the amplifier. This would save cost. However, the directly heated
rectifier would have a more efficient cathode system and would have a longer working
life expectancy.

Although there are several other reasons for this, such as the greater ease of manufacture,
by the use of modern techniques, of large value electrolytic capacitors, or the
contemporary requirement that there shall be no audible mains hum in the amplifier
output signal due to supply line AC ripple, it is apparent that the capacitance values
used in the smoothing, decoupling, and reservoir capacitors in traditional valve amplifier
circuits are much smaller than in contemporary systems, which operate at a lower output

voltage. The main reason for this is that the stored energy in a capacitor is defined by the relationship:

$$E_c = \tfrac{1}{2}CV^2$$

where $E_c$ is the stored energy, in joules; C is the capacitance, in farads; and V is the applied voltage. This means that there is as much energy stored in an 8-$\mu$F capacitor, charged to 450 V, as there is in a 400-$\mu$F capacitor charged only to 64 V. Because the effectiveness of a decoupling capacitor in avoiding the transmission of supply line rubbish, or a power supply reservoir capacitor in limiting the amount of ripple present on the output of a simple transformer/rectifier type of power supply, depends on the stored charge in the capacitor, its effectiveness is very dependent on the applied voltage, as is the discomfort of the electrical shock that the user would experience if he or she inadvertently discharged such a charged capacitor through his or her body.

## 5.2 Solid-State Rectifiers

The advent of solid state rectifiers—nowadays almost exclusively based on silicon bipolar junction technology—effectively caused the demise of valve rectifier systems, although for a short period, prior to the general adoption of semiconductor rectifiers, gas-filled rectifiers, such as the 0Z4, had been used, principally in car radios, in the interests of greater circuit convenience because, in these valves, the cathode was heated by reverse ionic bombardment so that no separate rectifier heater supply was required. The difficulties caused by the use of these gas-filled rectifiers were that they had a relatively short working life and that they generated a lot of radio frequency (RF) noise. This RF noise arose because of the very abrupt transition of the gas in the cathode/anode gap of the rectifier from a nonconducting to a conducting state. The very short duration high current spikes this caused shock excited the secondary windings of the transformer—and all its associated wiring interconnections—into bursts of RF oscillation, which caused a persistent 100- to 120-Hz rasping buzz called modulation hum to appear in the audio output.

The solution to this particular problem was the connection of a pair of capacitors, shown as C1 and C2 in Figure 5.1(a), across the transformer secondary windings to retune any shock-excited RF oscillation into a lower and less invasive frequency band. Sometimes these modulation hum prevention capacitors are placed across the rectifiers or across the mains transformer primary winding, but they are less effective in these positions.

With modern, low conduction resistance, semiconductor diodes, low equivalent series resistance (ESR) reservoir capacitors, and low winding resistance (e.g., toroidal) transformers, this problem can still arise, and the inclusion of these capacitors is a worthwhile and inexpensive precaution. The circuit layout shown in Figure 5.1(b) is the PSU arrangement used in most contemporary valve amplifiers. For lower voltages, a wider range of circuit layouts are commonly used, also shown in Figure 5.1.

## 5.3 Music Power

In their first flush of enthusiasm for solid-state audio amplifiers, manufacturers and advertising copy writers collectively made the happy discovery that most inexpensive audio amplifiers powered by simple supply circuits, such as that shown in Figure 5.1(b), would give a higher power output for short bursts of output signal, such as might quite reasonably be expected to arise in the reproduction of music, than they could give on a continuous sine-wave output. This short-duration, higher output power capability was therefore termed the music power rating, and, if based on a test in which perhaps only one channel was driven for a period of 100 ms every second, would allow a music power rating to be claimed that was double that of the power given on a continuous tone test in which both channels are driven simultaneously (the so-called rms output power rating).

## 5.4 Influence of Signal Type on Power Supply Design

Although this particular method of specification enhancement is no longer widely used, its echoes linger on in relation to modern expectations for the performance of hi-fi equipment. The reason for this is that in the earlier years of recorded music reproduction there were no such things as pop groups, and most of those interested in improving the quality of recording and replay systems were people such as Peter Walker of Quad or Gerald Briggs of Wharfedale Loudspeakers, whose spare-time musical activities were as an orchestral flautist and a concert pianist and whose interests, understandably, were almost exclusively concerned with the reproduction, as accurately as possible, of classical music. Consequently, when improvements in reproduction were attempted, they were in ways that helped enhance the perceived fidelity in the reproduction of classical music and the accuracy in the rendition of the tone of orchestral instruments. In general, this was easier to achieve if the electronic circuitry was fed from one or more accurately stabilized power supply sources, although this would nearly always mean that such power supplies would have, for reasons of circuit protection, a fixed maximum current output. While this would mean that the peak power and the rms power ratings would be the same, it also meant that there would be no reserve of power for sudden high-level signal demands—a penalty that the tonal purists were prepared to accept as a simple fact of life.

However, times change and hi-fi equipment has become easier to accommodate, less expensive in relative terms, and much more widely available. Also, there has been a considerable growth in the purchasing power of those within the relatively youthful age bracket, most of whose musical interests lie in the various forms of pop music—preferably performed at high signal levels—and it is for this large and relatively affluent group that most of the hi-fi magazines seek to cater.

The ways in which these popular musical preferences influence the design of audio amplifiers and their power supplies relate, in large measure, to the peak short-term output current that is available since one of the major instruments in any pop ensemble will be a string bass, whose sonic impact and attack will depend on the ability of the amplifier and power supply to drive large amounts of current into the LS load, and it must do this without causing any significant increase in the ripple on the DC supply lines or any loss of amplifier performance due to this cause.

A further important feature for the average listener to a typical pop ensemble is the performance of the lead vocalist, commonly a woman, the clarity of whose lyric must not be impaired by the high background signal level generated by the rest of the group. Indeed, with much pop music, with electronically enhanced instruments, the sound of the vocalist, although also enhanced electronically, is the nearest the listener will get to a recognizable reference sound. This clarity of the vocal line demands both low intermodulation distortion levels and a complete absence of peak-level clipping.

The designer of an amplifier that is intended to appeal to the pop music market must therefore ensure that the equipment can provide very large short-duration bursts of power; that the power supply line ripple level, at high output powers, must not cause problems to the amplifier; and that, when the amplifier is driven into overload, it copes gracefully with this condition. The use of large amounts of NFB, which causes hard clipping on overload, is thought to be undesirable. Similarly, the effects of electronic (i.e., fast acting) output transistor current limiting circuitry (used very widely in earlier transistor audio amplifiers) would be quite unacceptable for most pop music applications so alternative approaches, mainly based on more robust output transistors, must be used instead.

In view of the normal lack in much pop music of any identifiable reference sound source, such as would be provided by the orchestral or acoustic keyboard instruments in classical music forms, a variety of descriptive terms has emerged to indicate the success or otherwise of the amplifier system in providing attractive reproduction of the music. Terms such as "exciting," "giving precise image location," "vivid presence," "having full sound staging," "blurred," or transparent are colorful and widely used in performance reviews, but they do not help the engineer in his attempts to approach more closely to an ideal system performance—attempts that must rely on engineering intuition and trial and error.

## 5.5 High Current Power Supply Systems

In order for the power supply system to be able to provide high output currents for short periods of time, the reservoir capacitor, C3 in Figure 5.1(b), must be large and have a low ESR value. Ideally, the rectifier diodes used in the power supplies should have a low conducting resistance, the mains transformer should have low resistance windings and low leakage inductance, and all the associated wiring, including any PCB tracks, should have the lowest practicable path resistance. The output current drawn from the transformer secondary winding, to replace the charge lost from the reservoir capacitor during the previous half cycle of discharge, occurs in brief, high current bursts in the intervals between the points on the input voltage waveform labeled 1 and 2, 3 and 4, 5 and 6, and so on, shown in Figure 5.1(c). This leads to an output ripple pattern of the kind shown in Figure 5.1(d). Unfortunately, all of the measures that the designer can adopt to increase the peak DC output current capability of the power supply unit will reduce the interval of time during which the reservoir capacitor is able to recharge. This will increase the peak rectifier/reservoir capacitor recharge current and will shorten the duration of these high current pulses. This increases the transformer core losses, both the transformer winding and the lamination noise, and also the stray magnetic field radiated from the transformer windings. All of these factors increase the mains hum background, both electrical and acoustic, of the power supply unless steps are taken—in respect of the physical layout, and the placing of interconnections—to minimize it. The main action that can be taken is to provide a very large mains transformer, apparently excessively generously rated in relation to the output power it has to supply, in order that it can cope with the very high peak secondary current demand without mechanical hum or excessive electromagnetic radiation. Needless to say, the mains transformer should be mounted as far away as possible from regions of low signal level circuitry; its orientation should be chosen so that its stray magnetic field will be at right angles to the plane of the amplifier PCB.

## 5.6 Half-Wave and Full-Wave Rectification

Because the reservoir capacitor recharge current must replace the current drawn from it during the nonconducting portion of the input cycle, both the peak recharge current and the residual ripple will be twice as large if half-wave rectification is employed, such as that shown in the circuit of Figure 5.1(h), in which the rectifier diode only conducts during every other half cycle of the secondary output voltage rather than on both cycles, as would be the case in Figure 5.1(b). A drawback with the layouts of both Figures 5.1(a) and 5.1(b) is that the transformer secondary windings only deliver power to the load every other half cycle, which means that when they do conduct, they must pass twice the current they would have had to supply in, for example, the bridge rectifier circuit shown in Figure 5.1(e). The importance of this is that the winding losses are related to the square of the output current ($P = i \approx R$) so that the transformer copper losses would be four times as great in the circuit of Figure 5.1(b) as they would be for either of the bridge rectifier circuits of Figure 5.1(f). However, in the layout of Figure 5.1(b), during the conduction cycle in which the reservoir capacitor is recharged, only one conducting diode is in the current path, as compared with two in the bridge rectifier setups.

Many contemporary audio amplifier systems require symmetrical +ve and −ve power supply rails. If a mains transformer with a center-tapped secondary winding is available, such a pair of split-rail supplies can be provided by the layout of Figure 5.1(e) or, if component cost is of no importance, by the double bridge circuit of Figure 5.1(f). The half-wave voltage doubler circuit shown in Figure 5.1(g) is used mainly in low current applications where its output voltage characteristic is of value, such as perhaps a higher voltage, low-current source for a three-terminal voltage regulator.

## 5.7 Direct Current Supply Line Ripple Rejection

Avoidance of the intrusion of AC ripple or other unwanted signal components from the DC supply rails can be helped in two ways: by the use of voltage regulator circuitry to maintain these rails at a constant voltage or by choosing the design of the amplifier circuitry that is used so that there is a measure of inherent supply line signal rejection. In a typical audio power amplifier, there will be very little signal intrusion from the +ve supply line through the constant current source, Q6 and Q7, because this has a very high output impedance in comparison with the emitter impedance of Q1 and Q2, so any AC ripple passing down this path would be very highly attenuated. However, there would be no attenuation of rubbish entering the signal line via R5, so that, in a real-life amplifier, R5 would invariably be replaced by another constant current source, such as that arranged around Q7 and Q8.

For the negative supply rail, the cascode connection of Q10 would give this device an exceedingly high output impedance, so any signal entering via this path would be very heavily attenuated by the inevitable load impedance of the amplifier. Similarly, the output impedance of the cascode-connected transistors Q3 and Q4 would be so high that the voltage developed across the current mirror (Q5 and Q6) would be virtually independent of any −ve rail ripple voltage. In general, the techniques employed to avoid supply line intrusion are to use circuits with high output impedances wherever a connection must be made to the supply line rails. In order of effectiveness, these would be a cascode-connected field-effect transistor or bipolar device, a constant current source, a current mirror, or a decoupled output, such as a bootstrapped load. HT line decoupling, by means of an LF choke or a resistor and a shunt-connected capacitor, such as R2 and C2, was widely used in valve amplifier circuitry, mainly because there were few other options available to the designer. Such an arrangement is still a useful possibility if the current flow is low enough for the value of R2 to be high and if the supply voltage is high enough for the voltage drop across this component to be unimportant. It still suffers from the snag that its effectiveness decreases at low frequencies where the shunt impedance of C2 begins to increase.

## 5.8 Voltage Regulator Systems

Electronic voltage regulator systems can operate in two distinct modes, each with their own advantages and drawbacks: shunt and series. The shunt systems operate by drawing current from the supply at a level that is calculated to be somewhat greater than maximum value, which will be consumed by the load. A typical shunt regulator circuit is shown in Figure 5.2(a), in which the regulator device is an avalanche or Zener diode or a two-terminal band-gap element for low current, high stability requirements. Such simple circuits are normally only used for relatively low current applications, although high power avalanche diodes are available. If high power shunt regulators are needed, a better

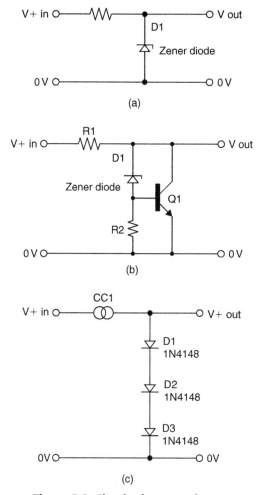

**Figure 5.2: Simple shunt regulators.**

approach is to use a combination of avalanche diode and power transistor, as shown in Figure 5.2(b). The obvious snag is that in order for such a system to work, there must be a continuous current drain that is rather larger than the maximum likely to be drawn by the load, which is wasteful. The main advantages of the shunt regulator system are that it is simple and that it can be used even when the available supply voltage is only a little greater than the required output voltage. Avalanche and Zener diodes are noisy, electrically speaking, although their noise can be lessened by connecting a low ESR capacitor in parallel with them. For applications where only a low voltage is needed and its actual value is not very important but a low circuit noise is essential, a simple arrangement is to use a string of silicon diodes, as shown in Figure 5.2(c). Each of these diodes will have a forward direction voltage drop of about 0.6 V, depending on the current flowing though them. Light-emitting diodes have also been recommended in this application, for which a typical forward voltage drop would be about 2.4 V, depending on the LED type and its forward current. All of these simple shunt regulator circuits will perform better if the input resistor (R1) is replaced by a constant current source, shown as CC1.

## 5.9  Series Regulator Layouts

The problem with the shunt regulator arrangement is that the circuit must draw a current that is always greater than would have been drawn by the load on its own. This is an acceptable situation if the total current levels are small, but this would not be tolerable if high output power levels were involved. In this situation it is necessary to use a series regulator arrangement, of which some simple circuit layouts are shown in Figure 5.3. The circuit of Figure 5.3(a) forms the basis for almost all of this type of regulator circuit, with various degrees of elaboration. Essentially, it is a fixed voltage source to which an emitter–follower has been connected to provide an output voltage (that of the Zener diode less the forward emitter bias of Q1) at a low output impedance. The main problem is that for the circuit to work, the input voltage must exceed the output voltage—the difference is termed the drop-out voltage—by enough voltage for the current flow through R1 to provide the necessary base current for Q1 and also enough current through D1 for D1 to reach its reference voltage. Practical considerations require that R1 shall not be too small. In a well-designed regulator of this kind, such as the 78xx series voltage regulator IC, the drop-out voltage will be about 2 V.

This drop-out voltage can be reduced by reversing the polarity of Q1, as shown in Figure 5.3(b), so that the required base input current for Q1 is drawn from the 0-V rail. This arrangement works quite well, except that the power supply output impedance is much higher than that of Figure 5.3(a), unless there is considerable gain in the NFB control loop. In this particular instance Q2 will conduct and feed current into the Q1 base until the voltage developed across R3 approaches the voltage on the base of Q2, when both Q2 and Q1 will be turned off. By augmenting Q2 with an op-amp, as shown in Figure 5.4, a very high performance can be obtained from this inverted type of regulator layout.

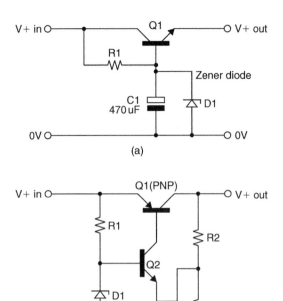

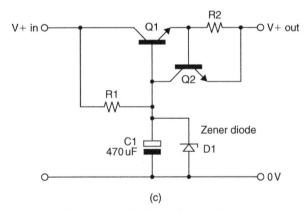

**Figure 5.3: Simple series regulators.**

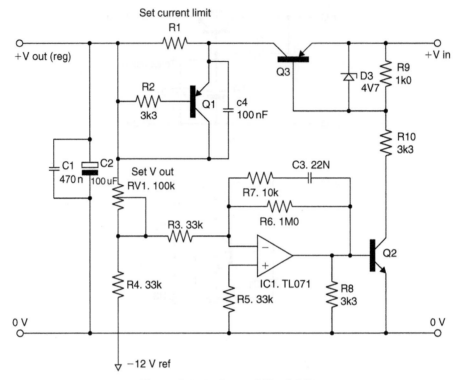

**Figure 5.4: Series-stabilized PSU.**

## 5.10 Overcurrent Protection

A fundamental problem with any kind of solid-state voltage regulator layout, such as that of Figure 5.3(a), is that if the output is short-circuited, the only limit to the current that can flow is the capacity of the input power supply, which could well be high enough to destroy the pass transistor (Q1). For such a circuit to be usable in the real world, where HT rail short-circuits can, and will, occur, some sort of overcurrent protection must be provided. In the case of Figure 5.3(c), this is done by putting a resistor (R2) in series with the regulator output and then arranging a further transistor (Q2) to monitor the voltage across this. If the output current demand is enough to develop a voltage greater than about 0.65 V across R2, Q2 will conduct and will progressively steal the base current from Q1.

In the inverted stabilizer circuit shown in Figure 5.4, R1 monitors the output current, and if this is large enough to cause Q1 to conduct, then the output voltage will progressively collapse, causing the PSU to behave as a constant current source at whatever output voltage causes the load to draw the current determined by R1. (I know this protection technique works because this is the circuit I designed for my workshop bench power supply 20 years ago,[1] which has been in use every working day since then, having endured countless inadvertent output short-circuits during normal use, as well as surviving my son having left it on overnight, at maximum current output, connected to a nickel-plating bath that he had hooked up, but which had inadvertently become short-circuited.) In the particular layout shown, the characteristics of the pass transistors used (Q3 and its opposite number) are such that no current/voltage combinations that can be applied will cause Q3 to exceed its safe operating area boundaries, but this is an aspect that must be borne in mind. Although I use this supply for the initial testing of nearly all my amplifier designs, it would not have an acceptable performance, for reasons given earlier, as the power supply for the output stage of a modern hi-fi amplifier.

However, there is no such demand for a completely unlimited supply current for voltage amplifier stages or preamplifier supply rails, and in these positions, a high-quality regulator circuit can be of considerable value in avoiding potential problems due to hum and distortion components breaking through from the PSU rails. Indeed, there is a trend in modern amplifier design to divide the power supplies to the amplifier into several separate groupings: one pair for the gain stages, a second pair for the output driver transistors, and a final pair of unregulated supplies to drive the output transistors themselves. Only this last pair of supplies normally needs to be fed directly from a simple high current rectifier/reservoir capacitor type of DC supply system.

A further possibility that arises from the availability of more than one power supply to the power amplifier is that it allows the designer, by the choice of the individual supply voltages provided, to determine whereabouts in the power amplifier the circuit will overload when driven too hard since, in general, it is better if it is not the output stage that clips. This was an option that I took advantage of in my 80-W power MOSFET design of 1984.[2]

## 5.11 Integrated Circuit (Three Terminals) Voltage Regulator ICs

For output voltages up to $\pm 24$ V and currents up to 5 A, depending on voltage rating, a range of highly developed IC voltage regulator packages are now offered, having overcurrent (s/c) and thermal overload protection, coupled with a very high degree of output voltage stability, and coupled with a typical >60-dB input/output line ripple rejection. They are available most readily in +5 V and +15V/−15V output voltages because of the requirements of 5-V logic ICs and of IC op-amps, widely used in preamplifier circuits, for which ±15-V supply rails are almost invariably specified. Indeed, the superlative performance of contemporary IC op-amps designed for use in audio applications is such an attractive feature that most audio power amplifiers are now designed so that the maximum signal voltage required from the pre amp is within the typical 9.5-V rms output voltage available from such IC op-amps.

Higher voltage regulator ICs, such as the LM337T and the LM317T, with output voltages up to −37 and +37 V, respectively, and output currents up to 1.5 A are available but where audio amplifier designs require higher voltage-stabilized supply rails, the most common approaches are either to extend the voltage and current capabilities of the standard IC regulator by adding on suitable discrete component circuitry, as shown in Figure 5.5, or by assembling a complete discrete component regulator of the kind shown in Figure 5.6.

In the circuit arrangement shown for a single channel in Figure 5.5, a small-power transistor, Q1, is used to reduce the 55- to 60-V output from the unregulated PSU to a level that is within the permitted input voltage range for the 7815 voltage regulator IC (IC2). This is one of a pair providing a ±15-V DC supply for a preamplifier. A similar 15-V regulator IC (IC1) has its input voltage reduced to the same level by the emitter–follower Q4 and is used to drive a resistive load (R7) via the control transistor, Q5. If the output voltage, and consequently the voltage at Q5 base, is too low, Q5 will conduct, current will be drawn from the regulator IC (IC1), and, via Q4, from the base of the pass transistor, Q2. This will increase the current through Q2 into the output load and will increase the output voltage. If, however, the output voltage tends to rise to a higher level than that set by RV1, Q5 will tend toward cutoff and the current drawn from Q2 base will be reduced to restore the target output voltage level.

Overcurrent protection is provided by the transistor Q3, which monitors the voltage developed across R4 and restricts the drive to Q2 if the output current is too high. Safe operating area conformity is ensured by the resistor R3, which monitors the voltage across the pass transistor and cuts off Q2 base current if this voltage becomes too high.

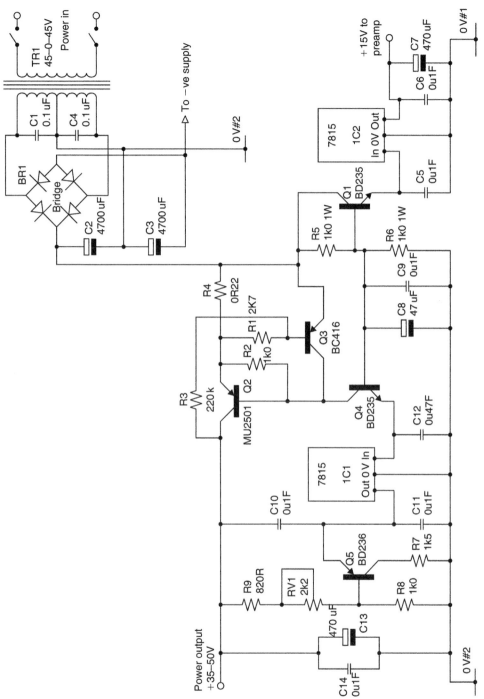

**Figure 5.5: Stabilized PSU (one-half only shown).**

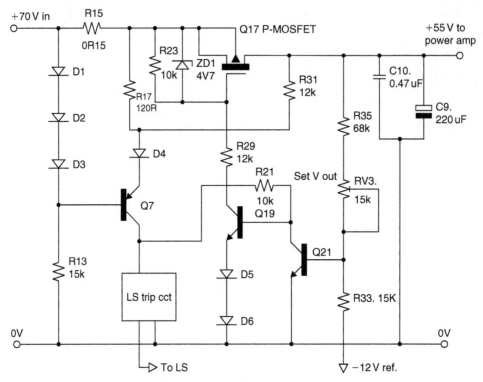

**Figure 5.6: S/C-protected PSU.**

In the circuit of Figure 5.6, which is used as the power supply for an 80- to 100-W power MOSFET audio amplifier—again only one channel is shown—a P-channel power MOSFET is used as the pass transistor and a circuit design based on discrete components is used to control the output voltage. In this, transistor Q21 is used to monitor the potential developed across R33 through the R35/RV3 resistor chain. If this is below the target value, current is drawn through Q19 and R29 to increase the current flow through the pass transistor (Q17). If either the output current or the voltage across Q17 is too high, Q7 is cut off and there is no current flow through Q18 into Q17 gate.

This regulator circuit allows electronic shut down of the power supply if an abnormal output voltage is detected across the LS terminals (due, perhaps, to a component failure). This monitoring circuit (one for each channel) is shown in Figure 5.7. This uses a pair of small-signal transistors, Q1 and Q2, in a thyristor configuration, which, if Q2 is turned

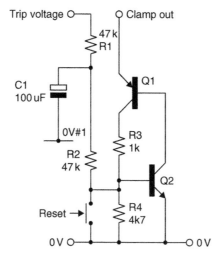

**Figure 5.7: Trip circuit.**

on, will connect Q1 base to the 0-V rail, which, in turn, causes current to be drawn from Q2 base, which causes Q2 to remain in conduction even if the original input voltage is removed. The trip voltage will arise if an excess DC signal (e.g., $>10\,$V) appears across the LS output for a sufficient length of time for Q1 to charge to $+5\,$V. Returning to Figure 5.6, when the circuit trips, the forward bias voltage present on Q19 base is removed and Q17 is cut off and remains cut off until the trip circuit is reset by shorting Q2 base to the 0-V rail. If the fault persists, the supply will cut out again as soon as the reset button is released. An electronic cut-out system like this avoids the need for relay contacts or fuses in the amplifier output lines. Relays can be satisfactory if they are sealed, inert gas-filled types, but fuse holders are, inevitably, crude, low-cost components, of poor construction quality, and with a variable and uncertain contact resistance. These are best eliminated from any signal line.

## 5.12 Typical Contemporary Commercial Practice

The power supply circuit used in the Rotel RHB10 330-W power amplifier is shown in Figure 5.8 as an example of typical modern commercial practice. In this design two separate mains power transformers are used, one for each channel (the drawing only shows the LH channel—the RH one is identical) and two separate bridge rectifiers are used to provide separate $\pm70$-V DC outputs for the power output transistors and the

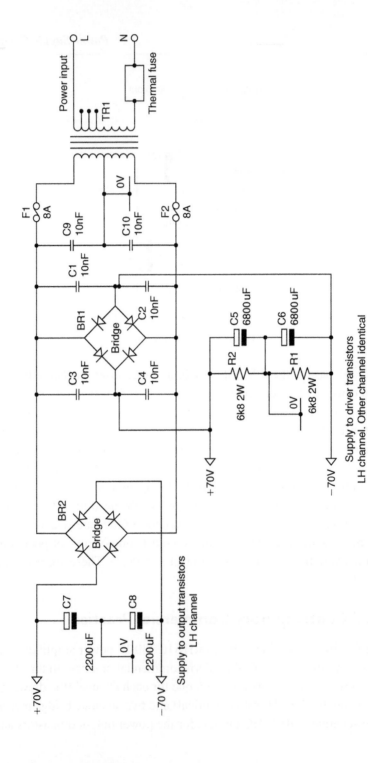

**Figure 5.8: Rotel rhb10 PSU (only one channel shown).**

driver transistors. This eliminates the distortion that might otherwise arise because of breakthrough of signal components from the output transistor supply rail into the low power signal channel. Similarly, use of a separate supply system for each channel eliminates any power supply line-induced L–R cross talk that might impair stereo image positioning.

## 5.13  Battery Supplies

An interesting new development is the use of internally mounted rechargeable batteries as the power supply source for sensitive parts of the amplifier circuitry, such as low input signal level gain stages. Provided that the unit is connected to a mains power line, these batteries will be recharged during the time the equipment is switched off, but will be disconnected automatically from the charger source as soon as the amplifier is switched on.

## 5.14  Switch-Mode Power Supplies

Switch-mode power supplies are widely used in computer power supply systems and offer a compact, high efficiency regulated voltage power source. They are not used in hi-fi systems because they generate an unacceptable level of HF switching noise due to the circuit operation. They would also fail the requirement to provide high peak output current levels.

## Reference

1. Linsley Hood, J. L., *Wireless World*, 43–45, January, 1975.
2. Linsley Hood, J. L., *Electronics Today International*, 24–31, June, 1984.

# PART 3

# Preamplifiers and Amplifiers

# CHAPTER 6

# Introduction to Audio Amplification

John Linsley Hood

In the field of audio amplifiers there has been great interest in techniques for making small electrical voltages larger ever since mankind first attempted to transmit the human voice along lengthy telephone cables. This quest received an enormous boost with the introduction of radio broadcasts and the resulting mass production of domestic radio receivers intended to operate a loudspeaker output. However, the final result, in the ear of the listener, although continually improved over the passage of the years, is still a relatively imperfect imitation of the real-life sounds that the engineer has attempted to copy. Although most of the shortcomings in this attempt at sonic imitation are not because of the electronic circuitry and the amplifiers that have been used, there are still some differences between them, and there is still some room for improvement.

I believe, very strongly, that the only way by which improvements in these things can be obtained is by making, analyzing, and recording, for future use, the results of instrumental tests of as many relevant aspects of the amplifier electrical performance as can be devised. Obviously, one must not forget that the final result will be judged in the ear of the listener so that when all the purely instrumental tests have been completed and the results judged to be satisfactory, the equipment should also be assessed for sound quality and the opinions in this context of as many interested parties as possible should be canvassed.

Listening trials are difficult to set up and hard to purge of any inadvertent bias in the way equipment is chosen or the tests are carried out. Human beings are also notoriously prone to believe that their preconceived views will prove to be correct. The tests must therefore be carried out on a double-blind basis, when neither the listening panel nor the persons selecting one or other of the items under test knows what piece of hardware is being tested.

If there is judged to be any significant difference in the perceived sound quality, as between different pieces of hardware that are apparently identical in their measured performance, the type and the scope of the electrical tests that have been made must be considered carefully to see if any likely performance factor has been left unmeasured or not given adequate weight in the balance of residual imperfections that exist in all real-life designs.

A further complicating factor arises because some people have been shown to be surprisingly sensitive to apparently insignificant differences in performance or to the presence of apparently trifling electrical defects—not always the same ones—so, because there are bound to be some residual defects in the performance of any piece of hardware, each listener is likely to have his or her own opinion of which of these sounds best or which gives the most accurate reproduction of the original sound—if this comparison is possible.

The most that the engineer can do, in this respect, is to try to discover where these performance differences arise or to help decide the best ways of getting the most generally acceptable performance.

It is simple to specify the electrical performance that should be sought. This means that for a signal waveform that does not contain any frequency components that fall outside the audio frequency spectrum, which may be defined as 10 Hz to 20 kHz, there should be no measurable differences, except in amplitude, between the waveform present at the input to the amplifier or other circuit layout (which must be identical to the waveform from the signal source before the amplifier or other circuit is connected to it) and that present across the circuit output to the load when the load is connected to it.

In order to achieve this objective, the following requirements must be met.

- The constant amplitude ($\pm 0.5$ dB) bandwidth of the circuit, under load and at all required gain and output amplitude levels, should be at least 20 Hz to 20 kHz.

- The gain- and signal-to-noise ratio of the circuit must be adequate to provide an output signal of adequate amplitude and the noise or other nonsignal-related components must be inaudible under all conditions of use.

- Both the harmonic and the intermodulation distortion components present in the output waveform, when the input signal consists of one or more pure sinusoidal waveforms within the audio frequency spectrum, should not exceed some agreed upon level. [In practice, this is very difficult to define because the tolerable magnitudes of such waveform distortion components depend on their frequency and also, in the case of harmonic distortion, on the order (i.e., whether they are second, third, fourth, or fifth as the case may be). Contemporary thinking is that all such distortion components should not exceed 0.02%, although, in the particular case of the second harmonic, it is probably undetectable below 0.05%.]

- The phase linearity and electrical stability of the circuit, with any likely reactive load, should be adequate to ensure that there is no significant alteration of the form of a transient or discontinuous waveform such as a fast square or rectangular wave, provided that this would not constitute an output or input overload. There should be no ringing (superimposed spurious oscillation) and, ideally, there should also be no waveform overshoot under square-wave testing in which the signal should recover to the undistorted voltage level, ±0.5%, within a settling time of 20 μs.

- The output power delivered by the circuit into a typical load—bearing in mind that this may be either higher or lower than the nominal impedance at certain parts of the audio spectrum—must be adequate for the purpose required.

- If the circuit is driven into overload conditions, it must remain stable. The clipped waveform should be clean and free from instability, and should recover to the normal signal waveform level with the least possible delay—certainly less than 20 μs.

In addition to these purely electrical specifications, which would probably be difficult to meet, even in a very high-quality solid-state design—and most unlikely to be satisfied in any transformer-coupled system—there are a number of purely practical considerations, such as that the equipment should be efficient in its use of electrical power; that its heat dissipation should not present problems in housing the equipment; and that the design should be cost-effective, compact, and reliable.

# CHAPTER 7

# Preamplifiers and Input Signals

John Linsley Hood

## 7.1 Requirements

Most high-quality audio systems are required to operate from a variety of signal inputs, including radio tuners, cassette or reel-to-reel tape recorders, compact disc players, and more traditional record player systems. It is unlikely at the present time that there will be much agreement between the suppliers of these ancillary units on the standards of output impedance or signal voltage that their equipment should offer.

Except where a manufacturer has assembled a group of such units, for which the interconnections are custom designed and there is in-house agreement on signal and impedance levels—and, sadly, such ready-made groupings of units seldom offer the highest overall sound quality available at any given time—both the designer and the user of the power amplifier are confronted with the need to ensure that their system is capable of working satisfactorily from all of these likely inputs.

For this reason, it is conventional practice to interpose a versatile preamplifier unit between the power amplifier and the external signal sources to perform the input signal switching and signal level adjustment functions.

This preamplifier either forms an integral part of the main power amplifier unit or, as is more common with the higher quality units, is a free-standing, separately powered unit.

## 7.2 Signal Voltage and Impedance Levels

Many different conventions exist for the output impedances and signal levels given by ancillary units. For tuners and cassette recorders, the output is either that of the German Deutsches Industrie Normal (DIN) standard, in which the unit is designed as a current source, which gives an output voltage of 1 mV for each 1000 ohms of load impedance, such that a unit with a 100-K input impedance would see an input signal voltage of 100 mV, or the line output standard, designed to drive a load of 600 ohms or greater, at a mean signal level of 0.775 V rms, often referred to in tape recorder terminology as OVU.

Generally, but not invariably, units having DIN type interconnections, of the styles shown in Figure 7.1, will conform to the DIN signal and impedance level convention, whereas those having "phono" plug/socket outputs, of the form shown in Figure 7.2, will not. In this case, the permissible minimum load impedance will be within the range 600 to 10,000 ohms, and the mean output signal level will commonly be within the range 0.25–1 V rms.

An exception to this exists regarding compact disc players, where the output level is most commonly 2 V rms.

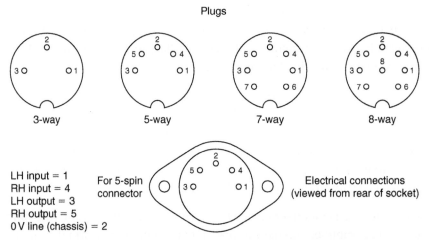

**Figure 7.1: Common DIN connector configurations.**

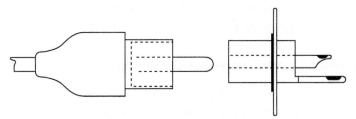

**Figure 7.2: The phono connector.**

## 7.3 Gramophone Pick-Up Inputs

Three broad categories of pick-up (PU) cartridges exist: the ceramic, the moving magnet or variable reluctance, and the moving coil. Each of these has different output characteristics and load requirements.

### 7.3.1 Ceramic Piezo-Electric Cartridges

These units operate by causing the movement of the stylus due to groove modulation to flex a resiliently mounted strip of piezo-electric ceramic, which then causes an electrical voltage to be developed across metallic contacts bonded to the surface of the strip. They are commonly found only on low-cost units and have a relatively high output signal level, in the range 100–200 mV at 1 kHz.

Generally, the electromechanical characteristics of these cartridges are tailored so that they give a fairly flat frequency response, although with some unavoidable loss of HF response beyond 2 kHz, when fed into a preamplifier input load of 47,000 ohms.

Neither the HF response nor the tracking characteristics of ceramic cartridges are particularly good, although circuitry has been designed with the specific aim of optimizing the performance obtainable from these units.[1] However, in recent years, the continuing development of PU cartridges has resulted in a substantial fall in the price of the less exotic moving magnet or variable reluctance types so that it no longer makes economic sense to use ceramic cartridges, except where their low cost and robust nature are of importance.

### 7.3.2 Moving Magnet and Variable Reluctance Cartridges

These are substantially identical in their performance characteristics and are designed to operate into a 47-K load impedance, in parallel with some 200–500 pF of anticipated lead capacitance. Since it is probable that the actual capacitance of the connecting leads will only be of the order of 50–100 pF, some additional input capacitance, connected across the phono input socket, is customary. This also will help reduce the probability of unwanted radio signal breakthrough.

PU cartridges of this type will give an output voltage that increases with frequency in the manner shown in Figure 7.3(a), following the velocity characteristics to which

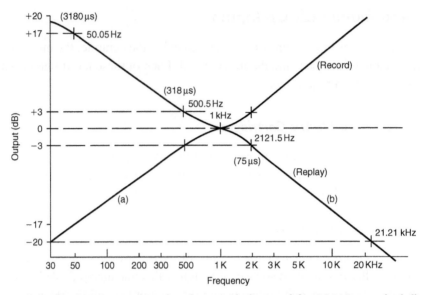

**Figure 7.3: The RIAA record/replay characteristics used for 33/45 rpm vinyl discs.**

LP records are produced, in conformity with the Recording Industry Association of America (RIAA) recording standards. The preamplifier will then be required to have a gain/frequency characteristic of the form shown in Figure 7.3(b), with the deemphasis time constants of 3180, 318, and 75 μs, as indicated in the figure.

The output levels produced by such PU cartridges will be of the order of 0.8–2 mV/cm/s of groove modulation velocity, giving typical mean outputs in the range of 3–10 mV at 1 kHz.

### 7.3.3 Moving Coil Pick-Up Cartridges

These low-impedance, low-output PU cartridges have been manufactured and used without particular comment for very many years. They have come into considerable prominence in the past decade because of their superior transient characteristics and dynamic range as the choice of those audiophiles who seek the ultimate in sound quality, even though their tracking characteristics are often less good than is normal for MM and variable reluctance types.

Typical signal output levels from these cartridges will be in the range 0.02–0.2 mV/cm/s into a 50- to 75-ohm load impedance. Normally, a very low-noise head amplifier circuit will be required to increase this signal voltage to a level acceptable at the input of the RIAA equalization circuitry, although some of the high output types will be capable of operating directly into the high-level RIAA input. Such cartridges will generally be designed to operate with a 47-K load impedance.

## 7.4 Input Circuitry

Most of the inputs to the preamplifier will merely require appropriate amplification and impedance transformation to match the signal and impedance levels of the source to those required at the input of the power amplifier. However, the necessary equalization of the input frequency response from a moving magnet, moving coil, or variable reluctance PU cartridge, when replaying an RIAA preemphasized vinyl disc, requires special frequency shaping networks.

Various circuit layouts have been employed in the preamplifier to generate the required RIAA replay curve for velocity sensitive PU transducers, and these are shown in Figure 7.4. Of these circuits, the two simplest are the "passive" equalization networks shown in Figures 7.4(a) and 7.4(b), although for accuracy in frequency response they require that the source impedance is very low and that the load impedance is very high in relation to $R_1$.

The required component values for these networks have been derived by Livy[2] in terms of $RC$ time constants and set out in a more easily applicable form by Baxandall[3] in his analysis of the various possible equalization circuit arrangements.

From the equations quoted, the component values required for use in the circuits of Figures 7.4(a) and 7.4(c) would be

$$R_1/R_2 = 6.818 \quad C_1 \cdot R_1 = 2187 \, \mu s \quad \text{and} \quad C_2 \cdot R_2 = 109 \, \mu s$$

For the circuit layouts shown in Figures 7.4(b) and 7.4(d), the component values can be derived from the relationships:

$$R_1/R_2 = 12.38 \quad C_1 \cdot R_1 = 2937 \, \mu s \quad \text{and} \quad C_2 \cdot R_2 = 81.1 \, \mu s$$

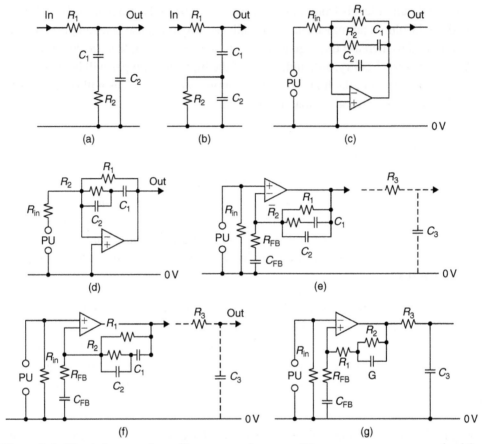

**Figure 7.4: Circuit layouts that will generate the type of frequency response required for RIAA input equalization.**

The circuit arrangements shown in Figures 7.4(c) and 7.4(d) use "shunt" type negative feedback (i.e., that type in which the negative feedback signal is applied to the amplifier in parallel with the input signal) connected around an internal gain block.

These layouts do not suffer from the same limitations with respect to source or load as the simple passive equalization systems of Figures 7.4(a) and 7.4(b). However, they do have the practical snag that the value of $R_{in}$ will be determined by the required PU input load resistor (usually 47k for a typical moving magnet or variable reluctance type of PU

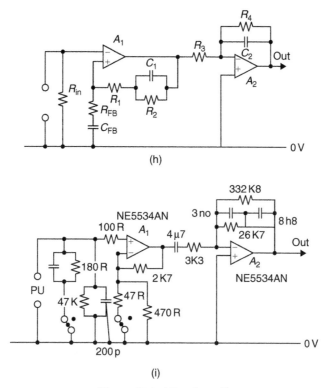

(h)

(i)

**Figure 7.4: (Continued)**

cartridge), and this sets an input "resistor noise" threshold, which is higher than desirable, as well as requiting inconveniently high values for $R_1$ and $R_2$.

For these reasons, the circuit arrangements shown in Figures 7.4(e) and 7.4(f) are found much more commonly in commercial audio circuitry. In these layouts, the frequency response shaping components are contained within a "series" type feedback network (i.e., one in which the negative feedback signal is connected to the amplifier in series with the input signal), which means that the input circuit impedance seen by the amplifier is essentially that of the PU coil alone and allows a lower midrange "thermal noise" background level.

The snag, in this case, is that at very high frequencies, where the impedance of the frequency-shaping feedback network is small in relation to $R_{FB}$, the circuit gain approaches unity, whereas both the RIAA specification and the accurate reproduction of transient waveforms require that the gain should asymptote to zero at higher audio frequencies.

This error in the shape of the upper half of the response curve can be remedied by the addition of a further *CR* network, $C_3/R_3$, on the output of the equalization circuit, as shown in Figures 7.4(e) and 7.4(f). This amendment is sometimes found in the circuit designs used by the more perfectionist of the audio amplifier manufacturers.

Other approaches to the problem of combining low input noise levels with accurate replay equalization are to divide the equalization circuit into two parts, in which the first part, which can be based on a low noise series feedback layout, is only required to shape the 20-Hz to 1-kHz section of the response curve. This can then be followed by either a simple passive *RC* roll-off network, as shown in Figure 7.4(g), or by some other circuit arrangement having a similar effect, such as that based on the use of a shunt feedback connected around an inverting amplifier stage, as shown in Figure 7.4(h), to generate that part of the response curve lying between 1 kHz and 20 kHz.

A further arrangement, which has attracted the interest of some Japanese circuit designers—as used, for example, in the Rotel RC-870BX preamp, of which the RIAA equalizing circuit is shown in a simplified form in Figure 7.4—simply employs one of the recently developed very low noise IC op-amps as a flat frequency response input buffer stage. This is used to amplify the input signal to a level at which circuit noise introduced by succeeding stages will only be a minor problem and also to convert the PU input impedance level to a value at which a straightforward shunt feedback equalizing circuit can be used, with resistor values chosen to minimize any thermal noise background rather than being dictated by the PU load requirements.

The use of "application-specific" audio ICs, to reduce the cost and component count of RIAA stages and other circuit functions, has become much less popular among the designers of higher quality audio equipment because of the tendency of the semiconductor manufacturers to discontinue the supply of such specialized ICs when the economic basis of their sales becomes unsatisfactory or to replace these devices by other, notionally equivalent, ICs that are not necessarily either pin or circuit function compatible.

There is now, however, a degree of unanimity among the suppliers of ICs as to the pin layout and operating conditions of the single and dual op-amp designs, commonly packaged in eight-pin dual-in-line forms. These are typified by the Texas Instruments TL071 and TL072 ICs or their more recent equivalents, such as the TL051 and TL052 devices; there is a growing tendency for circuit designers to base their circuits on the use of ICs of this type, and it is assumed that devices of this kind would be used in the circuits shown in Figure 7.4.

An incidental advantage of the choice of this style of IC is that commercial rivalry between semiconductor manufacturers leads to continuous improvements in the specification of these devices. Since these nearly always offer plug-in physical and electrical interchangeability, the performance of existing equipment can be upgraded easily, either on the production line or by the service department, by the replacement of existing op-amp ICs with those of a more recent vintage, which is an advantage to both manufacturer and user.

## 7.5 Moving Coil Pick-up Head Amplifier Design

The design of preamplifier input circuitry that will accept the very low signal levels associated with moving coil PUs presents special problems in attaining an adequately high signal-to-noise ratio, in respect to the microvolt level input signals, and in minimizing the intrusion of mains hum or unwanted radio frequency (RF) signals.

The problem of circuit noise is lessened somewhat with respect of such RIAA-equalized amplifier stages in that, because of the shape of the frequency response curve, the effective bandwidth of the amplifier is only about 800 Hz. The thermal noise due to amplifier input impedance, which is defined by the following equation, is proportional to the squared measurement bandwidth, other things being equal, so that the noise due to such a stage is less than would have been the case for a flat frequency response system. Nevertheless, the attainment of an adequate S/N ratio, which should be at least 60 dB, demands that the input circuit impedance should not exceed some 50 ohms.

$$\bar{V} = \sqrt{4KT\delta FR}$$

where $\delta F$ is the bandwidth, $T$ is the absolute temperature (room temperature being approximately 300°K), $R$ is resistance in ohms, and $K$ is Boltzmann's constant $(1.38 \times 10^{-23})$.

The moving coil PU cartridges themselves will normally have winding resistances that are only of the order of 5–25 ohms, except in the case of the high output units where the problem is less acute anyway, so the problem relates almost exclusively to the circuit impedance of the MC input circuitry and the semiconductor devices used in it.

## 7.6 Circuit Arrangements

Five different approaches are in common use for moving coil PU input amplification.

### 7.6.1 Step-Up Transformer

This was the earliest method to be explored and was advocated by Ortofon, which was one of the pioneering companies in the manufacture of MC PU designs. The advantage of this system is that it is substantially noiseless, in the sense that the only source of wide-band noise will be the circuit impedance of the transformer windings and that the output voltage can be high enough to minimize the thermal noise contribution from succeeding stages.

The principal disadvantages with transformer step-up systems, when these are operated at very low signal levels, are their proneness to mains "hum" pick up, even when well shrouded, and their somewhat less good handling of "transients" because of the effects of stray capacitances and leakage inductance. Care in their design is also needed to overcome the magnetic nonlinearities associated with the core, which are particularly significant at low signal levels.

### 7.6.2 Systems Using Paralleled Input Transistors

The need for a very low input circuit impedance to minimize thermal noise effects has been met in a number of commercial designs by simply connecting a number of small signal transistors in parallel to reduce their effective base-emitter circuit resistance. Designs of this type came from Ortofon, Linn/Naim, and Braithwaite and are shown in Figures 7.5–7.7.

If such small signal transistors are used without selection and matching—a time-consuming and expensive process for any commercial manufacturer—some means must be adopted to minimize the effects of the variation in base-emitter turn-on voltage that exist between nominally identical devices because of variations in the doping level in the silicon crystal slice or to other differences in manufacture.

This is achieved in the Ortofon circuit by individual collector-base bias current networks, for which the penalty is the loss of some usable signal in the collector circuit. In the

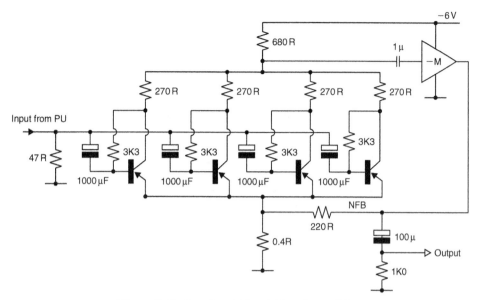

Figure 7.5: Ortofon MCA-76 head amplifier.

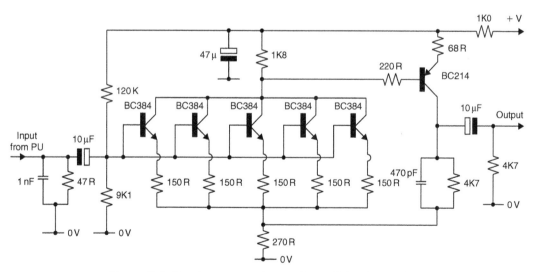

Figure 7.6: The Naim NAC 20 moving coil head amplifier.

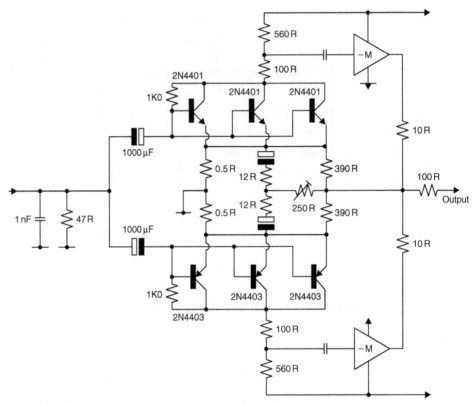

**Figure 7.7: Braithwaite RAI4 head amplifier. (The output stage is shown in a simplified form.)**

Linn/Naim and Braithwaite designs, this evening out of transistor characteristics in circuits having common base connections is achieved by the use of individual emitter resistors to swamp such differences in device characteristics. In this case, the penalty is that such resistors add to the base-emitter circuit impedance when the task of the design is to reduce this.

### 7.6.3 Monolithic Super-Matched Input Devices

An alternative method of reducing the input circuit impedance, without the need for separate bias systems or emitter circuit-swamping resistors, is to employ a monolithic (integrated circuit type) device in which a multiplicity of transistors has been

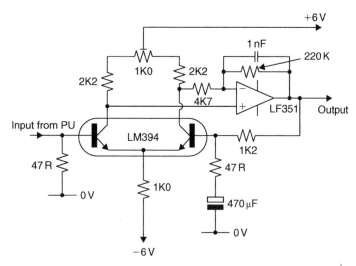

**Figure 7.8: Head amplifier using a LM394 multiple transistor array.**

simultaneously formed on the same silicon chip. Since these can be assumed to have virtually identical characteristics, they can be paralleled, at the time of manufacture, to give a very low impedance, low noise, matched pair.

An example of this approach is the National Semiconductors LM 194/394 super-match pair, for which a suitable circuit is shown in Figure 7.8. This input device probably offers the best input noise performance currently available, but is relatively expensive.

### 7.6.4 Small Power Transistors as Input Devices

The base-emitter impedance of a transistor depends largely on the size of the junction area on the silicon chip. This will be larger in power transistors than in small signal transistors, which mainly employ relatively small chip sizes. Unfortunately, the current gain of power transistors tends to decrease at low collector current levels, making them unsuitable for this application.

However, use of the plastic encapsulated medium power (3–4 A lc max.) styles, in T0126, T0127, and T0220 packages, at collector currents in the range of 1–3 mA, achieves a satisfactory compromise between input circuit impedance and transistor performance and allows the design of very linear low-noise circuitry. Two examples of MC head amplifier designs of this type, by the author, are shown in Figures 7.9 and 7.10.

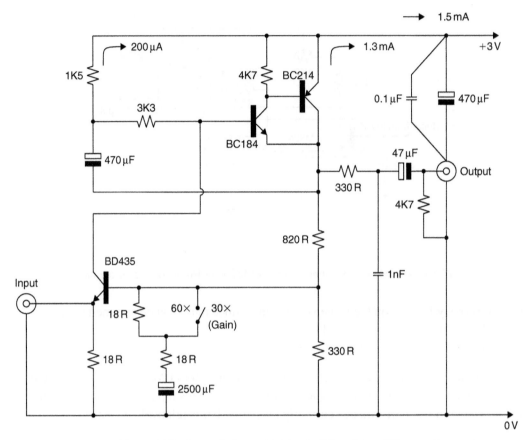

**Figure 7.9: Cascode input moving coil head amplifier.**

The penalty in this case is that, because such transistor types are not specified for low noise operation, some preliminary selection of the devices is desirable, although, in the writer's experience, the bulk of the devices of the types shown will be found to be satisfactory in this respect.

In the circuit shown in Figure 7.9, the input device is used in the common base (cascode) configuration so that the input current generated by the PU cartridge is transferred directly to the higher impedance point at the collector of this transistor so that the stage gain, prior to the application of negative feedback to the input transistor base, is simply the impedance transformation due to the input device.

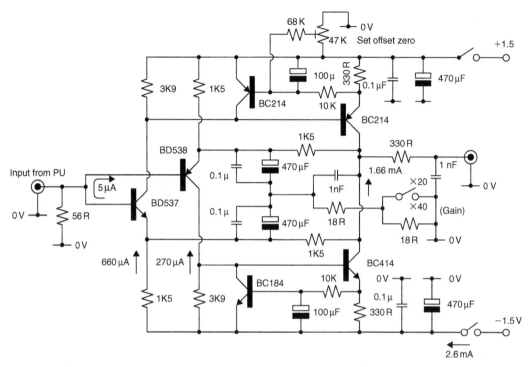

**Figure 7.10: Very low-noise, low-distortion, symmetrical MC head amplifier.**

In the circuit of Figure 7.10, the input transistors are used in a more conventional common-emitter mode, but the two input devices, although in a push–pull configuration, are effectively connected in parallel so far as the input impedance and noise figure are concerned. The very high degree of symmetry of this circuit assists in minimizing both harmonic and transient distortions.

Both of these circuits are designed to operate from 3-V DC "pen cell" battery supplies to avoid the introduction of mains hum due to the power supply circuitry or to earth loop effects. In mains-powered head amps, great care is always necessary to avoid supply line signal or noise intrusions in view of the very low signal levels at both the inputs and the outputs of the amplifier stage.

It is also particularly advisable to design such amplifiers with single point "0-V" line and supply line connections, which should be coupled by a suitable combination of good quality decoupling capacitors.

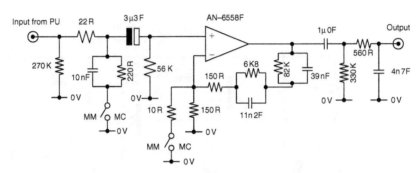

**Figure 7.11: Moving coil/moving magnet RIAA input stage in a Technics SU-V10 amplifier.**

### 7.6.5 Very Low Noise IC Op-Amps

The development, some years ago, of very low noise IC operational amplifiers, such as the Precision Monolithics OP-27 and OP-37 devices, has led to the proliferation of very high-quality, low-noise, low-distortion ICs aimed specifically at the audio market, such as the Signetics NE-5532/ 5534, the NS LM833, the PMI SSM2134/2139, and the TI TL051/052 devices.

With ICs of this type, it is a simple matter to design a conventional RIAA input stage in which the provision of a high-sensitivity, low-noise, moving coil PU input is accomplished by simply reducing the value of the input load resistor and increasing the gain of the RIAA stage in comparison with that needed for higher output PU types. An example of a typical Japanese design of this type is shown in Figure 7.11.

### 7.6.6 Other Approaches

A very ingenious, fully symmetrical circuit arrangement that allows the use of normal circuit layouts and components in ultralow noise (e.g., moving coil PU and similar signal level) inputs has been introduced by "Ouad" (Quad Electroacoustics Ltd.) and is employed in all their current series of preamps. This exploits the fact that, at low input signal levels, bipolar junction transistors will operate quite satisfactorily with their base and collector junctions at the same DC potential and permits the type of input circuit shown in Figure 7.12.

In the particular circuit shown, that used in the "Quad 44" disc input, a two-stage equalization layout is employed, using the type of structure illustrated in Figure 7.4(g),

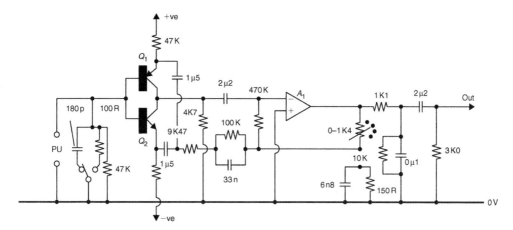

**Figure 7.12: The "Quad" ultralow noise input circuit layout.**

with the gain of the second stage amplifier (a TL071 IC op-amp) switchable to suit the type of input signal level available.

## 7.7 Input Connections

For all low-level input signals, care must be taken to ensure that the connections are of low contact resistance. This is obviously an important matter in the case of low-impedance circuits such as those associated with MC PU inputs, but is also important in higher impedance circuitry, as the resistance characteristics of poor contacts are likely to be nonlinear, and to introduce both noise and distortion.

In the better class modern units, the input connectors will invariably be of the "phono" type, and both the plugs and the connecting sockets will be gold plated to reduce the problem of poor connections as a consequence of contamination or tarnishing of the metallic contacts.

The use of separate connectors for L and R channels also lessens the problem of interchannel breakthrough due to capacitative coupling or leakage across the socket surface, a problem that can arise in the five- and seven-pin DIN connectors if they are fitted carelessly, particularly when both inputs and outputs are taken to that same DIN connector.

## 7.8 Input Switching

The comments made about input connections are equally true for the necessary switching of the input signal sources. Separate, but mechanically interlinked, switches of the push-on, push-off type are to be preferred to the ubiquitous rotary wafer switch, in that it is much easier, with separate switching elements, to obtain the required degree of isolation between inputs and channels than would be the case when the wiring is crowded around the switch wafer.

However, even with separate push switches, the problem remains that the input connections will invariably be made to the rear of the amplifier/preamplifier unit, whereas the switching function will be operated from the front panel so that the internal connecting leads must traverse the whole width of the unit.

Other switching systems, based on relays, or bipolar or field effect transistors, have been introduced to lessen the unwanted signal intrusions, which may arise on a lengthy connecting lead. The operation of a relay, which will behave simply as a remote switch when its coil is energized by a suitable DC supply, is straightforward, although for optimum performance it should either be hermetically sealed or have noble metal contacts to resist corrosion.

### 7.8.1 Transistor Switching

Typical bipolar and FET input switching arrangements are shown in Figures 7.13 and 7.14. In the case of the bipolar transistor switch circuit of Figure 7.13, the nonlinearity of the junction device when conducting precludes its use in the signal line; the circuit is therefore arranged so that the transistor is nonconducting when the signal is passing through the controlled signal channel, but acts as a short-circuit to shunt the signal path to the 0-V line when it is caused to conduct.

In the case of the FET switch, if $R_1$ and $R_2$ are high enough, the nonlinearity of the conducting resistance of the FET channel will be swamped, and the harmonic and other distortions introduced by this device will be negligible (typically less than 0.02% at 1 V rms and 1 kHz).

The CMOS bilateral switches of the CD4066 type are somewhat nonlinear and have a relatively high level of breakthrough. For these reasons they are generally thought to be

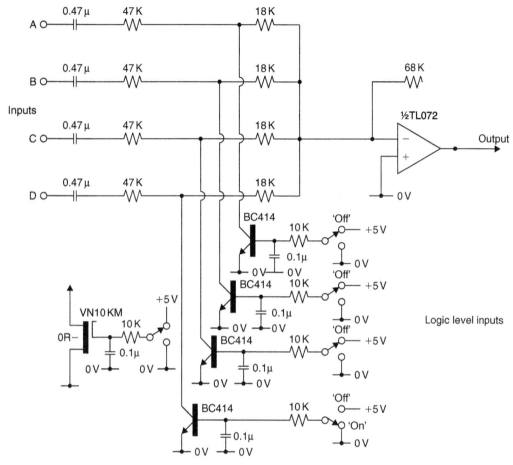

**Figure 7.13: Bipolar transistor-operated shunt switching. [Also suitable for
small-power metal–oxide–semiconductor field-effect transistor
(MOSFET) devices.]**

unsuitable for high-quality audio equipment where such remote switching is employed to
minimize cross talk and hum pick up.

However, such switching devices could well offer advantages in lower quality equipment
where the cost savings is being able to locate the switching element on the printed circuit
board, at the point where it was required, might offset the device cost.

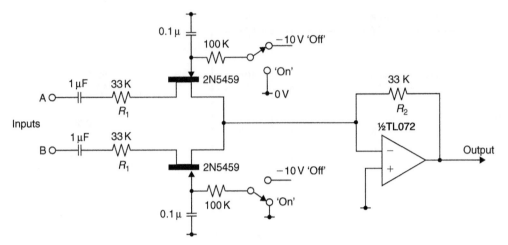

**Figure 7.14: Junction FET input switching circuit.**

### 7.8.2 Diode Switching

Diode switching of the form shown in Figure 7.15, while employed very commonly in RF circuitry, is unsuitable for audio use because of the large shifts in the DC level between the "on" and "off" conditions, which would produce intolerable "bangs" on operation.

For all switching, quietness of operation is an essential requirement, and this demands that care shall be taken to ensure that all of the switched inputs are at the same DC potential, preferably that of the 0-V line. For this reason, it is customary to introduce DC blocking capacitors on all input lines, as shown in Figure 7.16, and the time constants of the input RC networks should be chosen so that there is no unwanted loss of low-frequency signals due to this cause.

## VOLTAGE AMPLIFIERS AND CONTROLS

## 7.9  Preamplifier Stages

The popular concept of hi-fi attributes the major role in final sound quality to the audio power amplifier and the output devices or output configuration that it uses. Yet in reality the preamplifier system, used with the power amplifier, has at least as large an influence on the final sound quality as the power amplifier, and the design of the voltage gain stages within the pre- and power amplifiers is just as important as that of the power output

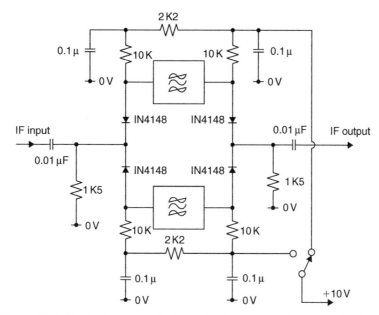

Figure 7.15: Typical diode switching circuit, as used in RF applications.

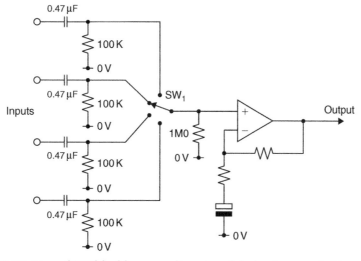

Figure 7.16: Use of DC blocking capacitors to minimize input switching noises.

stages. Moreover, developments in the design of such voltage amplifier stages have allowed continuing improvement in amplifier performance.

The developments in solid-state linear circuit technology that have occurred over the past 30 years seem to have been inspired in about equal measure by the needs of linear integrated circuits and by the demands of high-quality audio systems; engineers working in both of these fields have watched each other's progress and borrowed from each other's designs.

In general, the requirements for voltage gain stages in both audio amplifiers and integrated-circuit operational amplifiers are very similar. These are that they should be linear, which implies that they are free from waveform distortion over the required output signal range, have as high a stage gain as is practicable, have a wide AC bandwidth and a low noise level, and are capable of an adequate output voltage swing.

The performance improvements that have been made over this period have been due in part to the availability of new or improved types of semiconductor devices and in part to a growing understanding of the techniques for the circuit optimization of device performance. The interrelation of these aspects of circuit design is considered next.

## 7.10  Linearity

### 7.10.1  Bipolar Transistors

In the case of a normal bipolar (NPN or PNP) silicon junction transistor, for which the chip cross section and circuit symbol are shown in Figure 7.17, the major problem in

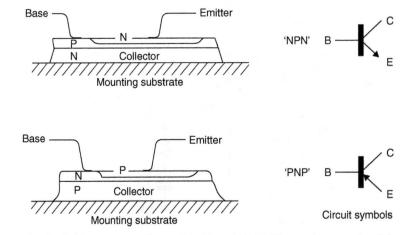

**Figure 7.17: Typical chip cross section of NPN and PNP silicon planar epitaxial transistors.**

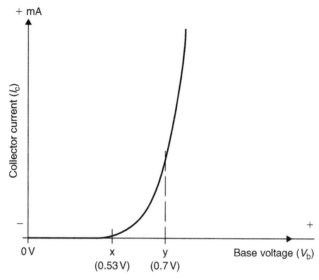

**Figure 7.18: Typical transfer characteristic of a silicon transistor.**

obtaining good linearity lies in the nature of the base voltage/collector current transfer characteristic, shown in the case of a typical "NPN" device (a "PNP" device would have a very similar characteristic, but with negative voltages and currents) in Figure 7.18.

In this, it can be seen that the input/output transfer characteristic is strongly curved in the region "X–Y" and that an input signal applied to the base of such a device, which is forward biased to operate within this region, would suffer from the very prominent (second harmonic) waveform distortion shown in Figure 7.19.

The way this type of nonlinearity is influenced by the signal output level is shown in Figure 7.20. It is normally found that the distortion increases as the output signal increases, and conversely.

There are two major improvements in the performance of such a bipolar amplifier stage that can be envisaged from these characteristics. First, because the nonlinearity is due to the curvature of the input characteristics of the device—the output characteristics, shown in Figure 7.21, are linear—the smaller the input signal that is applied to such a stage, the lower the nonlinearity, so that a higher stage gain will lead to reduced signal distortion at the same output level. Second, the distortion due to such a stage is very largely second harmonic in nature.

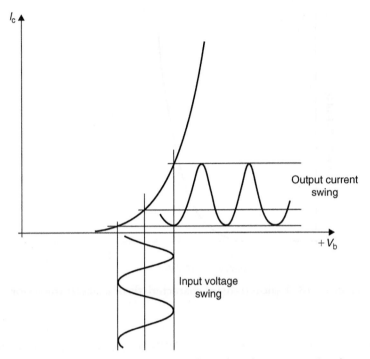

Figure 7.19: Transistor amplifier waveform distortion due to transfer characteristics.

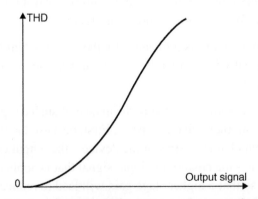

Figure 7.20: Relationship between signal distortion and output signal voltage in a bipolar transistor amplifier.

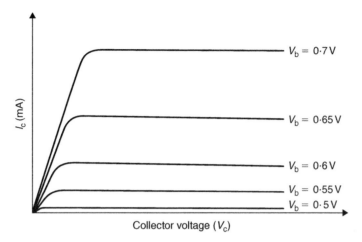

**Figure 7.21: Output current/voltage characteristics of a typical silicon bipolar transistor.**

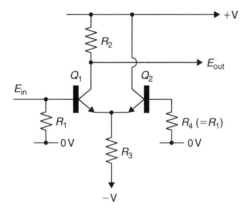

**Figure 7.22: Transistor voltage amplifier using a long-tailed pair circuit layout.**

This implies that a "push–pull" arrangement, such as the so-called "long-tailed pair" circuit shown in Figure 7.22, which tends to cancel second harmonic distortion components, will greatly improve the distortion characteristics of such a stage.

Also, because the output voltage swing for a given input signal (the stage gain) will increase as the collector load ($R_2$ in Figure 7.22) increases, the higher the effective impedance of this, the lower the distortion that will be introduced by the stage, for any given output voltage signal.

If a high value resistor is used as the collector load for $Q_1$ in Figure 7.22, either a very high supply line voltage must be applied, which may exceed the voltage ratings of the devices, or the collector current will be very small, which will reduce the gain of the device and therefore tend to diminish the benefit arising from the use of a higher value load resistor.

Various circuit techniques have been evolved to circumvent this problem by producing high dynamic impedance loads, which nevertheless permit the amplifying device to operate at an optimum value of collector current. These techniques are discussed later.

An unavoidable problem associated with the use of high values of collector load impedance as a means of attaining high stage gains in such amplifier stages is that the effect of "stray" capacitances, shown as $C_s$ in Figure 7.23, is to cause the stage gain to decrease at high frequencies as the impedance of the stray capacitance decreases and progressively begins to shunt the load. This effect is shown in Figure 7.24, in which the "transition" frequency, $f_o$ (the −3-dB gain point) is that frequency at which the shunt impedance of the stray capacitance is equal to that of the load resistor, or its effective equivalent, if the circuit design is such that an "active load" is used in its place.

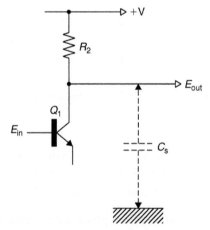

**Figure 7.23: Circuit effect of stray capacitance.**

### 7.10.2 Field Effect Devices

Other devices that may be used as amplifying components are field effect transistors and MOS devices. Both of these components are very much more linear in their transfer characteristics but have a very much lower mutual conductance ($G_m$).

This is a measure of the rate of change of output current as a function of an applied change in input voltage. For all bipolar devices, this is strongly dependent on collector current and is, for a small signal silicon transistor, typically of the order of 45 mA/V per mA collector current. Power transistors, operating at relatively high collector currents, for which a similar relationship applies, may therefore offer mutual conductances in the range of amperes/volt.

Because the output impedance of an emitter follower is approximately $1/G_m$, power output transistors used in this configuration can offer very low values of output impedance, even without externally applied negative feedback.

All field effect devices have very much lower values for $G_m$, which will lie, for small-signal components, in the range 2–10 mA/V, not significantly affected by drain currents. This means that amplifier stages employing field-effect transistors, although much more linear, offer much lower stage gains, with other things being equal.

The transfer characteristics of junction (bipolar) FETs, and enhancement and depletion mode MOSFETS are shown in Figure 7.25.

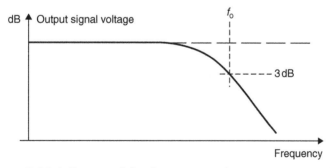

**Figure 7.24: Influence of circuit stray capacitances on stage gain.**

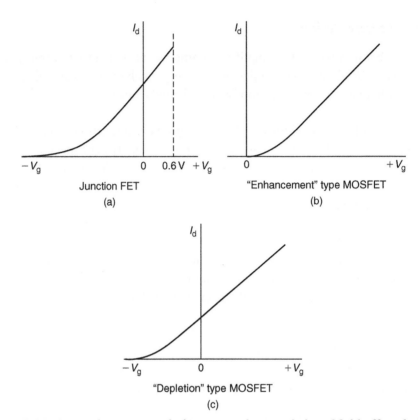

**Figure 7.25: Gate voltage versus drain current characteristics of field-effect devices.**

### 7.10.2.1 Metal–Oxide–Semiconductor Field-Effect Transistors

Metal–oxide–semiconductor field-effect transistors, in which the gate electrode is isolated from the source/drain channel, have very similar transfer characteristics to that of junction FETs. They have an advantage that, since the gate is isolated from the drain/source channel by a layer of insulation, usually silicon oxide or nitride, no maximum forward gate voltage can be applied—within the voltage breakdown limits of the insulating layer. In a junction FET the gate, which is simply a reverse biased PN diode junction, will conduct if a forward voltage somewhat in excess of 0.6 V is applied.

The chip constructions and circuit symbols employed for small signal lateral MOSFETs and junction FETs (known simply as FETs) are shown in Figures 7.26 and 7.27.

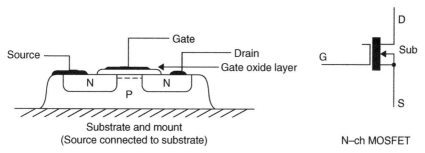

**Figure 7.26: Chip cross section and circuit symbol for lateral MOSFET (small signal type).**

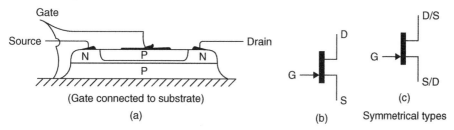

**Figure 7.27: Chip cross section and circuit symbols for (bipolar) junction FET.**

It is often found that the chip construction employed for junction FETs is symmetrical, so that the source and drain are interchangeable in use. For such devices the circuit symbol shown in Figure 7.27(c) should be used properly.

A practical problem with lateral devices, in which the current flow through the device is parallel to the surface of the chip, is that the path length from source to drain, and hence the device impedance and current carrying capacity, is limited by the practical problems of defining and etching separate regions that are in close proximity during the manufacture of the device.

### 7.10.2.2 V-MOS and T-MOS

This problem is not of very great importance for small signal devices, but is a major concern in high current ones such as those employed in power output stages. It has led to the development of MOSFETs in which the current flow is substantially in a direction that is vertical to the surface and in which the separation between layers is determined by diffusion processes rather than by photolithographic means.

Devices of this kind, known as V-MOS and T-MOS constructions, are shown in Figure 7.28.

Although these were originally introduced for power output stages, the electrical characteristics of such components are so good that these have been introduced, in smaller power versions, specifically for use in small signal linear amplifier stages. Their major advantages over bipolar devices, having equivalent chip sizes and dissipation ratings, are their high input impedance, their greater linearity, and their freedom from "hole storage" effects if driven into saturation.

These qualities are increasingly attracting the attention of circuit designers working in the audio field, where there is a trend toward the design of amplifiers having a very high intrinsic linearity rather than relying on the use of negative feedback to linearize an otherwise worse design.

### 7.10.2.3 Breakdown

A specific problem that arises in small signal MOSFET devices is that, because the gate-source capacitance is very small, it is possible to induce breakdown of the insulating

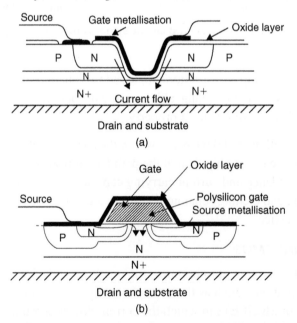

Figure 7.28: Power MOSFET constructions using (a) V and (b) T configurations. (Practical devices will employ many such cells in parallel.)

layer, which destroys the device, as a result of transferred static electrical charges arising from mishandling.

Although widely publicized and the source of much apprehension, this problem is actually very rarely encountered in use, as small signal MOSFETs usually incorporate protective zener diodes to prevent this eventuality, and power MOSFETs, where such diodes may not be used because they may lead to inadvertent "thyristor" action, have such a high gate-source capacitance that this problem does not normally arise.

In fact, when such power MOSFETs do fail, it is usually found to be because of circuit design defects, which have either allowed excessive operating potentials to be applied to the device, or have permitted inadvertent VHF oscillation, which has led to thermal failure.

## 7.11 Noise Levels

Improved manufacturing techniques have lessened the differences between the various types of semiconductor devices in respect to intrinsic noise level. For most practical purposes it can now be assumed that the characteristics of the device will be defined by the thermal noise figure of the circuit impedances. This relationship is shown in the graph of Figure 7.29.

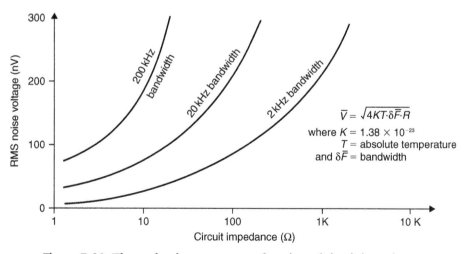

Figure 7.29: Thermal noise output as a function of circuit impedance.

For very low noise systems, operating at circuit impedance levels that have been deliberately chosen to be as low as practicable—such as in moving coil PU head amplifiers—bipolar junction transistors are still the preferred device. These will either be chosen to have a large base junction area or will be employed as a parallel-connected array, as, for example, in the LM194/394 "super-match pair" ICs, where a multiplicity of parallel-connected transistors are fabricated on a single chip, giving an effective input (noise) impedance as low as 40 ohms.

However, recent designs of monolithic-dual J-FETs, using a similar type of multiple parallel-connection system, such as the Hitachi 2SK389, can offer equivalent thermal noise resistance values as low as 33 ohms and a superior overall noise figure at input resistance values in excess of 100 ohms.

At impedance levels beyond about 1 kilohm there is little practical difference between any devices of recent design. Earlier MOSFET types were not so satisfactory because of excess noise effects arising from carrier-trapping mechanisms in impurities at the channel/gate interface.

## 7.12 Output Voltage Characteristics

Since it is desirable that output overload and signal clipping do not occur in audio systems, particularly in stages preceding the gain controls, much emphasis has been placed on the so-called "headroom" of signal handling stages, especially in hi-fi publications where the reviewers are able to distance themselves from the practical problems of circuit design.

While it is obviously desirable that inadvertent overload shall not occur in stages preceding signal level controls, high levels of feasible output voltage swing demand the use of high voltage supply rails, which, in turn, demand the use of active components that can support such working voltage levels.

Not only are such devices more costly, but they will usually have poorer performance characteristics than similar devices of lower voltage ratings. Also, the requirement for the use of high voltage operation may preclude the use of components having valuable characteristics, but which are restricted to lower voltage operation.

Practical audio circuit designs will therefore regard headroom simply as one of a group of desirable parameters in a working system whose design will be based on careful consideration of the maximum input signal levels likely to be found in practice.

Nevertheless, improved transistor or IC types, and new developments in circuit architecture, are welcomed as they occur and have eased the task of the audio design engineer, for whom the advent of new program sources, in particular the compact disc, and now digital audio tape systems, has greatly extended the likely dynamic range of the output signal.

### 7.12.1  Signal Characteristics

The practical implications of this can be seen from a consideration of the signal characteristics of existing program sources. Of these, in the past, the standard vinyl ("black") disc has been the major determining factor. In this, practical considerations of groove tracking have limited the recorded needle tip velocity to about 40 cm/s, and typical high-quality PU cartridges capable of tracking this recorded velocity will have a voltage output of some 3 mV at a standard 5-cm/s recording level.

If the preamplifier specification calls for maximum output to be obtainable at a 5-cm/s input, then the design should be chosen so that there is a "headroom factor" of at least $8\times$ in such stages preceding the gain controls.

In general, neither FM broadcasts, where the dynamic range of the transmitted signal is limited by the economics of transmitter power, nor cassette recorders, where the dynamic range is constrained by the limited tape overload characteristics, have offered such a high practicable dynamic range.

It is undeniable that the analogue tape recorder, when used at 15 in/s, twin-track, will exceed the LP record in dynamic range. After all, such recorders were originally used for mastering the discs. But such program sources are rarely found except among "live recording" enthusiasts. However, the compact disc, which is becoming increasingly common among purely domestic hi-fi systems, presents a new challenge, as the practicable dynamic range of this system exceeds 80 dB (10,000:1), and the likely range from mean (average listening level) to peak may well be as high as 35 dB (56:1) in comparison with the 18-dB (8:1) range likely with the vinyl disc.

Fortunately, because the output of the compact disc player is at a high level, typically 2 V rms, and requires no signal or frequency response conditioning prior to use, the gain control can be sited directly at the input of the preamp. Nevertheless, this still leaves the possibility that signal peaks may occur during use that are some 56× greater than the mean program level, with the consequence of the following amplifier stages being driven hard into overload.

This has refocused attention on the design of solid-state voltage amplifier stages having a high possible output voltage swing and upon power amplifiers that either have very high peak output power ratings or more graceful overload responses.

## 7.13 Voltage Amplifier Design

The sources of nonlinearity in bipolar junction transistors have already been referred to, in respect to the influence of collector load impedance, and push–pull symmetry in reducing harmonic distortion. An additional factor with bipolar junction devices is the external impedance in the base circuit, as the principal nonlinearity in a bipolar device is that due to its input voltage/output current characteristics. If the device is driven from a high impedance source, its linearity will be substantially greater, since it is operating under conditions of current drive.

This leads to the good relative performance of the simple, two-stage, bipolar transistor circuit of Figure 7.30 in that the input transistor, $Q_1$, is only required to deliver a very small voltage drive signal to the base of $Q_2$ so that the signal distortion due to $Q_1$ will

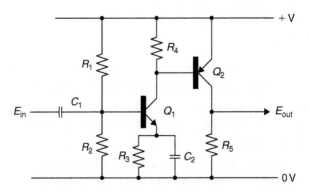

**Figure 7.30: A two-stage transistor voltage amplifier.**

be low. $Q_2$, however, which is required to develop a much larger output voltage swing, with a much greater potential signal nonlinearity, is driven from a relatively high source impedance, composed of the output impedance of $Q_1$, which is very high indeed, in parallel with the base-emitter resistor, $R_4$. $R_1$, $R_2$, and $R_3/C_2$ are employed to stabilize the DC working conditions of the circuit.

Normally, this circuit is elaborated somewhat to include both DC and AC negative feedback from the collector of $Q_2$ to the emitter of $Q_1$, as shown in the practical amplifier circuit of Figure 7.31.

This is capable of delivering a 14-V p-p output swing, at a gain of 100, and a bandwidth of 15 Hz to 250 kHz, at $-3$-dB points; largely determined by the value of $C_2$ and the output capacitances, with a THD figure of better that 0.01% at 1 kHz.

The practical drawbacks of this circuit relate to the relatively low value necessary for $R_3$—with the consequent large value necessary for $C_2$ if a good LF response is desired, and the DC offset between point 'X' and the output, due to the base-emitter junction potential of $Q_1$, and the DC voltage drop along $R_5$, which makes this circuit relatively unsuitable in DC amplifier applications.

An improved version of this simple two-stage amplifier circuit is shown in Figure 7.32, in which the single input transistor has been replaced by a "long-tailed pair" configuration

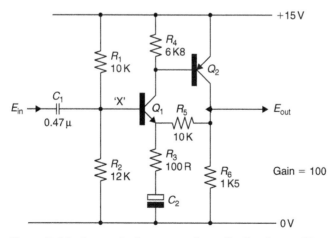

**Figure 7.31: A practical two-transistor feedback amplifier.**

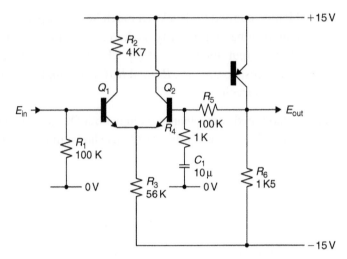

**Figure 7.32: Improved two-stage feedback amplifier.**

of the type shown in Figure 7.32. In this, if the two-input transistors are reasonably well matched in current gain and if the value of $R_3$ is chosen to give an equal collector current flow through both $Q_1$ and $Q_2$, the DC offset between input and output will be negligible, which will allow the circuit to be operated between symmetrical ($+$ and $-$) supply rails over a frequency range extending from DC to 250 kHz or more.

Because of the improved rejection of odd harmonic distortion inherent in the input "push–pull" layout, the THD due to this circuit, particularly at less than maximum output voltage swing, can be extremely low, which probably forms the basis of the bulk of linear amplifier designs. However, further technical improvements are possible, which are discussed next.

## 7.14 Constant-Current Sources and "Current Mirrors"

As mentioned earlier, the use of high-value collector load resistors in the interests of high stage gain and low inherent distortion carries with it the penalty that the performance of the amplifying device may be impaired by the low collector current levels that result from this approach.

Advantage can, however, be taken of the very high output impedance of a junction transistor, which is inherent in the type of collector current/supply voltage characteristics

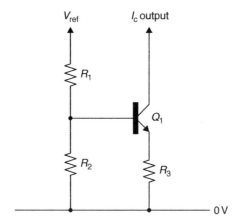

**Figure 7.33: Transistor constant current source.**

illustrated in Figure 7.21, where even at currents in the 1- to 10-mA region, dynamic impedances of the order of 100 kilohms may be expected.

A typical circuit layout that utilizes this characteristic is shown in Figure 7.33, in which $R_1$ and $R_2$ form a potential divider to define the base potential of $Q_1$, and $R_3$ defines the total emitter or collector currents for this effective base potential.

This configuration can be employed with transistors of either PNP or NPN types, which allow the circuit designer considerable freedom in their application.

An improved, two-transistor, constant current source is shown in Figure 7.34. In this, $R_1$ is used to bias $Q_2$ into conduction, and $Q_1$ is employed to sense the voltage developed across $R_2$, which is proportional to emitter current, and to withdraw the forward bias from $Q_2$ when that current level is reached at which the potential developed across $R_2$ is just sufficient to cause $Q_1$ to conduct.

The performance of this circuit is greatly superior to that of Figure 7.33 in that the output impedance is about $10\times$ greater and the circuit is insensitive to the potential, $+V_{ref.}$, applied to $R_1$, so long as it is adequate to force both $Q_2$ and $Q_1$ into conduction.

An even simpler circuit configuration makes use of the inherent very high output impedance of a junction FET under constant gate bias conditions. This employs the circuit layout shown in Figure 7.35, which allows a true "two-terminal" constant current

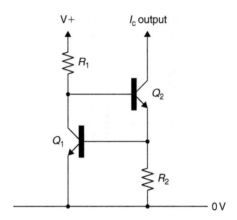

**Figure 7.34: Two-transistor constant current source.**

source, independent of supply lines or reference potentials, and which can be used at either end of the load chain.

The current output from this type of circuit is controlled by the value chosen for $R_1$, and this type of constant current source may be constructed using almost any available junction FET, provided that the voltage drop across the FET drain-gate junction does not exceed the breakdown voltage of the device. This type of constant current source is also available as small, plastic-encapsulated, two-lead devices, at a relatively low cost, and with a range of specified output currents.

All of these constant current circuit layouts share the common small disadvantage that they will not perform very well at low voltages across the current source element. In the case of Figures 7.33 and 7.34, the lowest practicable operating potential will be about 1 V. The circuit of Figure 7.35 may require, perhaps, 2–3 V, and this factor must be considered in circuit performance calculations.

The "boot-strapped" load resistor arrangement shown in Figure 7.36, and commonly used in earlier designs of audio amplifier to improve the linearity of the last class 'A' amplifier stage ($Q_1$), effectively multiplies the resistance value of $R_2$ by the gain which $Q_2$ would be deemed to have if operated as a common-emitter amplifier with a collector load of $R_3$ in parallel with $R_1$.

This arrangement is the best configuration practicable in terms of available rms output voltage swing as compared with conventional constant current sources, but has fallen into

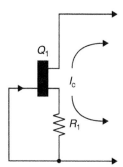

**Figure 7.35: Two-terminal constant current source.**

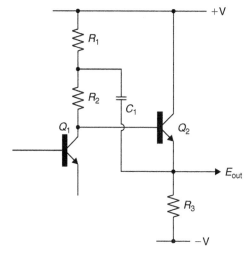

**Figure 7.36: Load impedance increase by boot-strap circuit.**

disuse because it leads to slightly lower quality THD figures than are possible with other circuit arrangements.

All these circuit arrangements suffer from a further disadvantage, from the point of view of the integrated circuit designer: they employ resistors as part of the circuit design, and resistors, although possible to fabricate in IC structures, occupy a disproportionately large area of the chip surface. Also, they are difficult to fabricate to precise resistance values without resorting to subsequent laser trimming, which is expensive and time-consuming.

Because of this, there is a marked preference on the part of IC design engineers for the use of circuit layouts known as "current mirrors," of which a simple form is shown in Figure 7.37.

### 7.14.1 IC Solutions

These are not true constant current sources in that they are only capable of generating an output current ($L_{out}$) that is a close equivalent of the input or drive current ($L_{in}$). However, the output impedance is very high, and if the drive current is held to a constant value, the output current will also remain constant.

A frequently found elaboration of this circuit, which offers improvements in respect to output impedance and the closeness of equivalence of the drive and output currents, is shown in Figure 7.38. Like the junction FET-based constant current source, these current mirror devices are available as discrete, plastic-encapsulated, three-lead components,

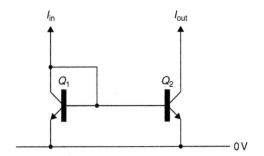

**Figure 7.37: Simple form of a current mirror.**

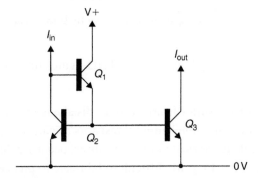

**Figure 7.38: Improved form of a current mirror.**

having various drive current/output current ratios, for incorporation into discrete component circuit designs.

The simple amplifier circuit of Figure 7.32 can be elaborated, as shown in Figure 7.39, to employ these additional circuit refinements, which would have the effect of increasing the open-loop gain, that is, that before negative feedback is applied, by $10-100\times$ and improving the harmonic and other distortions, and the effective bandwidth by perhaps $3-10\times$. From the point of view of the IC designer, there is also the advantage of a potential reduction in the total resistor count.

These techniques for improving the performance of semiconductor amplifier stages find particular application in the case of circuit layouts employing junction FETs and

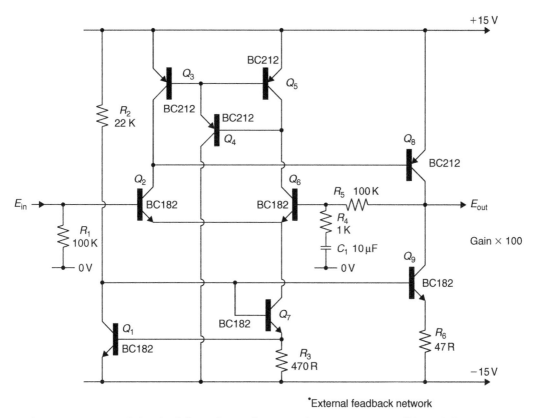

*External feedback network

**Figure 7.39: Use of circuit elaboration to improve the two-stage amplifier of Figure 7.32.**

MOSFETs, where the lower effective mutual conductance values for the devices would normally result in relatively poor stage gain figures.

This has allowed the design of IC operational amplifiers, such as the RCA CA3140 series or the Texas Instruments TL071 series, which employ, respectively, MOSFET and junction FET input devices. The circuit layout employed in the TL071 is shown, by way of example, in Figure 7.40.

Both of these op-amp designs offer input impedances in the million megohm range— in comparison with the input impedance figures of 5–10 kilohm, which were typical of early bipolar ICs—and the fact that the input impedance is so high allows the use of such ICs in circuit configurations for which earlier op-amp ICs were entirely inappropriate.

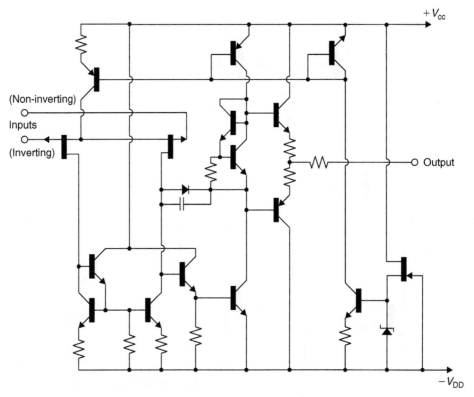

**Figure 7.40: Circuit layout of Texas Instruments TL071 op-amp.**

Although the RCA design employs MOSFET input devices, which offer, in principle, an input impedance that is perhaps 1000 times better than this figure, the presence of on-chip Zener diodes, to protect the device against damage through misuse or static electric charges, reduces the input impedance to roughly the same level as that of the junction FET device.

It is a matter for some regret that the design of the CA3140 series devices is now so elderly that the internal MOSFET devices do not offer the low level of internal noise of which more modern MOSFET types are capable. This tends to rule out the use of this MOSFET op-amp for high-quality audio use, although the TL071 and its equivalents, such as the LF351, have demonstrated impeccable audio behavior.

## 7.15 Performance Standards

It has always been accepted in the past, and is still held as axiomatic among a very large section of the engineering community, that performance characteristics can be measured and that improved levels of measured performance will correlate precisely, within the ability of the ear to detect such small differences, with improvements that the listener will hear in reproduced sound quality.

Within a strictly engineering context, it is difficult to do anything other than accept the concept that measured improvements in performance are the only things that should concern the designer.

However, the frequently repeated claim by journalists and reviewers working for periodicals in the hi-fi field—who, admittedly, are unlikely to be unbiased witnesses— that measured improvements in performance do not always go hand in hand with the impressions that the listener may form, tends to undermine the confidence of the circuit designer that the instrumentally determined performance parameters are all that matter.

It is clear that it is essential for engineering progress that circuit design improvements must be sought that lead to measurable performance improvements. However, there is now also the more difficult criterion that those things that appear to be better, in respect to measured parameters, must also be seen, or heard, to be better.

### 7.15.1 Use of ICs

This point is particularly relevant to the question of whether, in very high-quality audio equipment, it is acceptable to use IC operational amplifiers, such as the TL071, or some

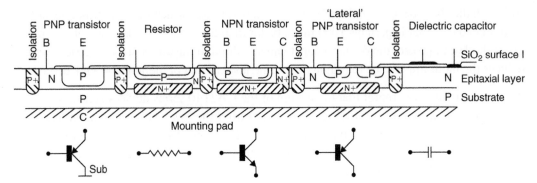

**Figure 7.41: Method of fabrication of components in a silicon-integrated circuit.**

of the even more exotic later developments such as the NE5534 or the OP27, as the basic gain blocks, around which the passive circuitry can be arranged, or whether, as some designers believe, it is preferable to construct such gain blocks entirely from discrete components.

Some years ago, there was a valid technical justification for this reluctance to use op-amp ICs in high-quality audio circuitry, as the method of construction of such ICs was as shown, schematically, in Figure 7.41, in which all the structural components were formed on the surface of a heavily 'P' doped silicon substrate, and relied for their isolation from one another or from the common substrate on the reverse-biased diodes formed between these elements.

This led to a relatively high residual background noise level, in comparison with discrete component circuitry, due to the effects of the multiplicity of reverse diode leakage currents associated with every component on the chip. Additionally, there were quality constraints in respect to the components formed on the chip surface—more severe for some component types than for others—that also impaired the circuit performance.

A particular instance of this problem arose in the case of PNP transistors used in normal ICs, where the circuit layout did not allow these to be formed with the substrate acting as the collector junction. In this case, it was necessary to employ the type of construction known as a "lateral PNP," in which all the junctions are diffused in, from the exposed chip surface, side by side.

In this type of device the width of the 'N' type base region, which must be very small for optimum results, depends mainly on the precision with which the various diffusion masking layers can be applied. The results are seldom very satisfactory. Such a lateral PNP device has a very poor current gain and HF performance.

In recent IC designs, considerable ingenuity has been shown in the choice of circuit layout to avoid the need to employ such unsatisfactory components in areas where their shortcomings would affect the end result. Substantial improvements, both in the purity of the base materials and in diffusion technology, have allowed the inherent noise background to be reduced to a level where it is no longer of practical concern.

### 7.15.2 Modern Standards

The standard of performance that is now obtainable in audio applications, from some of the recent IC op–amps, especially at relatively low closed-loop gain levels, is frequently of the same order as that of the best discrete component designs, but with considerable advantages in other respects, such as cost, reliability, and small size.

This has led to their increasing acceptance as practical gain blocks, even in very high-quality audio equipment.

When blanket criticism is made of the use of ICs in audio circuitry, it should be remembered that the 741, which was one of the earliest of these ICs to offer a satisfactory performance—although it is outclassed by more recent types—has been adopted with enthusiasm, as a universal gain block, for the signal handling chains in many recording and broadcasting studios.

This implies that the bulk of the program signals employed by the critics to judge whether or not a discrete component circuit is better than that using an IC will already have passed through a sizeable handful of 741-based circuit blocks, and if such ICs introduce audible defects, then their reference source is already suspect.

It is difficult to stipulate the level of performance that will be adequate in a high-quality audio installation. This arises partly because there is little agreement between engineers and circuit designers, on the one hand, and the hi-fi fraternity, on the other hand, about the characteristics that should be sought and partly because of the wide differences that exist between listeners in their expectations for sound quality or their sensitivity to distortions. These differences combine to make it a difficult and speculative task to attempt either to quantify or to specify the technical components of audio quality or to establish an acceptable minimum-quality level.

Because of this uncertainty, the designer of equipment in which price is not a major consideration will normally seek to attain standards substantially in excess of those that he supposes to be necessary, simply in order not to fall short. This means that the reason for the small residual differences in the sound quality, as between high-quality units, is the existence of malfunctions of types that are not currently known or measured.

## 7.16  Audibility of Distortion

### 7.16.1  Harmonic and Intermodulation Distortion

Because of the small dissipations that are normally involved, almost all discrete component voltage amplifier circuitry will operate in class 'A' (that condition in which the bias applied to the amplifying device is such as to make it operate in the middle of the linear region of its input/output transfer characteristic), and the residual harmonic components are likely to be mainly either second or third order, which are audibly much more tolerable than higher order distortion components.

Experiments in the late 1940s suggested that the level of audibility for second and third harmonics was of the order of 0.6 and 0.25%, respectively, which led to the setting of a target value, within the audio spectrum, of 0.1% THD, as desirable for high-quality audio equipment.

However, recent work aimed at discovering the ability of an average listener to detect the presence of low-order (i.e., second or third) harmonic distortions has drawn the uncomfortable conclusion that listeners, taken from a cross section of the public, may rate a signal to which 0.5% second harmonic distortion has been added as "more musical" than, and therefore preferable to, the original undistorted input. This discovery tends to cast doubt on the value of some subjective testing of equipment.

What is not in dispute is that the intermodulation distortion (IMD), which is associated with any nonlinearity in the transfer characteristics, leads to a muddling of the sound picture so that if the listener is asked not which sound he prefers, but which sound seems to him to be the clearer, he will generally choose that with the lower harmonic content.

The way in which IMD arises is shown in Figure 7.42, where a composite signal containing both high-frequency and low-frequency components, fed through a nonlinear

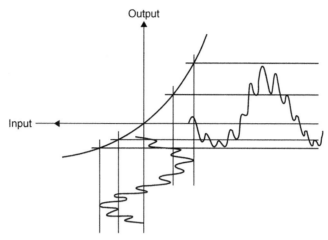

**Figure 7.42: Intermodulation distortions produced by the effect of a nonlinear input/output transfer characteristic on a complex tone.**

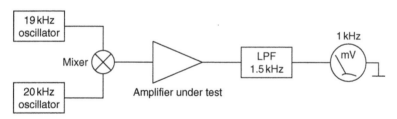

**Figure 7.43: Simple HF two-tone intermodulation distortion test.**

system, causes each signal to be modulated by the other. This is conspicuous in the drawing in respect to the HF component, but is also true for the LF one.

This can be shown mathematically to be due to the generation of sum and difference products, in addition to the original signal components, and provides a simple method, shown schematically in Figure 7.43, for the detection of this type of defect. A more formal IMD measurement system is shown in Figure 7.44.

With present circuit technology and device types, it is customary to design for total harmonic and IM distortions to be below 0.01% over the range 30 Hz–20 kHz, and at all signal levels below the onset of clipping. Linear IC op-amps, such as the TL071 and the LF351, will also meet this specification over the frequency range 30 Hz–10 kHz.

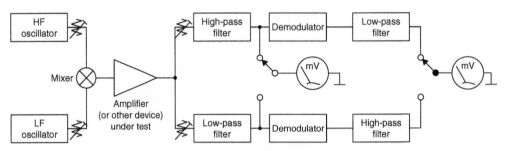

**Figure 7.44: Two-tone intermodulation distortion test rig.**

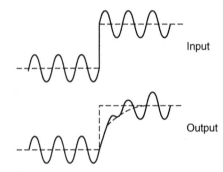

**Figure 7.45: Effect of amplifier slew-rate saturation or transient intermodulation distortion.**

### 7.16.2 Transient Defects

A more insidious group of signal distortions may occur when brief signals of a transient nature, or sudden step type changes in base level, are superimposed on the more continuous components of the program signal. These defects can take the form of slew-rate distortions, usually associated with a loss of signal during the period of the slew-rate saturation of the amplifier—often referred to as transient intermodulation distortion or TID.

This defect is illustrated in Figure 7.45 and arises particularly in amplifier systems employing substantial amounts of negative feedback when there is some slew-rate limiting component within the amplifier, as shown in Figure 7.46.

A further problem is that due to "overshoot," or "ringing," on a transient input, as illustrated in Figure 7.47. This arises particularly in feedback amplifiers if there is an inadequate stability margin in the feedback loop, particularly under reactive load

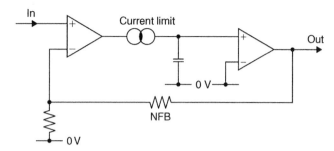

**Figure 7.46: Typical amplifier layout causing slew-rate saturation.**

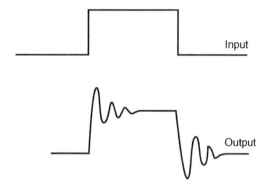

**Figure 7.47: Transient "ringing."**

conditions, but will also occur in low-pass filter systems if too high an attenuation rate is employed.

The ear is very sensitive to slew-rate induced distortion, which is perceived as a "tizziness" in the reproduced sound. Transient overshoot is normally noted as a somewhat overbright quality. The avoidance of both these problems demands care in the circuit design, particularly when a constant current source is used, as shown in Figure 7.48.

In this circuit, the constant current source, $CC_1$, will impose an absolute limit on the possible rate of change of potential across the capacitance, $C_1$ (which could well be simply the circuit stray capacitance), when the output voltage is caused to move in a positive-going direction. This problem is compounded if an additional current limit mechanism, $CC_2$, is included in the circuitry to protect the amplifier transistor ($Q_1$) from output current overload.

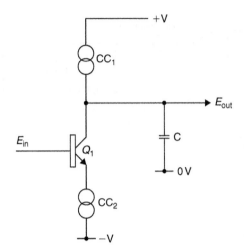

**Figure 7.48: Circuit design aspects that may cause slew-rate limiting.**

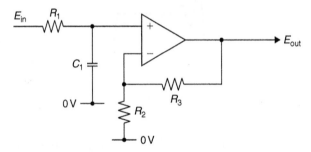

**Figure 7.49: Input HF limiting circuit to lessen slew-rate limiting.**

Since output load and other inadvertent capacitances are unavoidable, it is essential to ensure that all such current limited stages operate at a current level that allows potential slewing to occur at rates that are at least 10× greater than the fastest signal components. Alternatively, means may be taken, by way of a simple input integrating circuit, $(R_1 C_1)$, as shown in Figure 7.49, to ensure that the maximum rate of change of the input signal voltage is within the ability of the amplifier to handle it.

### 7.16.3 Spurious Signals

In addition to harmonic, IM, and transient defects in the signal channel, which will show up on normal instrumental testing, there is a whole range of spurious signals that may not arise in such tests. The most common of these is that of the intrusion of noise and alien signals, either from the supply line or by direct radio pick up.

This latter case is a random and capricious problem that can only be solved by steps appropriate to the circuit design in question. However, supply line intrusions, whether because of unwanted signals from the power supply or from the other channel in a stereo system, may be reduced greatly by the use of circuit designs offering a high immunity to voltage fluctuations on the DC supply.

Other steps, such as the use of electronically stabilized DC supplies or the use of separate power supplies in a stereo amplifier, are helpful, but the required high level of supply line signal rejection should be sought as a design feature before other palliatives are applied. Modern IC op-amps offer a typical supply voltage rejection ratio of 90 dB (30,000:1). Good discrete component designs should offer at least 80 dB (10,000:1).

This figure tends to degrade at higher frequencies, which has led to the growing use of supply line bypass capacitors having a low effective series resistance. This feature is either a result of the capacitor design or is achieved in the circuit by the designer's adoption of groups of parallel connected capacitors chosen so that the AC impedance remains low over a wide range of frequencies.

A particular problem in respect to spurious signals, which occurs in audio power amplifiers, is a consequence of the loudspeaker acting as a voltage generator, when stimulated by pressure waves within the cabinet, and injecting unwanted audio components directly into the negative feedback loop of the amplifier. This specific problem is unlikely to arise in small signal circuitry, but the designer must consider what effect output/line load characteristics may have, particularly in respect to reduced stability margin in a feedback amplifier.

In all amplifier systems there is a likelihood of microphonic effects due to vibration of the components. This is likely to be of increasing importance at the input of "low-level," high-sensitivity preamplifier stages and can lead to coloration of the signal when the equipment is in use, which is overlooked in the laboratory in a quiet environment.

### 7.16.4 Mains-Borne Interference

Mains-borne interference, as evidenced by noise pulses on switching electrical loads, is most commonly due to radio pick up problems and is soluble by the techniques (attention to signal and earth line paths, avoidance of excessive HF bandwidth at the input stages) that are applicable to these.

## 7.17 General Design Considerations

During the past three decades, a range of circuit design techniques has evolved to allow the construction of highly linear gain stages based on bipolar transistors whose input characteristics are, in themselves, very nonlinear. These techniques have also allowed substantial improvements in possible stage gain and have led to greatly improved performance from linear, but low gain, field-effect devices.

These techniques are used in both discrete component designs and in their monolithic integrated circuit equivalents, although, in general, the circuit designs employed in linear ICs are considerably more complex than those used in discrete component layouts.

This is partly dictated by economic considerations, partly by the requirements of reliability, and partly because of the nature of IC design.

The first two of these factors arise because both the manufacturing costs and the probability of failure in a discrete component design are directly proportional to the number of components used, so the fewer the better, whereas in an IC, both the reliability and the expense of manufacture are affected only minimally by the number of circuit elements employed.

In the manufacture of ICs, as indicated earlier, some of the components that must be employed are much worse than their discrete design equivalents. This has led the IC designer to employ fairly elaborate circuit structures, either to avoid the need to use a poor-quality component in a critical position or to compensate for its shortcomings.

Nevertheless, the ingenuity of the designers and the competitive pressures of the market-place have resulted in systems having a very high performance, usually limited only by their inability to accept differential supply line potentials in excess of 36 V unless nonstandard diffusion processes are employed.

For circuitry requiring higher output or input voltage swings than allowed by small signal ICs, the discrete component circuit layout is, at the moment, unchallenged. However, as every designer knows, it is a difficult matter to translate a design that is satisfactory at a low working voltage design into an equally good higher voltage system.

This is because:

- increased applied potentials produce higher thermal dissipations in the components for the same operating currents;

- device performance tends to deteriorate at higher interelectrode potentials and higher output voltage excursions; and,

- available high/voltage transistors tend to be more restricted in variety and less good in performance than lower voltage types.

## 7.18 Controls

These fall into a variety of categories:

- gain controls needed to adjust the signal level between source and power amplifier stages;

- tone controls used to modify the tonal characteristics of the signal chain; and,

- filters employed to remove unwanted parts of the incoming signal, and those adjustments used to alter the quality of the audio presentation, such as stereo channel balance or channel separation controls.

### 7.18.1 Gain Controls

These are the simplest in basic form and are often just a resistive potentiometer voltage divider of the type shown in Figure 7.50. Although simple, this component can generate a variety of problems. Of these, the first is due to the value chosen for $R_1$. Unless this is

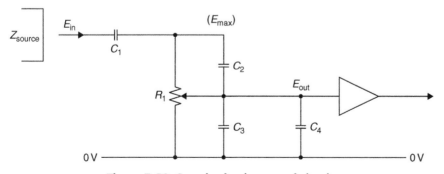

**Figure 7.50: Standard gain control circuit.**

infinitely high, it will attenuate the maximum signal voltage ($E_{max}$) obtainable from the source, in the ratio

$$E_{max} = \frac{E_n \times R_1}{(R_1 + Z_{source})}$$

where $Z_{source}$ is the output impedance of the driving circuit. This factor favors the use of a high value for $R_1$ to avoid loss of input signal.

However, the following amplifier stage may have specific input impedance requirements and is unlikely to operate satisfactorily unless the output impedance of the gain control circuit is fairly low. This will vary according to the setting of the control, between zero and a value, at the maximum gain setting of the control, due to the parallel impedances of the source and gain control.

$$Z_{out} = \frac{R_1}{(R_1 + Z_{source})}.$$

The output impedance at intermediate positions of the control varies as the effective source impedance and the impedance to the 0-V line are altered. However, in general, these factors would encourage the use of a low value for $R_1$.

An additional and common problem arises because the perceived volume level associated with a given sound pressure (power) level has a logarithmic characteristic. This means that the gain control potentiometer, $R_1$, must have a resistance value that has a logarithmic, rather than linear, relationship with the angular rotation of the potentiometer shaft.

### 7.18.1.1 Potentiometer Law

Since the most common types of control potentiometer employ a resistive composition material to form the potentiometer track, it is a difficult matter to ensure that the grading of conductivity within this material will follow an accurate logarithmic law.

On a single channel this error in the relationship between signal loudness and spindle rotation may be relatively unimportant. In a stereo system, having two ganged gain control spindles, intended to control the loudness of the two channels simultaneously, errors in following the required resistance law, existing between the two potentiometer sections, will cause a shift in the apparent location of the stereo image as the gain control is adjusted, which can be very annoying.

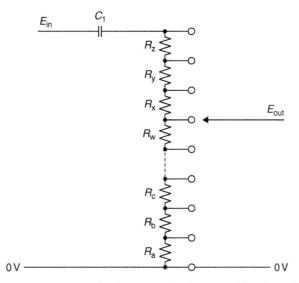

**Figure 7.51: Improved gain control using a multi-pole switch.**

In high-quality equipment, this problem is sometimes avoided by replacing $R_1$ by a precision resistor chain ($R_a - R_z$), as shown in Figure 7.51, in which the junctions between these resistors are connected to tapping points on a high-quality multiposition switch.

By this means, if a large enough number of switch tap positions is available, and this implies at least a 20-way switch to give a gentle gradation of sound level, a very close approximation to the required logarithmic law can be obtained, and two such channel controls could be ganged without unwanted errors in the differential output level.

### 7.18.1.2 Circuit Capacitances

A further practical problem, illustrated in Figure 7.50, is associated with circuit capacitances. First, it is essential to ensure that there is no standing DC potential across $R_1$ in normal operation, as this will cause an unwanted noise in the operation of the control. This imposes the need for a protective input capacitor, $C_1$, which will cause a loss of low-frequency signal components, with a $-3$-dB LF turnover point at the frequency at which the impedance of $C_m$ is equal to the sum of the source and gain control impedances. $C_1$ should therefore be of an adequate value.

Additionally, there are the effects of the stray capacitances, $C_2$ and $C_3$, associated with the potentiometer construction, and the amplifier input and wiring capacitances, $C_4$.

The effect of these is to modify the frequency response of the system, at the HF end, as a result of signal currents passing through these capacitances. The choice of a low value for $R_1$ is desirable to minimize this problem.

The use of the gain control to operate an on/off switch, which is fairly common in low-cost equipment, can lead to additional problems, especially with high resistance value gain control potentiometers, in respect to AC mains "hum" pick up. It also leads to a more rapid rate of wear of the gain control in that it is rotated at least twice whenever the equipment is used.

### 7.18.2 Tone Controls

These exist in the various forms shown in Figures 7.52–7.56, respectively, described as standard (bass and treble lift or cut), slope control, Clapham junction, parametric, and graphic equalizer types. The effect these will have on the frequency response of the equipment is shown in the drawings, and their purpose is to help remedy shortcomings in the source program material, the receiver or transducer, or in the loudspeaker and listening room combination.

To the hi-fi purist, all such modifications to the input signal tend to be regarded with distaste and are therefore omitted from some hi-fi equipment. However, they can be useful and make valuable additions to the audio equipment, if used with care.

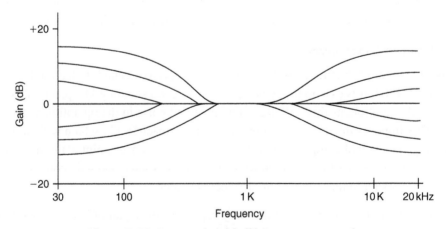

**Figure 7.52: Bass and treble lift/cut tone control.**

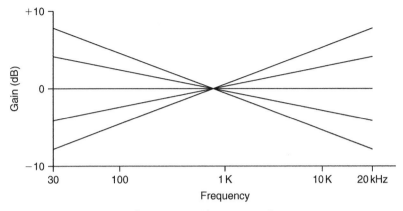

**Figure 7.53: Slope control.**

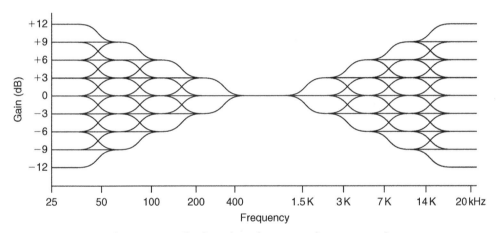

**Figure 7.54: Clapham junction type of tone control.**

### 7.18.2.1 Standard Tone Control Systems

These are either of the passive type, of which a typical circuit layout is shown in Figure 7.57, or are constructed as part of the negative feedback loop around a gain block using the general design due to Baxandall. A typical circuit layout for this kind of design is shown in Figure 7.58.

It is claimed that the passive layout has an advantage in quality over the active (feedback network) type of control in that the passive network merely contains resistors and

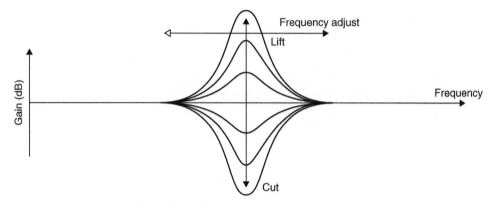

**Figure 7.55: Parametric equalizer control.**

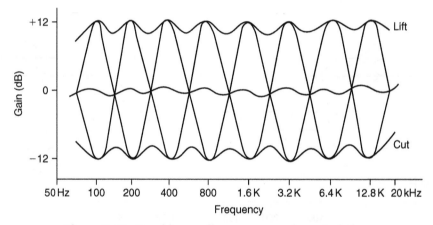

**Figure 7.56: Graphic equalizer response characteristics.**

capacitors and is therefore free from any possibility of introduced distortion, whereas the "active" network requires an internal gain block, which is not automatically above suspicion.

In reality, however, any passive network must introduce an attenuation, in its fiat response form, which is equal to the degree of boost sought at the maximum "lift" position, and some external gain block must therefore be added to compensate for this gain loss.

This added gain block is just as prone to introduce distortion as that in an active network, with the added disadvantage that it must provide a gain equal to that of the fiat-response

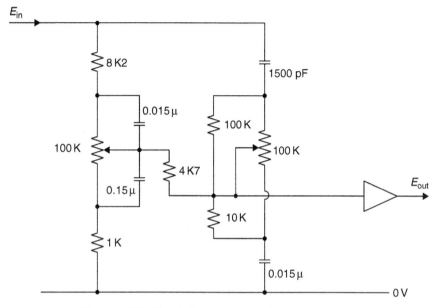

Figure 7.57: Circuit layout of passive tone control.

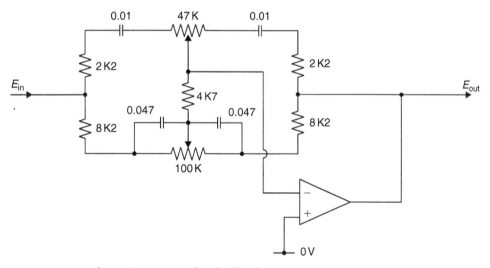

Figure 7.58: Negative feedback type tone control circuit.

network attenuation, whereas the active system gain block will typically have a gain of unity in the fiat response mode, with a consequently lower distortion level.

As a final point, it should be remembered that any treble lift circuit will cause an increase in harmonic distortion, simply because it increases the gain at the frequencies associated with harmonics, in comparison with that at the frequency of the fundamental.

The verdict of the amplifier designers appears to be substantially in favor of the Baxandall system in that this is the layout employed most commonly.

Both of these tone control systems—indeed this is true of all such circuitry—rely for their operation on the fact that the AC impedance of a capacitor will depend on the applied frequency, as defined by the equation:

$$Z_c = \frac{1}{(2\pi f_c)},$$

or, more accurately,

$$Z_c = \frac{1}{(2j\pi f_c)},$$

where $j$ is the square root of $-1$.

Commonly, in circuit calculations, the $2\pi f$ group of terms is lumped together and represented by the Greek symbol $\omega$.

The purpose of the $j$ term, which appears as a "quadrature" element in the algebraic manipulations, is to permit the circuit calculations to take account of the 90° phase shift introduced by the capacitative element. (The same is also true of inductors within such a circuit, except that the phase shift will be in the opposite sense.) This is important in most circuits of this type.

The effect of the change in impedance of the capacitor on the output signal voltage from a simple RC network, of the kind shown in Figures 7.59(a) and 7.60(a), is shown in Figures 7.59(b) and 7.60(b). If a further resistor, $R_2$, is added to the networks, the result is modified in the manner shown in Figures 7.61 and 7.62. This type of structure, elaborated by the use of variable resistors to control the amount of lift or fall of output as a function of frequency, is the basis of the passive tone control circuitry of Figure 7.57.

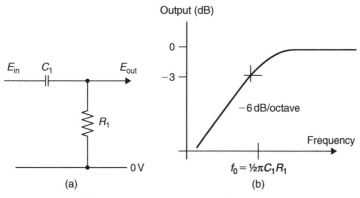

Figure 7.59: Layout and frequency response of a simple bass-cut circuit (high pass).

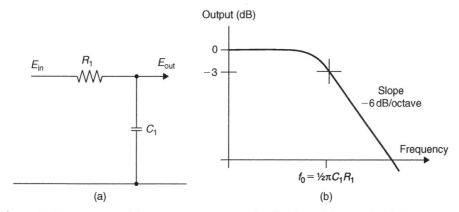

Figure 7.60: Layout and frequency response of a simple treble-cut circuit (low pass).

If such networks are connected across an inverting gain block, as shown in Figures 7.63(a) and 7.64(a), the resultant frequency response will be shown in Figures 7.63(b) and 7.64(b), since the gain of such a negative feedback configuration will be

$$\text{Gain} = \frac{Z_a}{Z_b}$$

assuming that the open-loop gain of the gain block is sufficiently high. This is the design basis of the Baxandall type of tone control, and a flat frequency response results when the impedance of the input and output limbs of such a feedback arrangement remains in equality as the frequency is varied.

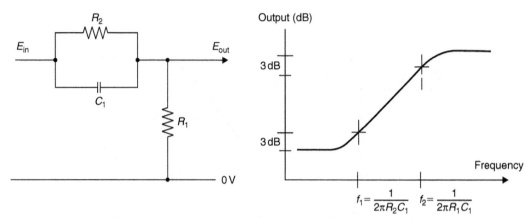

**Figure 7.61: Modified bass-cut (high-pass) RC circuit.**

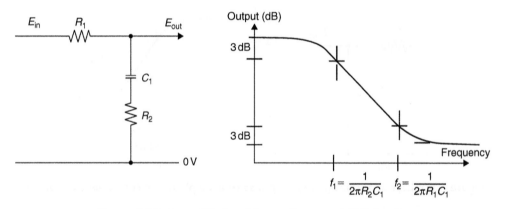

**Figure 7.62: A modified treble-cut (low-pass) RC circuit.**

### 7.18.2.2 Slope Controls

This is the type of tone control employed by Quad in its type 44 preamplifier and operates by altering the relative balance of the LF and HF components of the audio signal, with reference to some specified midpoint frequency, as is shown in Figure 7.53. A typical circuit for this type of design is shown in Figure 7.65.

The philosophical justification for this approach is that it is unusual for any commercially produced program material to be significantly in error in its overall frequency

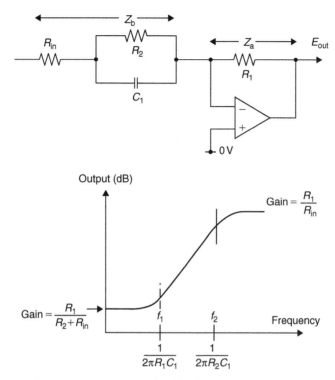

**Figure 7.63: Active RC treble-lift or bass-cut circuit.**

characteristics, but the tonal preferences of the recording or broadcasting balance engineer may differ from those of the listener.

In such a case, he might consider that the signal, as presented, was somewhat overheavy, in respect to its bass, or alternatively, perhaps, that it was somewhat light or thin in tone, and an adjustment of the skew of the frequency response could correct this difference in tonal preference without significantly altering the signal in other respects.

### 7.18.2.3 The Clapham Junction Type

This type of tone control, whose possible response curves are shown in Figure 7.54, was introduced by the author to provide a more versatile type of tonal adjustment than that offered by the conventional standard systems for remedying specific peaks or troughs in the frequency response, without the penalties associated with the graphic equalizer type of control, described later.

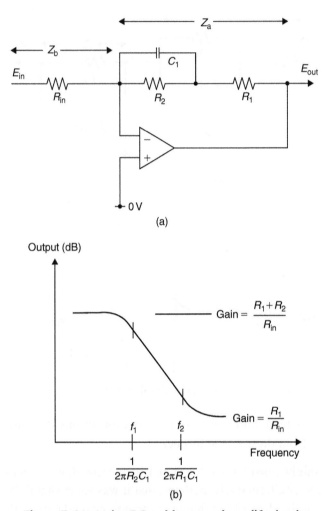

**Figure 7.64: Active RC treble-cut or bass-lift circuit.**

In the Clapham junction type system, so named because of the similarity of the possible frequency response curves to that of railway lines, a group of push switches is arranged to allow one or more of a multiplicity of RC networks to be introduced into the feedback loop of a negative feedback type tone control system, as shown in Figure 7.66, to allow individual ±3-dB frequency adjustments to be made, over a range of possible frequencies.

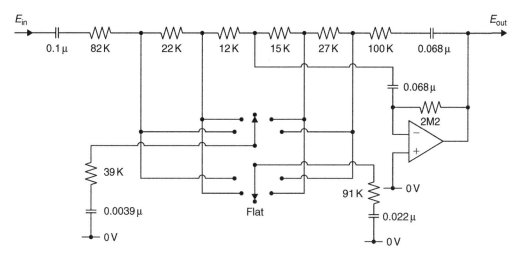

**Figure 7.65: The Quad tilt control.**

By this means it is possible, by combining elements of frequency lift or cut, to choose from a variety of possible frequency response curves without losing the ability to attain a linear frequency response.

### 7.18.2.4 Parametric Controls

This type of tone control, whose frequency response is shown in Figure 7.55, has elements of similarity to both the standard bass/treble lift/cut systems and the graphic equalizer arrangement in that while there is a choice of lift or cut in the frequency response, the actual frequency at which this occurs may be adjusted, up or down, in order to attain an optimal system frequency response.

A typical circuit layout is shown in Figure 7.67.

### 7.18.2.5 The Graphic Equalizer System

The aim of this type of arrangement is to compensate fully for the inevitable peaks and troughs in the frequency response of the audio system, including those due to deficiencies in the loudspeakers or the listening room acoustics, by permitting the individual adjustment of the channel gain, within any one of a group of eight single-octave segments of the frequency band, typically covering the range from 80 Hz to 20 kHz, although 10 octave equalizers covering the whole audio range from 20 Hz to 20 kHz have been offered.

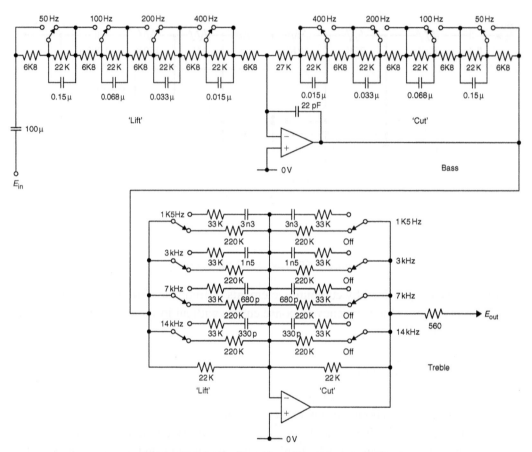

**Figure 7.66: Clapham junction tone control.**

Because the ideal solution to this requirement—that of employing a group of parallel connected amplifiers, each of which is filtered so that it covers a single octave band of the frequency spectrum, whose individual gains could be adjusted separately—would be excessively expensive to implement, conventional practice is to make use of a series of LC-tuned circuits, connected within a feedback control system, as shown in Figure 7.68.

This gives the type of frequency response curve shown in Figure 7.56. As can be seen, there is no position of lift or cut, or combination of control settings, that will permit a flat frequency response because of the interaction, within the circuitry, between the adjacent octave segments of the pass band.

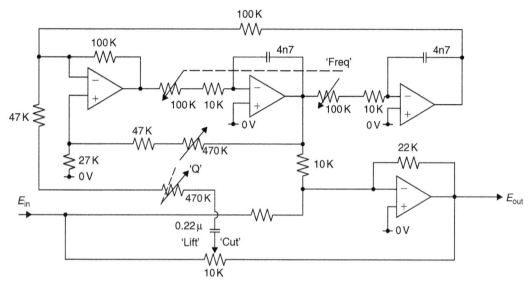

**Figure 7.67: Parametric equalizer circuit.**

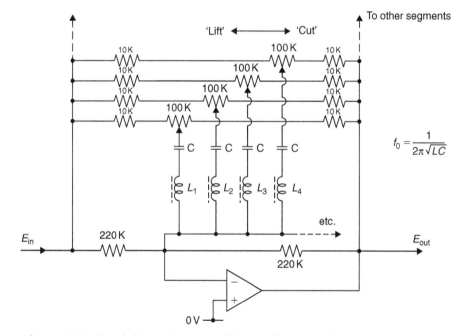

**Figure 7.68: Circuit layout for a graphic equalizer (only four sections shown).**

While such types of tone control are undoubtedly useful and can make significant improvements in the performance of otherwise unsatisfactory hi-fi systems, the inability to attain a flat frequency response when this is desired, even at the midposition of the octave-band controls, has given such arrangements a very poor status in the eyes of the hi-fi fraternity. This unfavorable opinion has been reinforced by the less than optimal performance offered by inexpensive, add-on units whose engineering standards have reflected their low purchase price.

### 7.18.3  Channel Balance Controls

These are provided in any stereo system to equalize the gain in the left- and right-hand channels and to obtain a desired balance in the sound image. (In a quadraphonic system, four such channel gain controls will ideally be provided.) In general, there are only two options available for this purpose: those balance controls that allow one or the other of the two channels to be reduced to zero output level and those systems, usually based on differential adjustment of the amount of negative feedback across controlled stages, in which the relative adjustment of the gain, in one channel with reference to the other, may only be about 10 dB.

This is adequate for all balance correction purposes, but does not allow the complete extinction of either channel.

The first type of balance control is merely a gain control, of the type shown in Figure 7.50. A negative feedback type of control is shown in Figure 7.69.

### 7.18.4  Channel Separation Controls

While the closest reproduction, within the environment of the listener, of the sound stage of the original performance will be given by a certain specific degree of separation between signals within the 'L' and 'R' channels, it is found that shortcomings in the design of the reproducing and amplifying equipment tend universally to lessen the degree of channel separation rather than the reverse.

Some degree of enhancement of channel separation is therefore often of great value, and electronic circuits for this purpose are available, such as that, due to the author, shown in Figure 7.70.

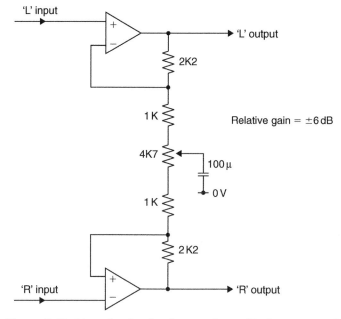

**Figure 7.69: Negative feedback type channel balance control.**

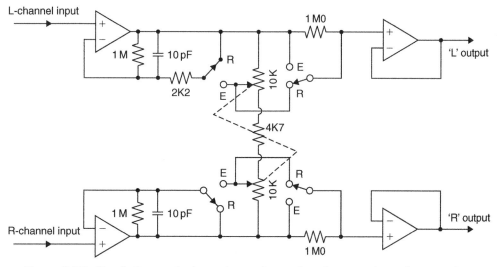

**Figure 7.70: Circuit for producing enhanced or reduced stereo channel separation.**

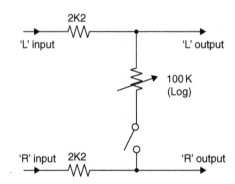

**Figure 7.71: Simple stereo channel blend control.**

There are also occasions when a deliberate reduction in the channel separation is advantageous, as, for example, in lessening "rumble" effects due to the vertical motion of a poorly engineered record turntable or in lessening the hiss component of a stereo FM broadcast. While this is also provided by the circuit of Figure 7.70, a much less elaborate arrangement, as shown in Figure 7.71, will suffice for this purpose.

A further, and interesting, approach is that offered by Blumlein, who found that an increase or reduction in the channel separation of a stereo signal was given by adjusting the relative magnitudes of the 'L + R' and 'L − R' signals in a stereo matrix, before these were added or subtracted to give the '2L' and '2R' components.

An electronic circuit for this purpose is shown in Figure 7.72.

### 7.18.5 Filters

While various kinds of filter circuits play a very large part in the studio equipment employed to generate the program material, both as radio broadcasts and as recordings on disc or tape, the only types of filters normally offered to the user are those designed to attenuate very low frequencies, below, say, 50 Hz and generally described as "rumble" filters, or those operating in the region above a few kHz, and generally described as "scratch" or "whistle" filters.

Three such filter circuits are shown in Figure 7.73. Of these, the first two are fixed frequency active filter configurations employing a bootstrap type circuit for use, respectively, in high-pass (rumble) and low-pass (hiss) applications, and the third is an

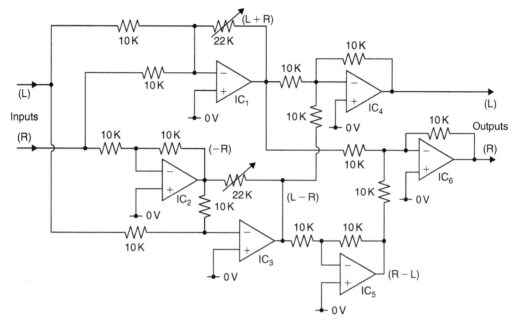

**Figure 7.72: Channel separation or blending using matrix addition or subtraction.**

inductor–capacitor passive circuit layout, which allows adjustment of the HF turnover frequency by variation of the capacitor value.

Such frequency adjustments are, of course, also possible with active filter systems, but require the simultaneous switching of a larger number of components. For such filters to be effective in their intended application, the slope of the response curve, as defined as the change in the rate of attenuation as a function of frequency, is normally chosen to be high—at least 20 dB/octave—as shown in Figure 7.74, and, in the case of the filters operating in the treble region, a choice of operating frequencies is often required, as is also, occasionally, the possibility of altering the attenuation rate.

This is of importance, as rates of attenuation in excess of 6 dB/octave lead to some degree of coloration of the reproduced sound, and the greater the attenuation rate, the more noticeable this coloration becomes. This problem becomes less important as the turnover frequency approaches the limits of the range of human hearing, but very steep rates of attenuation produce distortions in transient waveforms whose major frequency components are much lower than notional cut-off frequency.

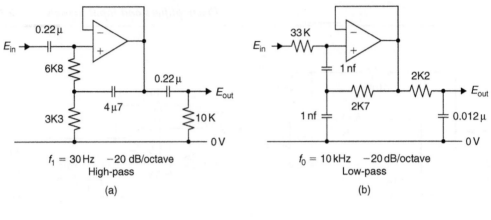

$f_1 = 30\,\text{Hz}$   $-20\,\text{dB/octave}$
High-pass

(a)

$f_0 = 10\,\text{kHz}$   $-20\,\text{dB/octave}$
Low-pass

(b)

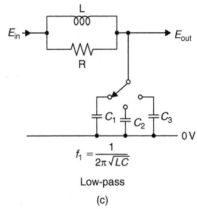

$$f_1 = \frac{1}{2\pi\sqrt{LC}}$$

Low-pass

(c)

**Figure 7.73: Steep-cut filter circuits.**

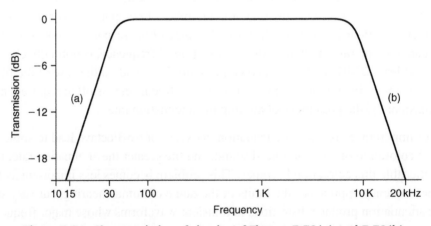

**Figure 7.74: Characteristics of circuits of Figures 7.73(a) and 7.73(b).**

It is, perhaps, significant in this context that recent improvements in compact disc players have all been concerned with an increase in the sampling rate, from 44.1 kHz to 88.2 kHz or 176.4 kHz, to allow more gentle filter attenuation rates beyond the 20-kHz audio pass band than that provided by the original 21-kHz "brick wall" filter.

The opinion of the audiophiles seems to be unanimous that such CD players, in which the recorded signal is two or four times "oversampled," which allows much more gentle "anti-aliasing" filter slopes, have a much preferable HF response and also have a more natural, and less prominent, high-frequency characteristic than that associated with some earlier designs.

## References

1. Linsley Hood, J., *Wireless World* (July 1969).
2. Livy, W. H., *Wireless World*, 29, (Jan. 1957).
3. Baxandall, P. J., '*Radio, TV, and audio reference book*', Chap. 14, S.W. Amos, Ed., Newnes-Butterworth Ltd.

# CHAPTER 8

# Interfacing and Processing

Ben Duncan

## 8.1 The Input

For the user, "the input" is often just a socket—often one groped for amidst a tangle of leads. This chapter untangles the details of the rarely recounted considerations that lie behind audio power amplifier input sockets that enable the signal source to connect to the amplifier (and maybe to many amps) with the least loss of fidelity and without introducing unwanted noise.

The amplifier is treated as a whole without considering the power capability or type of the output section.

### 8.1.1 Input Sensitivity and Gain Requirements

#### 8.1.1.1 Definition

Input sensitivity is the signal level at the input needed to drive an amplifier up to its full capability, to just before clip, into a stated, nominal impedance, often 8 ohms. Clip may be defined as the onset of visible waveform flattening or as a certain percentage THD+N distortion factor.

An older, less used definition (favored in the 1978 IHF standard) is the signal level needed to deliver I watt into a given nominal load, say 8f2. This is fine for comparing or normalizing drive levels between amps having different power ratings, but as input sensitivity *per se* has no particular merit, the usefulness, for real amplifiers and speakers of widely varying power capabilities and sensitivities, ends there.

#### 8.1.1.2 Description

Sensitivity is usually expressed as a voltage, either directly in volts or millivolts (1/1000ths of a volt), or in dBu. Mostly, sensitivity figures are assumed to be rms values (cf. peak) and also specified with a steady sine wave, and for power amps in particular, with loading—all unless stated otherwise. If a peak (or any other non-rms) voltage value is cited, the maximum output to which it is referred must also be cited likewise, so like is being compared with like.

### 8.1.1.3 Variables

The sensitivity of an amplifier depends (as defined earlier) on gain and swing. If an amp's output power rating, hence voltage swing capability into a given load impedance, were increased, maintaining the sensitivity requires more gain from the amplifier. This is a consideration for the maker and the installer who uses different sizes of a given design.

### 8.1.1.4 Do-It-Yourself Gain Resetting

For those uses with two or more different models and/or makes of amplifier, it is likely that sensitivities (however referred) will differ. Gain controls may not be present or it may be desired not to use them. If so, to align the system (ideally within a fraction of a dB), all the amps enter clip at about the same drive level and the gain(s) of one type of amp will need changing. Usually, any gain controls are assumed to be at maximum. Then any "accidental adjustments" can only cause reduced, not excess, gain.

In most well-designed, conventional high NFB power amps, gain may be changed up or down easily by changing one (global feedback) resistor per channel. The part being changed is usually in the output section. Changing gain by up to $+10\,dB$ or down by as much as $-6\,dB$ should have relatively little effect on sonic quality, assuming that RF stability is not upset. However, noise will be altered pro-rata.

In low- and zero-feedback designs, the availability of gain changing is far less, and the effect on both measured and sonic performances of even a modest 10-dB ($\times 3$) adjustment will be far more marked.

### 8.1.1.5 Gain Restriction

In some power amp designs, gain changes may be unavailable because they would upset RF stability, imperil a finally balanced gain/feedback structure, or violate some arbitrary %THD+N limit or other basic performance indication. Thus amplifiers from a product family spanning a range of output power ratings may have very similar gains ($+$ to $-3\,dB$); thus sensitivities (mV, V) almost commensurate with their ascending voltage swing. The upshot of this approach is (for example) a 2-kW 8fΩ amplifier, which only provides 100 W at normal drive levels (0 dBu say). The +13-dBu/3.5-V rms input drive needed for full output makes it safer and more likely that the high swing will be kept in reserve as an inviolate headroom.

In other words, in lieu of increased gain when output swing is increased, such an amplifier will need to be driven harder, that is, rated *less* sensitive. If the headroom achieved is ever used, then the higher input drive levels can cause increased distortion in the input stage. This effect will be noted most in esoteric amps with low feedback, but is still there in conventional high NFB amps.

### 8.1.1.6 Gain and Fidelity

As noted, the positive side of having high swing amplifiers desensitized, by *not* increasing gain commensurate with the increased voltage swing is that headroom occurs by default if the system's level/gain settings are not then altered. Reduced gain also reduces the risk of speaker damage by accidental loud blasts, dropped mics, styli, etc. Also, the audibility of the system's residual noise is lowered.

### 8.1.1.7 CM Stress

In conventional power amplifiers with high NFB, "common mode distortion," measurable as %THD+ N,[1] occurs because of common-mode voltage stress on the input stage, whether differential or single ended, with the latter suffering CM stress if, as is common, it is noninverting. The threshold voltage, 'Vth'—above which the input voltage to such an op-amp-type input becomes highly nonlinear when open loop may be sonically significant.[2,3] These setbacks may not be revealed with conventional tests, notably %THD+N, which can contrarily show lowered distortion at high input drive test levels, because the noise (+N) may "out-reduce" the rising common mode distortion.[1]

### 8.1.1.8 Real Figures

The sensitivity of every amplifier needs to match the zero (normal) levels of sources it is intended to be driven by. These vary. The upshot of all the factors is a spread of amplifier sensitivities that users know all too well (Table 8.1).

Ideally, there could be just one input sensitivity for all these uses. One that most could accept is the de facto professional standard of 0-dBu alias 775 mV. As a general rule, most lightweight domestic hi-fi and home studio equipment is likely to be more sensitive than 0 dBu, with pro equipment likewise less sensitive.

**Table 8.1: Range of Input Sensitivities**

| Category | In volts | In dBu |
|---|---|---|
| Home hi-fi | 30 mV to 2 V | –28 to + 8 |
| Home studios | 100 mV to I V | 18 to + 2 |
| Pro-audio | 775 mV to 5 V | 0 to + 16 |

However, as just discussed, a specific lower value, as low as 30 mV, may be best (at least in high NFB circuits) from the viewpoint of circuit and device physics for absolute best linearity.[2] However, the higher voltages that are mostly needed by desensitized high swing amplifiers (e.g., driving 2 V or +SdBu and above to clip) *confer the highest SNR,* hence dynamic range, *and also the highest RF EMI and CMV immunity.* So the best of both these worlds appears not to be immediately reconcilable.

As most amplifiers are not pure voltage sources, when driven with continuous, high-level test signals into a real (or simulated) loudspeaker load (as opposed to an ideal, simple resistive load), the sensitivity (for a given clip level) can appear to increase at some frequencies, as the maximum output voltage with a conventional amplifier having an unregulated supply is reduced by typically by –0.5 to –2 dB. The average shortfall is likely to be less with program, at least at mid- and high frequencies. *It follows that there is a complex frequency-conscious and dynamic peak-to-mean disparity in practical amplifiers' sensitivity ratings. The purer the voltage source, the less this can happen.*

### 8.1.1.9 Gain and Swing

Table 8.2 shows the gain requirements both in dB for some "round-figured" voltage swings, and how the nominal power then varies into 4 and 8 ohms.

For other sensitivities, gains are determined easily by appropriate subtraction or addition, for example, for +4 dBu, *subtract* 4 dB from the indicated gain(s) and for –10 dBu, *add* 10 dB to the indicated gain(s).

### 8.1.2 Input Impedance ($Z_{in}$)

### 8.1.2.1 Introduction

The amplifier's *input impedance* is the loading presented by the amplifier to the signal source *driving* (or "looking up" or "into") it.

**Table 8.2: Power Amp Gains for 0-dBu Sensitivity @ Clip ⇒ Means 'Into'**

| Gain (dB) | | rms voltage swing (V) | Average power ⇒ nom 8 Ω (W) | Average power ⇒ nom 4 Ω (W) |
|---|---|---|---|---|
| +24 | =×16 | 12.5 | 19 | 38 |
| +30 | =×32 | 25 | 78 | 156 |
| +33.5 | =×48 | 37.5 | 176 | 352 |
| +36 | =×65 | 50 | 312 | 624 |
| +40 | =×97 | 75 | 703 | 1406 |
| +42 | =×129 | 100 | 1250 | 2500 |
| +44 | =×161 | 125 | 1953 | 3906 |

Impedances (often abbreviated 'z') are rated in ohms ($\Omega$). As in this case, ohmic values are nearly always over 1000; the counting is usually in thousands (k). 10k or 10k$\Omega$ ("10k ohm") is easier to say than "ten thousand ohms." When near a million or over, 'M' for 'Mega' is used, for example, 1M$\Omega$ is 1000k$\Omega$.

### 8.1.2.2 Common Values

With ordinary, high NFB power amplifiers, high input impedances (high $Z_{in}$, say above 10k$\Omega$), to 1M$\Omega$ or more, are readily attained. For most sources, this is analogous to very light loading. However, in most cases, power amp input impedances are commonly at the low end of this range, at between 10 and 22k$\Omega$. This restricts noise and buzzes when (particularly *un*balanced) inputs are left open, unused, or floating, especially when cables are unplugged at the source end. This is less of a problem with short cables and in domestic environments.

The nominal values of amplifier input impedances vary widely. As a rule, professional equipment is defined in Table 8.3.

If balanced, $Z_{in}$ is the differential mode Z.

The input impedance of equipment may be described as the source's *load* impedance. This is true enough at frequencies below 1kHz. However, *load impedance* (since the signal source may be across a room, 100 yards down a hall, or even half-way across a field) is the *totality* of loading, namely including all the cable capacitance, which takes effect increasingly above 3kHz.

Table 8.3: Power Amplifier Input Impedances

| Type of power amplifier | $Z_{in}$ range |
|---|---|
| Domestic, seperated. and integrated | 10 k–200 kΩ |
| High-end domestic, esoteric | 600–2 MΩ |
| Professional | 5 k–20 kΩ |
| Vintage professional | 600 Ω |

### 8.1.2.3 Audio is Not RF

Precise "impedance matching," where specific impedances (often 50 or 75 ohms) must be adhered to, is correct for radio frequencies, where cables above a meter or so act as a transmission line.[4] But at the highest audible frequencies (20 kHz) even a 200-m-long input cable in a stadium PA system doesn't behave as a transmission line.

Where the wavelength (the dual of frequency) is a fair fraction, say 20 or 10 times greater than the cable, cables look mainly like the respective sums of their resistance, capacitance, and inductance. As the ratio falls, the cable begins to behave increasingly like a transmission line.

### 8.1.2.4 Voltage Matching

Since the widespread use of NFB (50 years ago), the majority of power amplifiers' inputs have been *voltage matched*. This means that the source impedance is low—much lower (at least 10 times less) than the total destination, or load impedance.[5,6] The intention is to transfer the signal, which is encoded as a voltage "wiggle," without significant loss of headroom, dynamic range, or detailing.

The source's impedance—whatever's feeding the amplifier(s)—also has to be *low enough* and remain so at hf to support a fiat hf response into the capacitative loading of likely cable lengths. Voltage matching is defined by de facto industry practice, in the IEC.268 standard. Here, recommended *input* impedances are 10 kΩ or over and equipment *source* impedances are 50 Ω or less. This is easily memorized as

Looking *back* from amp:                Looking *up* from amp:
$\leq 50\,\Omega \Leftarrow$                              $\Rightarrow \geq 10\,k\Omega$

With voltage matching there is no sharply defined "right" impedance. Except that in high common mode rejection (CMR) balanced systems and high resolution stereo systems alike, an amplifier's individual input impedances may be ultra-matched. Since with *voltage matched systems,* the wanted input signal is a voltage, the ideal, "noninvasive" amplifier input or load impedance would appear to be very high, say $1\,M\Omega$. Then only minuscule current would be taken from the source.

### 8.1.2.5 High Impedances

Some high-end hi-fi makers have taken the high impedance route, claiming better sonics. This may be inseparable from the circuitry used to create the high-Z conditions, and not necessarily down to the high-Z conditions *per se*.

In power amps with low (or zero) feedback, and using bipolar junction transistors (BJTs) in the input section, high input impedances (above $10\,k\Omega$) can be more difficult to implement consistently. On this basis, the early transistor amplifiers sometimes had their inputs rated in $\mu A$ of input current drawn! In contrast, there is usually no difficulty attaining impedances as high as $1\,M\Omega$ or more, when the input stage parts are valves, JFET or insulated gate FET (MOSFET) or any combination of these—whether loop or local feedback is zero, low, or high.

When *un*terminated, such high impedance circuits are noisier (hissier) and far more liable to allow parts to be microphonic than lower ("normal") impedance ones.[7] Demonstration is simple enough: try tapping the appropriate capacitors with an insulated tool while listening with full-range speaker(s) connected. High impedance inputs can also be the cause of difficulties and compromises with direct coupling. However, unless the input is direct coupled, or is at least coupled via very large capacitors, LF and subsonic microphony and electrostatic noise pick-up will not "see" the lower source impedance and will persist in accordance with the high impedance.

### 8.1.2.6 Low Impedances

As input impedance is lowered, there is less microphony and electrostatic noise pick-up when the amplifier inputs are disconnected, even with unshielded cabling. However, loading is increased, as is ultimately the *susceptibility to magnetic field noise pick-up, which is much, much harder to shield against.*

### 8.1.2.7 Loading

A single load of (say) $1k f\Omega$ may or may not compromise the source's performance. But two or a few of such loads almost certainly will, unless the source is rated appropriately (see later). Low impedance inputs are also the most easily damaged if one amp's output

**Table 8.4: The reciprocal pattern of conventional power amplifiers (with 10 kΩ input impedance)**

| No. of amps in tandem | Total $Z_{in}$ |
|:---:|:---:|
| ×1 | 10 kΩ |
| ×2 | 5 kΩ |
| ×3 | 3.3 kΩ |
| ×4 | 2.5 kΩ |
| ×5 | 2 kΩ |
| ×6 | 1.7 kΩ |
| ×10 | 1 kΩ |
| ×15 | 666 kΩ |
| ×20 | 500 kΩ |

is accidentally connected to another's input. Added protection would add complexity, increase the cost, and likely degrade sonics.

### 8.1.2.8 In Tandem

In professional (and even a few domestic) applications it is normal for each signal source to drive more than one amplifier input. The loading of amplifiers driven in tandem is cumulative: each added amplifier *reduces* the impedance (or *increases* the loading) pro-rata in accordance with its impedance. Assuming conventional power amplifiers with 10 kf2 input impedance, the reciprocal pattern is shown in Table 8.4.

Note that there are *very* few types and models of the likely sources (e.g., active crossovers, delay lines, preamps) that are rated and able to drive impedances of below 600 ohms without degraded performance. Much pro-gear is rated and even specified for 600 ohms, but still gives its best measured and sonic performance into 2 k or even higher.

For large tandem systems, existing equipment usually has to be retro-fitted with special *line-driver amplifiers,* or these are added as independent units, in line. Line drivers used in live sound practice do not expand the allowable loading by much, usually down to 300 ohms and possibly as low as 75 ohms. To be sure, only 50% of this rating would be used. The rest allows for tolerances, variables (see later), add ons, and the cable's capacitance loading at hf. In a major concert where 100 or more power amps have been required to handle just one frequency band alone,[8] the signal was split among up to 10 line drivers, all *daisy chained* off 1 line driver. This method is far preferable to having multiple crossovers, which might superficially simplify the signal path, but would also introduce near impossible set-up and band-matching demands.

### 8.1.2.9 Multiconnection

When one signal has to feed many amplifiers, it is normal to connect the amplifiers by *daisy chaining.* To permit this, amplifiers made for professional use have both female (input) and also male (output) XLR (or other, gendered or ungendered in/out) connectors, linked together in parallel. "Daisy chaining" means physically, as the name suggests, that a short cable "tail" carrying the input signal *loops* from one amplifier to the next in the rack or array. The signal being passed on is not really entering each amplifiers' input stage, but merely using the input sockets and case-work as a durable and shielded Y-splitting node. An alternative would be to make up a hydra-headed cable, that is, one splitting into *n* separate feeds. This would take up far more space and is far less flexible, but might prove the next best method if amplifiers without input "link-out" sockets have to be used.

### 8.1.2.10 Ramifications

Professional power amplifiers, which are the sort most likely to have long cables connected to their inputs and to reside in electrically noisy environments, mainly eschew impedances much above 10k. However, if they're to be usable for live sound, their makers also can't welcome any much lower impedance, as this would further limit the number of channels that can be daisy chained off a given line driver. In most multi-amp setups, the source that is being loaded is usually one of the band outputs of an active crossover, rated for 600 ohms with the NE5534 or 5532, 1977 IC technology that remains a de facto standard. In this common case, depending on the allowance for cable capacitance, between 10 and 15 amplifier channels (at most) should be driven.

### 8.1.2.11 Variables

As with other electronic equipment, input impedance is a function of electronic parts whose behavior almost inevitably varies with frequency and almost always depends on temperature. With unbalanced inputs, input impedance will also usually vary somewhat with the setting of the gain control (attenuator), if fitted.

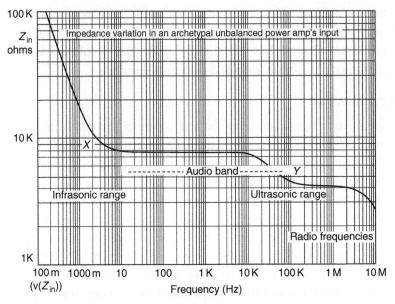

**Figure 8.1: Input impedance (load) variation in a typical, simple unbalanced power amplifier input stage.**

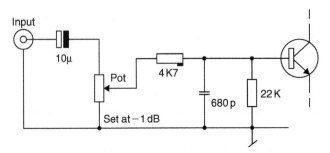

**Figure 8.2: A typical unbalanced input stage.**

Figure 8.1 shows how the input impedance of a typical, minimal power amplifier with an unbalanced input (Figure 8.2) varies across the frequency range. A 10kΩ gain control is assumed and is here backed off just 1dB. Note how the impedance in most of the audio band is almost constant at the scale used. Then notice how the impedance drops off (so the loading increases) at high audio frequencies, and more so at higher radio frequencies (Y). At low frequencies, if anything, the load impedance increases (X).

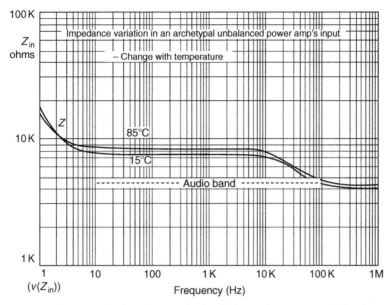

**Figure 8.3: Impedance variation in a typical unbalanced power amplifier input stage as the amplifier warms up.**

Figure 8.3 shows how the same input stage's impedance varies (without changing anything else) as temperature is changed from 15° to 85°. In other words, what can happen to the input impedance when an amplifier is "cooked?" For the most part, impedance increases, which will do no harm. However, in live work it *might* just alter a howl round threshold, as the higher load impedance allows the signal voltage to rise ever so slightly.

Figure 8.4 shows how the input impedance typically varies as the gain is adjusted. Because the change with each 30° rotation step is nonmonotonic, $Z_{in}$ goes up and then comes down, as you might expect. A 10kΩ log pot is assumed.

Ideally, an amp's input impedance would remain constant despite these changes. In unbalanced circuits, there is not much harm as long as any change in impedance is gradual and stays above certain limits, and anything that isn't like this happens well above (or even further below) the audio band. Staying constant is *far more* important in balanced circuits.

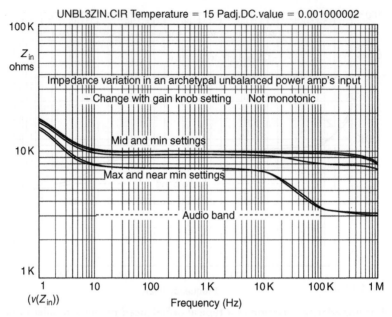

**Figure 8.4: Impedance variation in a typical unbalanced power amplifier's input stage as the gain control is swept.**

## 8.2 Radio Frequency Filtration

### 8.2.1 Introduction

Music starts out as air vibrations. These are not directly affected by electromagnetic (EM) waves, *except while* they are passing through an audio system in the form of electronic signals. Planet Earth has long had natural EMI, in the form of various electric and magnetic storms; both those occurring in the atmosphere and those occurring on the "surface" of the Sun and Jupiter in particular. Since 1900, the planet has increasingly abounded in human-made EMl babble, comprising electromagnetic energy fields and waves, some continuous, some pulsed, and others random. As stray signals nearly always have *nothing* to add to the music at hand, and most are profoundly unmusical, and as EMI permeates almost everywhere above ground unless guarded against, music signals require "*pro-active*" protection.

EM waves used for radio broadcasting and communications mainly start in earnest at 150 kHz (in the United Kingdom and continental Europe) and above, and continue to frequencies 10,000 times higher. However, special radio transmissions (for submerged submarines, national clocks, and caving) may use frequencies below 100 kHz and even those below 20 kHz.

### 8.2.2 Requirement

All active devices are potentially susceptible to EMI. BJTs, all kinds of field effect transistors (FETs), and also valves can all act as rectifiers at RF, demodulating radio transmissions. However, this is very much more likely with BJTs, as the nonlinearity of a BJT's forward biased base-emitter junction that gives rise to rectification is triggered by considerably lower levels of RF voltage or field strength. All kinds of FETs and valves are relatively "RF proof" in comparison. Oxidized copper, generally dirty contacts, crystalline soldered joints, or wrong metal-to-metal interfaces can all act as RF detectors as well, through rectification.

## 8.3 Balanced Input

Balanced inputs, when used properly, can clean up hums, buzzes, RFI, and general extraneous rubbish. When not used properly, the balanced-input's object may be partly defeated, but the connection will probably still improve the amplifier's and system's effective SNR.

### 8.3.1 Definition

To be truly balanced, a balanced input *and* the line coming in *and* the sending device must all have equal impedances to (signal) ground, to earth, and to everywhere else. Also, the signal must be exactly opposite in polarity but equal in magnitude, on each conductor.

### 8.3.2 Real Conditions

In practice, the signal is not of exactly opposite polarity. At high frequencies (and low frequencies in some poorly designed equipment), phase shifts add or subtract up to 90° or more, from the ideal 180° polarity difference. Otherwise the requirement for having a signal of opposite sign on each conductor is usually met. The exception is when one-half of a ground-referred, balanced source has been shorted to ground. Not surprisingly, this degrades the benefits of balancing.

### 8.3.3 Balancing Requirements

#### 8.3.3.1 Input Impedances

The norm in modem pro-audio equipment is $10\,k\Omega$ across the line. This is commonly known as a "bridging load." It is also the *differential input impedance.*

The *common mode impedance,* what any unwanted, induced noise signals will see, is often (but not always) half of this, for example, $5\,k\Omega$ in this case.

Considering the hum/RF noise rejection capability of an effective balanced input, input impedances much higher than $10\,k\Omega$, say, $500\,\Omega$, would seem feasible and useful in professional systems. However, if the input resistance is developed by the ubiquitous input bias path resistors connected from each input to the 0-V rail, then there are limits to the usable resistance, before the input stage's output offset voltage becomes unacceptably high. Although low $V_{oos}$ op-amps exist, a number of otherwise good ICs for audio have execrable DC characteristics, as IC designers do not appear to comprehend that good DC performance is a most helpful feature for high performance audio. In this case, input impedances above 15 to $100\,k\Omega$ are found to be impractical, depending on bias current.

A galvanically floating input (i.e., the primary of a suitably wired transformer) has no connection to signal 0 V (as it has no bias currents), so there can be a very high common-mode impedance, say, 1M or more, up to modest RF. This aids rejection.

Conversely, differential impedances of less than 1 Ok increase the influence of such random, external factors as mismatched cable core-to-shield capacitances.

### 8.3.4 Introducing Common Mode Rejection

Common mode rejection is an equipment and system specification that describes how well unwanted common mode signals, mainly hum and RF interference, are counteracted when using balanced connections.

#### 8.3.4.1 Minimum Requirements

At the very least, *all* the equipment in a system must have a balanced input (alias a "differential receiver"). CMR can be improved and made more rugged when balanced inputs are used in conjunction with balanced outputs (alias "differential transmitters"), but this is not essential.

### 8.3.4.2 What Does CMR Achieve?

Common mode rejection action prevents the egress and build-up of extraneous hum, buzzes, and RFI when analogue signals are conveyed down cables, and between equipment powered from different locations—all the more so in big or complex systems. CMR helps make shielding more effective by canceling the *attenuative residue*, the bit that any practical shield "lets through."

Sending the signal on a pair of twisted and parallel conductors ensures that this latter residue and any other stray signals that are picked up en route are literally coincident and appear "common mode," that is, equal to each other in size and polarity. A tight enough twist makes the conductors almost experience interfering fields as if they occupied the same space. This is true below high RF (200 MHz, say), when averaged out over a cable's length.

In contrast, the wanted, applied signal from both balanced and unbalanced output sockets is distinguished while being no less equal in size by appearing *opposite* in polarity on each input "leg," called *hot* and *cold*.

CMR also makes shielding more effective by freeing it from signal conveyance, enabling it to be connected at one end only, according solely to the dictates of optimum RF suppression and/or individual system practice. Breaking the shields through connection also prevents (or at least lessens) the build-up of the mesh of earth loops that causes most intractable hums and buzzes. CMR is also able to cancel differences between disparate, physically distant and electrically noisy ground points in a system.

Above 20 kHz, even a modest CMR lessens the immediacy of the need for aggressive RF filtering. RF filtering can take place at higher frequencies, and both the explicit and the component-level effects on the audio may be diminished accordingly.

Figure 8.5 shows the CMV that CMR helps the audio system ignore. Even when connection to mains safety earth is avoided by ground lifting (ground lift switch open) or by total isolation (switch open and ground lift R omitted), considerable capacitance frequently remains, through power transformers and wiring dress.

Overall, the rejection achieved (which is a ratio, not an absolute amount) is described in minus (−) dB. Often the minus is understood and omitted. In plain English, "CMR = 40 dB" means "all extraneous garbage entering this box will be made 100 times smaller."

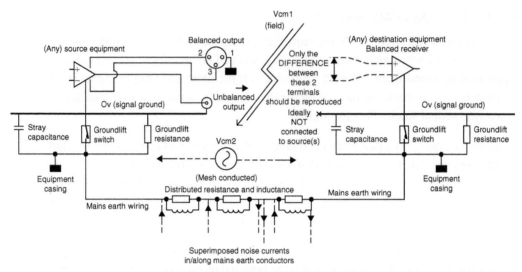

**Figure 8.5: Most of the common mode noise that CMR defends against is either RF and 50/60 Hz fundamental intercepted in cabling (Vcml) or 50/60 Hz hum + harmonics caused by magnetic loop, eddy, and leakage currents flowing in the safety ground wiring between any two equipment locations (Vcm2).**

### 8.3.4.3 What CMR Cannot Do

Like the stable door, the one thing CMR can't do is remove unwanted noises that are already embedded in with the music. It follows that just one piece of equipment with poor CMR, and in the wrong place, can determine the hum and RFI level in a complex studio or PA path.

The ingress of common mode noise, called *mode conversion*, is cumulative, as each unit in the chain lets some of it leak through. As a result, the CMR performance and/or interconnection standards of all the equipment in complex systems (e.g., multiroom studios and major live sets) must be doubly good.

The higher CMR of well-engineered equipment (80 dB or more) provides a safety factor of 100- to over 1000-fold over the minimum 40 dB that is common in more "cheerful" products. However, the higher CMRs are more likely to vary with temperature and aging, as with all finely tuned artifacts.

### 8.3.4.4  Relativity Rules

The size of common mode (noise) signals is not fixed or even very predictable; *they* may range from microvolts to tens of volts. CMR is just a layer of protection. Forty dB of protection is not much against 10 V of CMR, but it is definitely enough for 1 μV.

### 8.3.4.5  Sonic Effects of RF

Radio frequency interference is a common mode noise, and sources of RF go on increasing. In a competently wired system in premises away from radio transmitters and urban/industrial electrical hash, a modest rejection no better than 40 dB has previously seemed good enough to make inaudible induced 50/60-Hz hum and harmonics, and the "glazey" sound of RFI and RF intermodulation artifacts. Unfortunately, RFI artifacts aren't always blatant, and when any sound system is in use, they're the last thing that users are likely to be listening for the symptoms of. However, even if there are no blatant noises, inadequate CMR can allow ambient electrical hash to cover up ambient and reverberative detail.

### 8.3.4.6  System Reality

The CMRs discussed are those cited for power amplifier input stages. The actual *system* CMR is inevitably cumulatively degraded by the cabling and the source CMRs. However, it can be maintained by ensuring all three have individually high CMRs *and* have highly balanced leg impedances. Lines driven from unbalanced sources give numerically inferior results, but often quite adequate ones (subject to appropriate grounding and cable connections) in low-EMI domestic hi-fi and studio conditions, where equipment connections are also compact, and even in outdoor PA systems, in an open countryside.

### 8.3.4.7  Summary

Generally, 20 dB is a low, poor CMR, 40 to 70 dB is average to good, and 80 to 120 dB or more is very good and far harder to achieve in a real system. In a world where some audio measurements have had their credibility undermined, it's reassuring to know that with CMR, more dBs remain simply better.

## 8.4  Subsonic Protection and High-Pass Filtering

### 8.4.1  Rationale

All loudspeakers have a low-end limit; their bass response does not go endlessly deeper.

**Table 8.5: Loudspeaker Subsonic Handling (Infrasonic Handling)**

| More robust ⇑ | Transmission lines |
|---|---|
| | Differentially loaded cabs[a] |
| | Properly arrayed bass horns |
| | Sealed boxes |
| | Open-backed cabs |
| | Large cone-vented enclosures |
| Less robust ⇓ | Small cone-vented enclosures |
| [a]Alias band pass or push–pull. | |

*Subsonic* (*infrasonic*) *information*, comprising both music content and ambient information, may occur below the high-pass "turnover" frequency (or low-end roll off) of the bass loudspeaker(s). It will not be reproduced efficiently.

*Note:* While potentially within humans' aural perceptive range, subsonic signals are "below hearing" (strictly *infrasonic*) in the sense of being "out-of-band" to, and only faintly or at least reducingly reproducible by, the sound system.

Loudspeakers vary in their ability to handle large subsonic signals. Small ones may or may not be heard but won't ever cause damage. Large subsonic signals are more risky with some kinds of loading. An approximate ranking of subsonic signal handling robustness is shown in Table 8.5. *Individual designs can vary widely*, however.

### 8.4.2 Subsonic Stresses

Other than straining the speaker(s), if the amplitude of the subsonic (really infrasonic) signal(s) is large enough, then significant amplifier capability will be wasted. At the very least, the unrealizable portion will cause unnecessary amplifier heating and electricity consumption.

If the amplifier is also being driven hard, the presence of a large subsonic signal will reduce the threshold for clipping and also thermal shutdown. The amplifier will behave as if rated at only a fraction of its actual power capability. There are broadly two approaches to the problem.

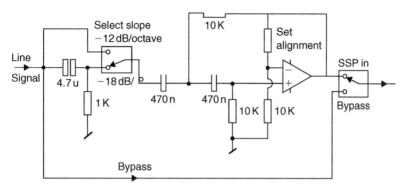

**Figure 8.6: Typical high-pass (subsonic protection) filter circuitry.**

### 8.4.3 The Pro Approach

Subsonic filtering may be regarded as an essential part of editing and sweetening in recording. "Subsonic" frequencies ("sub" here being rather loosely designated as any "out of context/too-low bass information") are usually removed before amplification by HP filters (HPF) with fixed, switchable, or sweepable roll-off frequencies, usually available on each channel or group of a mixing console. Alternatively, HP filtering may even be available "up front" as a switch on some microphones or on portable, location tape machines.

Generally, such filters are at least –12 dB/octave and, more usefully, the steeper –18-dB/octave (Figure 8.6) or even –24-dB/octave. They may be occasionally appended to professional power amplifiers, as well as to preceding active crossovers, on the basis of providing "maximum" (read: brute force) protection at all costs, in this guise they are described as "subsonic protection" (SSP). Often this facility is superfluous and repeated needlessly, as the mixer and active crossover already do or can provide subsonic filtering.

### 8.4.4 Logistics

The mixer can provide SSP most flexibly per channel, solely for those sources requiring filtration. The active crossover may provide overall back-up subsonic protection, in case a mic without HPF'ing on its channel is dropped.

When subsonic protection is fitted to and relied upon in amplifiers alone, there will be an enforced and unnecessary repetition and diversification of resources in any more than the simplest, two-channel PA. If subsonic filter provision is made in an amplifier, it should be switchable (or programmable or otherwise controllable) so that its action can be *removed positively* when not required.

### 8.4.5 Indication

A few power amplifiers have light-emitting diodes (LEDs) (often jointly error indicators) that indicate subsonic activity or protection shutdown arising from excess subsonic levels. This kind of protection is most common where the maker is also a speaker maker or where the amplifier is closely associated with a particular speaker, as the protection's frequency–amplitude envelope that will allow the most low frequency action is very specific to the cabinet and driver used.

Overall, in high performance professional power amplifier designs benefiting from modem knowledge, filtration and any HP filtering are avoided as far as possible or else minimized by adaptive circuitry.[9]

### 8.4.6 Hi-End Approach

In "hi-end" hi-fi *and* professional power amplifiers, high-pass filtering is (or should be) depreciated or at least kept to the bare minimum, for two reasons.

First, all practical HP filters progressively delay low frequencies relative to the rest of the music. Every added HP filter pole only adds to this "signal smearing."[10] Simulation in time and frequency domains shows this.[11]

Second, HP filters require the use of capacitors. Capacitors that are almost ideal for audio and not outrageously expensive and bulky are limited in type and values. Capacitors that are *faradically* large enough not to cause substantial "signal smearing" are, in practice, medium-type electrolytics, and not in practice nor in theory anywhere near so optimal for audio as other dielectric types.

For these reasons, even routine HP filtering (alias *DC blocking or ac coupling*) may be absent altogether. Figure 8.7 shows the points where HP filtering occurs in the majority of otherwise direct- and near-direct-coupled power amplifiers.

### 8.4.7 Low Approach

In "consumer-grade" audio power amplifiers, HP capacitors are made as small as possible in value while maintaining what is judged by casual listening or first-order theory to

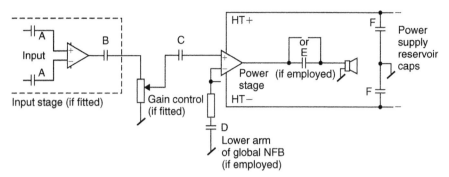

**Figure 8.7: High-pass filter capacitor positions. The potential locations of DC blocking/HPF capacitors in the signal path of conventional transistor power amplifiers, assuming that gain blocks (the triangles) are internally direct coupled.**

be an acceptable point for the bass response low cutoff frequency ($f3_L$). The result is considerable HP filtering, permanently engaged. Subsonic signals may then rarely pose a problem, but sonic quality may be degraded up into midfrequencies, while a great deal of the music's ambient cues is lost.

## 8.4.8 Direct Coupling

When all HP filtering is removed, a power amplifier becomes direct—or 'DC' (direct current)—coupled. 'DCC' would have been better, but that now means something else. Extending the response to zero frequency, that is, "down to DC," is achieved readily at the design stage with most transistor topologies. The advantages are sonic, and substantial, due to the excision of intrinsically imperfect parts *and* the removal of an intrinsically unnatural filtration, and the signal-delay and the possible charge accumulation on asymmetric music signals it brings. For this is the truth of all signal path HPF capacitors, both those in series and in NFB arms. Whether DC coupling is safe or workable in a particular amplifier is a separate design question.

With conventional valve amp topologies, the response to DC is not fully achievable, except in the few workable 'OTL' designs. However, it is still possible to direct couple the remainder of a valve amplifier, with global DC NFB taken before the transformer. In fact, the first precision DC amplifiers *were* valve op-amps.

### 8.4.8.1 Direct Current Management

With direct-coupled circuitry, unwanted DC "offset voltages" will be amplified by the power amplifier's respective stage gains. Excess DC is of great concern and must be avoided. It can be (i) produced internally, by mismatches in resistor or semiconductor values or by intrinsic topological asymmetry or (ii) introduced externally, from preceding DC-coupled signal sources.

Internally produced DC offsets may be kept to safe levels by precision in design and component selection. This requires matching of two or three apposite parameters of the *differential pair* at the front end of each stage, assuming some version of the conventional high NFB "op-amp" type of architecture. The "pair" might be BJTs, FETs, or valves. And to ensure that the source resistances (at DC) seen by each input leg are the same, or close, and not too high either, depending on bias current. If the resistor values then conflict with CMR, the latter should have priority, in view of EMC requirements, and the nonrecoverability of the CMR opportunity. Direct current balance may be restored by other means, for example, current injection.

Externally applied DC, appearing on the inputs, because of essentially healthy but imperfect preceding equipment, will usually be in the range of 0.1 to 100 mV. More than +/−100 mV would suggest a DC fault in the preceding source equipment. Assuming a gain of 30×, this would result in 3 V at the amplifier's output. Because such a steady *offset* will eat up headroom on one-half of the signal swing, the clip level is lowered asymmetrically. A direct coupled power amplifier should not be harmed by this and should also protect the speakers it is driving, but equally it is entitled to shut down to draw attention to such an unsatisfactory situation. In the most advanced designs of analogue path yet published,[9] DC coupling is adaptive: if DC above a problem level persists at the input, DC blocking capacitors are automatically installed and the user is informed by LED.

Some low-budget domestic power amplifiers have long offered part and manual direct coupling. The power stage may not be wholly direct coupled, but at least the DC blocking capacitor(s) at the input can be bypassed via a second "direct" or "laboratory" input. The user is expected to try this but revert to the ordinary ac-coupled inputs if DC on the source signal is enough to cause zits and plops. A blocking capacitor(s) at the input can be bypassed via a second "direct" or "laboratory" input.

### 8.4.8.2 Autonulling

Direct current offset may be continually forced to near zero volts by a *servo*, which is another name for brute-force VLF and DC feedback, applied around an amplifier overall, or just the input or output stage. Servos have been de rigueur in U.S. and U.S.-influenced high-end domestic power amplifiers for some years. Alas, those who have designed them into high-performance power amplifiers have clearly *not* thought through the consequences. Tellingly, servos are not usually nor likely to be found in amplifiers with truly accurate sounding bass.

The reasons are clear enough today: servos cause the same or even wilder distortions in LF frequency and/or phase response, and/or signal delay vs. frequency (group delay). Figure 8.8 shows this.

They also compromise the integrity of the circuitry they are wrapped around by increasing noise susceptibility, while the capacitor imperfections that DC coupling is supposed to overcome are reintroduced, as distortion-free DC servo action depends on an expensive, bulky, high-performance capacitor for integration. In this way, the DC servo returns us to *before* square one, with the added cost and complexity. Worse, the original thinking behind servo'ing was to save money (!) on input transistor and part matching, as a servo will "fix" any DC in its range, often up to $+/-5$ V, including DC appearing on the equipment input. This is neat, but like so many "smart" options, DC servo'ing is not *quite* suitable for audio.

## 8.5 Damage Protection

The input stages of most audio equipment are unprotected. This approach appears to save on parts cost, complexity, and sonic degradation; however, in reality, it may indeed *cause* costs and degraded sonics. The inputs of power amplifiers are certainly among those most likely to sustain input voltages that may be damaging to the active parts inside.

### 8.5.1 Causes

Typical culprits include first, large signals from line level sources, and from amplifier outputs, experienced through accidental connections (see Section 8.5.2). Here, excessive signal voltages that could be applied could range from a few volts, up to 230 V rms, and from below 10 Hz to above 30 kHz.

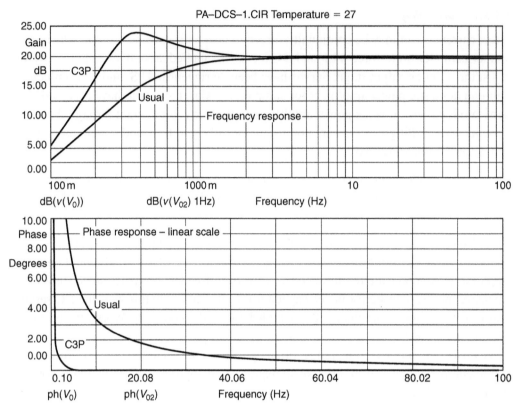

Figure 8.8: Direct current servo circuits cause at the very least the same phase and delay error as using a DC-blocking capacitor conventionally. The upper graph shows the frequency response of a standard two pole servo (2 × {1 M.O × 470 nF}). The lower graph shows the phase shift, which is clearly nonlinear below 85 Hz—place a ruler against the line. The curvature indicates a frequency-dependent signal delay, hence smearing (after Deane Jensen). An alternative, custom three-pole compensating type (C3P) is plotted. This overcomes the smearing, as the phase shift is much less than 0.1° above 5 Hz, but the amplitude (upper) is now peaking below 1 Hz.

Second, the outputs of crossovers or consoles, or misconnected amps, which are kaput and have DC faults, so the output voltage might range from +/–10 V to up to +/–30 V for line sources, and up to +/–160 V DC for power amplifiers, but more typically +/–30 to +/–90 V DC.

### 8.5.2 Scope

The parts most at risk from excess input voltages are the solid-state active devices, particularly discrete BJTs, and most monolithic IC op-amp input stages.

Valves are relatively immune to input voltage abuse. They are most likely to be harmed by gross overdrive conditions that bias the grid positive so a damagingly high current flows.

J-FETs and MOSFETs are next most rugged. MOSFETs are most susceptible to gate-source overvoltage, but gate-source protection is straightforward and effective.

IC input stages are the most fragile. Due to IC structure, even FETs, when monolithic, may have parasitic weak points. For long-term reliability, currents flowing into or out of IC op-amp pins[12] must always be kept below 5 mA.

### 8.5.3 Harmful Conditions

There are two kinds of potentially damaging input voltages: (1) common mode and (2) differential mode. Either may occur when a power amplifier is in (i) the on state or (ii) the off state, giving four possibilities.

### 8.5.4 On-State Risks

When an amplifier employing BJTs at the front of its input stage is on, powered up, and settled down, it can sustain relatively high differential (signal) voltages without damage. Generally, in high NFB op-amp and other dual-rail based designs, the max differential voltage is a volt below the supply rails, hence a maximum differential voltage ranges from +/–14 V for input stages working from +/–15-V supplies, up to +/–30 V or even over +/–100 V, where the input stage transistors operate from the same or else similarly high supplies, as the output stage.

Long before differential overload, the input stage will be driven strongly into clip. Provided the amplifier has clean recovery, an *overvoltaging* may pass unnoticed if the high differential voltage only lasts I mS. Yet this is plenty long enough to damage a semiconductor junction. In BJTs, the most vulnerable junction is the base emitter, when reverse biased.

Under the same powered-up conditions, common-mode voltages above +/–10 V can damage unprotected BJT input stages. In large systems, the common-mode voltage can be this high, commonly comprising 50/60-Hz AC and harmonics, and arising from differences in grounding or AC power potentials.

The input stage's supply rail voltage usually has a large bearing on the maximum safe DM and CM input voltages. Here, low supply voltages may do no favors.

### 8.5.5 Off-State Vulnerability

When an amplifier using BJTs is switched off, both differential and common-mode voltages as low as +/−0.5 V may be damaging. Users are advised to always power-up preceding equipment before the power amps. This is universal practice among informed users, both domestic and professional. However, if the prepowering of the source involves the passage of signals above 0.Sv peak to amplifier inputs, then unless the transistors behind the sockets are protected before the amp is powered-up, they may well be damaged. This mode of subtle, progressive damage and sonic degradation to analogue electronics has yet to be widely recognized. It can be overcome without changing otherwise sensible practices, by suitably designed input protection.

### 8.5.6 Occurrence Modes

Damage to input devices may be catastrophic if the overvoltage causes high currents to flow. This is rare.

Otherwise, with BJT inputs, damage may be subtle. Transistor parameters are degraded but NFB action initially hides the worst. Telltale signs would be changed or, reducing sonic quality, raised, increasing and/or intermittent noise, higher %THD, and possibly increased DC offset at the amp's output.

With ICs, damage may be cumulative, caused by peculiar metal migration effects occurring in ICs' microscopically thin conductors. This means an input stage can appear to handle abuse repeatedly until eventually the catastrophic failure occurs when all the conductor has migrated away!

### 8.5.7 Protection Circuitry

Power amps have been designed to survive likely levels of both CM and DM overvoltages by the use of some combination of the following.

1. Series input resistors, which may already be part of the input stage's RF filtering, will limit the current flowing into inputs. If the resistance between the input socket and the active device is 5 kΩ, then above 25 V DC or peak signal would be needed to get more than 51xA to flow.

2. Back-to-back zeners to 0 V, working in concert with series current-limiting resistors (which may already be part of the input stage's RF filtering). Both CM and DM voltages can be clamped to any available zener voltage. Designers must allow for quite wide variations with tolerance and temperature, and possible sonic degradation. Programmable zeners may also be used or zeners may be combined with BJTs.

3. Ordinary, fast diodes across the active differential inputs, in concert with series input resistors in both legs. Protects against DM overdrive only. Internal to some IC op-amps, for example, NE5534. External diodes with larger junctions may be used to enhance protection.

4. Clamping relays. Placed after the series input current limiting resistors, inputs are shorted to 0 V until power is up on all rails. With suitably rapid action and power sensing, relays in this configuration can provide complete protection against both DM and CM input signals.

5. Bin[13] describes a method developed at the BBC, using VDRs, zeners, and current sources, providing input protection to audio balanced line inputs (including power amps) up to 240 V ac. Alas, sonic quality may be detracted from.

## 8.6 What Are Process Functions?

When in use, an audio power amplifier is *always* but part of some greater system. In domestic audiophile and even recording studio systems, it is commonplace for power amplifiers to have no gain controls and to be devoid of any processing functions.

However, in professional music PA applications, by contrast, it is the exception to find power amplifiers *without* panel gain controls (really attenuators). This facility turns into a *system processing function* when the gain control element becomes remote controllable, most particularly when all the amplifiers in a system or grouping are so equipped and also when the rate of gain control change is fast enough for it to be used dynamically.

### 8.6.1 Common Gain Control (Panel Attenuator)

The most common, almost universal form of "gain control" is passive attenuation, set usually via a panel knob, with a rotary *pot* or *potentiometer.*

### 8.6.1.1 Characteristics

As "voltage matching" is the norm for modern audio, pots are nearly always wired in the voltage divider mode, where the wiper is the output. At this point, the source impedance seen varies, up to a maximum of a quarter (25%) of the pot's rated value (i.e., the end-to-end resistance) at half setting. At the pot's maximum and minimum settings, the source impedance reaches a few ohms above zero, which is usually much less than the preceding signal source's impedance.

### 8.6.1.2 Common Values

In audio power amplifiers, the pot's value is commonly 5 or 10 kΩ in professional and audiophile grade equipment and 20, 50, or 100 kΩ or even higher in "consumer" grade equipment. The lower pot values offer lower maximum impedances at half-setting, for example, just 2500 Ω (2.5 kΩ) for a (10kf) pot. This lessens the scope for noise pickup in the inevitably unbalanced and relatively sensitive part of the amplifier circuitry where the pot is placed.

### 8.6.1.3 Audio Taper

These considerations are true for ordinary pots with an audio taper, that is, those marked 'log' or 'B'. As shown wired in Figure 8.9(a), these normally sweep over the maximum possible range of level setting, from a purely nominal $-\infty$ (hard CCW or "shut off," really more like –60 to –70 dB) up to 0 dB (maximum level). The "audio taper" alias logarithmic resistance change per ° rotation makes the change in sound level reasonably constant with rotation. The full span and audio taper are relevant when a pot is needed to act sometimes as volume control, where output levels very much lower than the power

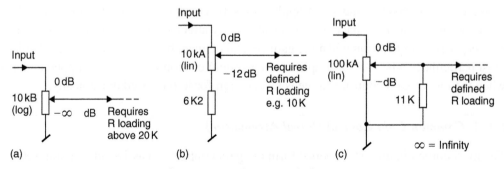

**Figure 8.9: Gain Pot Variations**

amplifier's capability are useful. It's also relevant where a quick sweep to $-\infty$ (infinite attenuation) may be needed as a mute—to turn off the signal in one speaker, say— without switching off or unplugging anything.

### 8.6.1.4 The Right Range

In many applications, the range offered by a raw pot is far too wide. In other industries employing pots, a vernier or a multiturn mechanism is added between the knob and shaft to aid fine settings. However, these are eschewed by modem professional audio operators, partly because of an ingrained fear of the loss of instant sweep control and because of relatively high cost versus relative fragility. There is also the false sense of alignment suggested by the verniers' 3 or 4 figure scale; scales on different amplifiers would be strictly incomparable, owing to most pots' poor tolerances, particularly good-sounding log pots. In the past 20 years, variations of 5 to 25% (or 0.5 dB to 3 dB) have remained the *norm for the resistance mismatch* between different pots at the same mechanical setting.

### 8.6.1.5 Linear Variants

Using a linear (A) pot and a fixed resistor, Figure 8.9(b) shows how adjustment range is restricted to the "top" 12 dB, that is, 0 dB to –12 dB. For system adjustment, this may be more usefully expressed as +/–6 dB. This range of adjustment is preferable for active crossover-based and arrayed systems, where the gain of individual amplifiers benefits from close adjustments and only needs this limited range. In practice, switched (say) –20 dB and $-\infty$ settings are then required. Note that the impedance vs. rotation relation is naturally slightly changed—the highest source impedance is here less at about 20% (rather than 25%) of the pot.

Returning to the full-scale mode, a linear pot may alternatively be used [Figure 8.9(c)], with a fixed resistor used for "law faking." This converts the linear law to a log-like curve, if the pot and resistor values are kept within tight limits; this approach can give approximations of an audio taper that are at least more consistent than most log pots, which are made by butting $n$ different-valued linear track segments together. Note that the pot's effective value is here a tenth of its rated value after the law faking resistor is included. As a result, the pot shown in Figure 8.9(c) looks like a 10 kf2 pot to the load. However, the maximum source resistance is, as with the audio taper, at the 50% attenuation point and is just about 10% from maximum.

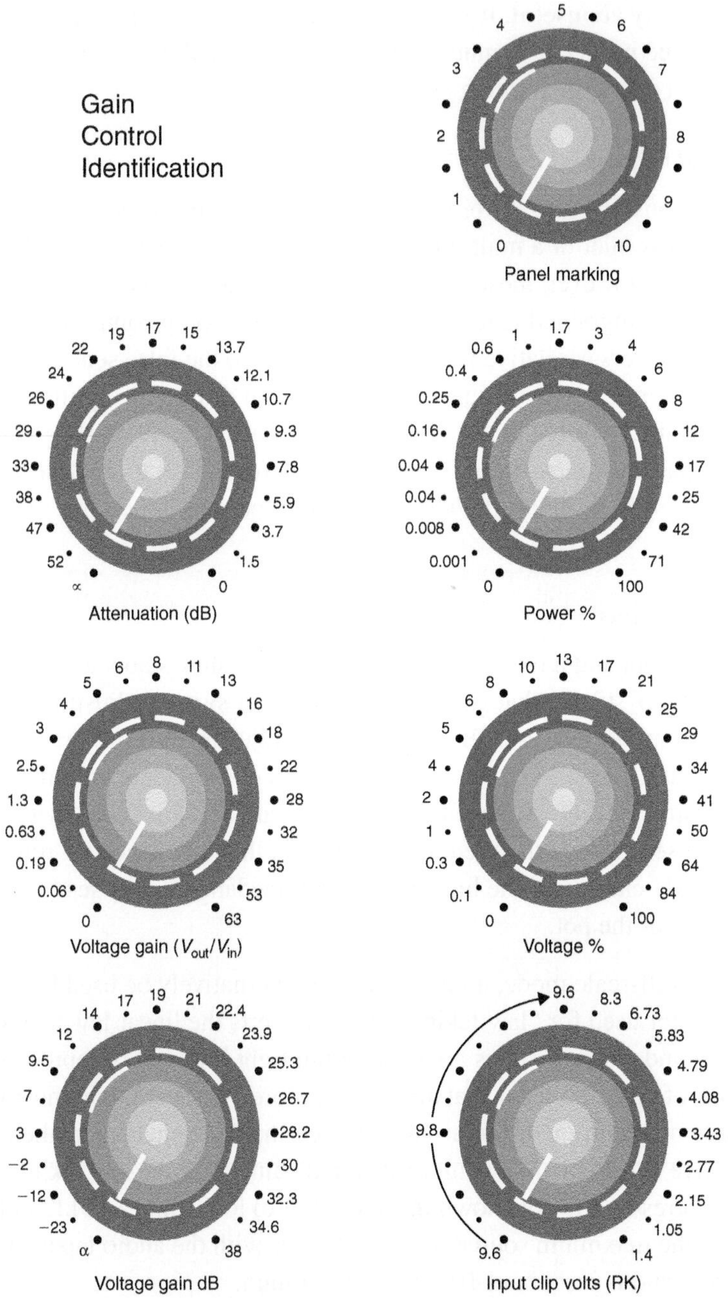
Gain
Control
Identification
0 1 2 3 4 5 6 7 8 9 10
Panel marking
∝ 52 47 38 33 29 26 24 22 19 17 15 13.7 12.1 10.7 9.3 7.8 5.9 3.7 1.5 0
Attenuation (dB)
0 0.001 0.008 0.04 0.04 0.16 0.25 0.4 0.6 1 1.7 3 4 6 8 12 17 25 42 71 100
Power %
0 0.06 0.19 0.63 1.3 2.5 3 4 5 6 8 11 13 16 18 22 28 32 35 53 63
Voltage gain ($V_{out}/V_{in}$)
0 0.1 0.3 1 2 4 5 6 8 10 13 17 21 25 29 34 41 50 64 84 100
Voltage %
α −23 −12 −2 3 7 9.5 12 14 17 19 21 22.4 23.9 25.3 26.7 28.2 30 32.3 34.6 38
Voltage gain dB
9.6 9.8 9.6 8.3 6.73 5.83 4.79 4.08 3.43 2.77 2.15 1.05 1.4
Input clip volts (PK)

### 8.6.1.6 Position

As in other analogue audio circuits, the placement of any gain control device requires careful considerations in regard to considering trade-offs in headroom and SNR. But in power amplifiers having a minimum path, there is not much choice for location. They all end up after the input is unbalanced but before it is raised far.

Placement couldn't be contemplated after the point of signal passing to the input of the power stage, for example, as pots having film tracks (cf. wire wound) that are suitable for audio by virtue of low rotation noise are unsuited to high dissipation. In any event, most power stage topologies don't have a place for inserting a single-ended, passive voltage divider, don't like having their gain widely changed, and are moreover wrapped around by NFB.

Adequate CMR (at the amplifier's input) demands good balancing, which in turn relies on resistance matching to better than *at least* 0.5%, and since even makers of very expensive, high specification pots have problems maintaining matching between two or more sections to even 2%, over the entire travel, pots passing audio have to be placed *after* the input signal has been converted to single ended, that is, after the debalancer (DTSEC). Virtually all power amplifier gain pots (or whatever other gain control devices) end up thusly sandwiched. A few are used in active mode, where the pot is used in the NFB loop, of either an added line-level stage or even a gain-change tolerant power stage. This seems smart but it has its own problems.

### 8.6.1.7 Fixed Install

In amps principally intended for fixed installation, whether for a home cinema or public venues, and where power amplifier gain trims are needed or helpful for setting up, "knobless" gain controls are welcomed. Here, shafts are normally recessed and can

---

**Figure 8.10: Gain pot settings. Shown are six ways of looking at any power amplifier's gain control; in this instance the simplest and most familiar "volume" control type. The final knob labeled "input clip volts (pk)" scale is for peak levels and is correct only for an amplifier that clips at 900 mV rms. In reality, the point would depend on speaker loading, mains voltage, the program, etc. The constant 9.6-V peak reached at lower levels shows where the input stage clips or where zener-based input-protection clamping is operating. Courtesy of Citronic Ltd.**

only be turned with a screwdriver. This avoids not just casual tampering, but knobs being moved (and settings lost) by accidental brushing, sweeping, or knocking. A collet nut may be included. When tightened, the setting will then be immune to attack by a screwdriver, as well as vibration creep. As a further discouragement to "let's turn this up," such controls may be placed on the rear panel of the amp or hidden behind cover plates.

### 8.6.2 Remotable Gain Controls (Machine Control)

Pots are mostly made to interface with human fingers via knobs. When a sound system moves past the point where a single driver in each band can handle the power required or where *Ambisonic* or other multichannel sound is contemplated, remote control opens the door to "intelligent" control of loudspeaker systems and clusters, including balancing and tweaking directivity, imaging, and focusing, by machines and via wires and radio links. The gain of an amp can be controlled by a variety of electronic means (Figure 8.11). The purely electronic means are fast enough to perform additional, true processing functions, for example, limiting.

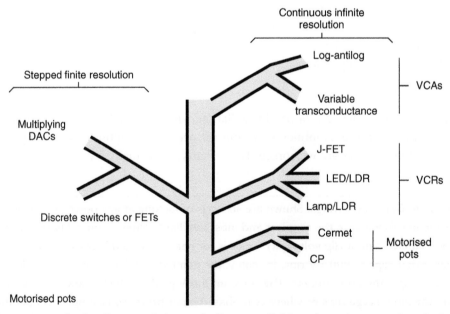

**Figure 8.11: The family tree of electronically controllable gain and attenuation devices.**

Usually a motor connects to the same shaft as a knob, but the latter via a slipping clutch. Either may override. This keeps the simplicity but shares the wide setting tolerance, sonic, and some of the mechanical limitations, of ordinary pots, for example, fragile shaft, relatively low setting speed. Which overrides the other depends on which way confidence most leans—toward human fingers or computers! Control circuitry is needed to decode remote command signals, which may be a variety of formats. Special driver ICs (e.g., BA series made by *Rohm* in Japan) make design and manufacture easy but might pose major replacement headaches to some owners in the future.

### 8.6.2.1 Voltage-Controlled Amplifiers

Commonly called a voltage-controlled amplifier, most are used as VC *attenuators*, usually as a solid state and always an analogue circuit. Most are ICs based around one of a limited number of proprietary schemes, which are made (or licensed, e.g., *That Corp.* in U.S. licenses, *National* in Japan) by one of three main patent holders, all in the United States.[14] Otherwise they are based on a discrete circuit or on a consumer grade 'OTA' IC. Gain is accurately settable to within a fraction a dB, down to at least −70 dB and even into positive gain with some parts. Gain is always defined by an analogue control voltage (or current) that may be derived locally after decoding from a digital line or buss. Refined VCAs introduce considerable added circuitry into the signal path, which may defeat its own purpose. The simplest parts add two stages. They may boast low noise but it is at the expense of exposing the unnatural distortion patterns they create. The best performers add as many as five sequential stages and more than 5 op-amps may be required. If part quality is not to be compromised, the added cost seems high. Operating speed with most types can be very high, under 1 ItS. In this way, VCAs *and all the following contrivances* are applicable to dynamic functions, up to the fastest meaningful audio *peak limiting*.

### 8.6.2.2 LED + LDRs

With this method, the control signal drives an LED so that full brightness is defined as either maximum level or full attenuation. An adjacent light-dependent resistor (LDR) acts as the upper or lower arm of a passive attenuator. The intrinsic circuit isolation and physical separation that is possible makes LED/LDRs attractive in systems where isolation (of both grounds and common-mode voltages to 2.5 kV or more) is important for safety or EMC. These parts provide remote control connections analogous to connecting digital feeds via opto-isolators.

Tolerance is an issue and is dependent on the constituent parts, both semiconductors. Because the tolerance of both LDRs and LEDs is rather wide, manufactured combination devices are likewise broadly specified. The performance of both devices also varies widely with temperature. Also, in many circuits, there is no negative feedback loop to keep these variables within limits. Thus LDR/LED combinations are unsuited to system gain control due to inconsistencies of say $+/-3\,$dB. They *are* fast enough to be used as limiters for bass and even midfrequencies in active crossover systems, and sonic quality is regarded as among the best. However, the above gain variation (in a population) would translate as a spectral imbalance, making overdriven conditions in a large system unsafe and/or uncomfortable, as well as drawing attention to the limiter action.

An LDR may also be partnered with an incandescent lamp. Even if small, the lamp is relatively slow to turn on and off, preventing its use for clean-cut dynamics processing, and lamp life span is more vibration sensitive and so not as certain as solid-state parts in road-going use.

### 8.6.2.3 Junction Field-Effect Transistors

JFETs are the lowest cost elements and can be made operative with little support circuitry. They are normally applied in the lower arm of an attenuator network. Without introducing complications of increased noise, noise pick-up, and other sonic degradation caused by introducing high ohmic value series resistors, attenuation is limited in range, and unless added circuitry can be justified, mild attenuation (around $-6\,$dB) produces high (1 to 10% but mainly benign, low order) harmonic distortion.[15] Low distortion control can be attained by placing the JFET in a control loop, comprising two or more op-amps and other active parts. However, as most JFETs' $R_{on}$ is in the order of a few tens of ohms, attenuation is still typically limited to $-20$ to $-30\,$dB, enough for limiting, but not as a VCA gain and mute control.

### 8.6.2.4 Multiplying Digital-to-Analogue Converters

Multiplying digital-to-analogue converters (M-DACs) involve a resistive ladder, usually binary, with semiconductor switches, usually small-signal MOSFETs. They are the solid-state equivalent of a relay-controlled attenuator ladder (see later). Types suitable for high-performance audio must have dB steps—awkward in binary format—and special MOSFETs for low distortion and absence of "zipper" noise. The latter undesired sonic effect occurs in low-grade M-DACs; it is caused by step changes in DC levels or feed through from the digital control signal. Unlike the previous elements, an M-DAC has discrete resolution—just like a stepped ("detented") pot. At low attenuations, step size must be no more than 1 dB for precise control; below –30 dB, larger steps (2 dB) are usually fine enough. To attenuate down to –70 dB in 1-dB steps, 12-bit M-DAC is required.

### 8.6.2.5 R&R Array

Comprising resistors and relays, this is the mechanical counterpart of the M-DAC, with relays opening and closing paths in a "ladder" or other array of (usually) discrete attenuator resistors. Only high reliability, ATE-grade, sealed reed relays are suited for high-performance audio on grounds of both reliability and sonics. Such relays can act in under 1 mS and have fast settling, but are still not really suited to dynamics processing! Getting dB steps to act binarily with a resistor array takes some lateral thinking. Although the relays required are relatively expensive, by ingenious network adaptation to increment in binary dB, a mere seven can offer a 60-dB range in 1-dB steps. With suitably well-specified resistors, this type can offer the highest transparency of any gain control device.

### 8.6.2.6 Summary

Motorized pots, lamp + LDRs, and relay/resistor arrays are good for remote- or machine-controlled gain trim and setting. The latter are the fastest and likely most reliable.

J-FETs and LED + LDRs are good for dynamics processing, but attaining accurate, noninvasive performance takes from the initial simplicity.

VCAs and M-DACs are elements that can do both kinds of jobs well.

## 8.6.3 Remote Control Considerations

Computers regularly feign precision that is only virtual. Until gain control elements become self-checking, self-calibrating, and self-aligning, they require careful specification.

### 8.6.3.1 Temperature

Pots (particularly conductive plastic), JFET, LDR, and particularly VCA elements are quite temperature sensitive. Unless designed with very low *tempco*, then when used in two or more channel amplifiers, they must be placed isothermally, that is, cosited to be independent of all the major temperature gradients, dependent on drive patterns, siting and even amplifier and rack orientation, as a hot gas usually rises upwards relative to the earth's surface. This is true even with amplifiers employing forced venting, when small signal parts are not in an air path and are left to cool by microconvection, conduction, and reradiation.

Without such precautions, differences in channel gains of 2 dB have been observed in an amplifier employing VCA-controlled gain when driven up to working temperatures. This is enough to cause howl round or upset spectral balance.

### 8.6.3.2 Repeatability

Remote gain settings must not drift or have *repeatability* errors, which can accumulate to cause more than (say) +/–0.15-dB total error. This may seem stringent, yet on top of an initial tolerance of another +/–0.15 dB, it allows a worst case total difference between speakers of 0.6 dB. Other errors (cable losses, driver mismatches) are of a similar order and add to the differencing toll so there is no room for complacency. Least is best.

### 8.6.3.3 Conclusion

M-DACs and relay-resistor-array attenuators have the highest stability against temperature and time. Other types may prove acceptable with ameliorative engineering. Setting precision should not be taken for granted.

### 8.6.4 Compression and Limiting

Compression and limiting (comp-lim) are gain reduction, alias dynamics processing techniques, that are employed (among other things) to protect speakers, ears, and amplifiers from excess, distorted signal levels. In professional, active crossover-based systems, they are usually embodied within the active crossover. This is the best position for logistics in traditional large systems, with only one comp-limper band to worry about. Positioned within the filter chain can also be the best location for sonics.

Where power amplifiers are driven full range or where active crossover filter sections are integral to the power stage, compression and limiting functions may take place within individual power amplifiers.

Compression must be used sparingly, as average power dissipation in the drivers will be increased, potentially part-defeating the object, as speakers may then suffer burnout.

Paradoxically, the compression threshold (at least for bass frequencies) should be increased if the gain reduction exceeds about 6 dB. Also, attack and release times require careful setting to avoid pumping on strong low bass.

Limiting is a higher ratio, more brute force (many dB-to- l) gain reduction. Its raison *d'etre* is to catch fast peaks, hence *"peak limiting."* Attack times that are useful for protecting most loudspeaker drivers are in the order of 10 μS. Faster rising peaks that "get through" rarely cause damage to hardware, but may be reproduced efficiently by metal-diaphragmed drive units (*cf*. paper cones) and perceived and found highly unpleasant by the ear. Hence faster-acting peak limiters may enhance sound quality under many real conditions of "operator abuse."

### 8.6.5 Clipping (Overload) Considerations

Driving any power amplifier with excessive input results in clipping because the output's excursion is finite. Amplifiers offering higher power into a given load impedance provide a higher voltage swing into that impedance so clipping for a given sound pressure level is less likely to arise. However, linear increases in power give only underproportionate, logarithmic increases in headroom (in dB) and cost linearly ascending amounts of money. At some point, whatever more swing could be afforded would make no difference, and a limit is set. Exceeding this is clipping. For short periods it can be benign but else it is unpleasant and potentially damaging to hearing and positively damaging to hf and bass drive units in particular. Moreover, considerable overdriving, into *hard* clip, as can happen at any time by accident, even with domestic systems, can heavily saturate and thus vaporize the BJT output stages of inadequately designed power amplifiers.

### 8.6.6 Clip Prevention

Destructive and antisocial clipping may be prevented with comparatively simple circuits performing like a dedicated, fast limiter. There are as many names as there are makers. Some examples are shown in Table 8.6.

In these and related schemes, clip prevention does not occur until a dB or so of clip. Using the 100-W analogy, the usual low % THD does not rise until the signal passes above about 50 to 70 W. If headroom is adequate, this point should hardly ever be reached with the majority of recorded sound. With live sound, it may be reached quite often, but the fact that the deeply unpleasant point only 1 dB higher is *not* crashed through is of far more importance.

**Table 8.6: Manufacturer and their products**

| ARX systems | *Anticlip* |
|---|---|
| Carver | Clipping eliminator |
| Crest Audio | IGM (Instantaneous Gain Modulation) |
| Crown (Amcron) | AGC in PSA2 (Automatic Gain Control) |
| Malcolm Hill | *Headlok* |

### 8.6.7 Soft Clip

"Soft clip" is a feature that aims to defeat the suddenness of the onset of hard distortion above the *clip level* in conventional, high NFB power amplifiers. It may be provided as a fixed or switchable option. Unlike compression and limiting, there are no time constants, no settings, and no attempt to avert serious distortion of a sine wave. However, the clipped waveform does not readily square off and retains some curvature ($dV/dt$) even with heavy overdrive (e.g., at +10dBvr). This greatly reduces the massed production of unpleasant, high harmonics and intermodulation products of hard clipping. One apparent (but not necessarily actual) snag is that because hard clipping is a real limit, soft clipping has to begin to occur up to –10dB below full output (–10dBvr). This is tantamount to saying that distortion (%THD say) with a 100-W amplifier begins rising from above about 10W, as opposed to rising very abruptly above exactly 100W, while remaining extremely low *up to* this point. Here is one difference between low and high *global* feedback amplifier behavior.

Soft clipping restores the more forgiving behavior of low feedback to a high NFB amplifier. The extent to which it undoes all the high feedback's other benefits is unqualified. At least the high NFB is in operation for most of the time, for with proper headroom allowance, most of the musical content should lie below the –10-dB threshold or so, whence the soft clip is inactive. Usually soft clipping is arranged to be symmetrical. This may not create the most consonant harmonic structure. Figure 8.12 shows a classic circuit.

## 8.7 Computer Control

Computer control of audio power amplifiers has been slow to develop. This is because amplifiers have not been a useful place, in most instances, for physical control surfaces.

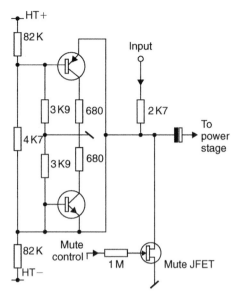

**Figure 8.12: A typical soft clip circuit as used in the Otis Power Station amplifier. Copyright Mead & Co. 1988.**

With a virtual control surface, the traditional limitation vanishes. In turn, installation setup, constant awareness of status, and troubleshooting of amplifiers in medium to large installations are all enhanced. One person can "be in six places at once."

The Dutch PA system manufacturer Stage Accompany was a pioneer of the computer-controlled and monitored PA system in the mid-1980s. However, the first widespread commercial system that wasn't a dedicated, integrated type was Crown's *IQ*, running on Apple Macintosh (1986). The second was Crest Audio's aptly named *Nexsys,* running on PC. Most subsequent systems have been IBM-PC-compatible types, running under Microsoft's *Windows.* Every system is different, yet offers similar, fairly predictable features; there is no clear-cut choice. At the time of writing (1996), some "future proofed" universal, nonpartisan, networkable system contenders that seem most likely to become industry standards appear to have priced themselves out of consideration. Instead, makers continue rolling their own. Recent examples include the IA (intelligent Amplifier) system by *C-Audio*, the MIDI-based interface used by MC 2 (UK), and QSC's *Dataport* system.

Today's computer-control systems theoretically offer:

1. the remote control of many of most of the facilities and controls considered here and in other chapters.

2. the flexible ganging, nesting, and prioritization of these controls.

3. the transmission of real-time signal, thermal, rail voltage, or PSU energy storage data, monitoring, logging, and alarming. May even include a measure of utilization, for example, if a particular amplifier's swing is largely unused as a consequence of overspecification.

4. the remote, even automatic, testing of amplifiers, speaker loads, and their connections.

Thus far, most computer-controlled power amplifiers require an interfacing card to be plugged in. Some types have integral microprocessors.

A well-designed computer control interface must not affect the analogue systems grounding or compromise mains safety. These requirements are met by the fiber-optic, opto-, or transformer-coupled interfacing, familiar enough in digital audio. Such systems must also not only meet EMC requirements, but also, in real world conditions, not radiate or introduce EMI to the power amplifiers. The system must also be able to recognize faults in its own connectivity to power amplifiers.

## References

1. Ball, Greg.M, Overlook THD at your peril, letters, *EW + WW*, August, 1993.

2. Cherry, Prof. Edward, Ironing out distortion, *EW + WW*, January 1995.

3. Jung, Walt, *Audio applications*, Section 8 of System Applications Guide, Analog Devices, 1993.

4. Penrose, H. E. and Boulding, R. H. S., *Principles and practice of radar*, 4th Ed., Newnes, 1953.

5. Duncan, B., 'Black box', *HFN/RR*, October, 1994.

6. Bohm, Dennis, 'Practical line driving current requirements', *Sound and Video Contractor*, September, 1991.

7. Duncan, Ben, 'AMP-O1 parts 3 and 4', *HFN/RR*, July and August, 1984.

8. Duncan, Ben, 'Building the world's biggest PA', *Lighting and Sound International*, October. 1988.

9. Duncan, Ben, A state of the art preamplifier: AMP-02, *Hi-Fi News*, March, 1990.

10. Duncan, Ben, Delayed audio signals, *EW + WW*, May 1995.

11. Duncan, Ben, Signal chain, *Studio Sound*, 1991.

12. Buxton, Joe, *Input overvoltage protection*, System Applications Guide, Analog Devices, Section 1, 156–173, 1993.

13. Bin, David, *Electronically balanced analogue line interfaces*, Proc. lOA, Vol. 12, Part 8, 1990.

14. Duncan, Ben, 'VCAs investigated, parts 1–4', *Studio Sound*, June to September, 1989.

15. (Nameless), *FETs as voltage controlled resistors*, FET data book, Siliconix, 1986.

## Further Reading

Augustadt, H. W., and Kannenberg, W. F., Longitudinal noise in audio circuits, *Audio Engineering*, 1950, reprinted J.AES, July, 1968.

Fletcher, T., 'Balanced or unbalanced?', *Studio Sound*, November, 1980.

Fletcher, T., 'Balanced or balanced?', *Studio Sound*, December, 1981.

Huber, M., 'Conceptual errors in microphone preamplification', *Studio Sound*, April, 1993.

Ott, H. W., *Noise reduction techniques in electronic systems*, Ch. 4, John Wiley, 1976.

Perkins, C., *Measurement techniques for debugging systems and their interconnection*, 11th AES conference, Oregon, May, 1992.

# CHAPTER 9

# Audio Amplifiers

**John Linsley Hood**

Solid-state device technologies, which are available to the amplifier designer, fall, broadly, into three categories: bipolar junction transistors (BJTs) and junction diodes; junction field effect transistors (FETs); and insulated gate FETs, usually referred to as MOSFETs (metal oxide silicon FETs), because of their method of construction. These devices are available in both P type—operating from a negative supply line—and N type—operating from a positive supply line. BJTs and MOSFETs are also available in small-signal and larger power versions, whereas FETs and MOSFETs are manufactured in both enhancement-mode and depletion-mode forms. Predictably, this allows the contemporary circuit designer very considerable scope for circuit innovation, by comparison with electronic engineers of the past, for whom there was only a very limited range of vacuum tube devices.

In addition, there is a very wide range of integrated circuits (ICs), which are complete functional modules in some (usually quite small) individual packages. These are designed both for general-purpose use, such as operational amplifiers, and for more specific applications, such as voltage regulator devices, current mirrors, current sources, phase-sensitive rectifiers, and an enormous variety of designs for digital applications, which mostly lie outside the scope of this book.

In the case of discrete devices, I think it is unnecessary for the purposes of audio amplifier design to understand the physical mechanisms by which the devices work, provided that their would-be user has a reasonable grasp of their operating characteristics and limitations and, above all, a knowledge of just what is available.

## 9.1 Junction Transistors

These are nearly always three-layer devices, fabricated by the multiple and simultaneous vapor phase diffusion and etching of small and intricate patterns on a large, thin slice of very high purity single crystal silicon. A few devices are still made in germanium, mainly for replacement purposes, and some VHF components are made in gallium arsenide, but these will not, in general, lie within the scope of this book. The fabrication techniques may be based on the use of a completely undoped (intrinsic) slice of silicon, into which carefully controlled quantities of impurities are diffused through an appropriate mask pattern from both sides of the slice. These are described in the manufacturers' literature as double diffused, triple diffused, and so on.

In a later technique, evolved by the Fairchild Instrument Corporation, all the diffusions were made from one side of the slice. These devices were called planar and had, normally, a better HF response and more precisely controlled characteristics than, for example, equivalent double-diffused devices. In a further, more recent, technique, also due to Fairchild, the silicon slice will have been made to grow a surface layer of uniformly doped silicon on the exposed side (which will usually form the base region of a transistor) and a single diffusion was then made into this doped layer to form the emitter junction. This technique was called epitaxial and led to transistors with superior characteristics, especially at HF. Since this is the least expensive BJT fabrication process, it will normally be used wherever it is practicable, and if no process is specified it may reasonably be supposed to be a planar-epitaxial type.

In contrast to a thermionic valve, which is a voltage-controlled device, the BJT is a current operated one. So while a change in the base voltage will result in a change in the collector current, this has a very nonlinear relationship to the applied base voltage. In comparison to this, the collector current changes with the input current to the base in a relatively linear manner. Unfortunately, this linear relationship between $I_c$ and $I_b$ tends to deteriorate at higher base current levels, as shown in Figure 9.1. This relationship between base and collector currents is called the current gain, and for AC operation is given the term $h_{fe}$, and its nonlinearity is an obvious source of distortion when the device is used as an amplifier. Alternatively, one could regard this lack of linearity as a change of $h_{FE}$ (this term is used to define the DC or LF characteristics of the device) as the base current is changed. A further problem of a similar kind is the change in $h_{fe}$ as a function of signal frequency, as shown in Figure 9.2.

However, as a current amplifier (which generally implies operation from a high impedance signal source) the behavior of a BJT is vastly more linear than when used as a voltage amplifying stage, for which the input voltage/output current relationships are

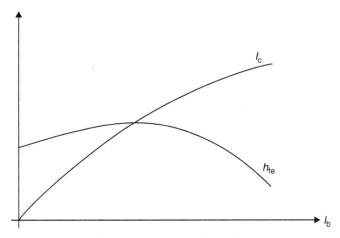

**Figure 9.1: BJT nonlinearity.**

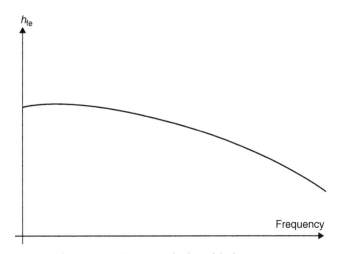

**Figure 9.2: Decrease in $h_{fe}$ with frequency.**

shown for an NPN silicon transistor as line 'a' in Figure 9.3. (I have included, as line 'b', for reference, the comparable characteristics for a germanium junction transistor, although this would normally be a PNP device with a negative base voltage, and a negative collector voltage supply line.) By comparison with, say, a triode valve, whose anode current/grid voltage relationships are also shown as line 'c' in Figure 9.3, the BJT

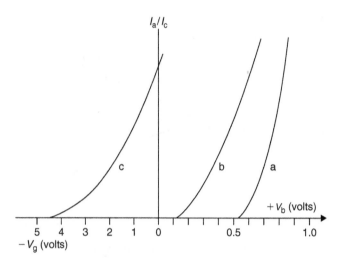

**Figure 9.3: Comparative characteristics of valve, germanium, and silicon based BJTs.**

is a grossly nonlinear amplifying device, even if some input (positive in the case of an NPN device) DC bias voltage has been chosen so that the transistor operates on a part of the curve away from the nonconducting initial region.

## 9.2  Control of Operating Bias

There are three basic ways of providing a DC quiescent voltage bias to a BJT, which is shown in Figure 9.4. In the first of these methods, shown in Figure 9.4(a), an arrangement that is fortunately seldom used, the method adopted is simply to connect an input resistor, $R_1$, between the base of the transistor and some suitable voltage source. This voltage can then be adjusted so that the collector current of the transistor is of the right order to place the collector potential near its desired operating voltage. The snag with this scheme is that transistors vary quite a lot from one to another of nominally the same type, so this would require to be set anew for each individual device. Also, if the operating temperature changes, the current gain of the device (which is temperature sensitive) will be altered and, with it, the collector current of $Q_1$ and its working potential. The arrangement shown in Figure 9.4(b) is somewhat preferable in that a high current gain transistor, or one working at a higher temperature, will pass more current, and this will lower the collector voltage of $Q_1$, which will, in turn, reduce the bias current flowing through $R_1$. However, this also provides NFB and will limit the stage gain to a value somewhat less than $R_1/Z_{in}$.

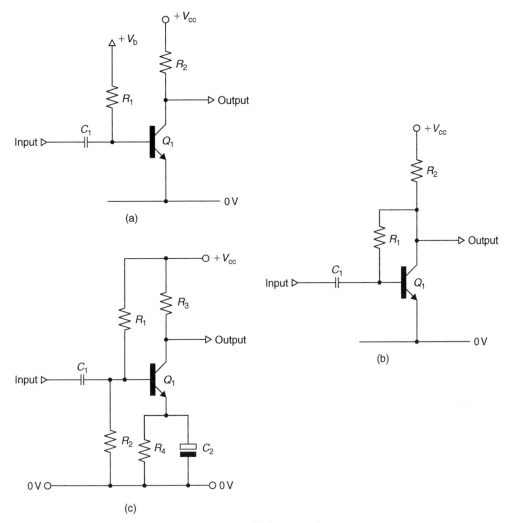

**Figure 9.4: Biasing circuits.**

The method almost invariably used in competently designed circuitry is that shown in Figure 9.4(c), or some equivalent layout. In this, a potential divider ($R_1,R_2$) having an output impedance low in relation to the base impedance of $Q_1$ is used to provide a fixed DC base potential. Since the emitter will, by emitter–follower action, sit at a potential, depending on emitter current, which is about 0.6 V below that of the base, the value of

$R_4$ will then determine the emitter and collector currents, and the operating conditions so provided will hold good for almost any broadly similar device used in this position. Since the emitter resistor would cause a significant reduction in stage gain, as seen in the equivalent analysis of valve cathode bias systems, it is customary to bypass this resistor with a capacitor, $C_2$, which is chosen to have an impedance low in relation to $R_4$ and $R_3$.

## 9.3  Stage Gain

The stage gain of a BJT, used as a simple amplifier, can be determined from the relationship:

$$\frac{V_{out}}{V_{in}} = \frac{h_{fe}R_L}{R_S + r_i}$$

where $R_s$ is the source resistance, $R_L$ is the collector load resistor, $h_{fe}$ is the small-signal (AC) current gain, and $r_i$ is the internal emitter-base resistance of the transistor. An alternative and somewhat simpler approach is similar to that used for a pentode valve gain stage in which

$$V_{out}/V_{in} = g_m R_L$$

where the $g_m$ of a typical modern planar epitaxial silicon transistor will be in the range of 25–40 mS/mA of collector current. Because the $g_m$ of the junction transistor is so high, high stage gains can be obtained with a relatively low value of load resistor. For example, a small-signal transistor with a supply voltage of 15 V, a 4k7 collector load resistor, and a collector current of 2 mA will have a low frequency stage gain, for a relatively low source resistance, of some 300˘. If some way can be found for increasing the load impedance, without also increasing the voltage drop across the load, very high gains indeed can be achieved—up to 2500 with a junction FET acting as a high impedance constant current load.[1]

A predictable, but interesting aspect of stage gain is that the higher the gain, which can be obtained from a circuit module, the lower the distortion in this which will be due to the input device. This is so because if increasingly small segments are taken from any curve, they will progressively approach more closely to a straight line in their form. This allows a very low THD figure, much less than 0.01% at 2 V rms output, over the frequency range 10 Hz–20 kHz, to be obtained from the simple NPN/PNP feedback pair shown in Figure 9.5, which would have an open loop gain of several thousand. The distortion contributed by

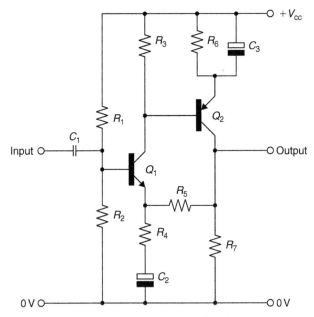

**Figure 9.5: NPN/PNP feedback pair.**

$Q_2$ will be relatively low because of the high effective source resistance seen by the $Q_2$ base. A similar low level of distortion is given by the amplifier layout (bipolar transistor with constant current load) described earlier because of the very high stage gain of the amplifying transistor and the consequent utilization of only a very small portion of its $I_c/V_b$ curve.

## 9.4 Basic Junction Transistor Circuit Configurations

As in the case of the thermionic valve, there are a number of layouts, in addition to the simple single transistor amplifier shown in Figure 9.4 or the two-stage amplifier of Figure 9.5, that can be used to provide a voltage gain or to perform an impedance transformation function. There is, for example, the grounded base layout of Figure 9.6, which has a very low input impedance, a high output impedance, and a very good HF response. This circuit is far from being only of academic interest in the audio field in that it can provide, for example, a very effective low input impedance amplifier circuit for a moving coil pick-up cartridge. I showed a circuit of this type, dating from about 1980, in an earlier book (*Audio Electronics*, Newnes, 1995, p. 133).

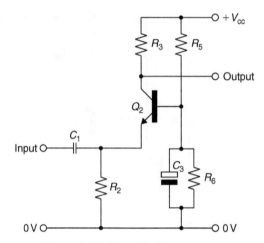

**Figure 9.6: Grounded base stage.**

The cascode layout is also used very widely as a voltage amplifier stage, using a circuit arrangement of the kind shown in Figure 9.7(a). As in the case of the valve amplifier stage, this circuit gives very good input/output isolation and an excellent HF performance due to its freedom from capacitative feedback from output to input. It can also be rearranged, as shown in Figure 9.7(b), so that the input stage acts as an emitter–follower, which gives a very high input impedance.

The long-tailed pair layout, shown in its simplest form in Figure 9.8(a), gives a very good input/output isolation; also, because it is of its nature a push–pull layout, it gives a measure of reduction in even-order harmonic distortion. Its principal advantage, and the reason why this layout is normally used, is that it allows, if the tail resistor ($R_1$) is returned to a −ve supply rail, both of the input signal ports to be referenced to the 0-V line—a feature that is enormously valuable in DC amplifying systems. The designer may sometimes seek to improve the performance of the circuit block by using a high impedance (active) tail in place of a simple resistor, as shown in Figure 9.8(b). This will lessen the likelihood of unwanted signal breakthrough from the −ve supply rail, as well as ensuring a greater degree of dynamic balance, and signal transfer, between the two halves.

Although like all solid-state amplifying systems it is free from the bugbears of hum and noise intrusion from the heater supply of a valve amplifying stage—likely in any valve

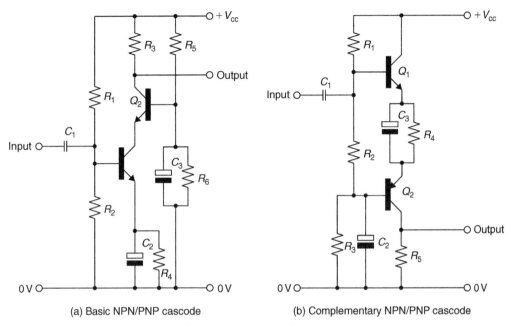

(a) Basic NPN/PNP cascode     (b) Complementary NPN/PNP cascode

**Figure 9.7: Cascode layouts.**

amplifier where there is a high impedance between cathode and ground—it is less good from the point of view of thermal noise than a similar single stage amplifier, partly because there is an additional device in the signal line and partly because the gain of a long-tailed pair layout will only be half that of a comparable single device gain stage. This arises because if a voltage increment is applied to the base of $Q_1$, then the $Q_1$ emitter will only rise half of that amount due to the constraint from $Q_2$, which will also see, but in opposite phase and halved in size, the same voltage increment. This allows, as in the case of the valve phase splitter, a very close similarity, but in opposite phase, of the output currents at $Q_1$ and $Q_2$ collectors.

## 9.5 Emitter–Follower Systems

These are the solid-state equivalent of the valve cathode follower layout, although offering superior performance and greater versatility. In the simple circuit shown in Figure 9.9 (the case shown is for an NPN transistor, but a virtually identical circuit, but with negative supply rails, could be made with a similar PNP transistor), the emitter will

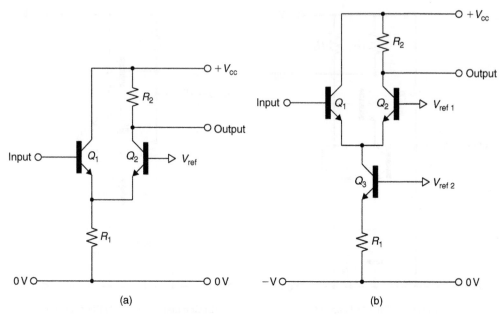

**Figure 9.8: Long-tailed pair layouts.**

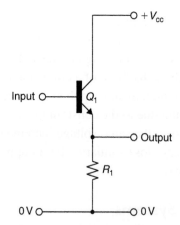

**Figure 9.9: Emitter–follower.**

sit at a quiescent potential about 0.6 V more negative than that of the base, and this will follow, quite accurately, any signal voltage excursions applied to the base. (There are some caveats in respect of capacitive loads; these potential problems will be explored under the heading of slew rate limiting.) The output impedance of this circuit is low because this is approximately equal to $1/g_m$, and the $g_m$ of a typical small-signal, silicon BJT is of the order of 35 mA/V (35 mS) per mA of emitter current. So, if $Q_1$ is operated at 5 mA, the expected output impedance, at low frequencies, will be 1/(5.35) kilohms, or 5.7 ohms, a value that is sufficiently smaller than any likely value for $R_1$, that the presence of this resistor will not greatly affect the output impedance of the circuit.

The output impedance of a simple emitter–follower can be reduced still further by the circuit elaboration shown in Figure 9.10, known as a compound emitter–follower. In this, the output impedance is lowered in proportion to the effective current gain of $Q_2$ in that, by analogy with the output impedance of an operational amplifier with overall NFB, any change in the potential of the $Q_1$ emitter, brought about by an externally impressed voltage, will result in an opposing change in the collector current of $Q_2$. This layout gives a comparable result to that of the Darlington pair, of two transistors, in cascade, connected as emitter–followers, shown in Figure 9.11, except that the arrangement of Figure 9.10 will only have an input/output DC offset equivalent to a single emitter-base

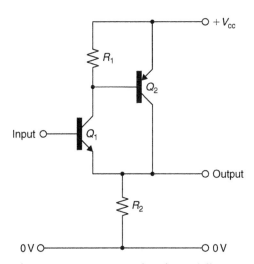

**Figure 9.10: Compound emitter–follower.**

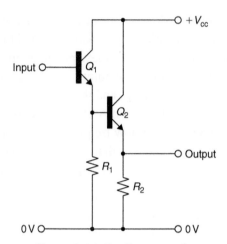

**Figure 9.11: Darlington pair.**

forward voltage drop, whereas the layout of Figure 9.11 will have two, giving a combined quiescent voltage offset of the order of 1.3–1.5 V. Nevertheless, in commercial terms, the popularity of power transistors, connected internally as a Darlington pair, mainly for use in the output stages of audio amplifiers, is great enough for a range of single chip Darlington devices to be offered by the semiconductor manufacturers.

## 9.6 Thermal Dissipation Limits

Unlike a thermionic valve, the active area of a BJT is very small, in the range of 0.5 mm for a small signal device to 4 mm or more for a power transistor. Because the physical area of the component is so small—this is a quite deliberate choice on the part of the manufacturer because it reduces the individual component cost by allowing a very large number of components to be fabricated on a single monocrystalline silicon slice—the slice thickness must also be kept as small as possible—values of 0.15–0.5 mm are typical—in order to assist the conduction of any heat evolved by the transistor action away from the collector junction to the metallic header on which the device is mounted.

Whereas in a valve, in which the internal electrode structure is quite massive and heat is lost by a combination of radiation and convection, the problem of overheating is usually the unwanted release of gases trapped in its internal metalwork, the problem

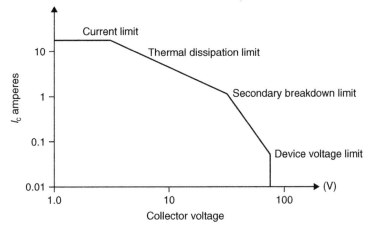

**Figure 9.12: Bipolar breakdown limits.**

in a BJT is the phenomenon known as thermal runaway. This can happen because the potential barrier of a P-N junction (that voltage that must be exceeded before current will flow in the forward direction) is temperature dependent and decreases with temperature. Because there will be unavoidable nonuniformities in the doping levels across the junction, this will lead to nonuniform current flow through the junction sandwich, with the greatest flow taking place through the hottest region. If the ability of the device to conduct heat away from the junction is inadequate to prevent the junction temperature rising above permissible levels, this process can become cumulative. This will result in the total current flow through the device being funneled through some very small area of the junction, which may permanently damage the transistor. This malfunction is termed secondary breakdown, and the operating limits imposed by the need to avoid this failure mechanism are shown in Figure 9.12. Field effect devices do not suffer from this type of failure.

## 9.7 Junction Field Effect Transistors (JFETs)

JFETs are, almost invariably, depletion mode devices, which means that there will be some drain current at a zero-applied gate-source potential. This current will decrease in a fairly linear manner as the reverse gate-source potential is increased, giving an operating characteristic, which is, in the case of an N-channel JFET, very similar to that of a triode valve, as shown in curve 'c' of Figure 9.3. Like a thermionic valve, the operation of the device is limited to the range between drain (or anode) current cutoff and gate (or grid) current. In the case of the JFET, this is because the gate-channel junction is effectively a silicon junction diode—normally operated under reverse bias conditions. If the gate source voltage exceeds some 0.6 V in the forward direction, it will conduct, which will prevent gate voltage control of the channel current.

P-channel JFETs are also made, although in a more limited range of types, and these have what is virtually a mirror image of the characteristics of their N-channel equivalents, although in this case the gate-source forward conduction voltage will be of the order of −0.6 V, and drain current cutoff will occur in the gate voltage range of +3 to +8 V. Although Sony did introduce a range of junction FETs for power applications, these are no longer available, and typical contemporary JFETs cover the voltage range (maximum) from 15 to 50 V, mainly limited by the gate-drain reverse breakdown potential, and with permitted dissipations in the range 250–400 mW. Typical values of $g_m$ (usually called $g_{fs}$ in the case of JFETs) fall in the range of 2–6 mS.

JFETs mainly have good high-frequency characteristics, particularly the N-channel types, of which there are some designs capable of use up to 500 MHz. Modern types can also offer very low noise characteristics, although their very high input impedance will lead to high values of thermal noise if their input circuitry is also of high impedance; however, this is within the control of the circuit designer. The internal noise resistance of a JFET, $R(n)$, is related to the $g_{fs}$ of the device by

$$R(n)\,\text{ohms} \approx 0.67/g_{fs}$$

and the value of $g_{fs}$ can be made higher by paralleling a number of channels within the chip. The Hitachi 2SK389 dual matched-pair JFET achieves a $g_{fs}$ value of 20 mS by this technique, with an equivalent channel thermal noise resistance of 33 ohms.[2]

Although JFETs will work in most of the circuit layouts shown for junction transistors, the most significant difference in the circuit structure is due to their different biasing needs. In the case of a depletion mode device it is possible to use a simple source bias arrangement, similar to the cathode bias used with an indirectly heated valve, of the kind shown in Figure 9.13. As before, the source resistor, $R_3$, will need to be bypassed with a capacitor, $C_2$, if the loss of stage gain due to local NFB is to be avoided. As with a

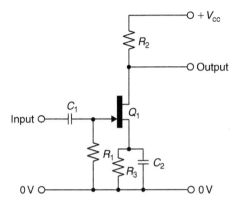

**Figure 9.13: Simple JFET biasing system.**

pentode valve, which the junction FET greatly resembles in its operational characteristics, the simplest way of calculating stage gain is by the relationship:

$$A \approx g_{fs}R_L.$$

The device manufacturers will frequently modify the structure of the JFET to linearize its $V_g/I_d$ characteristics, but, in an ideal device, these will have a square-law relationship, as defined by

$$g_{fs} = I_d/V_g \approx I_{dss}\left[1 - \left(V_{gs}/V_{gc}\right)\right] \approx /V_g.$$

For a typical JFET operating at $2\,mA$ drain current, the $g_{fs}$ value will be of the order of $1$–$4\,mS$, which would give a stage gain of up to 40 if $R_2$, in Figure 9.13, is $10\,k\Omega$. This is very much lower than would be given by a BJT and is the main reason why they are not often used as voltage-amplifying devices in audio systems unless their very high input impedance (typical values are of the order of $10^{12}\ \Omega$) or their high, and largely constant, drain impedance characteristics are advantageous.

The real value of the JFET emerges in its use with other devices, such as the bipolar/FET cascode shown in Figure 9.14 or the FET/FET cascode layout of Figure 9.15. In the first of these, use of the JFET in the cascode connection confers the very high output impedance of the JFET and the high degree of output/input isolation characteristic of the cascode

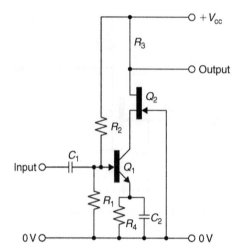

**Figure 9.14: Bipolar/FET cascode.**

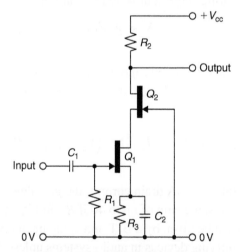

**Figure 9.15: FET/FET cascode.**

layout, coupled with the high stage gain of the BJT. The source potential of the JFET ($Q_2$) will be determined by the reverse bias appearing across the source/gate junction, and could typically be of the order of 2–5 V, which will define the collector potential applied to $Q_1$. A further common application of this type of layout is that in which the cascode FET ($Q_2$ in Figure 9.15) is replaced by a high-voltage BJT. The purpose of this

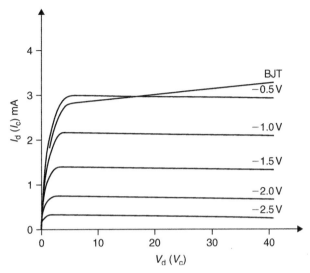

**Figure 9.16: Drain current characteristics of junction FET.**

arrangement is to allow a JFET amplifier stage to operate at a much higher rail voltage than would be allowable to the FET on its own; this layout is often found as the input stage of high-quality audio amps.

A feature that is very characteristic of the JFET is that for drain potentials above about 3 V, the drain current for a given gate voltage is almost independent of the drain voltage, as shown in Figure 9.16. BJTs have a high characteristic collector impedance, but their $I_c/V_c$ curve for a fixed base voltage, also shown, for comparison, in Figure 9.16, is not as flat as that of the JFET. The very high dynamic impedance of the JFET resulting from this very flat $I_d/V_d$ relationship encourages the use of these devices as constant current sources, shown in Figure 9.17. In this form the JFET can be treated as a true two-terminal device, from which the output current can be adjusted, with a suitable JFET, over the range of several milliamperes down to a few microamperes by means of RV1.

## 9.8  Insulated Gate FETs (MOSFETs)

Insulated gate FET devices, usually called MOSFETs, are by far the most widely available, and most widely used, of all the field effect transistors. They normally have a

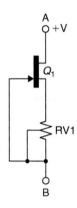

**Figure 9.17: Current source layout.**

rather worse noise figure than an equivalent JFET, but, on the plus side, they have rather more closely controlled operating characteristics. The range of types available covers the very small-signal, low-working voltage components used for VHF amplification in TVs and FM tuners (for which applications a depletion-mode dual-gate device has been introduced that has very similar characteristics to those of an RF pentode valve) to high-power, high-working voltage devices for use in the output stages of audio amplifiers, as well as many other high-power and industrial applications. They are made in both depletion- and enhancement-mode forms (the former having gate characteristics similar to that of the JFET, whereas the latter description refers to the style of device in which there is normally no drain current in the absence of any forward gate bias), in N-channel and P-channel versions, and, at the present time, in voltage and dissipation ratings of up to 1000 V and 600 W, respectively.

All MOSFETs operate in the same manner, in which a conducting electrode (the gate) situated in proximity to an undoped layer of very high purity single-crystal silicon (the channel), but separated from it by a very thin insulating layer, is caused to induce an electrostatic charge in the channel, which will take the form of a layer of mobile electrons or holes. In small-signal devices this channel is formed on the surface of the chip between two relatively heavily doped regions, which will act, respectively, as the source and the drain of the FET, while the conducting electrode will act as the current controlling gate.

Although modern photolithographic techniques are capable of generating exceedingly precise diffusion patterns, the length of the channel formed by surface-masking techniques in a lateral MOSFET will be too long to allow a low channel "on" resistance. For high current applications, the semiconductor manufacturers have therefore evolved a range of vertical MOSFETs. In these, very short channel lengths are achieved by sequential diffusion processes from the surface, which are then followed by etching a V- or U-shaped trough inward from the surface so that the active channel is formed across the exposed edge of a thin diffused region. Because this channel is short in length, its resistance will be low, and because the manufacturers generally adopt device structures that allow a multiplicity of channels to be connected electrically in parallel, channel "on" resistances as low as 0.008 $\Omega$ have been achieved.

Like a JFET, the MOSFET would, left to itself, have a square-law relationship between gate voltage and drain current. However, in practice, this is affected by the device geometry, and many modern devices have a quite linear $I_d/V_g$ characteristic, as shown in Figure 9.18 for an IRF520 power MOSFET.

The basic problem with the MOSFET is that of gate/channel overvoltage breakdown, in which the thin insulating layer of silicon oxide or silicon nitride between the gate electrode and the channel breaks down. If this happens the gate voltage will no longer

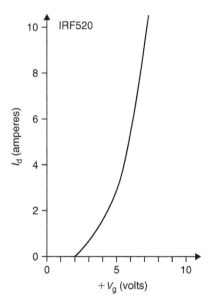

**Figure 9.18: Power MOSFET.**

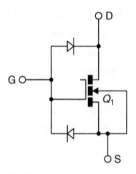

**Figure 9.19: Diode gate protection.**

control the drain current and the device is defunct. Because it is theoretically possible for an inadvertent electrostatic charge, such as might arise with respect to the ground if a user were to wear nylon or polyester fabric clothing and well-insulated shoes, it is common practice in the case of small-signal MOSFETs for protective diodes to be formed on the chip at the time of manufacture. These could be either zener diodes or simple junction diodes connected between the gate and the source or the source and drain, as shown in Figure 9.19.

In power MOSFETs, these protective devices are seldom incorporated into the chip. There are two reasons for this: (1) that the effective gate/channel area is so large that the associated capacitance is high, which would then require a relatively large inadvertently applied static charge to generate a destructive gate/channel voltage (typically >40 V), and (2) that such protective diodes could, if they were triggered into conduction, cause the MOSFET to act as a four-layer thyristor and become an effective electrical short circuit. However, there are usually no performance penalties that will be incurred by connecting some external protective zener diode in the circuit to prevent the gate/source or gate/drain voltage exceeding some safe value; this is a common feature in the output stages of audio power amplifiers using MOSFETs.

Apart from the possibility of gate breakdown, which, in power MOSFETs, always occurs at less than the maker's quoted voltage, except at zero drain current, MOSFETs are quite robust devices, and the safe operating area rating (SOAR) curve of these devices, shown for a typical MOSFET in Figure 9.20, is free from the threat of secondary breakdown whose limits are shown, for a power BJT, in Figure 9.12. The reason for this freedom

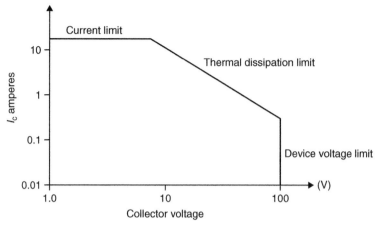

**Figure 9.20: Power MOSFET SOAR limits.**

from localized thermal breakdown in the MOSFET is that the mobility of the electrons (or holes) in the channel decreases as the temperature increases, which gives all FETs a positive temperature coefficient of channel resistance.

Although it is possible to propose a mathematical relationship between gate voltage and drain current, with MOSFETs as was done in the case of the JFET, the manufacturers tend to manipulate the diffusion pattern and construction of the device to linearize its operation, which leads to the type of performance (quoted for an actual device) shown in Figure 9.21. However, as a general rule, the $g_{fs}$ of a MOSFET will increase with drain current, and a forward transconductance (slope) of 10 S/A is quoted for an IRF140 at an ID value of 15 A.

## 9.9 Power BJTs vs Power MOSFETs as Amplifier Output Devices

Some rivalry appears to have arisen between audio amplifier designers over the relative merits of power BJTs, as compared with power MOSFETs. Predictably, this is a mixture of advantages and drawbacks. Because of the much more elaborate construction of the MOSFET, in which a multiplicity of parallel connected conducting channels is fabricated to reduce the conducting "on" resistance, the chip size is larger and the device is several times more expensive both to produce and to buy. The excellent HF characteristics of the MOSFET, especially the N-channel V and U MOS types, can lead to unexpected forms

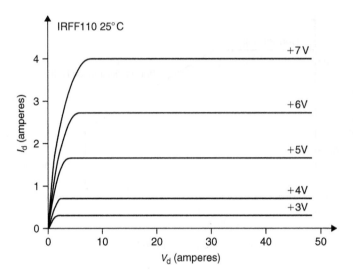

**Figure 9.21: MOSFET characteristics.**

of VHF instability, which can, in the hands of an unwary amplifier designer, lead to the rapid destruction of the output devices. However, this excellent HF performance, when handled properly, makes it much easier to design power amplifiers with good gain and phase margins in the feedback loop, where overall NFB is employed. In contrast, the relatively sluggish and complex characteristics of the junction power transistor can lead to difficulties in the design of feedback amplifiers with good stability margins.

Also, as has been noted, the power MOSFET is intrinsically free from the problem of secondary breakdown, and an amplifier based on these does not need the protective circuitry that is essential in amplifiers with BJT output devices if failure is to be avoided when they are used at high power levels with very low impedance or reactive loads. The problem here is that the protective circuitry may cut in during high-frequency signal level peaks during the normal use of the amplifier, which can lead to audible clipping. (Incidentally, the proponents of thermionic valve-based audio amplifiers have claimed that the superior audible qualities of these, by comparison with transistor-based designs, are due to the absence of any overload protection circuitry that could cause premature clipping and to their generally more graceful behavior under sporadic overload conditions).

A further benefit enjoyed by the MOSFET is that it is a majority carrier device, which means that it is free from the hole-storage effects that can impair the performance of power junction transistors and make them sluggish in their turn-off characteristics at high collector current levels. However, on the debit side, the slope of the $V_g/I_d$ curve of the MOSFET is less steep than that of the $V_b/I_c$ curve of the BJT, which means that the output impedance of power MOSFETs used as source followers is higher than that of an equivalent power BJT used as an emitter follower. Other things being equal, a greater amount of overall negative feedback (i.e., a higher loop gain) must therefore be used to achieve the same low amplifier output impedance with a power MOSFET design than would be needed with a power BJT one. If a pair of push–pull output source/emitter followers is to be used in a class AB output stage, more forward bias will be needed with the MOSFET than with the BJT to achieve the optimum level of quiescent operational current, and the discontinuity in the push–pull transfer characteristic will be larger in size, although likely to introduce, in the amplifier output signal, lower rather than higher order crossover harmonics.

## 9.10  U and D MOSFETs

I have, so far, lumped all power MOSFETs together in considering their performance. However, there are, in practice, two different and distinct categories of these, based on their constructional form, and these are illustrated in Figure 9.22. In the V or U MOS devices—these are just different names for what is essentially the same system, depending on the profile of the etched slot—the current flow, when the gate layer has been made sufficiently positive (in the case of an N-channel device) to induce a mobile electron layer, will be essentially vertical in direction, whereas in D-MOS or T-MOS construction the current flow is T shaped from the source metallization pads across the exposed face of the very lightly doped P region to the vertical N−/N+ drain sink. Because it is easier to manufacture a very thin diffused layer (=short channel) in the vertical sense than to control the lateral diffusion width, in the case of a T-MOS device, by surface masking, the U-MOS devices are usually much faster in response than the T-MOS versions, but the T-MOS equivalents are more rugged and more readily available in complementary (N-channel/P-channel) forms.

All power MOSFETs have a high input capacitance, typically in the range of 500–2500 pF, and because devices with a lower conducting resistance ($R_{ds/on}$) will have achieved this quality because of the connection of a large number of channels in parallel,

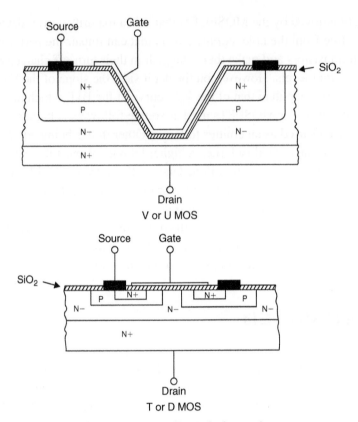

Figure 9.22: MOSFET design styles.

each of which will contribute its own element of capacitance, it is understandable that these low channel resistance types will have a larger input capacitance. Also, in general, P-channel devices will have a somewhat larger input capacitance than an N-channel one. The drain/gate capacitance—a factor that is very important if the MOSFET is used as a voltage amplifier—is usually in the range of 50–250 pF. The turn-on and turn-off times are about the same (in the range 30–100 nS) for both N-channel and P-channel types, mainly determined by the ease of applying or removing a charge from the gate electrode. If gate-stopper resistors are used—helpful in avoiding UHF parasitic oscillation and avoiding latch-up in audio amplifier output source followers—these will form a simple low-pass filter in conjunction with the device input capacitance and will slow down the operation of the MOSFET.

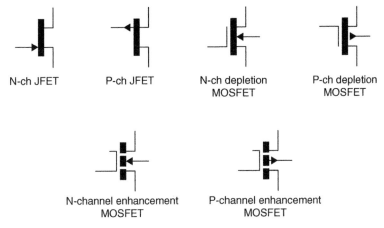

Figure 9.23: MOSFET symbols.

Although circuit designers tend to be rather lazy about using the proper symbols for the components in the designs they have drawn, enhancement-mode and depletion-mode MOSFETs should be differentiated in their symbol layout, as shown in Figure 9.23. As a personal idiosyncrasy, I also prefer to invert the symbol for P-channel field effect devices, as shown, to make this polarity distinction more obvious.

## 9.11 Useful Circuit Components

By comparison with the situation that existed at the time when most of the pioneering work was done on valve-operated audio amplifiers, the design of solid-state amplifier systems has been facilitated greatly by the existence of a number of circuit artifices, contrived with solid-state components, which perform useful functions in the design. This section shows a selection of the more common ones.

### 9.11.1 Constant Current Sources

A simple two-terminal constant current (CC) source is shown in Figure 9.17, and devices of this kind are made as single ICs with specified output currents. By comparison with the discrete JFET/resistor version, the IC will usually have a higher dynamic impedance and a rather higher maximum working voltage. In power amplifier circuits it is more common to use discrete component CC sources based on BJTs, as these are generally less expensive than JFETs and provide higher working voltages. The most obvious of these

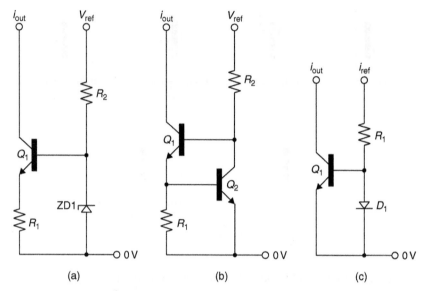

**Figure 9.24: Constant current sources.**

layouts is that shown in Figure 9.24(a), in which the transistor, $Q_1$, is fed with a fixed base voltage—in this case derived from a zener or avalanche diode, although any suitable voltage source will serve—and the current through $Q_1$ is constrained by the value chosen for $R_1$ in that if it grows too large, the base-emitter voltage of $Q_1$ will diminish and $Q_1$ output current will fall. Designers seeking economy of components will frequently operate several current source transistors and their associated emitter resistors (as $Q_1/R_1$) from the same reference voltage source.

In the somewhat preferable layout shown in Figure 9.24(b), a second transistor, $Q_2$, is used to monitor the voltage developed across $R_1$ due to the current through $Q_1$; when this exceeds the base emitter turn-on potential (about 0.6 V), $Q_2$ will conduct and will steal the base current to $Q_1$ provided from $V_{ref}$ through $R_2$. In the very simple layout shown in Figure 9.24(c), advantage is taken of the fact that the forward potential of a P-N junction diode, for any given junction temperature, will depend on the current flow through it. This means that if the base-emitter area and doping characteristics of $Q_1$ are the same as those for the P-N junction in $D_1$ (which would, obviously, be easy to arrange in the manufacture of ICs), then the current ($i_{out}$) through $Q_1$ will be caused to mimic that flowing through $R_1$,

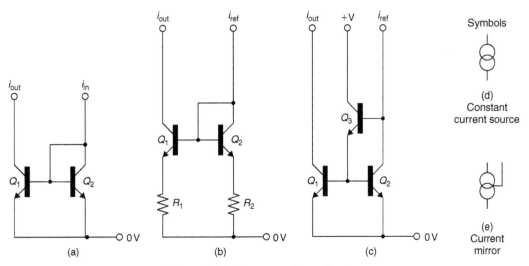

**Figure 9.25: Current mirror circuits.**

which is labeled $i_{ref}$. This particular action is called a current mirror, and several further versions of these are shown in Figure 9.25.

### 9.11.2 Current Mirror Layouts

Current mirror (CM) layouts allow, for example, the output currents from a long-tailed pair to be combined, which increases the gain from this circuit. In the version shown in Figure 9.25(a), two matched transistors are connected with their bases in parallel so that the current flow through $Q_2$ will generate a base-emitter voltage drop that will be precisely that which is needed to cause $Q_1$ to pass the same current. If any doubt exists about the similarity of the characteristics of the two transistors, as might reasonably be the case for randomly chosen devices, the equality of the two currents can be assisted by the inclusion of equal value emitter resistors $(R_1, R_2)$ as shown in Figure 9.25(b). For the perfectionist, an improved three-transistor current mirror layout is shown in Figure 9.25(c). Commonly used circuit symbols for these devices are shown in Figures 9.25(d) and 9.25(e).

## 9.12 Circuit Oddments

Several circuit modules have found their way into amplifier circuit design, and some of the more common of these are shown in Figure 9.26. Both the DC bootstrap, shown

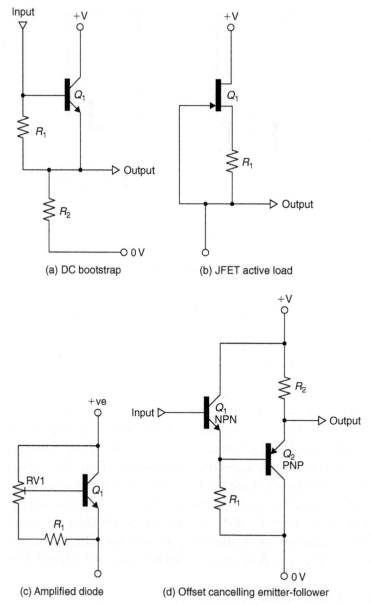

(a) DC bootstrap

(b) JFET active load

(c) Amplified diode

(d) Offset cancelling emitter-follower

Figure 9.26: Circuit oddments.

in Figure 9.26(a), and the JFET active load, shown in Figure 9.26(b), act to increase the dynamic impedance of $R_1$, although the DC bootstrap, which can, of course, be constructed using complementary devices, has the advantage of offering a low output impedance. The amplified diode, shown as Figure 9.26(c), is a device that is much used as a means of generating the forward bias required for the transistors used in a push–pull pair of output emitter followers, particularly if it is arranged so that $Q_1$ can sense the junction temperature of the output transistors. It can also be used, over a range of relatively low voltages, as an adjustable voltage source to complement the fixed voltage references provided by zener and avalanche diodes, band-gap references (IC stabilizers designed to provide extremely stable low voltage sources), and the wide range of voltage stabilizer ICs. Finally, when some form of impedance transformation is required, without the $V_{be}$ offset of an emitter follower, this can be contrived as shown by putting two complementary emitter followers in series. This layout will also provide a measure of temperature compensation.

## 9.13 Slew Rate Limiting

This is a potential problem that can occur in any voltage amplifier or other signal handling stage in which an element of load capacitance (which could simply be circuit stray capacitance) is associated with a drive circuit whose output current has a finite limit. The effect of this is shown in Figure 9.27. If an input step waveform is applied to network (a), then the output signal will have a waveform of the kind shown at 'a', and the slope of the curve will reflect the potential difference that exists, at any given moment, between the input and the output. Any other signal that is present at the same time will pass through this network, from input to output, and only the high-frequency components will be attenuated.

However, if the drive current is limited, the output waveform from circuit 9.27 (b) will be as shown at 'b' and the slope of the output ramp will be determined only by the current limit imposed by the source and the value of the load capacitance. This means that any other signal component that is present, at the time the circuit is driven into slew rate limiting, will be lost. This effect is noticeable, if it occurs, in any high-quality audio system and gives rise to a somewhat blurred sound—a defect that can be lessened or removed if the causes (such as too low a level of operating current for some amplifying stage) are remedied. It is prudent, therefore, for the amplifier designer to establish the possible voltage slew rates for the various stages in any new design and then to ensure

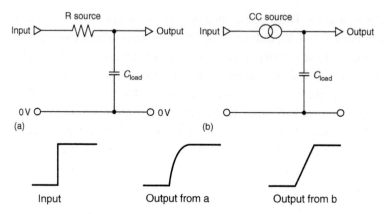

**Figure 9.27: Cause of slew rate limiting.**

that the amplifier does not receive any input signal that requires rates of change greater than the level that can be handled. A simple input integrating network of the kind shown in Figure 9.27(a) will often suffice.

## References

1. Linsley Hood, J., *Wireless World*, 437–441, September, 1971.

2. Linsley Hood, J., 'Low noise systems', *Electronics Today International*, 42–46, 1992.

# CHAPTER 10

# *Audio Amplifier Performance*

Douglas Self

## 10.1 A Brief History of Amplifiers

A full and detailed account of semiconductor amplifier design since its beginnings would be a book in itself, and a most fascinating volume it would be. This is not that book, but I still feel obliged to give a very brief account of how amplifier design has evolved in the last three or four decades.

Valve amplifiers, working in push–pull Class-A or AB1, and perforce transformer coupled to the load, were dominant until the early 1960s, when truly dependable transistors could be made at a reasonable price. Designs using germanium devices appeared first, but suffered severely from the vulnerability of germanium to even moderately high temperatures; the term *thermal runaway* was born. At first all silicon power transistors were NPN, and for a time most transistor amplifiers relied on input and output transformers for push–pull operation of the power output stage. These transformers were as always heavy, bulky, expensive, and nonlinear and added insult to injury as their LF and HF phase shifts severely limited the amount of negative feedback (NFB) that could be applied safely.

The advent of the transformerless Lin configuration,[1] with what became known as a quasi-complementary output stage, disposed of a good many problems. Because modestly capable PNP driver transistors were available, the power output devices could both be NPN and still work in push–pull. It was realized that a transformer was not required for impedance matching between power transistors and 8-$\Omega$ loudspeakers.

Proper complementary power devices appeared in the late 1960s, and full complementary output stages soon proved to give less distortion than their quasi-complementary predecessors. At about the same time, DC-coupled amplifiers began to take over from capacitor-coupled designs, as the transistor differential pair became a more familiar circuit element.

A much fuller and generally excellent history of power amplifier technology is given in Sweeney and Mantz.[2]

## 10.2 Amplifier Architectures

This grandiose title simply refers to the large-scale structure of the amplifier, that is, the block diagram of the circuit one level below that representing it as a single white block-labeled *power amplifier*. Almost all solid-state amplifiers have a three-stage architecture as described here, although they vary in the detail of each stage.

## 10.3 The Three-Stage Architecture

The vast majority of audio amplifiers use the conventional architecture, shown in Figure 10.1. There are three stages, the first being a transconductance stage (differential

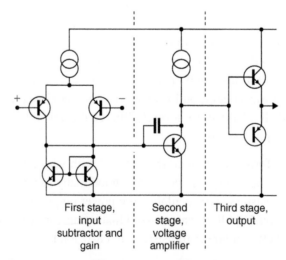

| First stage, input subtractor and gain | Second stage, voltage amplifier | Third stage, output |

**Figure 10.1: The three-stage amplifier structure. There is a transconductance stage, a transadmittance stage (the VAS), and a unity-gain buffer output stage.**

voltage in, current out), the second a transimpedance stage (current in, voltage out), and the third a unity-voltage-gain output stage. The second stage clearly has to provide all the voltage gain and I have therefore called it the voltage-amplifier stage or VAS. Other authors have called it the *predriver stage* but I prefer to reserve this term for the first transistors in output triples. This three-stage architecture has several advantages, not least being that it is easy to arrange things so that the interaction between stages is negligible. For example, there is very little signal voltage at the input to the second stage due to its current input (virtual-earth) nature, and therefore very little on the first stage output; this minimizes Miller phase shift and possible early effect in the input devices.

Similarly, the compensation capacitor reduces the second stage output impedance so that the nonlinear loading on it due to the input impedance of the third stage generates less distortion than might be expected. The conventional three-stage structure, familiar though it may be, holds several elegant mechanisms such as this. Since the amount of linearizing global NFB available depends on amplifier open-loop gain, how the stages contribute to this is of great interest. The three-stage architecture always has a unity-gain output stage—unless you really want to make life difficult for yourself—and so the total forward gain is simply the product of the transconductance of the input stage and the transimpedance of the VAS, the latter being determined solely by the Miller capacitor $C_{dom}$, except at very low frequencies. Typically, the closed-loop gain will be between +20 and +30 dB. The NFB factor at 20 kHz will be 25 to 40 dB, increasing at 6 dB per octave with falling frequency until it reaches the dominant pole frequency P1, when it flattens out. What matters for the control of distortion is the amount of NFB available, rather than the open-loop bandwidth, to which it has no direct relationship. In my Electronics World Class-B design, the input stage $g_m$ is about 9 mA/V, and $C_{dom}$ is 100 pF, giving an NFB factor of 31 dB at 20 kHz. In other designs I have used as little as 26 dB (at 20 kHz) with good results.

Compensating a three-stage amplifier is relatively simple; since the pole at the VAS is already dominant, it can be easily increased to lower the HF NFB factor to a safe level. The local NFB working on the VAS through $C_{dom}$ has an extremely valuable linearizing effect.

The conventional three-stage structure represents at least 99% of the solid-state amplifiers built, and I make no apology for devoting much of this book to its behavior. I doubt if I have exhausted its subtleties.

### 10.3.1 Two-Stage Amplifier Architecture

In contrast, the architecture shown in Figure 10.2 is a two-stage amplifier, with the first stage once again being more a transconductance stage, although now without a guaranteed low impedance to accept its output current. The second stage combines VAS and output stage in one block; it is inherent in this scheme that the VAS must double as a phase splitter as well as a generator of raw gain. There are then two quite dissimilar signal paths to the output, and it is not at all clear that trying to break this block down further will assist a linearity analysis. The use of a phase-splitting stage harks back to valve amplifiers; where it was inescapable as a complementary valve technology has, so far, eluded us.

Paradoxically, a two-stage amplifier is likely to be more complex in its gain structure than a three stage. The forward gain depends on the input stage $g_m$, the input stage collector load (because the input stage can no longer be assumed to be feeding a virtual earth), and the gain of the output stage, which will be found to vary in a most unsettling manner with bias and loading. Choosing the compensation is also more complex for a two-stage amplifier, as the VAS/phase splitter has a significant signal voltage on its input and so the usual pole-splitting mechanism that enhances Nyquist stability by increasing the pole

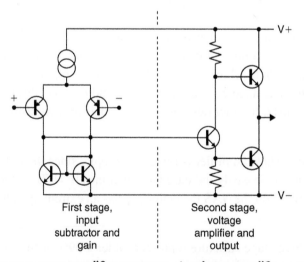

First stage, input subtractor and gain

Second stage, voltage amplifier and output

**Figure 10.2: The two-stage amplifier structure. A voltage-amplifier output follows the same transconductance input stage.**

frequency associated with the input stage collector will no longer work so well. (I have used the term Nyquist stability or Nyquist oscillation throughout this book to denote oscillation due to the accumulation of phase shift in a global NFB loop, as opposed to local parasitics, etc.)

The LF feedback factor is likely to be about 6 dB less with a 4-$\Omega$ load due to lower gain in the output stage. However, this variation is much reduced above the dominant pole frequency, as there is then increasing local NFB acting in the output stage.

Two-stage amplifiers are not popular; I can quote only two examples, Randi[3] and Harris.[4] The two-stage amplifier offers little or no reduction in parts cost, is harder to design, and, in my experience, invariably gives a poor distortion performance.

## 10.4 Power Amplifier Classes

For a long time the only amplifier classes relevant to high-quality audio were Class-A and Class-AB. This is because valves were the only active devices, and Class-B valve amplifiers generated so much distortion that they were barely acceptable, even for public address purposes. All amplifiers with pretensions to high fidelity operated in push–pull Class-A.

Solid state gives much more freedom of design; all of the following amplifier classes have been exploited commercially. Unfortunately, there will only be space to deal in detail in this book with A, AB, and B, although this certainly covers the vast majority of solid-state amplifiers. Plentiful references are given so that the intrigued can pursue matters further.

### 10.4.1 Class-A

In a Class-A amplifier, current flows continuously in all the output devices, which enables the nonlinearities of turning them on and off to be avoided. They come in two rather different kinds, although this is rarely explicitly stated, which work in very different ways. The first kind is simply a Class-B stage (i.e., two emitter–followers working back to back) with the bias voltage increased so that sufficient current flows for neither device to cut off under normal loading. The great advantage of this approach is that it cannot abruptly run out of output current; if the load impedance becomes lower than specified, then the amplifier simply takes brief excursions into Class-AB, hopefully with a modest increase in distortion and no seriously audible distress.

The other kind could be called a controlled-current source type, which is, in essence, a single emitter–follower with an active emitter load for adequate current sinking. If this latter element runs out of current capability, it makes the output stage clip much as if it had run out of output voltage. This kind of output stage demands a very clear idea of how low an impedance it will be asked to drive before design begins.

Valve textbooks contain enigmatic references to classes of operation called AB1 and AB2; in the former, grid current did not flow for any part of the cycle, but in the latter, it did. This distinction was important because the flow of output-valve grid current in AB2 made the design of the previous stage much more difficult.

AB1 or AB2 has no relevance to semiconductors, for base current in BJT always flows when a device is conducting, whereas gate current in power FET never does, apart from charging and discharging internal capacitances.

### 10.4.2 Class-AB

This is not really a separate class of its own, but a combination of A and B. If an amplifier is biased into Class-B and then the bias increased further, it will enter AB. For outputs below a certain level, both output devices conduct and operation is Class-A. At higher levels, one device will be turned completely off as the other provides more current, and the distortion jumps upward at this point as AB action begins. Each device will conduct between 50 and 100% of the time, depending on the degree of excess bias and the output level.

Class-AB is less linear than either A or B, and in my view its only legitimate use is as a fallback mode to allow Class-A amplifiers to continue working reasonably when faced with low-load impedance.

### 10.4.3 Class-B

Class-B is by far the most popular mode of operation, and probably more than 99% of the amplifiers currently made are of this type. Most of this book is devoted to it, so no more is said here.

### 10.4.4 Class-C

Class-C implies device conduction for significantly less than 50% of the time and is normally only usable in radio work, where an LC circuit can smooth out the current pulses and filters harmonics. Current-dumping amplifiers can be regarded as combining Class-A (the correcting amplifier) with Class-C (the current-dumping devices); however, it is hard to visualize how an audio amplifier using devices in Class-C only could be built.

### 10.4.5 Class-D

These amplifiers continuously switch the output from one rail to the other at a supersonic frequency, controlling the mark/space ratio to give an average representing the instantaneous level of the audio signal; this is alternatively called pulse width modulation. Great effort and ingenuity have been devoted to this approach, for the efficiency is, in theory, very high, but the practical difficulties are severe, especially so in a world of tightening EMC legislation, where it is not at all clear that a 200-kHz high-power square wave is a good place to start. Distortion is not inherently low[5] and the amount of global NFB that can be applied is severely limited by the pole due to the effective sampling frequency in the forward path. A sharp cutoff low-pass filter is needed between amplifier and speaker to remove most of the RF; this will require at least four inductors (for stereo) and will cost money, but its worst feature is that it will only give a flat frequency response into one specific load impedance. The technique now has a whole chapter of this book to itself. Other references to consult for further information are Goldberg and Sandler[6] and Hancock.[7]

### 10.4.6 Class-E

An extremely ingenious way to operate a transistor is to have either a small voltage across it or a small current through it almost all the time; in other words, the power dissipation is kept very low.[8] Regrettably, this is an RF technique that seems to have no sane application to audio.

### 10.4.7 Class-F

There is no Class-F, as far as I know. This seems like a gap that needs filling.

### 10.4.8 Class-G

This concept was introduced by Hitachi in 1976 with the aim of reducing amplifier power dissipation. Musical signals have a high peak/mean ratio, spending most of this at low levels, so internal dissipation is much reduced by running from low-voltage rails for small outputs, switching to higher rails current for larger excursions.

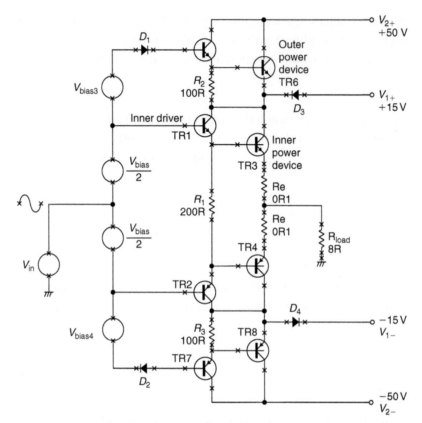

**Figure 10.3: Class-G-series output stage. When the output voltage exceeds the transition level, $D_3$ or $D_4$ turn off and power is drawn from the higher rails through the outer power devices.**

The basic series Class-G with two rail voltages (i.e., four supply rails, as both voltage are $\pm$) is shown in Figure 10.3.[9,11] Current is drawn from the lower $\pm V_1$ supply rails whenever possible; should the signal exceed $\pm V_1$, TR6 conducts and $D_3$ turns off, so the output current is now drawn entirely from the higher $\pm V_2$ rails, with power dissipation shared between TR3 and TR6. The inner stage TR3, TR4 is usually operated in Class-B, although AB or A is equally feasible if the output stage bias is suitably increased. The outer devices are effectively in Class-C as they conduct for significantly less than 50% of the time.

In principle, movements of the collector voltage on the inner device collectors should not significantly affect the output voltage, but in practice, Class-G is often considered to have poorer linearity than Class-B because of glitching due to charge storage in commutation diodes $D_3$, $D_4$. However, if glitches occur they do so at moderate power, well displaced from the crossover region, and so appear relatively infrequently with real signals.

An obvious extension of the Class-G principle is to increase the number of supply voltages. Typically the limit is three. Power dissipation is further reduced and efficiency increased as the average voltage from which the output current is drawn is kept closer to the minimum. The inner devices operate in Class-B/AB as before, and the middle devices are in Class-C. The outer devices are also in Class-C, but conduct for even less of the time.

To the best of my knowledge, three-level Class-G amplifiers have only been made in shunt mode, as described later, probably because in series mode the cumulative voltage drops become too great and compromise the efficiency gains. The extra complexity is significant, as there are now six supply rails and at least six power devices, all of which must carry the full output current. It seems most unlikely that this further reduction in power consumption could ever be worthwhile for domestic hi-fi.

A closely related type of amplifier is Class-G shunt.[10] Figure 10.4 shows the principle; at low outputs, only $Q_3$, $Q_4$ conduct, delivering power from the low-voltage rails. Above a threshold set by $V_{bias3}$ and $V_{bias4}$, $D_1$ or $D_2$ conduct and $Q_6$, $Q_8$ turn on, drawing current from the high-voltage rails, with $D_3$, 4 protecting $Q_3$, 4 against reverse bias. The conduction periods of the $Q_6$, $Q_8$ Class-C devices are variable, but inherently less than 50%. Normally the low-voltage section runs in Class-B to minimize dissipation. Such shunt Class-G arrangements are often called "commutating amplifiers."

Some of the more powerful Class-G shunt PA amplifiers have three sets of supply rails to further reduce the average voltage drop between rail and output. This is very useful in large PA amplifiers.

### 10.4.9 Class-H

Class-H is once more basically Class-B, but with a method of dynamically boosting the single supply rail (as opposed to switching to another one) in order to increase efficiency.[12] The usual mechanism is a form of bootstrapping. Class-H is used occasionally to describe Class-G as described earlier; this sort of confusion we can do without.

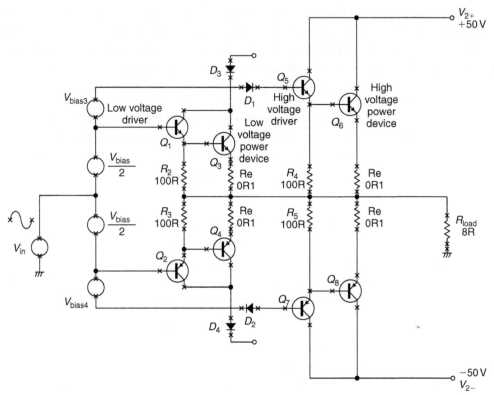

**Figure 10.4: A Class-G shunt output stage, composed of two EF output stages with the usual drivers. $V_{bias3,4}$ set the output level at which power is drawn from the higher rails.**

### 10.4.10 Class-S

Class-S, so named by Doctor Sandman,[13] uses a Class-A stage with very limited current capability, backed up by a Class-B stage connected so as to make the load appear as a higher resistance that is within the capability of the first amplifier.

The method used by the Technics SE-A100 amplifier is extremely similar.[14] I hope that that this necessarily brief catalogue is comprehensive; if anyone knows of other bona fide classes I would be glad to add them to the collection. This classification does not allow a completely consistent nomenclature; for example, quad-style current dumping can only be specified as a mixture of Class-A and -C, which says nothing about the basic principle of operation, which is error correction.

### 10.4.11  Variations on Class-B

The solid-state Class-B three-stage amplifier has proved both successful and flexible, so many attempts have been made to improve it further, usually by trying to combine the efficiency of Class-B with the linearity of Class-A. It would be impossible to give a comprehensive list of the changes and improvements attempted, so I give only those that have been either commercially successful or particularly thought provoking to the amplifier-design community.

### 10.4.12  Error-Correcting Amplifiers

This refers to error-cancellation strategies rather than the conventional use of NFB. This is a complex field, for there are at least three different forms of error correction, of which the best known is error feedforward as exemplified by the ground-breaking Quad 405.[15] Other versions include error feedback and other even more confusingly named techniques, some of which turn out on analysis to be conventional NFB in disguise. For a highly ingenious treatment of the feedforward method, see Giovanni Stochino.[16]

### 10.4.13  Nonswitching Amplifiers

Most of the distortion in Class-B is crossover distortion and results from gain changes in the output stage as the power devices turn on and off. Several researchers have attempted to avoid this by ensuring that each device is clamped to pass a certain minimum current at all times.[17] This approach has certainly been exploited commercially, but few technical details have been published. It is not intuitively obvious (to me, anyway) that stopping the diminishing device current in its tracks will give less crossover distortion.

### 10.4.14  Current-Drive Amplifiers

Almost all power amplifiers aspire to be voltage sources of zero output impedance. This minimizes frequency response variations caused by the peaks and dips of the impedance curve and gives a *universal* amplifier that can drive any loudspeaker directly.

The opposite approach is an amplifier with a sufficiently high output impedance to act as a constant-current source. This eliminates some problems, such as rising voice-coil resistance with heat dissipation, but introduces others, such as control of the cone resonance. Current amplifiers therefore appear to be only of use with active crossovers and velocity feedback from the cone.[18]

It is relatively simple to design an amplifier with any desired output impedance (even a negative one) and so any compromise between voltage and current drive is attainable. The snag is that loudspeakers are universally designed to be driven by voltage sources, and higher amplifier impedances demand tailoring to specific speaker types.[19]

### 10.4.15 The Blomley Principle

The goal of preventing output transistors from turning off completely was introduced by Peter Blomley in 1971[20]; here the positive/negative splitting is done by circuitry ahead of the output stage, which can then be designed so that a minimum idling current can be separately set up in each output device. However, to the best of my knowledge this approach has not yet achieved commercial exploitation.

### 10.4.16 Geometric Mean Class-AB

The classical explanations of Class-B operation assume that there is a fairly sharp transfer of control of the output voltage between the two output devices, stemming from an equally abrupt switch in conduction from one to the other. In practical audio amplifier stages this is indeed the case, but it is not an inescapable result of the basic principle. Figure 10.5 shows a conventional output stage, with emitter resistors Re1, Re2 included to increase quiescent-current stability and allow current sensing for overload protection; to a large extent, these emitter resistances make classical Class-B what it is.

However, if the emitter resistors are omitted and the stage biased with two matched diode junctions, then the diode and transistor junctions form a *translinear loop*[21] around which the junction voltages sum to zero. This links the two output transistor currents $I_p$, $I_n$ in the relationship $I_n * I_p = $ constant, which in op-amp practice is known as geometric-mean Class-AB operation. This gives smoother changes in device current at the crossover point, but this does not necessarily mean lower THD. Such techniques are not very practical for discrete power amplifiers; first, in the absence of the very tight thermal coupling between the four junctions that exists in an IC, the quiescent-current stability will be atrocious, with thermal runaway and spontaneous combustion a near certainty. Second, the output device bulk emitter resistance will probably give enough voltage drop to turn the other

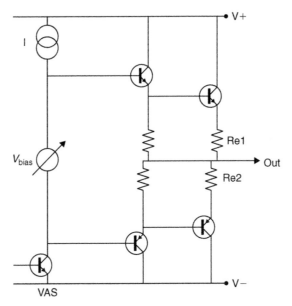

**Figure 10.5: A conventional double emitter–follower output stage with emitter resistors Re shown.**

device off anyway, when current flows. The need for drivers, with their extra junction drops, also complicates things.

A new extension of this technique is to redesign the translinear loop so that $1/I_n + 1/I_p =$ constant; this is known as harmonic-mean AB operation.[22] It is too early to say whether this technique (assuming it can be made to work outside an IC) will be of use in reducing crossover distortion and thus improving amplifier performance.

### 10.4.17 Nested Differentiating Feedback Loops

This is a most ingenious, but conceptually complex technique for significantly increasing the amount of NFB that can be applied to an amplifier (see Cherry[23]).

## 10.5 AC- and DC-Coupled Amplifiers

All power amplifiers are either AC coupled or DC coupled. The first kind have a single supply rail, with the output biased to be halfway between this rail and ground to give the maximum symmetrical voltage swing; a large DC-blocking capacitor is therefore used in series with the output. The second kind have positive and negative supply rails, and the output is biased to be at 0 V, so no output DC blocking is required in normal operation.

### 10.5.1 Advantages of AC Coupling

1. The output DC offset is always zero (unless the output capacitor is leaky).

2. It is very simple to prevent turn-on thump by purely electronic means. The amplifier output must rise up to half the supply voltage at turn on, but providing this occurs slowly, there is no audible transient. Note that in many designs, this is not simply a matter of making the input bias voltage rise slowly, as it also takes time for the DC feedback to establish itself, and it tends to do this with a snap action when a threshold is reached.

3. No protection against DC faults is required, providing that the output capacitor is voltage rated to withstand the full supply rail. A DC-coupled amplifier requires an expensive and possibly unreliable output relay for dependable speaker protection.

4. The amplifier should be easier to make short-circuit proof, as the output capacitor limits the amount of electric charge that can be transferred each cycle, no matter how low the load impedance. This is speculative; I have no data as to how much it really helps in practice.

5. AC-coupled amplifiers do not, in general, appear to require output inductors for stability. Large electrolytics have significant equivalent series resistance (ESR) and a little series inductance. For typical amplifier output sizes the ESR will be of the order of 100 mΩ; this resistance is probably the reason why AC-coupled amplifiers rarely had output inductors, as it is enough resistance to provide isolation from capacitive loading and so gives stability. Capacitor series inductance is very low and probably irrelevant, being quoted by one manufacturer as a few tens of nanoHenrys'. The output capacitor was often condemned in the past for reducing the low-frequency damping factor (DF), for its ESR alone is usually enough to limit the DF to 80 or so. As explained earlier, this is not a technical problem because "damping factor" means virtually nothing.

### 10.5.2 Advantages of DC Coupling

1. No large and expensive DC-blocking capacitor is required. However, the dual supply will need at least one more equally expensive reservoir capacitor and a few extra components such as fuses.

2. In principle, there should be no turn-on thump, as the symmetrical supply rails mean the output voltage does not have to move through half the supply voltage to reach its bias point—it can just stay where it is. In practice, the various filtering time constants used to keep the bias voltages free from ripple are likely to make various sections of the amplifier turn on at different times, and the resulting thump can be substantial. This can be dealt with almost for free, when a protection relay is fitted, by delaying the relay pull-in until any transients are over. The delay required is usually less than a second.

3. Audio is a field where almost any technical eccentricity is permissible, so it is remarkable that AC coupling appears to be the one technique that is widely regarded as unfashionable and unacceptable. DC coupling avoids any marketing difficulties.

4. Some potential customers will be convinced that DC-coupled amplifiers give better speaker damping due to the absence of output capacitor impedance. They will be wrong, as explained later, but this misconception has lasted at least 40 years and shows no sign of fading away.

5. Distortion generated by an output capacitor is avoided. This is a serious problem, as it is not confined to low frequencies, as is the case in small-signal circuitry. For a 6800-$\mu$F output capacitor driving 4 W into an 8-$\Omega$ load, there is significant midband third harmonic distortion at 0.0025%, as shown in Figure 10.6. This is at least five times more than the amplifier generates in this part of the frequency range. In addition, the THD rise at the LF end is much steeper than in the small-signal case, for reasons that are not yet clear. There are two cures for output capacitor distortion. The straightforward approach uses a huge output capacitor, far larger in value than required for a good low-frequency response. A 100,000-$\mu$F/40-V Aerovox from BHC eliminated all distortion, as shown in Figure 10.7. An allegedly "audiophile" capacitor gives some interesting results; a Cerafine Supercap of only moderate size

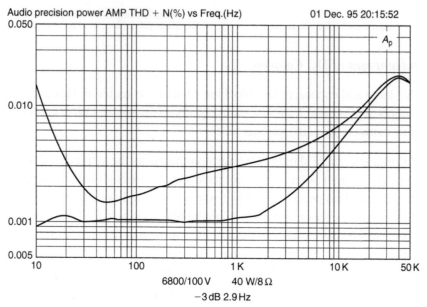

Figure 10.6: The extra distortion generated by an 6800-µF electrolytic delivering 40 W into 8 Ω. Distortion rises as frequency falls, as for the small-signal case, but at this current level there is also added distortion in the midband.

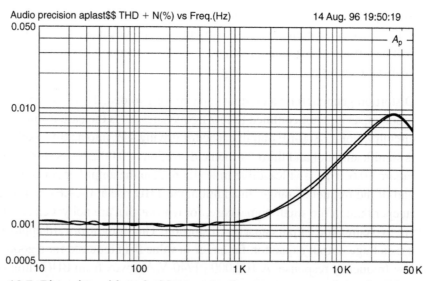

Figure 10.7: Distortion with and without a very large output capacitor, the BHC Aerovox 100,000 µF/40 V (40 watts/8 Ω). Capacitor distortion is eliminated.

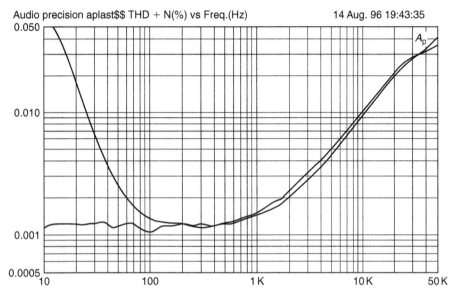

Audio precision aplast$$ THD + N(%) vs Freq.(Hz)                14 Aug. 96 19:43:35

**Figure 10.8: Distortion with and without an "audiophile" Cerafine 4700-μF/63-V capacitor. Midband distortion is eliminated but LF rise is much the same as the standard electrolytic.**

(4700 μF/63 V) gave Figure 10.8, where the midband distortion is gone, but the LF distortion rise remains. What special audio properties this component is supposed to have are unknown; as far as I know, electrolytics are never advertised as low midband THD, but that seems to be the case here. The volume of the capacitor case is about twice as great as conventional electrolytics of the same value, so it is possible the crucial difference may be a thicker dielectric film than is usual for this voltage rating.

Either of these special capacitors costs more than the rest of the amplifier electronics put together. Their physical size is large. A DC-coupled amplifier with protective output relay will be a more economical option.

A little-known complication with output capacitors is that their series reactance increases the power dissipation in the output stage at low frequencies. This is counterintuitive as it would seem that any impedance added in series must reduce the current drawn and hence the power dissipation. In fact, it is the load phase shift that increases the amplifier dissipation.

6. The supply currents can be kept out of the ground system. A single-rail AC amplifier has half-wave Class-B currents flowing in the 0-V rail, which can have a serious effect on distortion and cross talk performance.

## 10.6 Negative Feedback in Power Amplifiers

It is not the role of this book to step through elementary theory, which can be found easily in any number of textbooks. However, correspondence in audio and technical journals shows that considerable confusion exists regarding NFB as applied to power amplifiers; perhaps there is something inherently mysterious in a process that improves almost all performance parameters simply by feeding part of the output back to the input, but inflicts dire instability problems if used to excess. This chapter therefore deals with a few of the less obvious points here.

The main uses of NFB in amplifiers are the reduction of harmonic distortion, the reduction of output impedance, and the enhancement of supply-rail rejection. There are analogous improvements in frequency response and gain stability, and reductions in DC drift, but these are usually less important in audio applications.

By elementary feedback theory, the factor of improvement for all these quantities is

$$\text{Improvement ratio} = A \cdot \beta \tag{10-1}$$

where $A$ is the open-loop gain and $\beta$ is the attenuation in the feedback network, that is, the reciprocal of the closed-loop gain. In most audio applications the improvement factor can be regarded as simply open-loop gain divided by closed-loop gain.

In simple circuits you just apply NFB and that is the end of the matter. In a typical power amplifier, which cannot be operated without NFB, if only because it would be saturated by its own DC offset voltages, several stages may accumulate phase shift, and simply closing the loop usually brings on severe Nyquist oscillation at HF. This is a serious matter, as it will not only burn out any tweeters that are unlucky enough to be connected, but can also destroy the output devices by overheating, as they may be unable to turn off fast enough at ultrasonic frequencies.

The standard cure for this instability is compensation. A capacitor is added, usually in Miller-integrator format, to roll off the open-loop gain at 6 dB per octave, so it reaches unity loop gain before enough phase shift can build up to allow oscillation. This means that the NFB factor varies strongly with frequency, an inconvenient fact that many audio commentators seem to forget.

It is crucial to remember that a distortion harmonic, subjected to a frequency-dependent NFB factor as described earlier, will be reduced by the NFB factor corresponding to its own frequency, not that of its fundamental. If given a choice, generate low-order rather than high-order distortion harmonics, as the NFB deals with them much more effectively.

NFB can be applied either locally (i.e., to each stage, or each active device) or globally; in other words, right around the whole amplifier. Global NFB is more efficient at distortion reduction than the same amount distributed as local NFB, but places much stricter limits on the amount of phase shift that may be allowed to accumulate in the forward path.

Above the dominant pole frequency, the VAS acts as a Miller integrator and introduces a constant 90° phase lag into the forward path. In other words, the output from the input stage must be in quadrature if the final amplifier output is to be in phase with the input, which to a close approximation it is. This raises the question of how the 90° phase shift is accommodated by the NFB loop; the answer is that the input and feedback signals applied to the input stage are subtracted, and the small difference between two relatively large signals with a small phase shift between them has a much larger phase shift. This is the signal that drives the VAS input of the amplifier.

Solid-state power amplifiers, unlike many valve designs, are almost invariably designed to work at a fixed closed-loop gain. If the circuit is compensated by the usual dominant pole method, the HF open-loop gain is also fixed, and therefore so is the important NFB factor. This is in contrast to valve amplifiers, where the amount of NFB applied was regarded as a variable and often user-selectable parameter; it was presumably accepted that varying the NFB factor caused significant changes in input sensitivity. A further complication was serious peaking of the closed-loop frequency response at both LF and HF ends of the spectrum as NFB was increased due to the inevitable bandwidth limitations in a transformer-coupled forward path. Solid-state amplifier designers go cold at the thought of the customer tampering with something as vital as the NFB factor, and such an approach is only acceptable in cases such as valve amplification where global NFB plays a minor role.

### 10.6.1 Some Common Misconceptions About Negative Feedback

All of the comments quoted here have appeared many times in the hi-fi literature. All are wrong.

*NFB is a bad thing.* Some audio commentators hold that, without qualification, NFB is a bad thing. This is of course completely untrue and based on no objective reality. NFB is one of the fundamental concepts of electronics, and to avoid its use altogether is virtually impossible; apart from anything else, a small amount of local NFB exists in every common emitter transistor because of the internal emitter resistance. I detect here distrust of good fortune; the uneasy feeling that if something apparently works brilliantly then there must be something wrong with it.

*A low NFB factor is desirable.* Untrue; global NFB makes just about everything better, and the sole effect of too much is HF oscillation, or poor transient behavior on the brink of instability. These effects are painfully obvious on testing and not hard to avoid unless there is something badly wrong with the basic design.

In any case, just what does *low* mean? One indicator of imperfect knowledge of NFB is that the amount enjoyed by an amplifier is almost always baldly specified as *so many dB* on the very few occasions it is specified at all, despite the fact that most amplifiers have a feedback factor that varies considerably with frequency. A dB figure quoted alone is meaningless, as it cannot be assumed that this is the figure at 1 kHz or any other standard frequency.

My practice is to quote the NFB factor at 20 kHz, as this can normally be assumed to be above the dominant pole frequency and so in the region where open-loop gain is set by only two or three components. Normally the open-loop gain is falling at a constant 6-dB/octave at this frequency on its way down to intersect the unity-loop-gain line and so its magnitude allows some judgment as to Nyquist stability. Open-loop gain at LF depends on many more variables, such as transistor beta, and consequently has wide tolerances and is a much less useful quantity to know.

*NFB is a powerful technique and therefore dangerous when misused.* This bland truism usually implies an audio Rakes's progress that goes something like this: an amplifier has too much distortion and so the open-loop gain is increased to augment the NFB factor. This causes HF instability, which has to be cured by increasing the compensation capacitance. This is turn reduces the slew-rate capability, resulting in a sluggish, indolent, and generally bad amplifier.

The obvious flaw in this argument is that the amplifier so condemned no longer has a high NFB factor because the increased compensation capacitor has reduced the open-loop gain at HF; therefore feedback itself can hardly be blamed. The real problem in this situation is probably an unduly low standing current in the input stage; this is the other parameter determining slew rate.

*NFB may reduce low-order harmonics but increases the energy in the discordant higher harmonics.* A less common but recurring complaint is that the application of global NFB is a shady business because it transfers energy from low-order distortion harmonics—considered musically consonant—to higher order ones that are anything but. This objection contains a grain of truth, but appears to be based on a misunderstanding of one article in an important series by Peter Baxandall[24] in which he showed that if you took an amplifier with only second-harmonic distortion and then introduced NFB around it, higher order harmonics were indeed generated as the second harmonic was fed back round the loop. For example, the fundamental and the second harmonic intermodulate to give a component at third-harmonic frequency. Likewise, the second and third intermodulate to give the fifth harmonic. If we accept that high-order harmonics should be numerically weighted to reflect their greater unpleasantness, there could conceivably be a rise rather than a fall in the weighted THD when NFB is applied.

All active devices, in Class-A or -B (including FETs, which are often erroneously thought to be purely square law), generate small amounts of high-order harmonics. Feedback could and would generate these from nothing, but in practice they are already there.

The vital point is that if enough NFB is applied, all the harmonics can be reduced to a lower level than without it. The extra harmonics generated, effectively by the distortion of a distortion, are at an extremely low level, providing a reasonable NFB factor is used. This is a powerful argument against low feedback factors such as 6 dB, which are most likely to increase the weighted THD. For a full understanding of this topic, a careful reading of the Baxandall series is absolutely indispensable.

*A low open-loop bandwidth means a sluggish amplifier with a low slew rate.* Great confusion exists in some quarters between open-loop bandwidth and slew rate. In truth, open-loop bandwidth and slew rate have nothing to do with each other and may be altered independently. Open-loop bandwidth is determined by compensation $C_{dom}$, VAS β, and resistance at the VAS collector, whereas slew rate is set by the input stage standing current and $C_{dom} \bullet C_{dom}$ affects both, but all the other parameters are independent.

In an amplifier, there is a maximum amount of NFB you can safely apply at 20 kHz; this does not mean that you are restricted to applying the same amount at 1 kHz, or indeed 10 Hz. The obvious thing to do is to allow the NFB to continue increasing at 6 dB/octave—or faster if possible—as frequency falls so that the amount of NFB applied doubles with each octave as we move down in frequency, and we derive as much benefit as we can. This obviously cannot continue indefinitely, for eventually open-loop gain runs out, being limited by transistor beta and other factors. Hence the NFB factor levels out at a relatively low and ill-defined frequency; this frequency is the open-loop bandwidth and, for an amplifier that can never be used open loop, has very little importance.

## References

1. Lin, H. C., 'Transistor audio amplifier', *Electronics*, 173, September, 1956.

2. Sweeney, and Mantz., 'An informal history of amplifiers', *Audio*, 46, June, 1988.

3. Linsley-Hood., 'Simple class-A amplifier', *Wireless World*, 148, April, 1969.

4. Olsson, B., 'Better audio from non-complements?', *Electronics World*, 988, December, 1994.

5. Attwood, B., 'Design parameters important for the optimisation of PWM (class-D) amplifiers', *JAES,* (**31**) 842, November, 1983.

6. Goldberg, and Sandler., 'Noise shaping and pulse-width modulation for all-digital audio power amplifier', *JAES,* (**39**)449, February, 1991.

7. Hancock, J. A., 'Class-D amplifier using MOSFETS with reduced minority carrier lifetime', *JAES,* (**39**) 650, September, 1991.

8. Peters, A., 'Class E RF amplifiers IEEE', *J. Solid-State Circuits*,168, June, 1975.

9. Feldman, L., 'Class-G high-efficiency hi-fi amplifier', *Radio-Electronics*, 47, August, 1976.

10. Raab, F., 'Average efficiency of class-G power amplifiers', *IEEE Transactions on Consumer Electronics,* **CE-22**,145, May, 1986.

11. Sampei, *et al*, 'Highest efficiency and super quality audio amplifier using MOS-power FETs in class-G', *IEEE Transactions on Consumer Electronics,* **CE-24**, 300, August, 1978.

12. Buitendijk, P., *A 40W integrated car radio audio amplifier*, IEEE Conf. on Consumer Electronics, 1991 Session, THAM 12.4, 174, (Class-H), 1991.

13. Sandman, A., 'Class S: A novel approach to amplifier distortion', *Wireless World*, 38, September, 1982.

14. Sinclair, (ed.), *Audio and hi-fi handbook*, Newnes, 1993.

15. Walker, P. J., 'Current dumping audio amplifier', *Wireless World*, 560, December, 1975.

16. Stochino, G., 'Audio design leaps forward?', *Electronics World*, 818, October, 1994.

17. Tanaka, S. A., 'New biasing circuit for class-B operation', *JAES*, 27, January/February 1981.

18. Mills, and Hawksford., 'Transconductance power amplifier systems for current-driven loudspeakers', *JAES,* (**37**) 809, March, 1989.

19. Evenson, R., 1988. *Audio amplifiers with tailored output impedances*, Preprint for November, 1988 AES convention (Los Angeles).

20. Blomley, P., 'A new approach to class-B', *Wireless World*, 57, February, 1971.

21. Gilbert, B., 'Current mode circuits from a translinear viewpoint, chapter 2, Analogue IC design: The current-mode approach, Toumazou, Lidgey & Haigh, eds., *IEE* 1990.

22. *Thus compact bipolar class AB output stage IEEE Journal of Solid-State Circuits,* December, 1992.

23. Cherry, E., 'Nested differentiating feedback loops in simple audio power amplifiers', *JAES,* **30**(#5):295, May, 1982.

24. Baxandall, P., 'Audio power amplifier design: Part 5', *Wireless world*, 53, December, 1978. (This superb series of articles had six parts and ran on roughly alternate months, starting in January 1978.)

# CHAPTER 11

# Valve (Tube-Based) Amplifiers

John Linsley Hood

Although the bulk of modern electronic circuitry is based on "solid-state" components, for very good engineering reasons—one could not, for example, build a compact disc player using valves and still have room in one's house to sit down and listen to it—all the early audio amplifiers were based on valves, and it is useful to know how these worked and what the design problems and circuit options were in order to get a better understanding of the technology. Also, there is still interest on the part of some "hi-fi" enthusiasts in the construction and use of valve-operated audio amplifiers, and additional information on valve based circuitry may be welcomed by them.

## 11.1 Valves or Vacuum Tubes

The term thermionic valve (or valve for short) was given, by its inventor, Sir Ambrose Fleming, to the earliest of these devices, a rectifying diode. Fleming chose the name because of the similarity of its action in allowing only a one-way flow of current to that of a one-way air valve on an inflatable tire, and the way it operated was by controlling the internal flow of thermally generated electrons, which he called "thermions," hence the term thermionic valve. In the United States they are called "vacuum tubes." These devices consist of a heated cathode, mounted, in vacuum, inside a sealed glass or metal tube. Other electrodes, such as anodes or grids, are then arranged around the cathode so that various different functions can be performed.

The descriptive names given to the various types of valve are based on the number of its internal electrodes so that a valve with two electrodes (a cathode and an anode) is called a "diode," one with three electrodes (a cathode, a grid, and an anode) is called a "triode," one with four (a cathode, two grids, and an anode) is called a "tetrode," and so on.

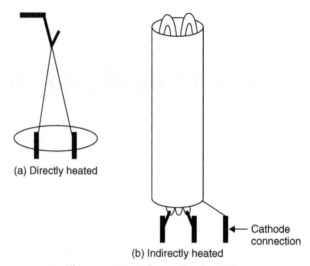

(a) Directly heated

Cathode connection

(b) Indirectly heated

**Figure 11.1: Valve cathode styles.**

It helps to understand the way in which valves work, and how to get the best performance from them, if one understands the functions of these internal electrodes and the way in which different groupings of them affect the characteristics of the valve, so, to this end, I have listed them and examined their functions separately.

### 11.1.1  The Cathode

This component is at the heart of any valve and is the source of the electrons with which it operates. It is made in one of two forms: either a short length of resistor wire, made of nickel, folded into a 'V' shape and supported between a pair of stiff wires at its base and a light tension spring at its top, as shown in Figure 11.1(a), or a metallic tube, usually made of nickel, with a bundle of nickel or tungsten heater wires gathered inside it, as shown in Figure 11.1(b). Whether the cathode is a directly heated "filament" or an indirectly heated metal cylinder, its function and method of operation are the same, although, other things being equal, the directly heated filament is much more efficient in terms of the available electron emission from the cathode in relation to the amount of power required to heat it to its required operating temperature (about 775°C for one having an oxide-coated construction).

It is possible to use a plain tungsten filament as a cathode, but it needs to be heated to some 2500°C to be usable, which requires quite a substantial amount of power and leads to other problems, such as fragility. Virtually all contemporary low to medium power valves use oxide-coated cathodes, which are made from a mixture of the oxides of calcium, barium, and strontium deposited on a nickel substrate.

In the manufacture of the valve, these chemicals are applied to the cathode as a paste composed of a binding agent, the metals in the form of their carbonates, and some small quantities of doping agents, typically of rare-earth origin. The metal carbonates are then reduced to their oxides by subsequent heating during the last stage in the process of evacuating the air from the valve envelope.

In use, a chemical reaction occurs between the oxide coating and the heated nickel cathode tube (or the directly heated filament), which causes the alkali metal oxides to be locally reduced to the free metal, which then slowly diffuses out to the cathode surface to form the electron-emitting layer. The extent of electronic emission from the cathode depends critically upon its temperature, and the value chosen for this in practice is a compromise between performance and life expectancy, as higher cathode temperatures lead to shorter cathode life due to the loss through evaporation of the active cathode metals, whereas a lower limit to the working temperature is set by the need to have an adequate level of electron emission.

When hot, the cathode will emit electrons, which form a cloud around it, a situation in which the thermal agitation of the electrons in the cathode body, which causes electrons to escape from its surface, is balanced by the growing positive charge that the cathode has acquired as the result of the loss of these electrons. This electron cloud is called the "space charge" and plays an important part in the operation of the valve; a matter that is discussed later.

### 11.1.2 The Anode

In the simplest form of valve, the diode, the cathode is surrounded by a metal tube or box, called the anode or plate. This is usually made of nickel and will attract electrons from the space charge if it is made positive with regard to the cathode. The amount of current that will flow depends on the closeness of the anode box to the cathode, the effective area of the cathode, the voltage on the anode, and the cathode temperature. For a fixed cathode temperature and anode voltage, the ratio of anode voltage to current flow determines the anode current resistance, $R_a$, which is measured by the current flow for a given applied voltage— as shown in the equation

$$R_a = dV_a/dI_a.$$

Because the anode is bombarded by electrons accelerated toward it by the applied anode voltage, when they collide with the anode their kinetic energy is converted into heat, which raises the anode temperature. This heat evolution is normally unimportant, except in the case of power rectifiers or power output valves, when care should be taken to ensure that the makers' current and voltage ratings are not exceeded. In particular, there is an inherent problem that if the anode becomes too hot, any gases that have been trapped in pores within its structure will be released, and this will impair the vacuum within the valve, which can lead to other problems.

### 11.1.3 The Control Grid

If the cathode is surrounded by a wire grid or mesh—in practice, this will usually take the form of a spiral coil, spot welded between two stiff supporting wires, of the form shown in Figure 11.2—the current flow from the cathode to the anode can be controlled by the voltage applied to the grid, such that if the grid is made positive, more negatively charged electrons will be attracted away from the cathode and encouraged to continue on their way to the anode. However, if the grid is made negative, it will repel the electrons emitted by the cathode and reduce the current flow to the anode.

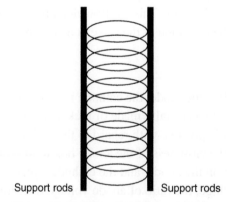

Support rods          Support rods

**Figure 11.2: Control grid construction.**

It is this quality that is the most useful aspect of a valve in that a quite large anode current flow can be controlled by a relatively small voltage applied to the grid, and so long as the grid is not allowed to swing positive with respect to the cathode, no current will flow in the grid circuit, and its effective input impedance at low frequencies will be almost infinite. This ability to regulate a large current at a high voltage by a much smaller control voltage allows the valve to amplify small electrical signals, and since the relationship between grid voltage and anode current is relatively linear, as shown in Figure 11.3, this amplification will cause relatively little distortion in the amplified signal. The theoretical amplification factor of a valve, operating into an infinitely high impedance anode load, is denoted by the Greek symbol $\mu$.

Although there may be several grids between the cathode and the anode in more complex valves, the grid closest to the cathode will have the greatest influence on the anode current flow, which is therefore usually called the control grid.

The effectiveness of the grid in regulating the anode current depends on the relative proximity of the grid and the anode to the cathode, in that, if the grid is close to the cathode, but the anode is relatively remote, the effectiveness of the grid in determining the anode current will be much greater and will therefore give a higher value of $\mu$ than if the anode is closer to the grid and cathode. Unfortunately, there is a snag in that the

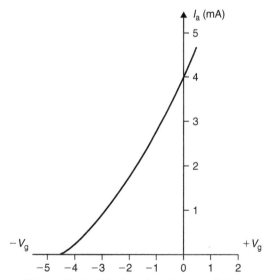

**Figure 11.3: Triode valve characteristics.**

anode current resistance of the valve, $R_a$, is also related to the anode/cathode spacing and becomes higher as the anode/cathode spacing is increased. The closeness of the pitch of the wire spiral that forms the grid also affects the anode current resistance in that a close spacing will lead to a high $R_a$, and vice versa.

The stage gain (M) of a simple valve amplifier, of the kind shown in Figure 11.4, is given by the equation

$$M = \frac{\mu R}{R + R_a}$$

so that a low impedance valve, such as a 6SN7 (typical $I_a = 9\,mA$, $R_a = 7.7\,K$, $\mu = 20$), which has close anode–grid and grid–cathode spacings, and a relatively open pitch in the grid wire spiral, will have a high possible anode current but a low amplification factor, while a high impedance valve such as a 6SL7 (typical $I_a = 2.3\,mA$, $R_a = 44\,K$, $\mu = 70$) will have a low stage gain unless the circuit used has a high value of anode load resistance ($R$), which, in turn, will demand a high value of HT voltage.

### 11.1.4 The Space Charge

Although a cloud of electrons will surround any heated cathode mounted in a vacuum and will act as a reservoir of electrons when these are drawn off as anode current, their

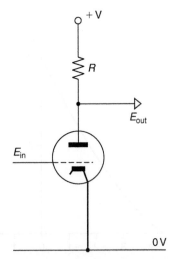

Figure 11.4: A simple valve amplifier.

presence becomes of particular importance when a negatively charged control grid is introduced into the system, in that the electron cloud will effectively fill the space between the cathode and the grid and will act as the principal source of electrons.

The presence of this electron cloud—known as the space charge—has several important operational advantages. Of these the first is that, by acting as an electron reservoir, it allows larger, brief-duration, current flows than would be available from the cathode on its own and that it acts as a measure of protection to the cathode against the impacts of positive ions created by electronic collisions with the residual gases in the envelope, as these ions will be attracted toward the more negatively charged cathode. Finally, left to itself, the electronic emission from the cathode suffers from both "shot" and "flicker" noise, a current fluctuation that is averaged out if the anode current is drawn from the space charge.

This random emission of electrons from a space charge-depleted cathode is used to advantage in a "noise diode," a wide-band noise source that consists of a valve in which the cathode is deliberately operated at a low temperature to prevent a space charge from forming so that a resultant noisy current can be drawn off by the anode.

In the case of a triode used as an output power valve, where large anode currents are needed, the grid mesh must be coarse and the grid–cathode spacing must be close. This limits the formation of an adequate space charge in the grid–cathode gap, and, in its absence, the cathode must have higher emission efficiency than would be practicable with an indirectly heated system, which means that a directly heated filament must be used instead. Usually, the filament voltage will be low to minimize cathode-induced "hum" and the filament current will be high because of the size of the filament (2.5 A at 2.5 V in the case of the 2A3 valve).

Directly heated cathodes are also commonly used in valve HT rectifiers, such as the 5U4 or the 5Y3, because the higher cathode emission reduces the voltage drop across the valve and increases the available HT output voltage by comparison with a similar power supply using an indirectly heated cathode type.

### 11.1.5 Tetrodes and Pentodes

Although the triode valve has a number of advantages as an amplifier, such as a low noise and low distortion factor, it suffers from the snag that there will be a significant capacitance, typically of the order of 2.5 pF, between the grid and the anode. In itself, this latter capacitance would seem to be too small to be troublesome, but, in an amplifying stage with a gain of, say, 100, the Miller effect will increase the capacitance by a factor of 101, increasing the effective input capacitance to 252.5 pF, which could influence the performance of the stage.

When triode valves were used as RF amplifiers, in the early years of radio, this anode–grid capacitance caused unwanted RF instability, and the solution adopted was the introduction of a "screening" grid between the triode control grid and its anode, which reduced this anode–grid capacitance, in the case of a screened grid or tetrode valve, to some 0.025 pF.

A further effect that the inclusion of a screening grid had upon the valve characteristics was to make the anode current, in its linear region, almost independent of the anode voltage, which led to very high values for $R_a$ and $\mu$. Unfortunately, the presence of this grid caused a problem that when the anode voltage fell, during dynamic conditions, to less than that of the screening grid, electrons hitting the anode could cause secondary electrons to be ejected from its surface, especially if the anode was hot or its surface had been contaminated by cathode material, and these would be collected by the screening grid, which would cause a kink in the anode current/voltage characteristics. While this might not matter much in an RF amplifier, it would cause an unacceptable level of distortion if used in an audio amplifier stage.

Two solutions were found for this problem, of which the simplest was to interpose an additional, open mesh grid between the anode and the screening grid. This grid will normally be connected to the cathode, either externally or within the valve envelope, and is called the suppressor grid because it acts to suppress the emission of secondary electrons from the anode.

Since this type of valve had five electrodes it was called a "pentode." A typical small-signal pentode designed specifically for use in audio systems is the EF86, in which steps have also been taken to reduce the problem of microphony when the valve is used in the early stages of an amplifying system. The EF86 also has a wire mesh screen inside the glass envelope and surrounding the whole of the electrode structure. This is connected to pins 2 and 7 and is intended to lessen the influence of external voltage fields on the electron flow between the valve electrodes.

In use, a small-signal pentode amplifying stage will give a much higher stage gain than a medium impedance triode valve (250ˇ in comparison with, say, 30ˇ). It will also have a better HF gain due to its lower effective anode–grid capacitance. However, a triode gain stage will probably have a distortion figure, other things being equal, which is about half that of a pentode.

The second solution to the problem of anode current nonlinearity in tetrodes, particularly suited to the output stages of audio amplifiers, was alignment of the wires of the control grid and screening grid so that they constrained the electron flow into a series of beams, which served to sweep any secondary electrons back toward the anode—a process that was helped by the inclusion within the anode box of a pair of "beam-confining electrodes," which modified the internal electrostatic field pattern. These are connected to the cathode internally and take the form shown in Figure 11.5. These valves were called beam tetrodes or kinkless tetrodes and had a lower distortion than output pentodes. Valves of this type, such as the 6L6, the 807, and the KT66 and KT88, were widely employed in the output stages of the high-quality audio amplifiers of the 1950s and early 1960s.

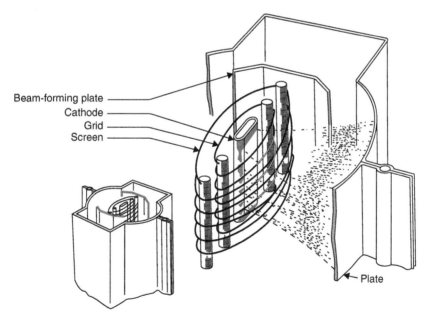

Beam-forming plate
Cathode
Grid
Screen

Plate

**Figure 11.5: Construction of a beam tetrode.**

(Courtesy of RCA.)

Both pentodes and beam tetrodes can be used with their screen grids connected to their anodes. In this mode their characteristics will resemble a triode having similar grid–cathode and grid–anode spacings to the grid–cathode and grid–screen grid spacings of the pentode. The most common use of this form of connection is in power output stages, where a triode connected beam tetrode will behave much like a power triode, without the need for a directly heated (and hum-inducing) cathode.

### 11.1.6 Valve Parameters

In addition to the anode current resistance, $R_a$, and the amplification factor, $\mu$, mentioned earlier, there is also the valve slope or mutual conductance ($g_m$), which is a measure of the extent to which the anode current will be changed by a change in grid voltage. Traditionally, this would be quoted in milliamperes per volt (mA/V or milli-Siemens, written as mS) and would be a useful indication of the likely stage gain given by the valve in an amplifying circuit.

This would be particularly helpful in the case of a pentode amplifying stage, where the value of $R_a$ would probably be very high in comparison with the likely value of load resistance. (For example, in the case of the EF86, $R_a$ is quoted as $2.5\,M\Omega$ and the $g_m$ is $2\,mA/V$.) In this case, the stage gain (M) can be determined, approximately, by the relationship $M \approx -g_m \cdot R_L$, which, for a 100k anode load would be $\approx -200^\vee$.

The various valve characteristics are defined mathematically as

$$R_a = dV_a/dI_a \qquad \text{at a constant grid voltage,}$$
$$g_m = -dI_a/dV_g \qquad \text{at a constant anode voltage, and}$$
$$\mu = -dV_a/dV_g \qquad \text{at a constant anode current.}$$

In these equations the negative sign takes account of the phase inversion of the signal. These parameters are related to one another by the further equation,

$$g_m = \mu/R_a \qquad \text{or} \qquad \mu = g_m R_a.$$

### 11.1.7 Gettering

Preservation of a high vacuum within its envelope is essential to the life expectancy and proper operation of the valve. However, it is difficult to remove all traces of residual gas on the initial pumping out of the envelope, quite apart from the small but continuing gas evolution from the cathode, or any other electrodes that may become hot in use. The solution to this problem is the inclusion of a small container, known as a boat, mounted somewhere within the envelope, but facing away from the valve electrodes, which contains a small quantity of reactive material, such as metallic calcium and magnesium.

The boat is positioned so that after the pumping out of the envelope has been completed, and the valve had been sealed off, the getter could be caused to evaporate on to the inner face of the envelope by heating the boat with an induction heating coil. Care is taken to ensure that as little as possible of the getter material finds its way on to the inner faces of the valve electrodes, where it may cause secondary emission, or on to the mica spacers, where it may cause leakage currents between the electrodes.

While this technique is reasonably effective in cleaning up the gas traces that arise during use of the valve, the vacuum is never absolute, and evidence of the residual gas can sometimes be seen as a faint, deep blue glow in the space within the anode envelope of a power output valve. If, however, there is a crack in the glass envelope, or some other cause of significant air leakage into the valve interior, this will become apparent because of a whitening of the edges of the normally dark, mirror-like surface of the getter deposit on the inside of the valve envelope. A further sign of the ingress of air into the valve envelope is the presence of a pinkish-violet glow that extends beyond the confines of the anode box. By this time the valve must be removed and discarded to prevent damage to other circuit components through an increasing and uncontrolled current flow.

### 11.1.8 Cathode and Heater Ratings

For optimum performance, the cathode temperature should be maintained, when in use, at its optimum value, which requires that the heater or filament voltages should be set at the correct levels. Since the voltage of the domestic AC power supply is not constant, the design ratings for the heater or filament supply must take account of this. However, this is not as difficult to do as it might appear. For example, Brimar, a well-known valve manufacturer, makes the following recommendations in their *Valve and Teletube Manual*: "the heater supply voltages should be within ±5% of the rated value when the heater transformer is fed with its nominal input voltage, provided that the mains power supply is within ±10% of its declared value."

An additional requirement is that, because of inevitable cathode-heater leakage currents, the voltages between these electrodes should be kept as low as possible and should not exceed 200 V. Moreover, there must always be a resistive path, not exceeding 250 kΩ, between the cathode and heater circuits.

As a practical point, the wiring of the heater circuit, which is usually operated at 6.3 V AC, will normally be installed as a twisted pair to minimize the induction of mains hum into sensitive parts of the system, as will the heater wiring inside the cathode tube of low noise valves, such as the EF86. With modern components, such as silicon diodes and low-cost regulator ICs, there is no good reason why the heater supplies to high-quality valve amplifiers should not be derived from smoothed and stabilized DC sources.

It has been suggested that the cathodes of valves can be damaged by reverse direction ionic bombardment if the HT voltage is applied before the cathode has had a chance to warm up and form a space charge, and that the valve heaters should be left on to avoid this problem. In practice, this problem does not arise because gaseous ions are only formed by collisions between residual gas molecules and the electrons in the anode current stream. If the cathode has not reached operating temperature there will be little or no anode current and, consequently, no gaseous ions produced as a result of it. Brimar specifically warns against leaving the cathode heated, in the absence of anode current, in that this may lead to cathode poisoning because of chemical reactions occurring between the exposed reactive metal of the cathode surface and any gaseous contaminants present within the envelope. Unfortunately, the loss of electron emissivity as the cathode temperature is reduced occurs more rapidly than the reduction in the chemical reactivity of the cathode metals.

Indirectly heated HT rectifier valves have been used, despite their lower operating efficiency, to ensure that the full HT voltage was not applied to the equipment before the other valves had warmed up. This was done to avoid the HT rail overvoltage surge that would otherwise occur and allow the safe use of lower working voltage and less expensive components, such as HT reservoir, smoothing, or intervalve coupling capacitors.

### 11.1.9 Microphony

Any physical vibration of the grid (or filament, in the case a directly heated cathode) will, by altering the grid–cathode spacing, cause a fluctuation of the anode current, which will cause an audible ringing sound when the envelope is tapped—an effect known as microphony in the case of a valve used in audio circuitry. Great care must therefore be taken in the manufacture of valves to maintain the firmness of the mounting of the grids and other electrodes. This is done by the use of rigid supporting struts whose ends are located in holes punched in stiff mica disc-shaped spacers, which, in turn, are a tight fit within the valve envelope.

Since a microphonic valve will pick up vibration from any sound source, such as a loudspeaker system in proximity to it, and convert these sounds into (inevitably distorted) electrical signals, which will be added to the amplifier output, this can be a significant, but unsuspected, source of signal distortion, which will not be revealed during laboratory testing on a resistive dummy load. Because it is difficult to avoid valve microphony completely, and it is equally difficult to sound proof amplifiers, this type of distortion will always occur unless such valve amplifier systems are operated at a low volume level or the amplifier is located in a room remote from the loudspeakers.

## 11.2 Solid-State Devices

### 11.2.1 Bipolar Junction Transistors

#### 11.2.1.1 'N'- and 'P'-Type Materials

Most materials can be grouped in one or other of three classes, insulators, semiconductors, or conductors, depending on the ease or difficulty with which electrons can pass through them. In insulators, all of the electrons associated with the atomic structure will be firmly bound in the valency bands of the material, whereas in good, usually metallic, conductors many of the atomic electrons will only be loosely bound and will be free to move within the body of the material.

In semiconductors, at temperatures above absolute zero (0°K or –273.15°C), electrons will exist both in the valency levels where they are not free to leave the atoms with which they are associated and in the conduction band in which they are free to travel within the body of the material. This characteristic is influenced greatly by the "doping" of the material, which is normally done during the manufacture of the semiconductor material by introducing carefully controlled amounts of specific impurities into the molten mass from which the single semiconductor crystal is grown. The most common semiconductor material in normal use is silicon because it is inexpensive, readily available, and has good thermal properties. Germanium, the material from which all early transistors were made, has electrical characteristics that are influenced greatly by its temperature, which is inconvenient in use. Also, it does not lend itself at all well to contemporary mass-production techniques.

In the case of silicon, which has very little conductivity in its undoped "intrinsic" form, the most common dopants are boron or aluminium, which give rise to a semiconductor with a deficiency of valency electrons, usually referred to as holes—called a 'P'-type material—or phosphorus, which will cause the silicon to have a surplus of valency electrons, which forces some of them into the conduction band. Such a semiconductor material would be termed 'N' type. Both P-type and N-type silicon can be quite highly conductive, depending on the doping levels used.

### 11.2.1.2 Fermi Levels

The electron energy distribution in single-crystal P- and N-type materials is shown in Figure 11.6, and the mean electron energy levels, known as the Fermi levels, are shown.

## 11.3 Valve Audio Amplifier Layouts

In its simplest form, shown in Figure 11.6, an audio amplifier consists of an input voltage amplifier stage (A) whose gain can be varied to provide the desired output signal level, an impedance converter stage (ZC) to adjust the output impedance of the amplifier to suit the load, which could be a loudspeaker, a pair of headphones, or the cutting head in a vinyl disc manufacturing machine.

In the case of headphones, their load impedance could be high enough for them to be driven directly by the voltage amplifier stage without a serious impedance mismatch, but with other types of output load it will be necessary to interpose some sort of impedance conversion device; in valve-operated audio systems this is most commonly an iron-cored

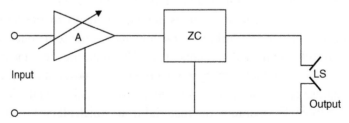

**Figure 11.6: An audio amplifier block diagram.**

audio frequency transformer. This is a difficult component to incorporate within a high-fidelity system, and much thought must be given both to its design and the way it is used in the circuit.

A very simple circuit layout embodying the structure outlined in Figure 11.6, using directly heated (battery operated) valves, is shown in Figure 11.7. This is the type of design that might have been built some 50 years ago by a technically minded youngster who wanted some means of driving a loudspeaker from a simple piezo-electric gramophone pick-up.

For the maximum transfer of power from an amplifier to its load it is necessary that both of these should have the same impedance, and since the anode resistance ($R_a$) of the output valve is of the order of $10\,\text{k}\Omega$, and the most common speech coil impedance of an inexpensive moving coil loudspeaker is $3\,\Omega$, there would be a drastic loss of available power unless some impedance converting output transformer was employed.

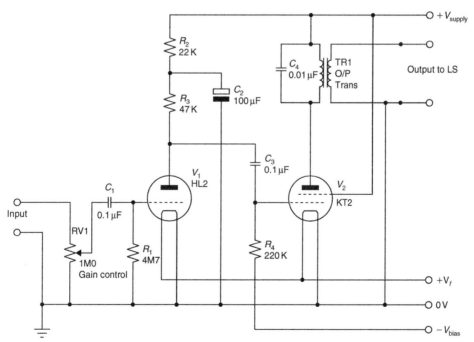

**Figure 11.7: A simple valve amplifier.**

The primary:secondary turns ratio of this component would need to be $\sqrt{(10k/3)} = 58:1$. It is difficult to design transformers having such high turns ratios without losses in performance; consequently, when higher audio quality was required, the LS manufacturers responded by making loudspeaker drive units with higher impedance speech coils. Before the advent of transistor-operated audio amplifiers the most common LS driver impedance was $15\,\Omega$.

With regard to the amplifier design shown, the input stage ($V_1$) uses a simple directly heated triode, with grid-current bias developed across the $4M7\,\Omega$ grid resistor, $R_1$. This is resistor/capacitor coupled to $V_2$, a small-power beam tetrode or pentode, operated with fixed bias derived from an external DC voltage source.

Because both $V_1$ and $V_2$ will contribute some distortion to the signal—in the case of $V_1$, this will mainly be second harmonic, but in the output valve ($V_2$) there will also be a substantial third harmonic component—the output signal will sound somewhat shrill due to the presence of these spurious high signal frequency components in the output. The simplest and most commonly adopted remedy for this defect was to connect a capacitor ($C_4$) across the primary of the output transformer (TR1) to roll-off the high-frequency response of the amplifier as a whole to give it the required mellow sound. The HT line decoupling capacitor ($C_2$) serves to reduce the amount of spurious and distorted audio signal, present on the $+V$ supply line, which will be added to the wanted signal present on the $V_2$ grid. An amplifier of this type would have an output power of, perhaps, $0.5\,W$, a bandwidth, mainly depending on the quality of the output transformer, that could be $150\,Hz - 6\,kHz$, $\pm 6\,dB$, and a harmonic distortion, at $1\,kHz$ and $0.4\,W$ output, of $10\%$.

The amplifier shown in Figure 11.7 uses a circuit of the kind that would allow operation from batteries, and it was accepted that such designs would have a low output power and a relatively poor performance in respect to its audio quality: this was the price paid for the low current drain on its power source. If, however, the amplifier was to be powered from an AC mains supply, the constraints imposed by the need to keep the total current demand low no longer applied, which gave the circuit designer much greater freedom. The other consideration in the progress toward higher audio power outputs was the type of output stage layout, in that this influenced the output stage efficiency, as examined later.

## 11.4 Single-Ended Versus Push–Pull Operation

These two options are shown schematically in Figure 11.8, in which $Q_1$ and $Q_2$ are notional amplifier blocks, simplified to the extent that they are only considered as being

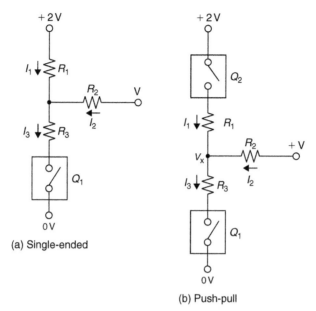

**Figure 11.8: Output arrangements.**

either open-circuit (O/C) or short-circuit (S/C), but with some internal resistance, shown as $R^3$ [or $R_1$ in the case of Figure 11.8(b)]. I have also adopted the convention that the current flow into the load resistor ($R_2$) is deemed to be positive when the amplifier circuit is feeding current (as a current source) into the load and to be negative when the amplifier is acting as a current sink and drawing current from $R_2$ and its associated power supply. I have also labeled the voltage at the junction of these three resistors as $V_x$. The efficiency of the system can be considered as related to the extent of the change in the current through $R_2$ brought about by the change from O/C to S/C in $Q_1$ or $Q_2$.

If we consider first the single-ended layout of Figure 11.8(a), when $Q_1$ is O/C, the current flow into $R_2$ is only through $R_1$ and $i_2 = V/(R_1 + R_2)$. If, however, $Q_1$ is short circuited, S/C, then, from inspection,

$$i_3 = i_1 + i_2 \tag{11.1}$$

$$\text{but } i_2 = \frac{V - V_x}{R_2} \tag{11.2}$$

$$i_1 = \frac{2V - V_x}{R_1} \tag{11.3}$$

and $\quad i_3 = \dfrac{V_x}{R_3}$ 
$$\tag{11.4}$$

from Equations (11.1), (11.2), and (11.3) we have

$$i_3 = \frac{2V - V_x}{R_1} + \frac{V - V_x}{R_2} \tag{11.5}$$

but it has been seen from Equation (11.4) that $i_3 = V_x/R_3$

$$\text{therefore } \frac{2V - V_x}{R_1} + \frac{V - V_x}{R_2} = \frac{V_x}{R_3}. \tag{11.6}$$

If we insert the actual values for $R_1$, $R_2$, and $R_3$, we can discover the difference in output current flow in the load resistor ($R_2$) between the O/C and S/C conditions of $Q_1$. For example, if all resistors are $10\,\Omega$ in value, when $Q_1$ is S/C, $V_x$ will be equal to $V$, and there will be no current flow in $R_2$ and the change on making $Q_1$ O/C will be (V/20)A. If $R_1$ and $R_2$ are $10\,\Omega$ in value and $R_3$ is $5\,\Omega$, then the current flow in $R_2$, when $Q_1$ is O/C, will still be (V/20)A, whereas when $Q_1$ is S/C, the current will be (–0.25 V/10)A and the change in current will be (3 V/40)A. By comparison, for the push–pull system of Figure 11.8(b), the change in current through $R_2$, when this is $10\,\Omega$ and both $R_1$ and $R_3$ are $5\,\Omega$ in value, on the alteration in the conducting states of $Q_1$ and $Q_2$, will be (2 V/15)A, which is nearly twice as large.

The increase in available output power from similar output valves when operated at the same V+ line voltage in a push–pull rather than in a single-ended layout is the major advantage of this arrangement, although if the output devices have similar distortion characteristics, and the output transformer is well made, the even harmonic distortion components will tend to cancel. Also, the magnetization of the core of the output transformer due to the valve anode currents flowing in the two halves of the primary winding will be reduced substantially because the induced fields will be in opposition.

In addition, an increase in the drive voltage to the grids of the output valves, provided that it is not large enough to drive them into grid current, will, by reducing their equivalent series resistance ($R_1$, $R_3$ in the calculations given earlier), increase the available output power, whereas in the single-ended layout the dynamic drive current cannot be increased beyond twice the quiescent level without running into waveform clipping. However, there are other problems, which are discussed later.

## 11.5 Phase Splitters

In order to drive a pair of output valves in push–pull it is necessary to generate a pair of AC control grid drive voltages that are equal in magnitude but in phase opposition. The simplest way of doing this is to use a transformer as the anode load for an amplifier stage, but with a center-tapped secondary winding.

Figure 11.9 shows a typical center-tapped, transformer-coupled 20W audio amplifier, of the kind that would have been common in the period spanning the late 1930s to early 1940s. Because there are two coupling transformers in the signal path from the input and the LS output, which would cause substantial phase shifts at the ends of the audio spectrum, it would be impractical to try to clean up the amplifier's relatively poor performance by applying overall negative feedback (from the LS output to $V_1$ cathode) to the system.

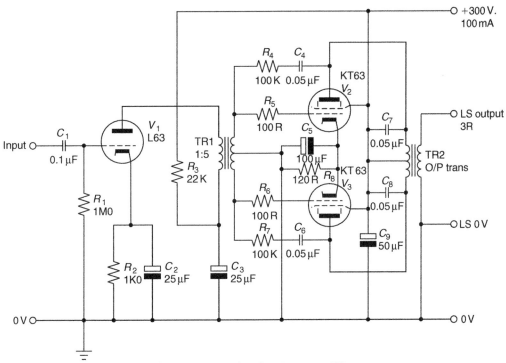

**Figure 11.9: A simple 20-W amplifier.**

Some local negative feedback from anodes to grids in $V_2$ and $V_3$ is applied by way of $C_4/R_4$ and $C_6/R_7$ in an attempt to reduce the third, and other odd-order, harmonic distortion components generated by the output valves. Since the designer expected that the output sound quality could still be somewhat shrill, a pair of 0.05-$\mu$F capacitors, $C_7$ and $C_8$, has been added across the two halves of the output transformer primary windings to reduce the high-frequency performance. These would also have the effect of lessening the tendency of the output valves to flash over if the amplifier was driven into an open-circuited LS load—an endemic problem in designs without the benefit of overall negative feedback to stabilize the output voltage.

The anode voltage decoupling circuit ($R_3$, $C_3$), shown in Figure 11.9, is essential to prevent the spurious signal voltages from the $+V$ supply line to the output valves being introduced to the output valve grid circuits. This would, in the absence of the supply line decoupling circuit, cause the amplifier to oscillate continuously at some low frequency—a problem that was called motorboating, from the sound produced in the loudspeakers.

Various circuit arrangements have been proposed as a means of generating a pair of low distortion, low phase shift, push–pull drive voltages. Of these, the phase inverter circuit of Figure 11.10 is the simplest, but does not offer a very high-quality performance. It is, in principle, a bad thing to attenuate and then to amplify again, as is done in this

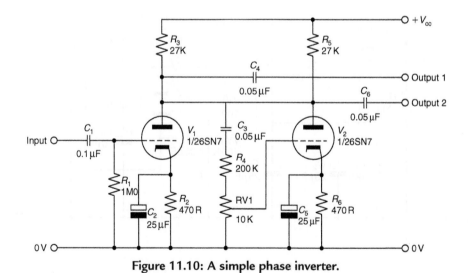

**Figure 11.10: A simple phase inverter.**

arrangement, because this simply adds just another increment of waveform distortion, due to $V_2$, to that contributed by $V_1$.

A much more satisfactory arrangement is that shown in Figure 11.11, in which $V_2$ is operated as an anode follower, which, like the cathode follower, employs 100% negative feedback, although in this case derived from the anode. This stage contributes very little waveform distortion. Also, because both valves operate as normal amplifier stages, the available voltage from either output point will be largely unaffected by the operation of the circuit. An additional advantage over the circuit shown in Figure 11.10 is that the two antiphase output voltages are equal in magnitude, without the need to adjust the preset gain control, RV1.

Another satisfactory circuit is that based on the long-tailed pair layout, in which, provided that the tail resistor is large in relation to the cathode source resistance ($1/g_m$), the two antiphase anode currents will be closely similar in magnitude. The advantage of this circuit is that it can be direct coupled (i.e., without the need for a DC blocking coupling capacitor) to the output of the preceding stage, which minimizes circuit phase shifts, especially at the LF end of the passband. By comparison with the two preceding

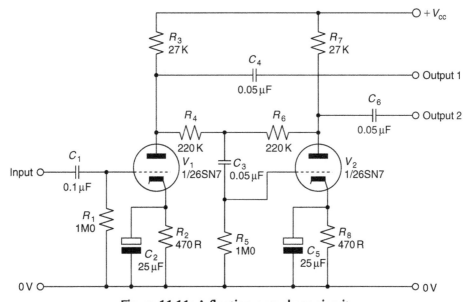

**Figure 11.11: A floating paraphase circuit.**

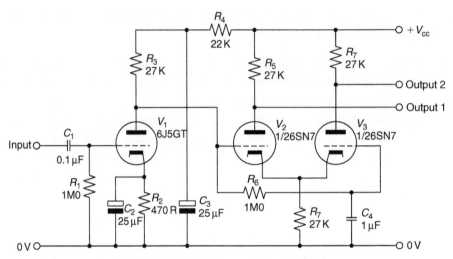

**Figure 11.12: A long-tailed pair circuit.**

phase-splitter circuits, it has the disadvantage that the available AC output swing, at either anode, is reduced greatly by the fact that the cathode voltages of $V_2$ and $V_3$ are considerably positive in relation to the 0-V line, which will almost certainly require an additional amplifier stage between its output and the input of any succeeding triode or beam-tetrode output stage.

This disadvantage is shared by the circuit layout shown in Figure 11.13, in which a direct-coupled triode amplifier is operated with identical value resistive loads in both its anode and cathode circuits. Because of the very high level of negative feedback due to the cathode resistor, both the distortion and the unwanted phase shifts introduced by this stage are very low. Significantly, this was the type of phase splitter adopted by D. T .N. Williamson in his classic 15-W audio amplifier design.

## 11.6 Output Stages

The basic choice of output valves will lie between a triode, a beam tetrode, or a pentode. If large output powers are required—say, in excess of 2 W—triode output valves are unsuitable because the physical spacing between the control grid and the anode must be small, and the grid mesh must be relatively widely spaced, in order to achieve a low anode current resistance and a high practicable anode current level. This closely packed

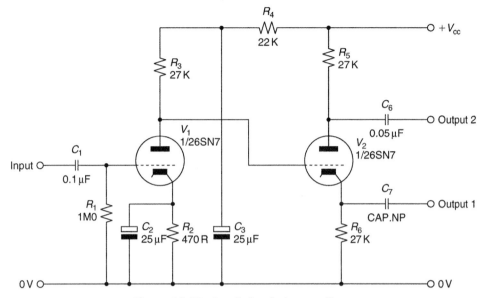

**Figure 11.13: A split load phase splitter.**

type of construction will lead to the almost complete stripping of the space charge from the region between the cathode and the grid. Experience shows that the life expectancy of cathodes operated under such conditions is short, and the only way by which this problem can be avoided is by the use of a directly heated (filament type) cathode construction, which is much more prolific as a source of electrons, and this leads to other difficulties such as hum intrusion from the AC heater supplies, and the awkwardness of arranging cathode bias systems.

So, if it is required to use a triode output stage, at anything greater than the 50-mA anode current obtainable from a parallel connected 6N7 double triode (the 6SN7 has a smaller envelope and, in consequence, a lower permissible anode dissipation), a directly heated valve such as the now long obsolete 6B4 or PX25 would need to be found. Therefore, in practice, the choice for output valves will be between output beam tetrodes or pentodes. Although a fairly close simulation of a triode characteristic can be obtained in both of these valve types if the anode and $G_2$ are connected together, this approach works better with a beam tetrode than a power output pentode because the presence of the suppressor grid in the pentode somewhat disturbs the anode current flow.

The required grid drive voltage for typical pentode or beam tetrode output valves, at $V_a = 300\,V$, will be in the range of 20–50 $V_{p-p}$ for each output valve, and whether or not the valve is triode connected has little effect on this requirement. The triode connection does, however, greatly affect the anode current impedance, which is reduced, in the case of the KT88, from $12\,k\Omega$ to $670\,\Omega$, and the need for a lower turns ratio greatly simplifies the design of the necessary, load-matching, output transformer with low half-primary to half-primary and primary to secondary leakage inductances.

## 11.7  Output (Load-Matching) Transformer

This component is probably the most important factor in determining the quality of the sound given by a valve-operated audio amplifier, and the performance of this component is influenced by a number of factors, both mechanical and electrical, which will become of critical importance if an attempt is made to apply negative feedback (NFB) over the whole amplifier. However, for a low power system, such as might be used as a headphone amplifier, it is possible to make a quite decent sounding system without the need for

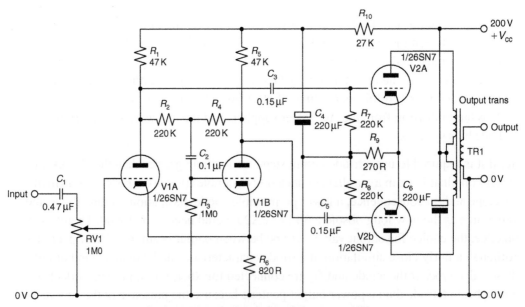

**Figure 11.14: A simple headphone amplifier.**

much in the way of exotic components, circuit complexity, or very high-quality output impedance-matching transformers, and I have sketched out in Figure 11.14 a typical circuit for a two-valve, 1 W headphone amplifier based on a pair of 6SN7s or equivalents.

In this design the input pair of valves acts as a floating paraphase phase-splitter circuit, which provides the drive for the output valves. Since the cathode currents from the two input valves are substantially identical, but opposite in phase, it is unnecessary to provide a cathode bypass capacitor to avoid loss of stage gain. Also, since this cathode resistor is common to both valves, it assists in reducing any differences between the two output signals, as the arrangement acts, in part, as a long-tailed pair circuit such as that shown in Figure 11.12. Since the total harmonic distortion from a push–pull pair of triodes will probably be less than 0.5% and will decrease as the output power is reduced, provided a reasonable quality output transformer is used, I have not included any overall NFB, which avoids any likely instability problems. To match the output impedances of V2A and V2B to a notional load impedance of $100\,\Omega$, a transformer turns ratio, from total primary to secondary, of 12:1 is required.

In more ambitious systems, in which NFB is used to improve the performance of the amplifier and reduce the distortion introduced by the output transformer, much more care is needed in the design of the circuit. In particular, the phase shifts in the signal that are introduced by the output transformer become very important if a voltage is to be derived from its output and fed back in antiphase to the input of the amplifier, in that to avoid instability the total phase angle within the feedback loop must not exceed $180°$ at any frequency at which the loop gain is greater than unity. This requirement can be met by both limiting the amount of NFB that is applied, which would, of course, limit its effectiveness, and controlling the gain/frequency characteristics of the system.

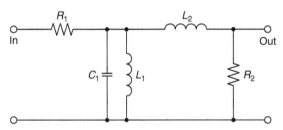

Figure 11.15: Equivalent circuits of idealized coupling transformer.

Although there are a number of factors that determine the phase shifts within the transformer, the two most important are the inductance of the primary winding and the leakage inductance between primary and secondary; a simple analysis of this problem, based on an idealized, loss-free transformer, can be made by reference to Figure 11.15. In this, $R_1$ is the effective input resistance seen by the transformer, made up of the anode current resistance of the valve, in parallel with the effective load resistance, and $L_1$ is the inductance of the transformer primary winding. When the signal frequency is lowered, a frequency will be reached at which there will be an attenuation of 3 dB and a phase shift of 60°. This will occur when $R_1 = j\omega L_1$, where $\omega$ is the frequency in radians per seconds.

$R_2$ is the secondary load resistance, which is the sum of the resistance reflected through the transformer and the anode resistance, and $L_2$ is the primary leakage inductance—a term that denotes the lack of total inductive coupling between primary and secondary windings—which behaves like an inductance between the output and the load and introduces an attenuation, and associated phase shift, at the HF end of the passband. The HF –3-dB gain point, at which the phase shift will be 60°, will occur at a frequency at which $R_2 = j\omega L_2$.

To see what these figures mean, consider the case of a 15-$\Omega$ resistive load, driven by a triode-connected KT66 that has an anode current resistance of 1000 ohms. Let us assume that, in order to achieve a low anode current distortion figure, it has been decided to provide an anode load of 5000 ohms. The turns ratio required will be $\sqrt{(5000/15)} = 18.25{:}1$ and the effective input resistance ($R_1$) due to the output load reflected through the transformer will be 833 ohms. If it is decided that the transformer shall have an LF −3-dB point at 10 Hz, then the primary inductance would need to be $833/2\pi10 = 833/62.8 = 13.26$ H. If it is also decided that the HF –3-dB point is to be 50 kHz, then the leakage inductance must be $833/2\pi50{,}000 = 2.7$ mH. The interesting feature here is that if an output pentode is used, which has a much higher value of $R_a$ than a triode, not only will a higher primary inductance be required, but the leakage inductance can also be higher for the same HF phase error.

Unfortunately, a number of other factors affect the performance of the transformer. The first of these is the dependence of the permeability of the core material on the magnetizing flux density, as shown in Figure 11.16. Since the current through the windings in any audio application is continually changing, so therefore is the permeability, and with it the winding inductances and the phase errors introduced into the feedback loop. Williamson

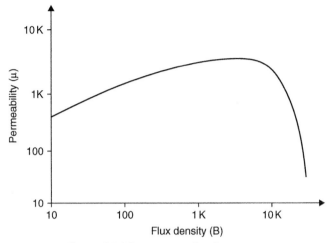

**Figure 11.16: A magnetization curve.**

urged that, for good LF stability, the value of permeability, μ, for low values of *B* should be used for primary inductance calculations.

Second, this change in inductance, as a function of current in the windings, is a source of transformer waveform distortion, as are—especially at high frequencies—the magnetic hysteresis of the core material and the eddy current losses in the core. These problems are exacerbated by the inevitable DC resistance of the windings and provide another reason, in addition to that of improved efficiency, for keeping the winding resistance as low as possible.

The third problem is that the permeability of the core material falls dramatically, as seen in Figure 11.16, if the magnetization force exceeds some effective core saturation level. This means that the cross-sectional area of the core (and the size and weight of the transformer) must be adequate if a distortion-generating collapse in the transformer output voltage is not to occur at high signal levels. The calculations here are essentially the same as those made to determine the minimum turns per volt figure permissible for the windings of a power transformer.[1]

In practical terms, the requirements of high primary inductance and low leakage inductance are conflicting and require that primary winding is divided into a number of sections between which portions of the secondary winding are interleaved. Williamson proposed that eight secondary segments should be placed in the gaps left between 10 primary windings. This increases the stray capacitance, $C_1$, across the primary winding and between primary and secondary coils. However, the HF phase errors introduced by these will probably be unimportant within the design frequency spectrum.

## 11.8 Effect of Output Load Impedance

This is yet another area in which there is a conflict in design requirements, between output power and output stage distortion. Figure 11.17(a), shows the output power

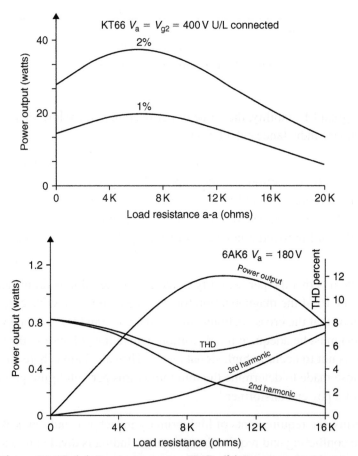

Figure 11.17: (a) Power output vs THD. (b) Power output curve.

given for 1 and 2% THD values by a push–pull pair of U/L-connected KT88s in relation to the anode to anode load impedance chosen by the designer. These data are courtesy of the GEC[2] Since the distortion can also alter in its form as a function of load impedance, Figure 11.17(b) shows the way these circuit characteristics change as the load resistance changes. The figures given for a single-ended 6AK6 output pentode are due to Langford-Smith.[3]

## 11.9  Available Output Power

The power available from an audio amplifier, for a given THD figure, is an important aspect of the design. Although there are a number of factors that will influence this, such as the maximum permitted anode voltage or the maximum allowable cathode current, the first of these that must be considered is the permissible thermal dissipation of the anode of the valve. These limiting values are quoted in the manufacturers' handbooks, and from these it is possible to draw a graph of the kind shown in Figure 11.18, where the maximum permitted combinations of anode current and anode voltage result in the curved (dashed) line indicating the dissipation limits for the valve, and the load line for its particular operating conditions can then be superimposed on this graph to confirm that the proposed working conditions will be within these thermal limits.

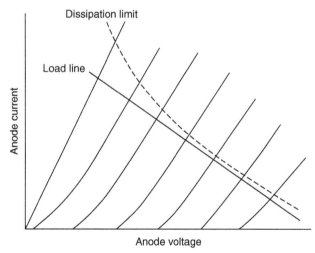

**Figure 11.18: An anode dissipation curve.**

## References

1. Linsley Hood, J., *The art of linear electronics*, Butterworth-Heinemann, pp. 29–31, 1993.

2. 'GEC', *Audio Amplifier Design*, 1957, p. 41.

3. Langford-Smith, *Radio designers handbook*, 4th Ed., 1954, p. 566.

# CHAPTER 12

# Negative Feedback

Douglas Self

It is difficult to convince people that this frequency is of no relevance whatsoever to the speed of amplifiers and that it does not affect the slew rate. Nonetheless, it is so, and any first-year electronics textbook will confirm this. High-gain op-amps with sub-1-Hz bandwidths and blindingly fast slewing are as common as the grass (if somewhat less expensive) and if that does not demonstrate the point beyond doubt then I really do not know what will.

*Limited open-loop bandwidth prevents the feedback signal from immediately following the system input, so the utility of this delayed feedback is limited.* No linear circuit can introduce a pure time delay; the output must begin to respond at once, even if it takes a long time to complete its response. In the typical amplifier the dominant-pole capacitor introduces a 90° phase shift between input pair and output at all but the lowest audio frequencies, but this is not a true time delay. The phrase *delayed feedback* is often used to describe this situation, and it is a wretchedly inaccurate term; if you really delay the feedback to a power amplifier (which can only be done by adding a time constant to the feedback network rather than the forward path) it will quickly turn into the proverbial power oscillator as sure as night follows day.

## 12.1 Amplifier Stability and Negative Feedback

In controlling amplifier distortion, there are two main weapons. The first is to make the linearity of the circuitry as good as possible before closing the feedback loop. This is unquestionably important, but it could be argued that it can only be taken so far before the complexity of the various amplifier stages involved becomes awkward. The second is to apply as much negative feedback (NFB) as possible while maintaining amplifier stability. It is well known that an amplifier with a single time constant is always stable, no matter how high the feedback factor. The linearization of the VAS by local Miller feedback is a good example. However, more complex circuitry, such as the generic three-stage power amplifier, has more than one time constant, and these extra poles will cause poor transient response or instability if a high feedback factor is maintained up to the higher frequencies where they start to take effect. It is therefore clear that if these higher poles can be eliminated or moved upward in frequency, more feedback can be applied and distortion will be less for the same stability margins. Before they can be altered—if indeed this is practical at all—they must be found and their impact assessed.

The dominant pole frequency of an amplifier is, in principle, easy to calculate; the mathematics are very simple. In practice, two of the most important factors, the effective beta of the VAS and the VAS collector impedance, are only known approximately, so the dominant pole frequency is a rather uncertain thing. Fortunately, this parameter in itself has no effect on amplifier stability. What matters is the amount of feedback at high frequencies.

Things are different with the higher poles. To begin with, where are they? They are caused by internal transistor capacitances and so on, so there is no physical component to show where the roll-off is. It is generally regarded as fact that the next poles occur in the output stage, which use power devices that are slow compared with small-signal transistors. Taking the Class-B design, the TO-92 MPSA06 devices have an Ft of 100 MHz, the MJE340 drivers about 15 MHz (for some reason this parameter is missing from the data sheet), and the MJ802 output devices an Ft of 2.0 MHz. Clearly the output stage is the prime suspect. The next question is at what frequencies these poles exist. There is no reason to suspect that each transistor can be modeled by one simple pole.

There is a huge body of knowledge devoted to the art of keeping feedback loops stable while optimizing their accuracy; this is called control theory, and any technical bookshop will yield some intimidatingly fat volumes called things like "control system design." Inside, system stability is tackled by Laplace-domain analysis, eigenmatrix methods, and joys such as the Lyapunov stability criterion. I think that makes it clear that you need to be pretty good at mathematics to appreciate this kind of approach.

Even so, it is puzzling that there seems to have been so little application of control theory to audio amplifier design. The reason may be that so much control theory assumes that you know fairly accurately the characteristics of what you are trying to control, especially in terms of poles and zeros.

One approach to appreciating NFB and its stability problems is SPICE simulation. Some SPICE simulators have the ability to work in the Laplace or s-domain, but my own experiences with this have been deeply unhappy. Otherwise respectable simulator packages output complete rubbish in this mode. Quite what the issues are here I do not know, but it does seem that s-domain methods are best avoided. The approach suggested here instead models poles directly as poles, using RC networks to generate the time constants. This requires minimal mathematics and is far more robust. Almost any SPICE simulator, evaluation versions included, should be able to handle the simple circuit used here.

Figure 12.1 shows the basic model, with SPICE node numbers. The scheme is to idealize the situation enough to highlight the basic issues and exclude distractions such as nonlinearities or clipping. The forward gain is simply the transconductance of the input stage multiplied by the transadmittance of the VAS integrator. An important point is that with correct parameter values, the current from the input stage is realistic, as are all the voltages.

The input differential amplifier is represented by G. This is a standard SPICE element—the VCIS, or voltage-controlled current source. It is inherently differential, as the output current from Node 4 is the scaled difference between the voltages at Nodes 3 and 7. The scaling factor of 0.009 sets the input stage transconductance ($g_m$) to 9 mA/V, a typical

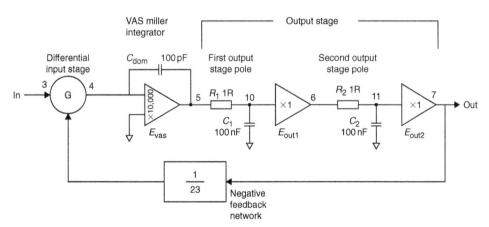

**Figure 12.1: Block diagram of system for SPICE stability testing.**

figure for a bipolar input with some local feedback. Stability in an amplifier depends on the amount of NFB available at 20 kHz. This is set at the design stage by choosing the input $g_m$ and $C_{dom}$, which are the only two factors affecting the open-loop gain. In simulation it would be equally valid to change $g_m$ instead; however, in real life it is easier to alter $C_{dom}$ as the only other parameter this affects is slew rate. Changing input stage transconductance is likely to mean altering the standing current and the amount of local feedback, which will in turn impact input stage linearity.

The VAS with its dominant pole is modeled by the integrator $E_{vas}$, which is given a high but finite open-loop gain, so there really is a dominant pole $P_1$ created when the gain demanded becomes equal to that available. With $C_{dom} = 100\,pF$, this is below 1 Hz. With infinite (or as near-infinite as SPICE allows) open-loop gain, the stage would be a perfect integrator. As explained elsewhere, the amount of open-loop gain available in real versions of this stage is not a well-controlled quantity, and $P_1$ is liable to wander about in the 1- to 100-Hz region; fortunately, this has no effect at all on HF stability. $C_{dom}$ is the Miller capacitor that defines the transadmittance, and since the input stage has a realistic transconductance, $C_{dom}$ can be set to 100 pF, its usual real-life value. Even with this simple model we have a nested feedback loop. This apparent complication here has little effect, as long as the open-loop gain of the VAS is kept high.

The output stage is modeled as a unity-gain buffer, to which we add extra poles modeled by $R_1$, $C_1$ and $R_2$, $C_2$. $E_{out1}$ is a unity-gain buffer internal to the output stage model, added so that the second pole does not load the first. The second buffer $E_{out2}$ is not strictly necessary as no real loads are being driven, but it is convenient if extra complications are introduced later. Both are shown here as a part of the output stage but the first pole could equally well be due to input stage limitations instead; the order in which the poles are connected makes no difference to the final output. Strictly speaking, it would be more accurate to give the output stage a gain of 0.95, but this is so small a factor that it can be ignored.

The component values here are of course completely unrealistic and chosen purely to make the math simple. It is easy to appreciate that $1\,\Omega$ and $1\,\mu F$ make up a 1-$\mu s$ time constant. This is a pole at 159 kHz. Remember that the voltages in the latter half of the circuit are realistic, but the currents most certainly are not.

The feedback network is represented simply by scaling the output as it is fed back to the input stage. The closed-loop gain is set to 23 times, which is representative of most power amplifiers.

Note that this is strictly a linear model, so that the slew-rate limiting associated with Miller compensation is not modeled here. It would be done by placing limits on the amount of current that can flow in and out of the input stage.

Figure 12.2 shows the response to a 1-V step input, with the dominant pole the only time element in the circuit. (The other poles are disabled by making $C_1$, $C_2$ 0.00001 pF, because this is quicker than changing the actual circuit.) The output is an exponential rise to an asymptote of 23 V, which is exactly what elementary theory predicts. The exponential shape comes from the way that the error signal that drives the integrator becomes less as the output approaches the desired level. The error, in the shape of the output current from G, is the smaller signal shown; it has been multiplied by 1000 to get mA onto the same scale as volts. The speed of response is inversely proportional to the size of $C_{dom}$ and is shown here for values of 50 and 220 pF as well as the standard 100 pF. This simulation technique works well in the frequency domain, as well as the time domain. Simply tell SPICE to run an AC simulation instead of a TRANS (transient) simulation. The frequency response in Figure 12.3 exploits this to show how the closed-loop gain in an

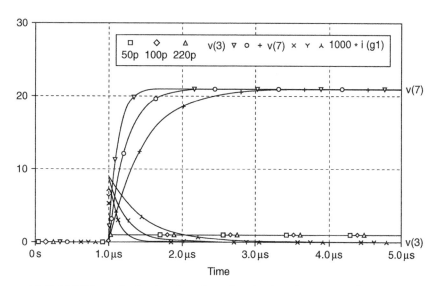

**Figure 12.2: SPICE results in the time domain. As $C_{dom}$ increases, the response V(7) becomes slower and the error (g1) declines more slowly. The input is the step-function V(3) at the bottom.**

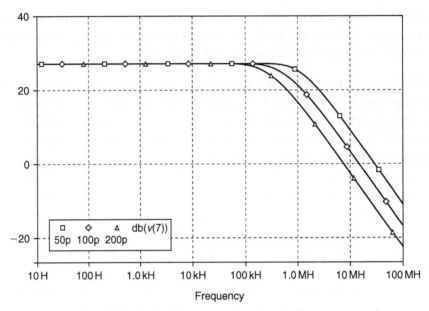

**Figure 12.3: SPICE simulation in the frequency domain. As the compensation capacitor is increased, the closed-loop bandwidth decreases proportionally.**

NFB amplifier depends on the open-loop gain available. Once more elementary feedback theory is brought to life. The value of $C_{dom}$ controls the bandwidth, and it can be seen that the values used in the simulation do not give a very extended response compared with a 20-kHz audio bandwidth.

In Figure 12.4, one extra pole $P_2$ at 1.59 MHz (a time constant of only 100 ns) is added to the output stage, and $C_{dom}$ stepped through 50, 100, and 200 pF as before. 100 pF shows a slight overshoot that was not there before; with 50 pF there is a serious overshoot that does not bode well for the frequency response. Actually, it's not that bad; Figure 12.5 returns to the frequency-response domain to show that an apparently vicious overshoot is actually associated with a very mild peaking in the frequency domain.

From here on $C_{dom}$ is left set to 100 pF, which is its real value in most cases. In Figure 12.6, $P_2$ is stepped instead, increasing from 100 ns to 5 µs, and while the response gets slower and shows more overshoot, the system does not become unstable. The reason is simple: sustained oscillation (as opposed to transient ringing) in a feedback loop requires

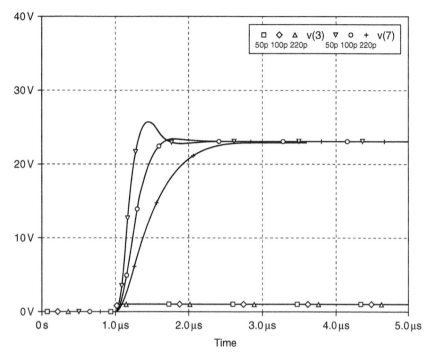

**Figure 12.4: Adding a second pole $P_2$ causes overshoot with smaller values $C_{dom}$, but cannot bring about sustained oscillation.**

positive feedback, which means that a total phase shift of 180° must have accumulated in the forward path and reversed the phase of the feedback connection. With only two poles in a system the phase shift cannot reach 180°. The VAS integrator gives a dependable 90° phase shift above $P_1$, being an integrator, but $P_2$ is instead a simple lag and can only give a 90° phase lag at infinite frequency. So, even this very simple model gives some insight. Real amplifiers do oscillate if $C_{dom}$ is too small, so we know that the frequency response of the output stage cannot be modeled meaningfully with one simple lag.

A certain president of the United States is alleged to have said: "Two wrongs don't make a right—so let's see if three will do it." Adding in a third pole, $P_3$, in the shape of another simple lag gives the possibility of sustained oscillation.

Stepping the value of $P_2$ from 0.1 to 5 μs with $P_3 = 500$ ns shows sustained oscillation starting to occur at $P_2 = 0.45$ μs. For values such as $P_2 = 0.2$ μs, the system is stable

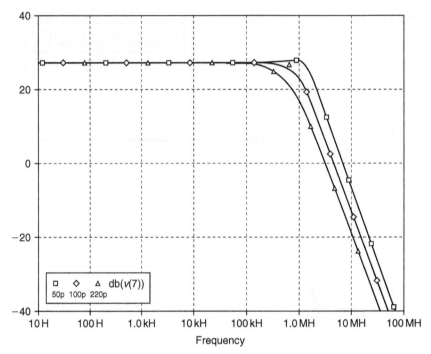

**Figure 12.5: The frequency responses that go with the transient plots of Figure 12.4. The response peaking for $C_{dom} = 50\,pF$ is very small compared with the transient overshoot.**

and shows only damped oscillation. Figure 12.7 shows over 50 μs what happens when the amplifier is made very unstable (there are degrees of this) by setting $P_2 = 5\,\mu s$ and $P_3 = 500\,ns$. It still takes time for the oscillation to develop, but exponentially diverging oscillation like this is a sure sign of disaster. Even in the short time examined here the amplitude has exceeded a rather theoretical half a kilovolt. In reality, oscillation cannot increase indefinitely, if only because the supply rail voltages would limit the amplitude. In practice, slew rate limiting is probably the major controlling factor in the amplitude of high-frequency oscillation.

We have now modeled a system that will show instability. But does it do it right? Sadly, no. The oscillation is about 200 kHz, which is a rather lower frequency than is usually seen when an amplifier misbehaves. This low frequency stems from the low $P_2$ frequency we have to use to provoke oscillation; apart from anything else this seems out of line with the

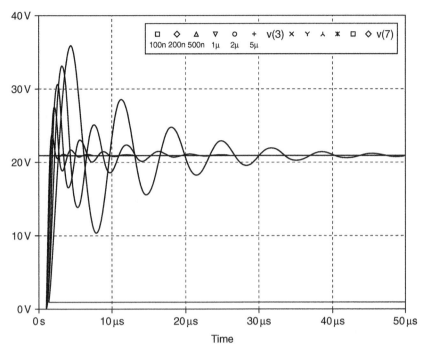

**Figure 12.6: Manipulating the $P_2$ frequency can make ringing more prolonged but it is still not possible to provoke sustained oscillation.**

known Ft of power transistors. Practical amplifiers are likely to take off at around 500 kHz to 1 MHz when $C_{dom}$ is reduced, which seems to suggest that a phase shift is accumulating quickly at this sort of frequency. One possible explanation is that there are a large number of poles close together at a relatively high frequency.

A fourth pole can be simply added to Figure 12.1 by inserting another RC–buffer combination into the system. With $P_2 = 0.5\,\mu s$ and $P_3 = P4 = 0.2\,\mu s$, instability occurs at 345 kHz, which is a step toward a realistic frequency of oscillation. This is case B in Table 12.1.

When a fifth output stage pole is grafted on, so that $P_3 = P_4 = P_5 = 0.2\,\mu s$, the system just oscillates at 500 kHz with $P_2$ set to 0.01 $\mu s$. This takes us close to a realistic frequency of oscillation. Rearranging the order of poles so that $P_2 = P_3 = P_4 = 0.2\,\mu s$, while $P_5 = 0.01\,\mu s$, is tidier and the stability results are of course the same; this is a linear system so the order does not matter. This is case C in Table 12.1.

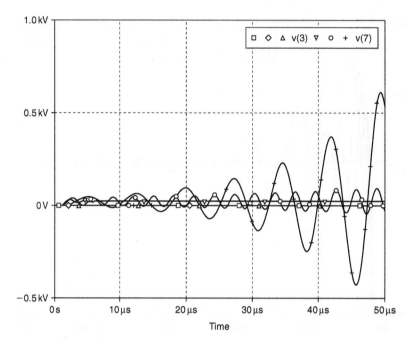

**Figure 12.7: Adding a third pole makes possible true instability with an exponentially increasing amplitude of oscillation. Note the unrealistic voltage scale on this plot.**

**Table 12.1: Instability Onset: $P_2$ is Increased Until Sustained Oscillation Occurs**

| Case | $C_{dom}$ | $P_2$ | $P_3$ | $P_4$ | $P_5$ | $P_6$ | |
|------|-----------|-------|-------|-------|-------|-------|---------|
| A | 100p | 0.45 | 0.5 | – | – | | 200 kHz |
| B | 100p | 0.5 | 0.2 | 0.2 | – | | 345 kHz |
| C | 100p | 0.2 | 0.2 | 0.2 | 0.01 | | 500 kHz |
| D | 100p | 0.3 | 0.2 | 0.1 | 0.05 | | 400 kHz |
| E | 100p | 0.4 | 0.2 | 0.1 | 0.01 | | 370 kHz |
| F | 100p | 0.2 | 0.2 | 0.1 | 0.05 | 0.02 | 475 kHz |

Having $P_2$, $P_3$, and $P_4$ all at the same frequency does not seem very plausible in physical terms, so case D shows what happens when the five poles are staggered in frequency. $P_2$ needs to be increased to 0.3 μs to start the oscillation, which is now at 400 kHz. Case E is another version with five poles, showing that if $P_5$ is reduced, $P_2$ needs to be doubled to 04 μs for instability to begin.

In the final case F, a sixth pole is added to see if this permitted sustained oscillation is above 500 kHz. This seems not to be the case; the highest frequency that could be obtained after a lot of pole twiddling was 475 kHz. This makes it clear that this model is of limited accuracy (as indeed are all models—it is a matter of degree) at high frequencies and that further refinement is required to gain further insight.

## 12.2 Maximizing Negative Feedback

Having freed ourselves from fear of feedback, and appreciating the dangers of using only a little of it, the next step is to see how much can be used. It is my view that the amount of NFB applied should be maximized at all audio frequencies to maximize linearity, and the only limit is the requirement for reliable HF stability. In fact, global or Nyquist oscillation is not normally a difficult design problem in power amplifiers; the HF feedback factor can be calculated simply and accurately, and set to whatever figure is considered safe. (Local oscillations and parasitics are beyond the reach of design calculations and simulations and cause much more trouble in practice.)

In classical control theory, the stability of a servomechanism is specified by its phase margin, the amount of extra phase shift that would be required to induce sustained oscillation, and its gain margin, the amount by which the open-loop gain would need to be increased for the same result. These concepts are not very useful in amplifier work, where many of the significant time constants are known only vaguely. However, it is worth remembering that the phase margin will never be better than 90° because of the phase lag caused by the VAS Miller capacitor; fortunately, this is more than adequate.

In practice, the designer must use his judgment and experience to determine an NFB factor that will give reliable stability in production. My own experience leads me to believe that when the conventional three-stage architecture is used, 30 dB of global feedback at 20 kHz is safe, providing an output inductor is used to prevent capacitive loads from eroding the stability margins. I would say that 40 dB was distinctly risky, and I would not care to pin it down any more closely than that.

The 30-dB figure assumes simple dominant-pole compensation with a 6-dB/octave roll-off for the open-loop gain. The phase and gain margins are determined by the angle at which this slope cuts the horizontal unity loop-gain line. (I am deliberately terse here; almost all textbooks give a very full treatment of this stability criterion.) An intersection of 12 dB/octave is definitely unstable. Working within this, there are two basic ways in which to maximize the NFB factor.

1. While a 12-dB/octave gain slope is unstable, intermediate slopes greater than 6 dB/octave can be made to work. The maximum usable slope is normally considered to be 10 dB/octave, which gives a phase margin of 30°. This may be acceptable in some cases, but I think it cuts it a little fine. The steeper fall in gain means that more NFB is applied at lower frequencies and so less distortion is produced. Electronic circuitry only provides slopes in multiples of 6 dB/octave, so 10 dB/octave requires multiple overlapping time constants to approximate a straight line at an intermediate slope. This gets complicated, and this method of maximizing NFB is not popular.

2. The gain slope varies with frequency so that maximum open-loop gain and hence NFB factor is sustained as long as possible as frequency increases; the gain then drops quickly, at 12 dB/octave or more, but flattens out to 6 dB/octave before it reaches the critical unity loop-gain intersection. In this case the stability margins should be relatively unchanged compared with the conventional situation.

## 12.3  Maximizing Linearity Before Feedback

*Make your amplifier as linear as possible before applying NFB* has long been a cliché. It blithely ignores the difficulty of running a typical solid-state amplifier without any feedback to determine its basic linearity.

Virtually no dependable advice on how to perform this desirable linearization has been published. The two factors are the basic linearity of the forward path and the amount of NFB applied to further straighten it out. The latter cannot be increased beyond certain limits or else high-frequency stability is put in peril, whereas there seems no reason why open-loop linearity could not be improved without limit, leading us to what in some senses must be the ultimate goal—a distortionless amplifier. This book therefore takes as one of its main aims the understanding and improvement of open-loop linearity.

# Further Reading

Attwood, B., Design parameters important for the optimisation of PWM (Class-D) amplifiers, *JAES,* **31**, p. 842, November 1983.

Baxandall, P., Audio power amplifier design: Part 5, *Wireless World*, p. 53, Dec 1978. (This superb series of articles had six parts and ran on roughly alternate months, starting in January 1978.)

Blomley, P., A new approach to Class-B, *Wireless World*, p. 57, February 1971.

Buitendijk, P., A 40W Integrated car radio audio amplifier, *IEEE Conf on Consumer Electronics*, 1991 Session THAM 12.4, p. 174. (Class-H)

Cherry, E., Nested differentiating feedback loops in simple audio power amplifiers, *JAES,* **30**(5): 295, May 1982.

Evenson, R., *Audio amplifiers with tailored output impedances*. Preprint for November 1988 AES convention (Los Angeles).

Feldman, L., Class-G high-efficiency hi-fi amplifier, *Radio-Electronics*, p. 47, August 1976.

Gilbert, B., Current mode circuits from a translinear viewpoint Ch 2, Analogue IC design: The current-mode approach ed Toumazou, Lidgey & Haigh, IEE, 1990.

Goldberg and Sandler, Noise shaping and pulse-width modulation for all-digital audio power amplifier, *JAES,* **39**, p. 449, February 1991.

Hancock, J., A Class-D amplifier using MOSFETS with reduced minority carrier lifetime, *JAES,* **39**, p. 650, September 1991.

Lin, H. C., Transistor audio amplifier, *Electronics*, p. 173, September 1956.

Linsley-Hood, Simple Class-A amplifier, *Wireless World*, p. 148, April 1969.

Mills and Hawksford, Transconductance power amplifier systems for current-driven loudspeakers, *JAES,* **37**, p. 809, March 1989.

Olsson, B., Better audio from non-complements?, *Electronics World*, p. 988, December 1994.

Peters, A., Class E RF amplifiers, *IEEE J. Solid-State Circuits*, p. 168, June 1975.

Raab, F., Average efficiency of Class-G power amplifiers, *IEEE Transactions on Consumer Electronics,* **CE-22**, p. 145, May 1986.

Sampei, *et al.*, Highest efficiency & super quality audio amplifier using MOS-power FETs in Class-G, *IEEE Transactions on Consumer Electronics,* **CE-24**, p. 300, August 1978.

Sandman, A Class S: A novel approach to amplifier, *Distortion Wireless World*, September 1982, p. 38

Sinclair, Audio and hi-fi handbook, Newnes, p. 541, 1993.

Stochino, G., Audio design leaps forward?, *Electronics World*, p. 818, October 1994.

Sweeney,  and Mantz, , An informal history of amplifiers, *Audio*, p. 46, June 1988.

Tanaka, S. A., New biasing circuit for Class-B operation, *JAES*, p. 27, January/February 1981.

Thus, Compact bipolar Class AB output stage, *IEEE Journal of Solid-State Circuits*, p. 1718, December 1992.

Walker, P. J., Current dumping audio amplifier, *Wireless World*, p. 560, December 1975.

# CHAPTER 13

# Noise and Grounding

Douglas Self

## 13.1 Audio Amplifier Printed Circuit Board Design

This section addresses the special printed circuit board (PCB) design problems presented by power amplifiers, particularly those operating in Class-B. All power amplifier systems contain the power-amp stages themselves, and usually associated control and protection circuitry; most also contain small-signal audio sections such as balanced input amplifiers, subsonic filters, output meters, and so on.

Other topics related to PCB design, such as grounding, safety, and reliability, are also dealt with.

The performance of an audio power amplifier depends on many factors, but in all cases the detailed design of the PCB is critical because of the risk of inductive distortion due to cross talk between the supply rails and the signal circuitry; this can very easily be the ultimate limitation on amplifier linearity, and it is hard to overemphasize its importance. The PCB design will, to a great extent, define both the distortion and the cross talk performance of the amplifier.

Apart from these performance considerations, the PCB design can have considerable influence on ease of manufacture, ease of testing and repair, and reliability. All of these issues are addressed here.

Successful audio PCB layout requires enough electronic knowledge to fully appreciate the points set out here so that layout can proceed smoothly and effectively. It is common in many electronic fields for PCB design to be handed over to draftspersons, who, while very skilled in the use of CAD, have little or no understanding of the details of circuit operation. In some fields this works fine; in power amplifier design it will not because basic parameters such as cross talk and distortion are so strongly layout dependent. At the very least the PCB designer should understand the points that follow.

### 13.1.1 Cross Talk

All cross talk has a transmitting end (which can be at any impedance) and a receiving end, usually either at high impedance or at virtual earth. Either way, it is sensitive to the injection of small currents. When interchannel cross talk is being discussed, the transmitting and receiving channels are usually called speaking and nonspeaking channels, respectively.

Cross talk comes in various forms:

- Capacitive cross talk is a consequence of the physical proximity of different circuits and may be represented by a small notional capacitor joining the two circuits. It usually increases at the rate of 6 dB/octave, although higher dB/octave rates are possible. Screening with any conductive material is a complete cure, but physical distance is usually less expensive.

- Resistive cross talk usually occurs simply because ground tracks have a nonzero resistance. Copper is not a room-temperature superconductor. Resistive cross talk is constant with frequency.

- Inductive cross talk is rarely a problem in general audio design; it might occur if you have to mount two uncanned audio transformers close together, but otherwise you can usually forget it. The notable exception to this rule is the Class-B audio power amplifier, where the rail currents are halfwave sines that seriously degrade the distortion performance if allowed to couple into the input, feedback, or output circuitry.

In most line-level audio circuitry the primary cause of cross talk is unwanted capacitive coupling between different parts of a circuit, and in most cases this is defined solely by the PCB layout. Class-B power amplifiers, in contrast, should suffer very low or negligible levels of cross talk from capacitive effects, as circuit impedances tend to be low and the physical separation large; a much greater problem is inductive coupling between the supply-rail currents and the signal circuitry. If coupling occurs to the same channel, it manifests itself as distortion and can dominate amplifier nonlinearity. If it occurs to the other (nonspeaking) channel it will appear as cross talk of a distorted signal. In either case it is thoroughly undesirable, and precautions must be taken to prevent it.

The PCB layout is only one component of this, as cross talk must be both emitted and received. In general, the emission is greatest from internal wiring due to its length and extent; wiring layout will probably be critical for best performance and needs to be fixed by cable ties, etc. The receiving end is probably the input and feedback circuitry of the amplifier, which will be fixed on the PCB. Designing these sections for maximum immunity is critical to good performance.

### 13.1.2 Rail Induction Distortion

The supply rails of a Class-B power-amp carry large and very distorted currents. As outlined previously, if these are allowed to cross talk into the audio path by induction, the distortion performance will be severely degraded. This applies to PCB conductors just as much as cabling, and it is sadly true that it is easy to produce an amplifier PCB that is absolutely satisfactory in every respect but this one, and the only solution is another board iteration. The effect can be completely prevented, but in the present state of knowledge I cannot give detailed guidelines to suit every constructional topology. The best approach is to minimize radiation from the supply rails by running the $V+$ and $V-$ rails as close together as possible. Keep them away from the input stages of the amplifier and the output connections; the best method is to bring the rails up to the output stage from one side, with the rest of the amplifier on the other side. Then run tracks from the output to power the rest of the amp; these carry no halfwave currents and should cause no problems.

Minimize pick-up of rail radiation by keeping the area of the input and feedback circuits to a minimum. These form loops with the audio ground and these loops must be as small in area as possible. This can often best be done by straddling the feedback and input networks across the audio ground track, which is taken across the center of the PCB from input ground to output ground.

Induction of distortion can also occur into the output and output-ground cabling, and even the output inductor. The latter presents a problem as it is usually difficult to change its orientation without a PCB update.

### 13.1.3 Mounting of Output Devices

The most important decision is whether to mount the power output devices directly on the main amplifier PCB. There are strong arguments for doing so, but it is not always the best choice.

### 13.1.3.1 Advantages

- The amplifier PCB can be constructed so as to form a complete operational unit that can be tested thoroughly before being fixed into the chassis. This makes testing much easier, as there is access from all sides; it also minimizes the possibility of cosmetic damage (scratches, etc.) to the metalwork during testing.

- It is impossible to connect the power devices wrongly, providing you get the right devices in the right positions. This is important for such errors usually destroy both output devices and cause other domino-effect faults that are very time-consuming to correct.

- The output device connections can be very short. This seems to help stability of the output stage against HF parasitic oscillations.

### 13.1.3.2 Disadvantages

- If the output devices require frequent changing (which obviously indicates something very wrong somewhere) then repeated resoldering will damage the PCB tracks. However, if the worst happens, the damaged track can usually be bridged out with short sections of wire so that the PCB need not be scrapped; make sure this is possible.

- The output devices will probably get fairly hot, even if run well within their ratings; a case temperature of 90°C is not unusual for a TO3 device. If the mounting method does not have a degree of resilience, then thermal expansion may set up stresses that push the pads off the PCB.

- Because the heat sink will be heavy, there must be a solid structural fixing between this and the PCB. Otherwise the assembly will flex when handled, putting stress on soldered connections.

### 13.1.4 Single- and Double-Sided Printed Circuit Boards

Because of their lower cost, single-sided PCBs are the usual choice for power amplifiers; however, the price differential between single- and double-sided plated-through-hole (PTH) PCBs is much less than it used to be. It is not usually necessary to go double-sided for reason of space or convoluted connectivity, as power amplifier components tend to be physically large, determining the PCB size, and in typical circuitry there are a large number of discrete resistors, etc., that can be used for jumping tracks.

Bear in mind that single-sided boards need thicker tracks to ensure adhesion in case desoldering is necessary. Adding one or more ears to pads with only one track leading to them gives much better adhesion and is highly recommended for pads that may need resoldering during maintenance; unfortunately, it is a very tedious task with most CAD systems.

The advantages of double-sided PTH for power amplifiers are as follow:

- No links are required.

- Double-sided PCBs may allow one side to be used primarily as a ground plane, minimizing cross talk and EMC problems.

- Much better pad adhesion on resoldering as the pads are retained by the through-hole plating.

- There is more total room for tracks so they can be wider, giving less volt drop and PCB heating.

- The extra cost is small.

### 13.1.5 Power Supply Printed Circuit Board Layout

Power supply subsystems have special requirements due to the very high capacitor-charging currents involved.

- Tracks carrying the full supply-rail current must have generous widths. The board material used should have not less than 2-oz copper. Four-ounce copper can be obtained but it is expensive and has long lead times; it is not really recommended.

- Reservoir capacitors must have the incoming tracks going directly to the capacitor terminals; likewise the outgoing tracks to the regulator must leave from these terminals. In other words, do not run a tee off to the cap. Failure to observe this puts sharp pulses on the DC and tends to worsen the hum level.

- The tracks to and from the rectifiers carry charging pulses that have a considerably higher peak value than the DC output current. Conductor heating is therefore much greater due to the higher value of $I^2R$. Heating is likely to be especially severe at PC-mount fuse holders. Wire links may also heat up and consideration should be given to two links in parallel; this sounds crude but actually works very effectively.

Track heating can usually be detected simply by examining the state of the solder mask after several hours of full-load operation; the green mask materials currently in use discolor to brown on heating. If this occurs then as a very rough rule the track is too hot. If the discoloration tends to dark brown or black then the heating is serious and must definitely be reduced.

- If there are PCB tracks on the primary side of the mains transformer, and this has multiple taps for multicountry operation, then remember that some of these tracks will carry much greater currents at low voltage tappings; mains current drawn on 90 V input will be nearly three times that at 240 V.

Be sure to observe the standard safety spacing of 60 thou between mains tracks and other conductors for creepage and clearance. (This applies to all track-track, track-PCB edge, and track-metal-fixings spacings.)

In general, PCB tracks carrying mains voltages should be avoided, as presenting an unacceptable safety risk to service personnel. If it must be done, then warnings must be displayed very clearly on both sides of the PCB. Mains-carrying tracks are unacceptable in equipment intended to meet UL regulations in the United States, unless they are fully covered with insulating material that is nonflammable and can withstand at least 120°C (e.g., polycarbonate).

### 13.1.6  Power Amplifier Printed Circuit Board Layout Details

A simple unregulated supply is assumed.

- Power amplifiers have heavy currents flowing through the circuitry, and all of the requirements for power supply design also apply here. Thick tracks are essential, and 2-oz copper is highly desirable, especially if the layout is cramped.

If attempting to thicken tracks by laying solder on top, remember that ordinary 60:40 solder has a resistivity of about six times that of copper, so even a thick layer may not be very effective.

- The positive and negative rail reservoir caps will be joined together by a thick earth connection; this is called reservoir ground (RG). *Do not* attempt to use any point on this track as the audio-ground starpoint, as it carries heavy charging pulses and will induce ripple into the signal. Instead, take a thick tee from the center of this track (through which the charging pulses will not flow) and use the end of this as the starpoint.

- Low-value resistors in the output stage are likely to get very hot in operation— possibly up to 200°C. They must be spaced out as much as possible and kept from contact with components such as electrolytic capacitors. Keep them away from sensitive devices such as the driver transistors and the bias-generator transistor.

- Vertical power resistors. The use of these in power amplifiers appears attractive at first because of the small amount of PCB area they take up. However, the vertical construction means that any impact on the component, such as might be received in normal handling, puts a very great strain on the PCB pads, which are likely to be forced off the board. This may result in it being scrapped. Single-sided boards are particularly vulnerable, having much lower pad adhesion due to the absence of vias.

- Solderable metal clips to strengthen the vertical resistors are available in some ranges (e.g., Vitrohm) but this is not a complete solution, and the conclusion must be that horizontal-format power resistors are preferable.

- Rail decoupler capacitors must have a separate ground return to the reservoir ground. This ground must *not* share any part of the audio ground system, and must *not* be returned to the starpoint.

- The exact layout of the feedback takeoff point is criticial for proper operation. Usually the output stage has an *output rail* that connects the emitter power resistors together. This carries the full output current and must be substantial. Take a tee from this track for the output connection and attach the feedback takeoff point to somewhere along this tee. *Do not* attach it to the track joining the emitter resistors.

- The input stages (usually a differential pair) should be at the other end of the circuitry from the output stage. Never run input tracks close to the output stage. Input stage ground and the ground at the bottom of the feedback network must be the same track running back to the starpoint. No decoupling capacitors may be connected to this track, but it seems to be permissible to connect input bias resistors that pass only very small DC currents.

- Put the input transistors close together. The closer the temperature match, the less the amplifier output DC offset due to VBE mismatching. If they can both be hidden from *seeing* the infra-red radiation from the heat sink (e.g., by hiding them behind a large electrolytic), then DC drift is reduced.

- Most power amplifiers will have additional control circuitry for muting relays, thermal protection, etc. Grounds from this must take a separate path back to the reservoir ground, and *not* the audio starpoint.

- Unlike most audio boards, power amps will contain a mixture of sensitive circuitry and a high-current power supply. Be careful to keep bridge rectifier connections and so on away from input circuitry.

- Mains/chassis ground will need to be connected to the power amplifier at some point. Do not do this at the transformer center tap as this is spaced away from the input ground voltage by the return charging pulses and will create severe ground-loop hum when the input ground is connected to mains ground through another piece of equipment.

Connecting mains ground to starpoint is better, as the charging pulses are excluded, but the track resistance between input ground and star will carry any ground-loop currents and induce a buzz.

Connecting mains ground to the input ground gives maximal immunity against ground loops.

- If capacitors are installed the wrong way round the results are likely to be explosive. Make every possible effort to put all capacitors in the same orientation to allow efficient visual checking. Mark polarity clearly on the PCB, positioned so it is still visible when the component is fitted.

- Drivers and the bias generator are likely to be fitted to small vertical heat sinks. Try to position them so that the transistor numbers are visible.

- All transistor positions should have emitter, base, and collector or whatever marked on the top print to aid fault finding. TO3 devices also need to be identified on the copper side, as any screen printing is covered up when the devices are installed.

- Any wire links should be numbered to make it easier to check that they have all been fitted.

### 13.1.7 Audio Printed Circuit Board Layout Sequence

PCB layout must be considered from an early stage of amplifier design. For example, if a front-facial layout shows the volume control immediately adjacent to a loudspeaker routing switch, then a satisfactory cross talk performance will be difficult to obtain because of the relatively high impedance of the volume control wipers. Shielding metalwork may be required for satisfactory performance, which adds cost. In many cases the detailed electronic design has an effect on cross talk quite independently from physical layout.

a. Consider implications of facia layout for PCB layout.

b. Circuitry designed to minimize cross talk. At this stage, try to look ahead to see how op-amp halves, switch sections, and so on should be allocated to keep signals away from sensitive areas. Consider cross talk at above-PCB level; for example, when designing a module made up of two parallel double-sided PCBs, it is desirable to place signal circuitry on the inside faces of the boards, and power and grounds on the outside, to minimize cross talk and maximize RF immunity.

c. Facia components (pots, switches, etc.) placed to partly define available board area.

d. Other fixed components, such as power devices, driver heat sinks, input and output connectors, and mounting holes placed. The area left remains for the purely electronic parts of the circuitry that do not have to align with metalwork and so may be moved about fairly freely.

e. Detailed layout of components in each circuit block, with consideration toward manufacturability.

f. Make efficient use of any spare PCB area to fatten grounds and high-current tracks as much as possible. It is not wise to fill in every spare corner of a prototype board with copper as this can be time-consuming (depending on the facilities of your PCB CAD system) and some of it will probably have to be undone to allow modifications.

Ground tracks should always be as thick as practicable. Copper is free.

### 13.1.8 Miscellaneous Points

- On double-sided PCBs, copper areas should be solid on the component side for minimum resistance and maximum screening, but will need to be cross-hatched on the solder side to prevent distortion if the PCB is flow soldered. A common standard is 10 thou wide noncopper areas, that is, mostly copper with small square holes; this is determined in the CAD package. If in doubt, consult those doing the flow soldering.

- Do not bury component pads in large areas of copper, as this causes soldering difficulties.

- There is often a choice between running two tracks into a pad or taking off a tee so that only one track reaches it. The former is better because it holds the pad more firmly to the board if desoldering is necessary. This is *particularly important* for components such as transistors that are relatively likely to be replaced; for single-sided PCBs it is absolutely vital.

- If two parallel tracks are likely to cross talk, then it is beneficial to run a grounded screening track between them. However, the improvement is likely to be disappointing, as electrostatic lines of force will curve over the top of the screen track.

- Jumper options must always be clearly labeled. Assume that everyone loses the manual the moment they get it.

- Label pots and switches with their function on the screen-print layer, as this is a great help when testing. If possible, also label circuit blocks, for example, *DC offset detect*. The labels must be bigger than component ident text to be clearly readable.

## 13.2  Amplifier Grounding

The grounding system of an amplifier must fulfil several requirements, among which are:

- The definition of a *starpoint* as the reference for all signal voltages.

- In a stereo amplifier, grounds must be suitably segregated for good cross talk performance. A few inches of wire as a shared ground to the output terminals will probably dominate the cross talk behavior.

- Unwanted AC currents entering the amplifier on the signal ground, due to external ground loops, must be diverted away from the critical signal grounds, that is, the input ground and the ground for the feedback arm. Any voltage difference between these last two grounds appears directly in the output.

- Charging currents for the power supply unit (PSU) reservoir capacitors must be kept out of all other grounds.

Ground is the point of reference for all signals, and it is vital that it is made solid and kept clean; every ground track and wire must be treated as a resistance across which signal currents will cause unwanted voltage drops. The best method is to keep ground currents apart by means of a suitable connection topology, such as a separate ground return to the starpoint for the local HT decoupling, but when this is not practical it is necessary to make every ground track as thick as possible and fattened up with copper at every possible point. It is vital that the ground path has no necks or narrow sections, as it is no stronger than the weakest part. If the ground path changes board side then a single via hole may be insufficient, and several should be connected in parallel. Some CAD systems make this difficult, but there is usually a way to fool them.

Power amplifiers rarely use double-insulated construction and so the chassis and all metalwork must be grounded permanently and solidly for safety. One result of permanent chassis grounding is that an amplifier with unbalanced inputs may appear susceptible to ground loops. One solution is to connect audio ground to chassis only through a 10-$\Omega$ resistor, which is large enough to prevent loop currents becoming significant. This is not very satisfactory as:

- The audio system as a whole may not be grounded solidly.

- If the resistor is burnt out due to misconnected speaker outputs, the audio circuitry is floating and could become a safety hazard.

- The RF rejection of the power amplifier is likely to be degraded. A 100-nF capacitor across the resistor may help.

A better approach is to put the audio-chassis ground connection at the input connector so that in Figure 13.1, ground-loop currents must flow through A–B to the protected earth at B and then to mains ground via B–C. They cannot flow through the audio path E–F. This

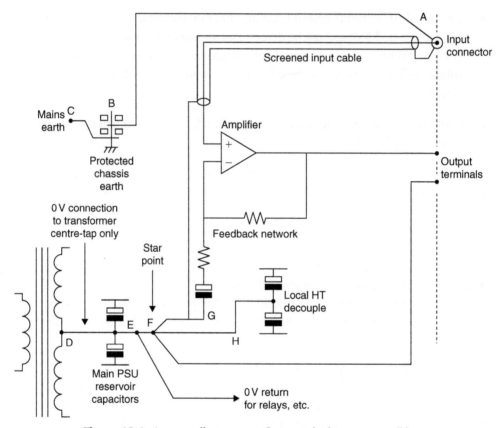

**Figure 13.1: A grounding system for a typical power amplifier.**

topology is very resistant to ground loops, even with an unbalanced input; the limitation on system performance in the presence of a ground loop is now determined by the voltage drop in the input cable ground, which is outside the control of the amplifier designer. A balanced input could, in theory, cancel out this voltage drop completely.

Figure 13.1 also shows how the other grounding requirements are met. The reservoir charging pulses are confined to the connection D–E and do not flow E–F, as there is no other circuit path. E–F–H carries ripple, etc., from the local HT decouplers, but likewise cannot contaminate the crucial audio ground A–G.

# 13.3  Ground Loops: How They Work and How to Deal with Them

A ground loop is created whenever two or more pieces of mains-powered equipment are connected together so that mains-derived AC flows through shields and ground conductors, degrading the noise floor of the system. The effect is the worst when two or more units are connected through mains ground as well as audio cabling, and this situation is what is normally meant by the term "ground loop." However, ground currents can also flow in systems that are not grounded galvanically; they are of lower magnitude but can still degrade the noise floor, so this scenario is also considered here.

Ground currents may either be inherent in the mains supply wiring (see Section 13.3.1) or generated by one or more of the pieces of equipment that make up the audio system (see Sections 13.3.2 and 13.3.3).

Once flowing in the ground wiring, these currents will give rise to voltage drops that introduce hum and buzzing noises. This may occur either in the audio interconnects or inside the equipment itself if it is not well designed.

Here I have used the word "ground" for conductors and so on, whereas "earth" is reserved for the damp crumbly stuff into which copper rods are thrust.

### 13.3.1  Hum Injection by Mains Grounding Currents

Figure 13.2 shows what happens when a so-called "technical ground," such as a buried copper rod, is attached to a grounding system that is already connected to "mains ground" at the power distribution board. The latter is mandatory both legally and technically, so one might as well accept this and denote as the reference ground. In many cases this "mains ground" is actually the neutral conductor, which is only grounded at the remote transformer substation. AB is the cable from substation to consumer, which serves many houses from connections tapped off along its length. There is substantial current flowing down the N+E conductor, so point B is often 1-V rms or more above earth. From B onward, in the internal house wiring, neutral and ground are always separate (in the United Kingdom anyway).

Two pieces of audio equipment are connected to this mains wiring at C and D and are joined to each other through an unbalanced cable F–G. Then an ill-advised connection is made to earth at D; the 1-V rms is now impressed on the path B–C–D, and substantial

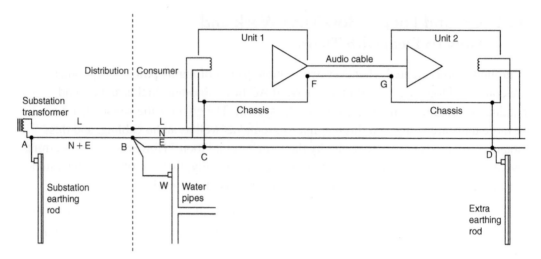

**Figure 13.2: Pitfalls of adding a "technical ground" to a system that is already grounded via the mains.**

current is likely to flow through it, depending on the total resistance of this path. There will be a voltage drop from C to D, with its magnitude depending on what fraction of the total BCDE resistance is made up by the section C–D. The earth wire C–D will be of at least $1.5\,mm^2$ cross section, and so the extra connection FG down the audio cable is unlikely to reduce the interfering voltage much.

To get a feel for the magnitudes involved, take a plausible ground current of 1 A. The 1.5-mm² ground conductor will have a resistance of $0.012\,\Omega/m$, so if the mains sockets at C and D are 1 m apart, the voltage C–D will be 12 mV rms. Almost all of this will appear between F and G and will be indistinguishable from wanted signal to the input stage of unit 2, so the hum will be severe, probably only 30 dB below the nominal signal level.

The best way to solve this problem is not to create it in the first place. If some ground current is unavoidable, then the use of balanced inputs (or ground-cancel outputs—it is not necessary to use both) should give at least 40 dB of rejection at audio frequencies.

Figure 13.2 also shows a third earthing point, which fortunately does not complicate the situation. Metal water pipes are bonded to the incoming mains ground for safety reasons, and because they are usually connected electrically to an incoming water supply, current

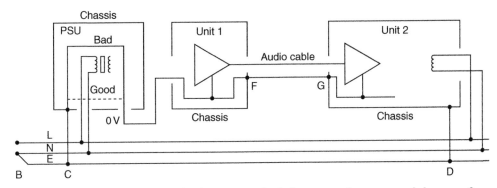

**Figure 13.3: A poor cable layout in the PSU at the left wraps a loop around the transformer and induces ground currents.**

flows through B–W in the same way as it does through the copper rod link D–E. This waterpipe current does not, however, flow through C–D and cannot cause a ground-loop problem. It may, however, cause the pipes to generate an AC magnetic field, which is picked up by other wiring.

### 13.3.2 Hum Injection by Transformer Stray Magnetic Fields

Figure 13.3 shows a thoroughly bad piece of physical layout that will cause ground currents to flow even if the system is grounded correctly to just one point.

Here unit 1 has an external DC power supply; this makes it possible to use an inexpensive frame-type transformer, which will have a large stray field. However, note that the wire in the PSU that connects mains ground to the outgoing 0 V takes a half-turn around the transformer, and significant current will be induced into it, which will flow round the loop C–F–G–D, and give an unwanted voltage drop between F and G. In this case, reinforcing the ground of the audio interconnection is likely to be of some help, as it directly reduces the fraction of the total loop voltage that is dropped between F and G.

It is difficult to put any magnitudes to this effect because it depends on many imponderables, such as the build quality of the transformer and the exact physical arrangement of the ground cable in the PSU. If this cable is rerouted to the dotted position in the diagram, the transformer is no longer enclosed in a half-turn, and the effect will be much smaller.

www.newnespress.com

### 13.3.3 Hum Injection by Transformer Stray Capacitance

It seems at first sight that the adoption of Class II (double-insulated) equipment throughout an audio system will give inherent immunity to ground-loop problems. Life is not so simple, although it has to be said that when such problems do occur they are likely to be much less severe. This problem afflicts all Class II equipment to a certain extent.

Figure 13.4 shows two Class II units connected together by an unbalanced audio cable. The two mains transformers in the units have stray capacitance from both live and neutral to the secondary. If these capacitances were all identical, no current would flow, but in practice they are not, so 50-Hz currents are injected into the internal 0-V rail and flow through the resistance of F–G, adding hum to the signal. A balanced input or ground-canceling output will remove or render negligible the ill effects.

Reducing the resistance of the interconnect ground path is also useful—more so than with other types of ground loop, because the ground current is essentially fixed by the small stray capacitances, and so halving the resistance F–G will dependably halve the interfering voltage. There are limits to how far you can take this; while a simple balanced input will give 40 dB of rejection at low cost, increasing the cross-sectional area of copper in the ground of an audio cable by a factor of 100 times is not going to be either easy or inexpensive. Figure 13.4 shows equipment with metal chassis connected to the 0 V (this is quite acceptable for safety approvals—what counts is the isolation between mains and everything else, not between low-voltage circuitry and touchable metalwork); note that the

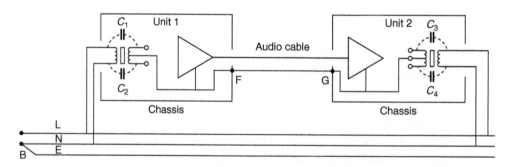

**Figure 13.4: The injection of mains current into the ground wiring via transformer interwinding capacitance.**

chassis connection, however, has no relevance to the basic effect, which would still occur even if the equipment enclosure was completely nonconducting.

The magnitude of ground current varies with the details of transformer construction and increases as the size of the transformer grows. Therefore, the more power a unit draws, the larger the ground current it can sustain. This is why many systems are subjectively hum free until the connection of a powered subwoofer, which is likely to have a larger transformer than other components of the system.

| Equipment type | Power consumption | Ground current |
|---|---|---|
| Turntable, CD, cassette deck | 20 W or less | 5 μA |
| Tuners, amplifiers, small TVs | 20–100 W | 100 μA |
| Big amplifiers, subwoofers, large TVs | More than 100 W | 1 mA |

### 13.3.4 Ground Currents Inside Equipment

Once ground currents have been set flowing, they can degrade system performance in two locations: outside the system units, by flowing in the interconnect grounds, or inside the units, by flowing through internal PCB tracks, etc. The first problem can be dealt with effectively by the use of balanced inputs, but the internal effects of ground currents can be much more severe if the equipment is poorly designed.

Figure 13.5 shows the situation. There is, for whatever reason, ground current flowing through the ground conductor C–D, causing an interfering current to flow round the loop

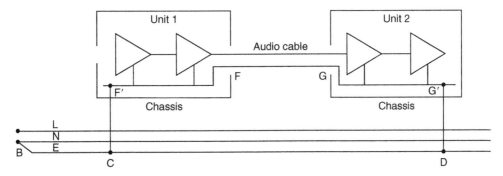

**Figure 13.5: If ground current flows through the path F'FGG', then the relatively high resistance of the PCB tracks produces voltage drops between the internal circuit blocks.**

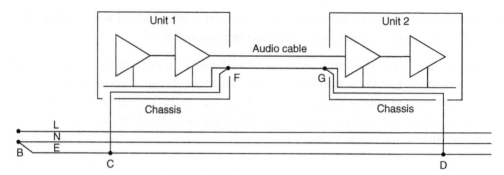

**Figure 13.6: The correct method of dealing with ground currents; they are diverted away from internal circuitry.**

C–F–G–D as before. Now, however, the internal design of unit 2 is such that the ground current flowing through F–G also flows through G–G′ before it encounters the ground wire going to point D. G–G′ is almost certain to be a PCB track with higher resistance than any of the cabling and so the voltage drop across it can be relatively large and the hum performance correspondingly poor. Exactly similar effects can occur at signal outputs; in this case the ground current is flowing through F–F′.

Balanced inputs will have no effect on this; they can cancel out the voltage drop along F–G, but if internal hum is introduced further down the internal signal path, there is nothing they can do about it.

The correct method of handling this is shown in Figure 13.6. The connection to mains ground is made right where the signal grounds leave and enter the units and are made as solidly as possible. The ground current no longer flows through the internal circuitry. It does, however, still flow through the interconnection at F–G, so either a balanced input or a ground-canceling output will be required to deal with this.

### 13.3.5 Balanced Mains Power

There has been speculation in recent times as to whether a balanced mains supply is a good idea. This means that instead of live and neutral (230 and 0 V) you have live and the other live (115 V–0–115 V) created by a center-tapped transformer with the tap connected to neutral (see Figure 13.7).

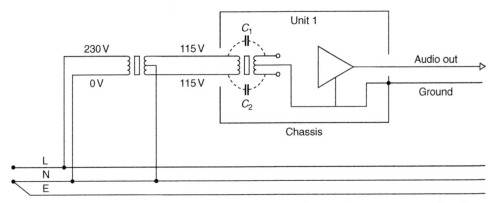

**Figure 13.7: Use of a balanced mains supply to cancel ground currents stemming from interwinding capacitance in the mains transformer. This is an expensive solution.**

It has been suggested that balanced mains has miraculous effects on sound quality, makes the sound stage ten dimensional, etc. This is obviously nonsense. If a piece of gear is that fussy about its mains (and I do not believe any such gear exists) then dispose of it.

If there is severe radio frequency interference (RFI) on the mains, an extra transformer in the path may tend to filter it out. However, a proper mains RFI filter will almost certainly be more effective—it is designed for the job, after all—and will definitely be less expensive.

Where you might gain a real benefit is in a Class II (i.e., double-insulated) system with very feeble ground connections. Balanced mains would tend to cancel out the ground currents caused by transformer capacitance (see Figure 13.4 and previous discussion for more details on this) and so reduce hum. The effectiveness of this will depend on $C_1$ being equal to $C_2$ in Figure 13.7, which is determined by the details of transformer construction in the unit being powered. I think that the effect would be small with well-designed equipment and reasonably heavy ground conductors in interconnects. Balanced audio connections are a much less expensive and better way of handling this problem, but if none of the equipment has them then beefing up the ground conductors should give an improvement. If the results are not good enough, then, as a last resort, balanced mains may be worth considering.

Finally, bear in mind that any transformer you add must be able to handle the maximum power drawn by the audio system at full throttle. This can mean a large and expensive component.

I would not be certain about the whole of Europe, but to the best of my knowledge it is the same as the United Kingdom, that is, not balanced. The neutral line is at earth potential, give or take a volt, and the live is 230 V above this. The three-phase 11-kV distribution to substations is often described as "balanced" but this just means that power delivered by each phase is kept as near equal as possible for the most efficient use of the cables.

It has often occurred to me that balanced mains 115 V–0–115 V would be a lot safer. Since I am one of those people that put their hands inside live equipment a lot, I do have a kind of personal interest here.

## 13.4  Class I and Class II

Mains-powered equipment comes in two types: grounded and double insulated. These are officially called Class I and Class II, respectively.

Class I equipment has its external metalwork grounded. Safety against electric shock is provided by limiting the current the live connection can supply with a fuse. Therefore, if a fault causes a short-circuit between live and metalwork, the fuse blows and the metalwork remains at ground potential. A reasonably low resistance in the ground connection is essential to guarantee the fuse blows. A three-core mains lead is mandatory. Two-core IEC mains leads are designed so that they cannot be plugged into three-pin Class I equipment. Class I mains transformers are tested to 1.5 kV rms.

Class II equipment is not grounded. Safety is maintained not by interrupting the supply in case of a fault, but by preventing the fault happening in the first place. Regulations require double insulation and a generally high standard of construction to prevent any possible connection between live and the chassis. A two-core IEC mains lead is mandatory; it is not permitted to sell a three-core lead with a Class II product. This would present no hazard in itself, but is presumably intended to prevent confusion as to what kind of product is in use. Class II mains transformers are tested to 3 kV rms to give greater confidence against insulation breakdown.

Class II is often adopted in an attempt to avoid ground loops. Doing so eliminates the possibility of major problems, at the expense of throwing away all hope of fixing minor ones. There is no way to prevent capacitance currents from the mains transformer flowing through the ground connections (see Section 13.3). It is also no longer possible to put a grounded electrostatic screen between primary and secondary windings. This is serious as it deprives you of your best weapon against mains noise coming in and circuit RF emissions getting out. In Class II the external chassis may be metallic and connected to signal 0 V as often as you like.

If a Class II system is not connected to ground at any point, then the capacitance between primaries and secondaries in the various mains transformers can cause its potential to rise well above ground. If it is touched by a grounded human, then current will flow, and this can sometimes be perceptible, although not directly, as a painful shock such as static electricity. The usual complaint is that the front panel of equipment is "vibrating" or that it feels "furry." The maximum permitted touch current (flowing to ground through the human body) permitted by current regulations is 700 μA, but currents well below this are perceptible. It is recommended, although not required, that this limit be halved in the tropics where fingers are more likely to be damp. The current is measured through a 50 k resistance to ground.

When planning new equipment, remember that the larger the mains transformer, the greater the capacitance between primary and secondary and the more likely this is to be a problem. To put the magnitudes into perspective, I measured a 500 VA toroid (intended for Class II usage and with no interwinding screen) and found 847 pF between the windings. At 50 Hz and 230 V, this implies a maximum current of 63 μA flowing into the signal circuitry, with the actual figure depending on precisely how the windings are arranged. A much larger 1500 VA toroidal transformer had 1.3 nF between the windings, but this was meant for Class I use and had a screen, which was left floating to get the figure above.

### 13.4.1 Warning

Please note that the legal requirements for electrical safety are always liable to change. This book does not attempt to give a complete guide to what is required for compliance. The information given here is correct at the time of writing, but it is the designer's responsibility to check for changes to compliance requirements. The information is given here in good faith but the author accepts no responsibility for loss or damage under any circumstances.

## 13.5 Mechanical Layout and Design Considerations

The mechanical design adopted depends very much on the intended market and production and tooling resources, but I offer a few purely technical points that need to be taken into account.

### 13.5.1  Cooling

All power amplifiers will have a heat sink that needs cooling, usually by free convection, and the mechanical design is often arranged around this requirement. There are three main approaches to the problem.

    a.  The heat sink is entirely internal and relies on convected air entering the bottom of the enclosure and leaving near the top (passive cooling).

#### 13.5.1.1  Advantages

The heat sink may be connected to any voltage, which may eliminate the need for thermal washers between power device and sink. However, some sort of conformal material is still needed between transistor and heat sink. A thermal washer is much easier to handle than the traditional white oxide-filled silicone compound, so you will be using them anyway. There are no safety issues as to heat sink temperatures.

#### 13.5.1.2  Disadvantages

Because of the limited fin area possible inside a normal-sized box and the relatively restricted convection path, this system is not suitable for large dissipations.

    b.  The heat sink is partly internal and partly external, as it forms one or more sides of the enclosure. Advantages and disadvantages are much as just described; if any part of the heat sink can be touched, then the restrictions on temperature and voltage apply. Greater heat dissipation is possible.

    c.  The heat sink is primarily internal, but is fan cooled (active cooling). Fans always create some noise, which increases with the amount of air they are asked to move. Fan noise is most unwelcome in a domestic hi-fi environment, but is of little importance in PA applications. This allows maximal heat dissipation, but requires an inlet filter to prevent the build-up of dust and fluff internally. Persuading people to clean such filters regularly is near impossible.

Efficient passive heat removal requires extensive heat sinking with a free convective air flow, and this indicates putting the sinks on the side of the amplifier; the front will carry at least the mains switch and power indicator light, while the back carries the in/out and mains connectors so that only the sides are completely free.

The internal space in the enclosure will require some ventilation to prevent heat build-up; slots or small holes are desirable to keep foreign bodies out. Avoid openings on the top surface as these will allow the entry of spilled liquids and increase dust entry. BS415 is a good starting point for this sort of safety consideration, and this specifies that slots should be no more than 3 mm wide.

Reservoir electrolytics, unlike most capacitors, suffer significant internal heating due to ripple current. Because the electrolytic capacitor life is very sensitive to temperature, mount them in the coolest position available and, if possible, leave room for air to circulate between them to minimize the temperature rise.

### 13.5.2 Convection Cooling

It is important to realize that the buoyancy forces that drive natural convection are very small and that even small obstructions to flow can seriously reduce the rate of flow, and hence the cooling. If ventilation is by slots in the top and bottom of an amplifier case, then the air must be drawn under the unit and then execute a sharp right-angle turn to go up through the bottom slots. This change of direction is a major impediment to air flow, and if you are planning to lose a lot of heat then it feeds into the design of something so humble as the feet the unit stands on; the higher the better for air flow. In one instance the amplifier feet were made 13 mm taller and all the internal amplifier temperatures dropped by 5°C. Standing such a unit on a thick-pile carpet can be a really bad idea, but someone is bound to do it (and then drop their coat on top of it); hence the need for overtemperature cutouts if amplifiers are to be fully protected.

### 13.5.3 Mains Transformers

A toroidal transformer is useful because of its low external field. It must be mounted so that it can be rotated to minimize the effect of what stray fields it does emit. Most suitable toroids have single-strand secondary leadouts, which are too stiff to allow rotation; these can be cut short and connected to suitably large flexible wire such as 32/02, with carefully sleeved and insulated joints. One prototype amplifier I have built had a sizeable toroid mounted immediately adjacent to the TO3 end of the amplifier PCB; however, complete cancellation of magnetic hum (hum and ripple output level below −90 dBu) was possible on rotation of the transformer.

A more difficult problem is magnetic radiation caused by reservoir charging pulses (as opposed to the ordinary magnetization of the core, which would be essentially the same if the load current was sinusoidal), which can be picked up by either the output connections or cabling to the power transistors if these are mounted off board. For this reason, the transformer should be kept physically as far away as possible from even the high-current section of the amplifier PCB.

As usual with toroids, ensure that the bolt through the middle cannot form a shorted turn by contacting the chassis in two places.

### 13.5.4 Wiring Layout

There are several important points about the wiring for any power amplifier:

- Keep the + and − HT supply wires to the amplifiers close together. This minimizes the generation of distorted magnetic fields that may otherwise couple into the signal wiring and degrade linearity. Sometimes it seems more effective to include the 0-V line in this cable run; if so, it should be tightly braided to keep the wires in close proximity. For the same reason, if the power transistors are mounted off the PCB, the cabling to each device should be configured to minimize loop formation.

- The rectifier connections should go directly to the reservoir capacitor terminals and then away again to the amplifiers. Common impedance in these connections superimposes charging pulses on the rail ripple waveform, which may degrade amplifier PSRR.

- Do not use the actual connection between the two reservoir capacitors as any form of starpoint. It carries heavy capacitor-charging pulses that generate a significant voltage drop even if thick wire is used. As Figure 13.1 shows, the *starpoint* is teed off from this connection. This is a starpoint only insofar as the amplifier ground connections split off from here, so do not connect the input grounds to it, as distortion performance will suffer.

### 13.5.5 Semiconductor Installation

- Driver transistor installation. These are usually mounted onto separate heat sinks that are light enough to be soldered into the PCB without further fixing. Silicone thermal washers ensure good thermal contact, and spring clips are used to hold the package firmly against the sink. Electrical isolation between device and heat sink is not normally essential, as the PCB need not make any connection to the heat sink fixing pads.

- TO3P power transistor installation. These large flat plastic devices are usually mounted on to the main heat sink with spring clips, which are not only rapid to install, but also generate less mechanical stress in the package than bolting the device down by its mounting hole. They also give a more uniform pressure onto the thermal washer material.

- TO3 power transistor installation. The TO3 package is extremely efficient at heat transfer, but notably more awkward to mount.

My preference is for TO3s to be mounted on an aluminium thermal coupler bolted against the component side of the PCB. The TO3 pins may then be soldered directly on the PCB solder side. The thermal coupler is drilled with suitable holes to allow M3.5 fixing bolts to pass through the TO3 flange holes, through the flange, and then be secured on the other side of the PCB by nuts and crinkle washers, which will ensure good contact with the PCB mounting pads. For reliability, the crinkle washers must cut through the solder tinning into the underlying copper; a solder contact alone will creep under pressure and the contact force will decay over time.

Insulating sleeves are essential around the fixing bolts where they pass through the thermal coupler; nylon is a good material for these as it has a good high-temperature capability. Depending on the size of the holes drilled in the thermal coupler for the two TO3 package pins (and this should be as small as practicable to maximize the area for heat transfer), these are also likely to require insulation; silicone rubber sleeving carefully cut to length is very suitable.

An insulating thermal washer must be used between TO3 and flange; these tend to be delicate and the bolts must not be overtightened. If you have a torque wrench, then 10 Nw/m is an approximate upper limit for M3.5 fixing bolts. *Do not* solder the two transistor pins to the PCB until the TO3 is mounted firmly and correctly, fully bolted down, and checked for electrical isolation from the heat sink. Soldering these pins and *then* tightening the fixing bolts is likely to force the pads from the PCB. If this should happen, then it is quite in order to repair the relevant track or pad with a small length of stranded wire to the pin; 7/02 size is suitable for a very short run.

Alternatively, TO3s can be mounted off PCB (e.g., if you already have a large heat sink with TO3 drillings) with wires taken from the TO3 pads on the PCB to the remote devices. These wires should be fastened together (two bunches of three is fine) to prevent loop formation; see earlier discussion. I cannot give a maximum safe length for such cabling, but certainly 8 inches causes no HF stability problems. The emitter and collector wires should be substantial, for example, 32/02, but the base connections can be as thin as 7/02.

# PART 4

# Digital Audio

# CHAPTER 14

# Digital Audio Fundamentals

John Watkinson

## 14.1 Audio as Data

The most exciting aspects of digital technology are the tremendous possibilities that were not available with analog technology. Many processes that are difficult or impossible in the analog domain are straightforward in the digital domain. Once audio is in the digital domain, it becomes data, and only differs from generic data in that it needs to be reproduced with a certain time base.

The worlds of digital audio, digital video, communication, and computation are closely related, and that is where the real potential lies. The time when audio was a specialist subject that could evolve in isolation from other disciplines has gone. Audio has now become a branch of information technology (IT); a fact that is reflected in the approach of this book.

Systems and techniques developed in other industries for other purposes can be used to store, process, and transmit audio, video, or both at once. IT equipment is available at low cost because the volume of production is far greater than that of professional audiovisual equipment. Disk drives and memories developed for computers can be put to use in such products. Communications networks developed to handle data can happily carry audiovisual data over indefinite distances without quality loss.

As the power of processors increases, it becomes possible to perform under software control processes that previously required dedicated hardware. This allows a dramatic reduction in hardware cost. Inevitably the very nature of audiovisual equipment and the ways in which it is used is changing along with the manufacturers who supply it. The computer industry is competing with traditional manufacturers, using the economics of mass production.

Tape is a linear medium and it is necessary to wait for the tape to wind to a desired part of the recording. In contrast, the head of a hard disk drive can access any stored data in milliseconds. This is known in computers as direct access and in audio production as nonlinear access. As a result, the nonlinear editing workstation based on hard drives has eclipsed the use of tape for editing.

Digital broadcasting uses coding techniques to eliminate the interference, fading, and multipath reception problems of analog broadcasting. At the same time, more efficient use is made of available bandwidth. The hard drive-based consumer audio recorder gives the consumer more power.

Figure 14.1 shows what the home audio system of the future may look like. MPEG-compressed signals may arrive in real time by terrestrial or satellite broadcast, via

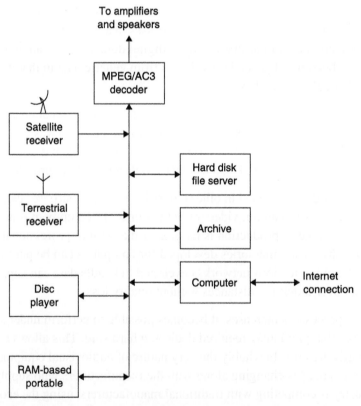

Figure 14.1: Audio system of the future based on data technology.

the Internet, or as the soundtrack of media such as DVD. Media such as compact disc supply uncompressed data for higher quality. The heart of the system is a hard drive-based server. This can be used to time shift broadcast programs, to skip commercial breaks, or to assemble requested audio material transmitted in nonreal time at low bit rates. If equipped with a Web browser, the server may explore the Web looking for material that is of the same kind the user normally wants. As the cost of storage falls, the server may download this material speculatively.

For portable use, the user may download compressed audio files into memory-based devices, which act as audio players, yet have no moving parts. On playback the bit stream is recovered from memory, decoded, and converted typically to a signal that can drive headphones.

Ultimately, digital technology will change the nature of broadcasting out of recognition. Once the viewer has nonlinear storage technology and electronic program guides, the traditional broadcaster's transmitted schedule is irrelevant. Increasingly, consumers will be able to choose what is played and when, rather than the broadcaster deciding for them. The broadcasting of conventional commercials will cease to be effective when viewers have the technology to skip them. Anyone with a Web site that can stream audio data can become a broadcaster.

## 14.2 What is an Audio Signal?

An analog audio signal is an electrical waveform that is a representation of the velocity of a microphone diaphragm. Such a signal is two dimensional in that it carries a voltage changing with respect to time. In analog systems, these waveforms are conveyed by some infinite variation of a continuous parameter. In a recorder, distance along the medium is a further, continuous analog of time. It does not matter at what point a recording is examined along its length, a value will be found for the recorded signal. That value can itself change with infinite resolution within the physical limits of the system.

Those characteristics are the main weakness of analog signals. Within the allowable bandwidth, *any* waveform is valid. If the speed of the medium is not constant, one valid waveform is changed into another valid waveform; a problem that cannot be detected in an analog system and that results in wow and flutter. In addition, a voltage error simply changes one valid voltage into another; noise cannot be detected in an analog signal. Noise might be suspected, but how is one to know what proportion of the received signal is noise and what is the original? If the transfer function of a system is not linear, distortion results, but the distorted waveforms are still valid; an analog system cannot detect distortion. Again distortion might be suspected, but it is impossible to tell how much of the energy at a given frequency is due to distortion and how much was actually present in the original signal.

It is a characteristic of analog systems that degradations cannot be separated from the original signal, so nothing can be done about them. At the end of a system a signal carries the sum of all degradations introduced at each stage through which it passed. This sets a limit to the number of stages through which a signal can be passed before it is useless. Alternatively, if many stages are envisaged, each piece of equipment must be far better than necessary so that the signal is still acceptable at the end. The equipment will naturally be more expensive.

Digital audio is simply an alternative means of carrying an audio waveform. Although there are a number of ways in which this can be done, there is one system, known as pulse code modulation (PCM), that is in virtually universal use.[1] Figure 14.2 shows how PCM works.

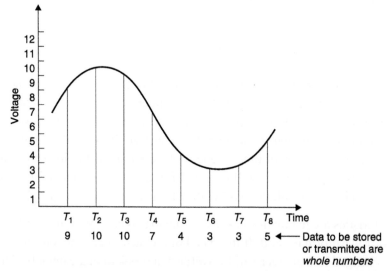

Figure 14.2: In pulse code modulation the analog waveform is measured periodically at the sampling rate. The voltage (represented here by the height) of each sample is then described by a whole number. The whole numbers are stored or transmitted rather than the waveform itself.

Instead of being continuous, the time axis is represented in a discrete or stepwise manner. The audio waveform is not carried by continuous representation, but by measurement at regular intervals. This process is called sampling, and the frequency with which samples are taken is called the sampling rate or sampling frequency $F_s$. Each sample still varies infinitely as the original waveform did. To complete the conversion to PCM, each sample is then represented to finite accuracy by a discrete number in a process known as quantizing.

At the analog-to-digital convertor (ADC), every effort is made to rid the sampling clock of jitter, or time instability, so every sample is taken at an exactly even time step. Clearly, if there is any subsequent time base error, the instants at which samples arrive will be changed and the effect can be detected. If samples arrive at some destination with an irregular time base, the effect can be eliminated by temporarily storing the samples in a memory and reading them out using a stable, locally generated clock. This process is called time base correction and all properly engineered digital audio systems will use it.

Those who are not familiar with digital principles often worry that sampling takes away something from a signal because it appears not to be taking notice of what happened between the samples. This would be true in a system having infinite bandwidth, but no analog signal can have infinite bandwidth. All analog signal sources from microphones and so on have a resolution or frequency response limit, as indeed do devices such as loudspeakers and human hearing. When a signal has finite bandwidth, the rate at which it can change is limited, and the way in which it changes becomes predictable. When a waveform can only change between samples in one way, it is then only necessary to convey the samples and the original waveform can be unambiguously reconstructed from them.

As stated, each sample is also discrete or represented in a stepwise manner. The magnitude of the sample, which will be proportional to the voltage of the audio signal, is represented by a whole number. This process is known as quantizing and results in an approximation, but the size of the error can be controlled until it is negligible. The advantage of using whole numbers is that they are not prone to drift.

If a whole number can be carried from one place to another without numerical error, it has not changed at all. By describing audio waveforms numerically, the original information has been expressed in a way that is more robust.

Essentially, digital audio carries the sound numerically. Each sample is a numerical analog of the voltage at the corresponding instant in the sound.

## 14.3 Why Binary?

Arithmetically, the binary system is the simplest numbering scheme possible.

Figure 14.3(a) shows that there are only two symbols: 1 and 0. Each symbol is a binary digit, abbreviated to *bit*. One bit is a datum and many bits are data. Logically, binary allows a system of thought in which statements can only be true or false.

The great advantage of binary systems is that they are the most resistant to misinterpretation. In information terms they are *robust*. Figures 14.3(b) and 14.3(c) show some binary terms and some nonbinary terms, respectively, for comparison. In all real processes, the wanted information is disturbed by noise and distortion, but with only two possibilities to distinguish, binary systems have the greatest resistance to such effects.

Figure 14.4(a) shows an ideal binary electrical signal is simply two different voltages: a high voltage representing a true logic state or a binary 1 and a low voltage representing a false logic state or a binary 0. The ideal waveform is also shown in Figure 14.4(b) after

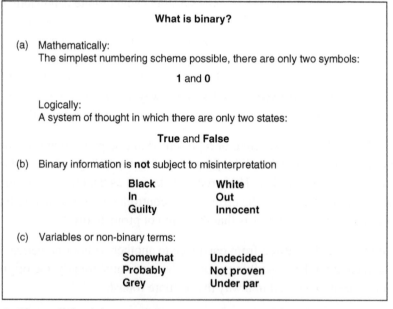

**Figure 14.3: Binary digits (a) can only have two values. At (b) some everyday binary terms are shown, whereas (c) shows some terms that cannot be expressed by a binary digit.**

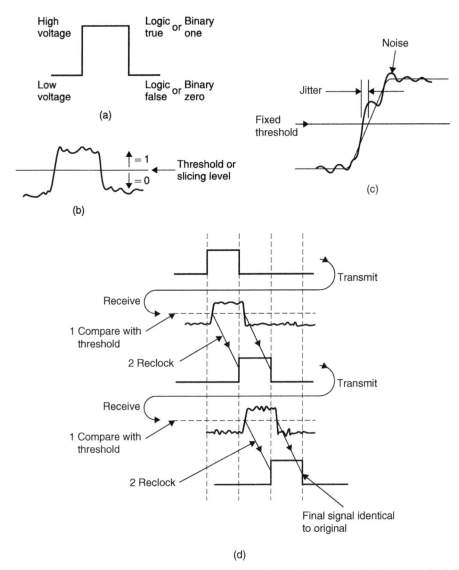

(d)

Figure 14.4: An ideal binary signal (a) has two levels. After transmission it may look like (b), but after slicing the two levels can be recovered. Noise on a sliced signal can result in jitter (c), but reclocking combined with slicing makes the final signal identical to the original as shown in (d).

it has passed through a real system. The waveform has been considerably altered, but the binary information can be recovered by comparing the voltage with a threshold that is set half way between the ideal levels. In this way any received voltage above the threshold is considered a 1 and any voltage below is considered a 0. This process is called slicing and can reject significant amounts of unwanted noise added to the signal. The signal will be carried in a channel with finite bandwidth, which limits the slew rate of the signal; an ideally upright edge is made to slope.

Noise added to a sloping signal [Figure 14.4(c)] can change the time at which the slicer judges that the level passed through the threshold. This effect is also eliminated when the output of the slicer is reclocked. Figure 14.4(d) shows that however many stages the binary signal passes through, the information is unchanged except for a delay. Of course, excessive noise could cause a problem. If it had sufficient level and an appropriate polarity, noise could force the signal to cross the threshold and the output of the slicer would then be incorrect. However, as binary has only two symbols, if it is known that the symbol is incorrect, it need only be set to the other state and a perfect correction has been achieved. Error correction really is as trivial as that, although determining which bit needs to be changed is somewhat harder.

Figure 14.5 shows that binary information can be represented by a wide range of real phenomena. All that is needed is the ability to exist in two states. A switch can be open

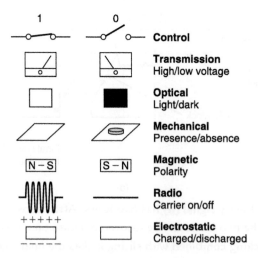

**Figure 14.5: A large number of real phenomena can be used to represent binary data.**

or closed and so represent a single bit. This switch may control the voltage in a wire that allows the bit to be transmitted. In an optical system, light may be transmitted or obstructed. In a mechanical system, the presence or absence of some feature can denote the state of a bit. The presence or absence of a radio carrier can signal a bit. In a random access memory (RAM), the state of an electric charge stores a bit.

Figure 14.5 also shows that magnetism is naturally binary as two stable directions of magnetization are easily arranged and rearranged as required. This is why digital magnetic recording has been so successful: it is a natural way of storing binary signals.

The robustness of binary signals means that bits can be packed more densely onto storage media, increasing the performance or reducing the cost. In radio signaling, lower power can be used.

In decimal systems, the digits in a number (counting from the right, or least significant end) represent ones, tens, hundreds, thousands, and so on. Figure 14.6 shows that in binary, the bits represent one, two, four, eight, sixteen, and so on. A multidigit binary number is commonly called a word, and the number of bits in the word is called the wordlength. The right-hand bit is called the least significant bit (LSB), whereas the bit on

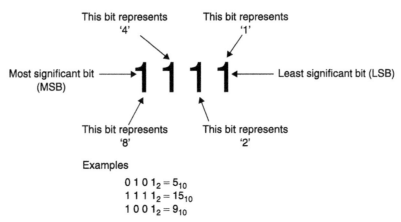

Figure 14.6: In a binary number, the digits represent increasing powers of two from the LSB. Also defined here are MSB and wordlength. When the wordlength is eight bits, the word is a byte. Binary numbers are used as memory addresses, and the range is defined by the address wordlength. Some examples are shown here.

the left-hand end of the word is called the most significant bit (MSB). Clearly more digits are required in binary than in decimal, but they are handled more easily. A word of eight bits is called a byte, which is a contraction of "by eight."

Figure 14.6 also shows some binary numbers and their equivalent in decimal. The radix point has the same significance in binary: symbols to the right of it represent one-half, one-quarter, and so on.

Binary words can have a remarkable range of meanings. They may describe the magnitude of a number such as an audio sample or an image pixel or they may specify the address of a single location in a memory. In all cases the possible range of a word is limited by the wordlength. The range is found by raising two to the power of the wordlength. Thus a 4-bit word has 16 combinations and could address a memory having 16 locations. A 16-bit word has 65,536 combinations. Figure 14.7(a) shows some examples of wordlength and resolution.

The capacity of memories and storage media is measured in bytes, but to avoid large numbers, kilobytes, megabytes, and gigabytes are often used. A 10-bit word has 1024 combinations, which is close to 1000. In digital terminology, 1 K is defined as 1024, so a kilobyte of memory contains 1024 bytes. A megabyte (1 MB) contains 1024 kilobytes and would need a 20-bit address. A gigabyte contains 1024 megabytes and would need a 30-bit address. Figure 14.7(b) shows some examples.

## 14.4  Why Digital?

There are two main answers to this question, and it is not possible to say which is the most important, as it will depend on one's standpoint.

  a.  The quality of reproduction of a well-engineered digital audio system is independent of the medium and depends only on the quality of the conversion processes and of any compression scheme.

  b.  The conversion of audio to the digital domain allows tremendous opportunities that were denied to analog signals.

Someone who is only interested in sound quality will judge the former the most relevant. If good-quality convertors can be obtained, all the shortcomings of analog recording and transmission can be eliminated to great advantage. An extremely good signal-to-noise ratio is possible, coupled with very low distortion. Timing errors between channels can be

The wordlength determines the possible range of values:

| Wordlength | | Range |
|:---:|:---:|:---:|
| 1 | | $2\ (2^1)$ |
| 2 | | $4\ (2^2)$ |
| 3 | | $8\ (2^3)$ |
| • | | |
| • | | |
| • | | |
| 8 | | $256\ (2^8)$ |
| • | | |
| 10 | | $1024\ (2^{10})$ |
| • | | |
| • | | |
| • | | |
| • | | |
| • | | |
| 16 | | $65\,536\ (2^{16})$ |

(a)

**Round numbers in binary**

$10000000000_2$ = 1024 = 1 K (Kilo in computers)
1 K × 1 K    = 1 M  (Mega)
1 M × 1 K    = 1 G  (Giga)
1 M × 1 M    = 1 T  (Tera)

(b)

**Figure 14.7: The wordlength of a sample controls the resolution as shown in (a). The ability to address memory locations is also determined in the same way as in (b).**

eliminated, making for accurate stereo images. One's greatest effort is expended in the design of convertors, whereas those parts of the system that handle data need only be workmanlike. When a digital recording is copied, the same numbers appear on the copy: it is not a dub, it is a clone. If the copy is undistinguishable from the original, there has been no generation loss. Digital recordings can be copied indefinitely without loss of quality. This is, of course, wonderful for the production process, but when the technology becomes available to the consumer, the issue of copyright becomes of great importance.

In the real world everything has a cost, and one of the greatest strengths of digital technology is low cost. When the information to be recorded consists of discrete numbers, they can be packed densely on the medium without quality loss. Should some bits be in error because of noise or dropout, error correction can restore the original value. Digital recordings take up less space than analog recordings for the same or better quality. Digital circuitry costs less to manufacture because more functionality can be put in the same chip.

Digital equipment can have self-diagnosis programs built in. The machine points out its own failures so the cost of maintenance falls. A small operation may not need maintenance staff at all; a service contract is sufficient. A larger organization will still need maintenance staff, but they will be fewer in number and their skills will be oriented more to systems than to devices.

## 14.5  Some Digital Audio Processes Outlined

While digital audio is a large subject, it is not necessarily a difficult one. Every process can be broken down into smaller steps, each of which is relatively easy to follow. The main difficulty with study is to appreciate where the small steps fit into the overall picture. Subsequent chapters of this book will describe the key processes found in digital technology in some detail, whereas this chapter illustrates why these processes are necessary and shows how they are combined in various ways in real equipment. Once the general structure of digital devices is appreciated, other chapters can be put in perspective.

Figure 14.8(a) shows a minimal digital audio system. This is no more than a point-to-point link that conveys analog audio from one place to another. It consists of a pair of convertors and hardware to serialize and deserialize the samples. There is a need for standardization in serial transmission so that various devices can be connected together.

Analog audio entering the system is converted in the ADC to samples that are expressed as binary numbers. A typical sample would have a wordlength of 16 bits. The sample is connected in parallel into an output register that controls the cable drivers. The cable also carries the sampling rate clock. Data are sent to the other end of the line where a slicer rejects noise picked up on each signal. Sliced data are then loaded into a receiving register by the clock and sent to the digital-to-analog convertor (DAC), which converts the sample back to an analog voltage.

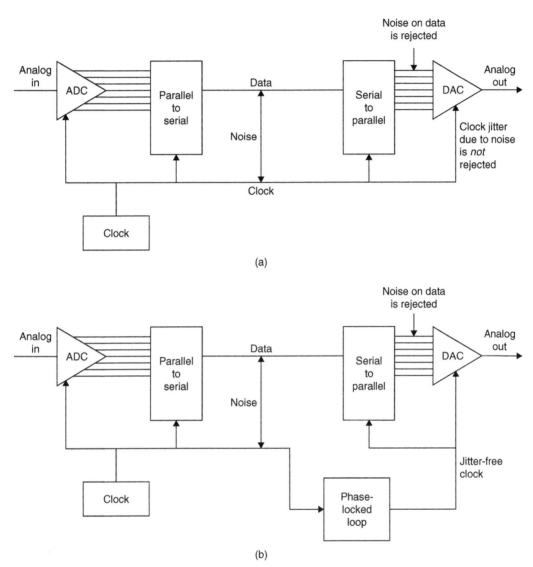

Figure 14.8: In (a) two convertors are joined by a serial link. Although simple, this system is deficient because it has no means to prevent noise on the clock lines causing jitter at the receiver. In (b) a phase-locked loop is incorporated, which filters jitter from the clock.

As Figure 14.4 showed, noise can change the timing of a sliced signal. While this system rejects noise that threatens to change the numerical value of the samples, it is powerless to prevent noise from causing jitter in the receipt of the sample clock. Noise on the clock means that samples are not converted with a regular time base and the impairment caused will be audible.

The jitter problem is overcome in Figure 14.8(b) by the inclusion of a phase-locked loop, which is an oscillator that synchronizes itself to the *average* frequency of the clock but which filters out the instantaneous jitter.

The system of Figure 14.8 is extended in Figure 14.9 by the addition of some RAM. What the device does is determined by the way in which the RAM address is controlled. If the RAM address increases by one every time a sample from the ADC is stored in the RAM, an audio recording can be made for a short period until the RAM is full. The recording can be played back by repeating the address sequence at the same clock rate but reading the memory into the DAC. The result is generally called a sampler. If the memory capacity is increased, the device can be used for general recording. RAM recorders are replacing dictating machines and the tape recorders used by journalists. In general they will be restricted to a fairly short playing time because of the high cost of memory in comparison with other storage media.

Using compression, the playing time of a RAM-based recorder can be extended. For unchanging sounds such as test signals and station IDs, read only memory can be used instead as it is nonvolatile.

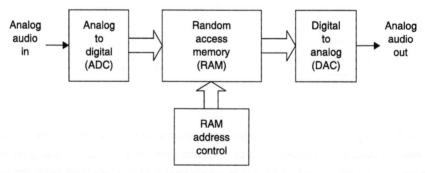

**Figure 14.9: In the digital sampler, the recording medium is a RAM. Recording time available is short compared with other media, but access to the recording is immediate and flexible as it is controlled by addressing the RAM.**

## 14.6 Time Compression and Expansion

Data files such as computer programs are simply lists of instructions and have no natural time axis. In contrast, audio and video data are sampled at a fixed rate and need to be presented to the viewer at the same rate. In audiovisual systems the audio also needs to be synchronized to the video. Continuous bit streams at a fixed bit rate are difficult for generic data recording and transmission systems to handle. Such systems mostly work on blocks of data that can be addressed and/or routed individually. The bit rate may be fixed at the design stage at a value that may be too low or too high for the audio or video data to be handled.

The solution is to use time compression or expansion. Figure 14.10 shows a RAM that is addressed by binary counters that periodically overflow to zero and start counting again, giving the RAM a ring structure. If write and read addresses increment at the same speed, the RAM becomes a fixed data delay as the addresses retain a fixed relationship. However, if the read address clock runs at a higher frequency but in bursts, output data are assembled into blocks with spaces in between. Data are now time compressed. Instead of being an unbroken stream, which is difficult to handle, data are in blocks with convenient pauses in between them. Numerous processes can take place in these pauses. A hard disk might move its heads to another track. In all types of recording and

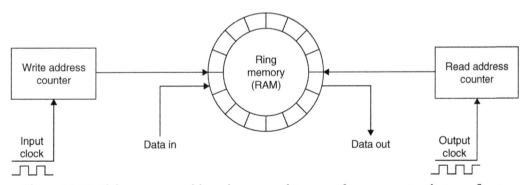

**Figure 14.10: If the memory address is arranged to come from a counter that overflows, the memory can be made to appear circular. The write address then rotates endlessly, overwriting previous data once per revolution. The read address can follow the write address by a variable distance (not exceeding one revolution) and so a variable delay takes place between reading and writing.**

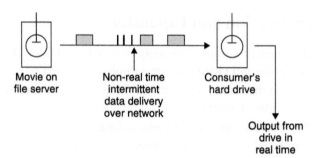

Figure 14.11: In nonreal-time transmission, data are transferred slowly to a storage medium, which then outputs real-time data. Recordings can be downloaded to the home in this way.

transmission, the time compression of the samples allows time for synchronizing patterns, subcode, and error-correction words to be inserted.

Subsequently, any time compression can be reversed by time expansion. This requires a second RAM identical to the one shown. Data are written into the RAM in bursts, but read out at the standard sampling rate to restore a continuous bit stream. In a recorder, the time-expansion stage can be combined with the time base correction stage so that speed variations in the medium can be eliminated at the same time. The use of time compression is universal in digital recording and is widely used in transmission. In general the *instantaneous* data rate in the channel is not the same as the original rate, although clearly the *average* rate must be the same.

Where the bit rate of the communication path is inadequate, transmission is still possible, but not in real time. Figure 14.11 shows that data to be transmitted will have to be written in real time on a storage device such as a disk drive, and the drive will then transfer data at whatever rate is possible to another drive at the receiver. When the transmission is complete, the second drive can then provide data at the correct bit rate.

In the case where the available bit rate is higher than the correct data rate, the same configuration can be used to copy an audio data file faster than in real time. Another application of time compression is to allow several streams of data to be carried along the same channel in a technique known as *multiplexing*. Figure 14.12 shows some examples. In Figure 14.12(a), multiplexing allows audio and video data to be recorded on the same heads in a digital video recorder such as DVC. In Figure 14.12(b), several radio or television channels are multiplexed into one MPEG transport stream.

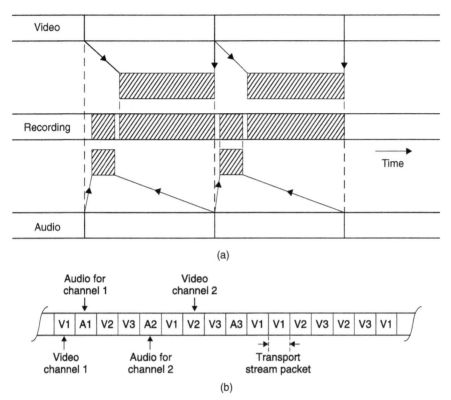

Figure 14.12: (a) Time compression is used to shorten the length of track needed by the video. Heavily time-compressed audio samples can then be recorded on the same track using common circuitry. In MPEG, multiplexing allows data from several TV channels to share one bit stream (b).

## 14.7 Error Correction and Concealment

All practical recording and transmission media are imperfect. Magnetic media, for example, suffer from noise and dropouts. In a digital recording of binary data, a bit is either correct or wrong, with no intermediate stage. Small amounts of noise are rejected, but inevitably, infrequent noise impulses cause some individual bits to be in error. Dropouts cause a larger number of bits in one place to be in error. An error of this kind is called a burst error. Whatever the medium and whatever the nature of the mechanism responsible, data are either recovered correctly or suffer some combination of bit errors and burst errors. In optical disks, random errors can be caused by imperfections in the moulding process, whereas burst errors are due to contamination or scratching of the disk surface.

The audibility of a bit error depends on which bit of the sample is involved. If the LSB of one sample was in error in a detailed musical passage, the effect would be totally masked and no one could detect it. Conversely, if the MSB of one sample was in error during a pure tone, no one could fail to notice the resulting click. Clearly a means is needed to render errors from the medium inaudible. This is the purpose of error correction.

In binary, a bit has only two states. If it is wrong, it is only necessary to reverse the state and it must be right. Thus the correction process is trivial and perfect. The main difficulty is in identifying the bits that are in error. This is done by coding data by adding redundant bits. Adding redundancy is not confined to digital technology, airliners have several engines and cars have twin braking systems. Clearly the more failures that have to be handled, the more redundancy is needed.

In digital recording, the amount of error that can be corrected is proportional to the amount of redundancy. Consequently, *corrected* samples are undetectable. If the amount of error exceeds the amount of redundancy, correction is not possible, and, in order to allow graceful degradation, concealment will be used. Concealment is a process where the value of a missing sample is estimated from those nearby. The estimated sample value is not necessarily exactly the same as the original, and so under some circumstances concealment can be audible, especially if it is frequent. However, in a well-designed system, concealments occur with negligible frequency unless there is an actual fault or problem.

Concealment is made possible by rearranging the sample sequence prior to recording. This is shown in Figure 14.13 where odd-numbered samples are separated from even-numbered samples prior to recording. The odd and even sets of samples may be recorded in different places on the medium so that an uncorrectable burst error affects only one set. On replay, the samples are recombined into their natural sequence, and the error is now split up so that it results in every other sample being lost in two different places. In those places, the waveform is described half as often, but can still be reproduced with some loss of accuracy. This is better than not being reproduced at all even if it is not perfect. Most tape-based digital audio recorders use such an odd/even distribution for concealment. Clearly, if any errors are fully correctable, the distribution is a waste of time; it is only needed if correction is not possible.

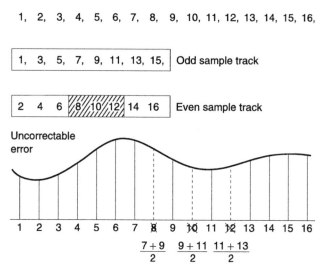

Figure 14.13: In cases where error correction is inadequate, concealment can be used, provided that the samples have been ordered appropriately in the recording. Odd and even samples are recorded in different places as shown here. As a result, an uncorrectable error causes incorrect samples to occur singly, between correct samples. In the example shown, sample 8 is incorrect, but samples 7 and 9 are unaffected and an approximation to the value of sample 8 can be had by taking the average value of the two. This interpolated value is substituted for the incorrect value.

The presence of an error-correction system means that the audio quality is independent of the medium/head quality within limits. There is no point in trying to assess the health of a machine by listening to the audio, as this will not reveal whether the error rate is normal or within a whisker of failure. The only useful procedure is to monitor the frequency with which errors are being corrected and to compare it with normal figures.

Digital systems such as broadcast channels, optical disks, and magnetic recorders are prone to burst errors. Adding redundancy equal to the size of expected bursts to every code is inefficient. Figure 14.14(a) shows that the efficiency of the system can be raised using interleaving. Sequential samples from the ADC are assembled into codes, but these are not recorded/transmitted in their natural sequence. A number of sequential codes are assembled along rows in a memory. When the memory is full, it is copied to the medium by reading down columns.

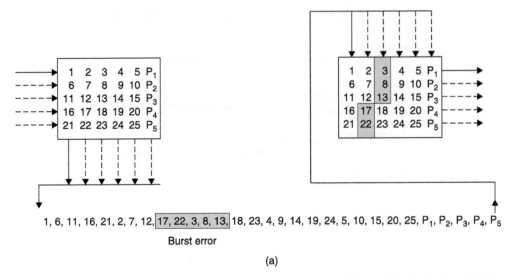

1, 6, 11, 16, 21, 2, 7, 12, 17, 22, 3, 8, 13, 18, 23, 4, 9, 14, 19, 24, 5, 10, 15, 20, 25, $P_1$, $P_2$, $P_3$, $P_4$, $P_5$

Burst error

(a)

**Figure 14.14(a): Interleaving is essential to make error-correction schemes more efficient. Samples written sequentially in rows into a memory have redundancy P added to each row. The memory is then read in columns and data are sent to the recording medium. On replay the nonsequential samples from the medium are deinterleaved to return them to their normal sequence. This breaks up the burst error (shaded) into one error symbol per row in the memory, which can be corrected by the redundancy P.**

Subsequently, the samples need to be deinterleaved to return them to their natural sequence. This is done by writing samples from tape into a memory in columns, and when it is full, the memory is read in rows. Samples read from the memory are now in their original sequence so there is no effect on the information. However, if a burst error occurs, as is shown shaded on the diagram, it will damage sequential samples in a vertical direction in the deinterleave memory. When the memory is read, a single large error is broken down into a number of small errors whose sizes are exactly equal to the correcting power of the codes and the correction is performed with maximum efficiency.

An extension of the process of interleave is where the memory array has not only rows made into code words but also columns made into code words by the addition of vertical redundancy. This is known as a product code. Figure 14.14(b) shows that in a product code the redundancy calculated first and checked last is called the outer code, and the redundancy calculated second and checked first is called the inner code. The inner code

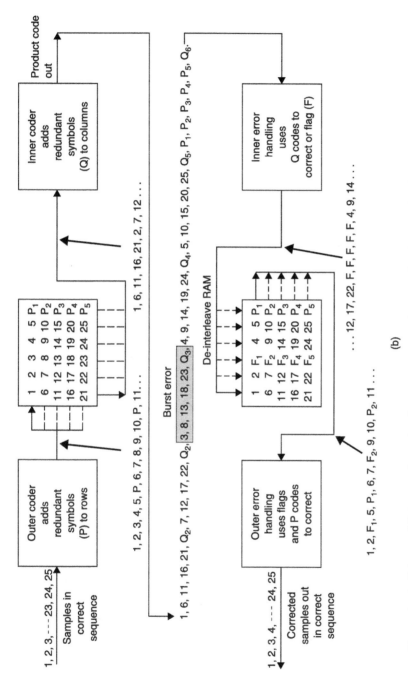

Figure 14.14(b): In addition to the redundancy P on rows, inner redundancy Q is also generated on columns. On replay, the Q code checker will pass on flag F if it finds an error too large to handle itself. Flags pass through the deinterleave process and are used by the outer error correction to identify which symbol in the row needs correcting with P redundancy. The concept of crossing two codes in this way is called a product code.

is formed along tracks on the medium. Random errors due to noise are corrected by the inner code and do not impair the burst-correcting power of the outer code. Burst errors are declared uncorrectable by the inner code, which flags the bad samples on the way into the deinterleave memory. The outer code reads the error flags in order to locate erroneous data. As it does not have to compute the error locations, the outer code can correct more errors.

The interleave, deinterleave, time-compression, and time base-correction processes inevitably cause delay.

## 14.8 Channel Coding

In most recorders used for storing digital information, the medium carries a track that reproduces a single waveform. Clearly, data words representing audio samples contain many bits and so they have to be recorded serially, a bit at a time. Some media, such as optical or magnetic disks, have only one active track, so it must be totally self-contained. Tape-based recorders may have several tracks read or written simultaneously. At high recording densities, physical tolerances cause phase shifts, or timing errors, between tracks and so it is not possible to read them in parallel. Each track must still be self-contained until the replayed signal has been time base corrected.

Recording data serially is not as simple as connecting the serial output of a shift register to the head. In digital audio, samples may contain strings of identical bits. For example, silence in digital audio is represented by samples in which all the bits are zero. If a shift register is loaded with such a sample and shifted out serially, the output stays at a constant level for the period of the identical bits, and nothing is recorded on the track. On replay there is nothing to indicate how many bits were present or even how fast to move the medium. Clearly, serialized raw data cannot be recorded directly, they must be modulated into a waveform that contains an embedded clock irrespective of the values of the bits in the samples. On replay, a circuit called a data separator can lock to the embedded clock and use it to separate strings of identical bits.

The process of modulating serial data to make them self-clocking is called channel coding. Channel coding also shapes the spectrum of the serialized waveform to make it more efficient. With a good channel code, more data can be stored on a given medium. Spectrum shaping is used in optical disks to prevent data from interfering with the focus and tracking servos and in hard disks and in certain tape formats to allow rerecording without erase heads.

Channel coding is also needed to broadcast digital signals where shaping of the spectrum is an obvious requirement to avoid interference with other services.

## 14.9 Audio Compression

In its native form, high-quality digital audio requires a high data rate, which may be excessive for certain applications. One approach to the problem is to use compression, which reduces that rate significantly with a moderate loss of subjective quality. Because the human hearing system is not equally sensitive to all frequencies, some coding gain can be obtained using fewer bits to describe the frequencies that are less audible.

While compression may achieve considerable reduction in bit rate, it must be appreciated that compression systems reintroduce the generation loss of the analog domain to digital systems.

One of the most popular compression standards for audio and video is known as MPEG. Figure 14.15 shows that the output of a single MPEG compressor is called an *elementary*

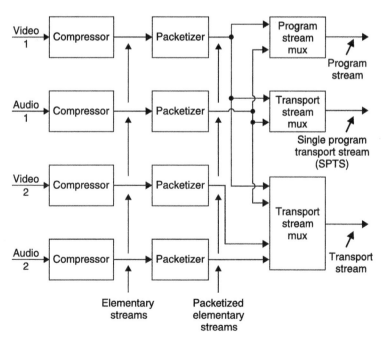

**Figure 14.15: The bit stream types of MPEG-2. See the text for details.**

*stream.* In practice, audio and video streams of this type can be combined using multiplexing. The *program stream* is optimized for recording and is based on blocks of arbitrary size. The *transport stream* is optimized for transmission and is based on blocks of constant size.

It should be appreciated that many successful products use non-MPEG compression.

Compression and the corresponding decoding are complex processes and take time, adding to existing delays in signal paths. Concealment of uncorrectable errors is also more difficult on compressed data.

## 14.10  Disk-Based Recording

The magnetic disk drive was perfected by the computer industry to allow rapid random access to data, and so it makes an ideal medium for editing. The heads do not touch the disk, but are supported on a thin air film, which gives them a long life but which restricts the recording density. Thus disks cannot compete with tape for archiving, but for work such as compact disc production they have no equal.

The disk drive provides intermittent data transfer owing to the need to reposition the heads. Figure 14.16 shows that disk-based devices rely on a quantity of RAM acting as a buffer between the real-time audio environment and the intermittent data environment.

Figure 14.17 shows the block diagram of an audio recorder based on disks and compression. The recording time and sound quality will not compete with full bandwidth tape-based devices, but following acquisition the disks can be used directly in an edit system, allowing a useful time saving in electronic news-gathering applications.

Development of the optical disk was stimulated by the availability of low-cost lasers. Optical disks are available in many different types, some which can only be recorded once, whereas others are erasable. Optical disks have in common the fact that access is generally slower than with magnetic drives and that it is difficult to obtain high data rates, but most of them are removable and can act as interchange media.

## 14.11  Rotary Head Digital Recorders

The rotary head recorder has the advantage that the spinning heads create a high head-to-tape speed, offering a high bit rate recording without high linear tape speed.

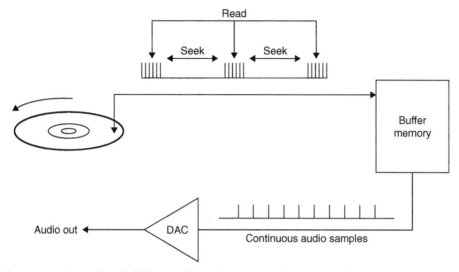

**Figure 14.16: In a hard disk recorder, a large-capacity memory is used as a buffer or time base corrector between the convertors and the disk. The memory allows the convertors to run constantly, despite the interruptions in disk transfer caused by the head moving between tracks.**

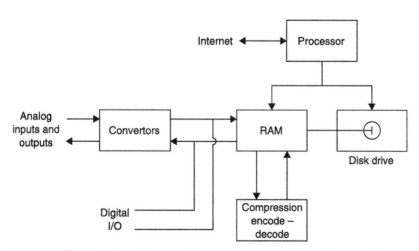

**Figure 14.17: A disk-based audio recorder can capture audio and transmit compressed audio files over the Internet.**

While mechanically complex, rotary head transport has been raised to a high degree of refinement and offers the highest recording density and thus lowest cost per bit of all digital recorders.

Figure 14.18 shows a representative block diagram of a rotary head machine. Following the convertors, a compression process may be found. In an uncompressed recorder, there will be distribution of odd and even samples for concealment purposes. An interleaved product code will be formed prior to the channel coding stage, which produces the recorded waveform. On replay the data separator decodes the channel code and the inner and outer codes perform correction as in Section 14.7. Following this the data channels are recombined and any necessary concealment will take place. Any compression will be decoded prior to the output convertors.

## 14.12 Digital Audio Broadcasting

Although it has given good service for many years, analog broadcasting is an inefficient use of bandwidth. Using compression, digital modulation, and error-correction techniques, acceptable sound quality can be obtained in a fraction of the bandwidth of analog. Pressure on spectrum use from other uses, such as cellular telephones, will only increase, which may result in a rapid changeover to digital broadcasts.

In addition to conserving spectrum, digital transmission is (or should be) resistant to multipath reception and gives consistent quality throughout the service area. Resistance to multipath means that omnidirectional antennae can be used, essential for mobile reception.

## 14.13 Networks

Communications networks allow transmission of data files whose content or meaning is irrelevant to the transmission medium. These files can therefore contain digital audio. Production systems can be based on high bit rate networks instead of traditional routing techniques. Contribution feeds between broadcasters and station output to transmitters no longer require special-purpose links. Audio delivery is also possible on the Internet. As a practical matter, most Internet users suffer from a relatively limited bit rate and compression will have to be used until greater bandwidth becomes available. While the quality does not compare with that of traditional broadcasts, this is not the point.

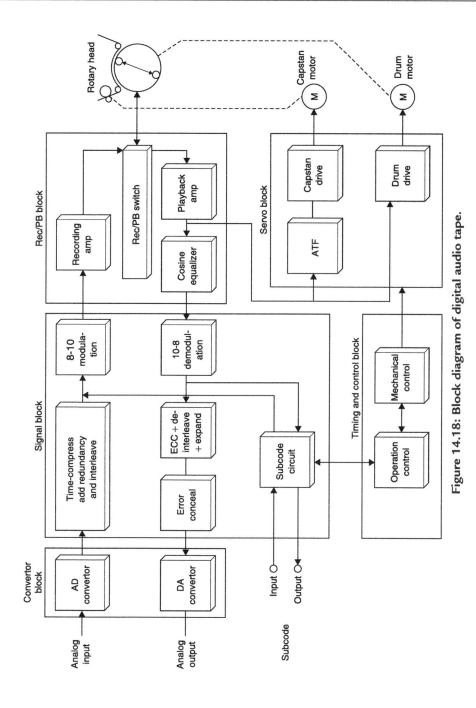

**Figure 14.18: Block diagram of digital audio tape.**

Internet audio allows a wide range of services that traditional broadcasting cannot provide and phenomenal growth is expected in this area.

## Reference

1. Devereux, V. G., 'Pulse code modulation of video signals: 8 bit coder and decoder,' *BBC Res. Dept. Rept.*, EL-42, No. 25, 1970.

# CHAPTER 15

# Representation of Audio Signals

Ian Sinclair

The impact that digital methods have made on audio has been at least as remarkable as it was on computing. Ian Sinclair uses this chapter to introduce the digital methods that seem so alien to anyone trained in analogue systems.

## 15.1 Introduction

The term digital audio is used so freely by so many that you could be excused for thinking there was nothing much new to tell. It is easy in fast conversation to present the impression of immense knowledge on the subject but it is more difficult to express the ideas concisely yet readably. The range of topics and disciplines that need to be harnessed in order to cover the field of digital audio is very wide and some of the concepts may appear paradoxical at first sight. One way of covering the topics would be to go for the apparent precision of the mathematical statement but, although this has its just place, a simpler physical understanding of the principles is of greater importance here. Thus in writing this chapter we steer between excessive arithmetic precision and ambiguous oversimplified description.

## 15.2 Analogue and Digital

Many of the physical things that we can sense in our environment appear to us to be part of a continuous range of sensation. For example, throughout the day much of coastal England is subject to tides. The cycle of tidal height can be plotted throughout the day. Imagine a pen plotter marking the height on a drum in much the same way as a barograph is arranged (Figure 15.1). The continuous line that is plotted is a feature of analogue signals in which the information is carried as a continuous infinitely fine variation of a voltage, current, or, as in this case, height of the sea level.

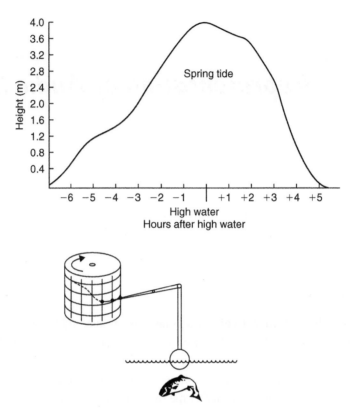

**Figure 15.1: The plot of tidal height versus diurnal time for Portsmouth, United Kingdom, time in hours. Mariners will note the characteristically distorted shape of the tidal curve for the Solent. We could mark the height as a continuous function of time using the crude arrangement shown.**

When we attempt to take a measurement from this plot we will need to recognize the effects of limited measurement accuracy and resolution. As we attempt greater resolution we will find that we approach a limit described by the noise or random errors in the measurement technique. You should appreciate the difference between resolution and accuracy since inaccuracy gives rise to distortion in the measurement due to some nonlinearity in the measurement process. This facility of measurement is useful. Suppose, for example, that we wished to send the information regarding the tidal heights we had measured to a colleague in another part of the country. One, admittedly crude, method might involve turning the drum as we traced out the plotted shape while at the far end an

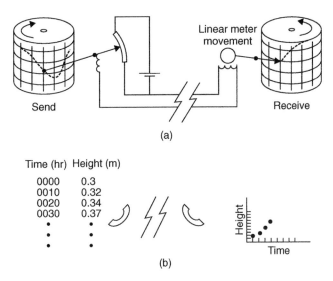

**Figure 15.2: Sending tidal height data to a colleague in two ways: (a) by tracing out the curve shape using a pen attached to a variable resistor and using a meter driven pen at the far end and (b) by calling out numbers, having agreed what the scale and resolution of the numbers will be.**

electrically driven pen wrote the same shape onto a second drum [Figure 15.2(a)]. In this method we would be subject to the nonlinearity of both the reading pen and the writing pen at the far end. We would also have to come to terms with the noise that the line, and any amplifiers, between us would add to the signal describing the plot. This additive property of noise and distortion is characteristic of handling a signal in its analogue form and, if an analogue signal has to travel through many such links, then it can be appreciated that the quality of the analogue signal is abraded irretrievably.

As a contrast consider describing the shape of the curve to your colleague by measuring the height of the curve at frequent intervals around the drum [Figure 15.2(b)]. You'll need to agree first that you will make the measurement at each 10-min mark on the drum, for example, and you will need to agree on the units of the measurement. Your colleague will now receive a string of numbers from you. The noise of the line and its associated amplifiers will not affect the accuracy of the received information since the received information should be a recognizable number. The distortion and noise performance of the line must be gross for the spoken numbers to be garbled and thus you are very well assured of correctly conveying the information requested. At the receiving end the numbers are plotted on to the chart and, in the simplest approach, they can be simply joined up with straight lines. The result will be a curve looking very much like the original.

Let's look at this analogy a little more closely. We have already recognized that we have had to agree on the time interval between each measurement and on the meaning of the units we will use. The optimum choice for this rate is determined by the fastest rate at which the tidal height changes. If, within the 10-minute interval chosen, the tidal height could have ebbed and flowed then we would find that this nuance in the change of tidal height would not be reflected in our set of readings. At this stage we would need to recognize the need to decrease the interval between readings. We will have to agree on the resolution of the measurement, since, if an arbitrarily fine resolution is requested, it will take a much longer time for all of the information to be conveyed or transmitted. We will also need to recognize the effect of inaccuracies in marking off the time intervals at both the transmit or coding end and the receiving end since this is a source of error that affects each end independently.

In this simple example of digitizing a simple wave shape we have turned over a few ideas. We note that the method is robust and relatively immune to noise and distortion in the transmission and we note also that, provided we agree on what the time interval between readings should represent, small amounts of error in the timing of the reception of each piece of data will be completely removed when the data are plotted. We also note that greater resolution requires a longer time and that the choice of time interval affects our ability to resolve the shape of the curve. All of these concepts have their own special terms and we will meet them slightly more formally.

In the example just given we used implicitly the usual decimal base for counting. In the decimal base there are 10 digits (0 through 9). As we count beyond 9 we adopt the convention that we increment our count of the number of tens by one and recommence counting in the units column from 0. The process is repeated for the count of hundreds, thousands, and so on. Each column thus represents the number of powers of 10 ($10 = 10^1$, $100 = 10^2$, $1000 = 10^3$, and so on). We are not restricted to using the number base of 10 for counting. Among the bases in common use these days are base 16 (known more commonly as the hexadecimal base), base 8 (known as octal), and the simplest of them all, base 2 (known as binary). Some of these scales have been, and continue to be, in common use. We recognize that the old coinage system in the United Kingdom used the base of 12 for pennies, as, indeed, the old way of marking distance still uses the unit of 12 inches to a foot.

The binary counting scale has many useful properties. Counting in the base of 2 means that there can only be two unique digits, 1 and 0. Thus each column must represent a power of 2 ($2 = 2^1$, $4 = 2^2$, $8 = 2^3$, $16 = 2^4$, and so on) and, by convention, we use a 1 to mark the presence of a power of 2 in a given column. We can represent any number by adding up an appropriate collection of powers of 2 and, if you try it, remember that $2^0$ is equal to 1. We refer to each symbol as a bit (actually a contraction of the words binary digit). The bit that appears in the units column is referred to as the least significant bit ( LSB), and the bit position that carries the most weight is referred to as the most significant bit (MSB).

Binary arithmetic is relatively easy to perform since the result of any arithmetic operation on a single bit can only be either 1 or 0.

We have two small puzzles at this stage. The first concerns how we represent numbers that are smaller than unity and the second is how negative numbers are represented. In the everyday decimal (base of 10) system we have adopted the convention that numbers which appear to the right of the decimal point indicate successively smaller values. This is in exactly the opposite way in which numbers appearing to the left of the decimal point indicated the presence of increasing powers of 10. Thus successive columns represent $0.1 = 1/10 = 10^{-1}$, $0.01 = 1/100 = 10^{-2}$, $0.001 = 1/1000 = 10^{-3}$, and so on. We follow the same idea for binary numbers and thus the successive columns represent $0.5 = 1/2 = 2^{-1}$, $0.25 = 1/4 = 2^{-2}$, $0.125 = 1/8 = 2^{-3}$, and so on.

One of the most useful properties of binary numbers is the ease with which arithmetic operations can be carried out by simple binary logic. For this to be viable there has to be a way of including some sign in the number itself since we have only the two symbols 0 and 1. Here are two ways it can be done. We can add a 1 at the beginning of the number to indicate that it was negative or we can use a more flexible technique known as two's complement. Here the positive numbers appear as we would expect but the negative numbering is formed by subtracting the value of the intended negative number from the largest possible positive number incremented by 1. Table 15.1 shows both of these approaches. The use of a sign bit is only possible because we will arrange that we will use the same numbering and marking convention. We will thus know the size of the largest

### Table 15.1: Four-Bit Binary Number Coding Methods

| | Binary number representation | | |
|---|---|---|---|
| Decimal number | Sign plus magnitude | Two's complement | Offset binary |
| 7 | 0011 | 0111 | 1111 |
| 6 | 0110 | 0110 | 1110 |
| 5 | 0101 | 0101 | 1101 |
| 4 | 0100 | 0100 | 1100 |
| 3 | 0011 | 0011 | 1011 |
| 2 | 0010 | 0010 | 1010 |
| 1 | 0001 | 0001 | 1001 |
| 0 | 0000 | 0000 | 1000 |
| −0 | 1000 | (0000) | (1000) |
| −1 | 1001 | 1111 | 0111 |
| −2 | 1010 | 1110 | 0110 |
| −3 | 1011 | 1101 | 0101 |
| −4 | 1100 | 1100 | 0100 |
| −5 | 1101 | 1011 | 0011 |
| −6 | 1110 | 1010 | 0010 |
| −7 | 1111 | 1001 | 0001 |
| −8 | | 1000 | 0000 |

positive or negative number we can count to. The simple use of a sign bit leads to two values for zero, which is not elegant or useful. One of the advantages of two's complement coding is that it makes subtraction simply a matter of addition. Arithmetic processes are at the heart of digital signal processing and thus hold the key to handling digitized audio signals.

There are many advantages to be gained by handling analogue signals in digitized form and, in no particular order, they include:

- great immunity from noise since the digitized signal can only be 1 or 0;
- exactly repeatable behavior;
- ability to correct for errors when they do occur;
- simple arithmetic operations, very easy for computers;

- more flexible processing possible and easy programmability;

- low cost potential; and

- processing can be independent of real time.

## 15.3 Elementary Logical Processes

We have described an outline of a binary counting scale and shown how we may implement a count using it but some physical method of performing this is needed. We can represent two states, a 1 and an 0 state, by using switches since their contacts will be either open or closed and there is no half-way state. Relay contacts also share this property but there are many advantages in representing the 1 and 0 states by the polarity or existence of a voltage or a current, not least of which is the facility of handling such signals at very high speed in integrated circuitry. Manipulation of the 1 and 0 signals is referred to as logic and, in practice, is usually implemented by simple logic circuits called gates. Digital integrated circuits comprise collections of various gates, which can number from a single gate (as in the eight input NAND gate exemplified by the 74LS30 part number) to many millions (as can be found in some microprocessors).

All logic operations can be implemented by the appropriate combination of just three operations:

- the AND gate, circuit symbol &, arithmetic symbol '.';

- the OR gate, circuit symbol |, arithmetic symbol '+'; and

- the inverter or NOT gate, circuit and arithmetic symbol '−'.

From this primitive trio we can derive the NAND, NOR, and EXOR (exclusive-OR gate). Gates are characterized by the relationship of their output to combinations of their inputs (Figure 15.3). Note how the NAND (literally negated AND gate) performs the same logical function as OR gate fed with inverted signals and, similarly, note the equivalent duality in the NOR function. This particular set of dualities is known as De Morgan's theorem.

Practical logic systems are formed by grouping many gates together and naturally there are formal tools available to help with the design, the most simple and common of which is known as Boolean algebra. This is an algebra that allows logic problems to be expressed in symbolic terms. These can then be manipulated and the resulting expression can be directly interpreted as a logic circuit diagram. Boolean expressions cope best with

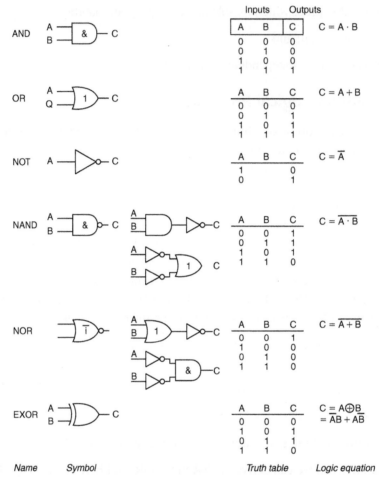

Figure 15.3: Symbols and truth tables for the common basic gates. For larger arrays of gates it is more useful to express the overall logical function as a set of sums (the OR function) and products (the AND function); this is the terminology used by gate array designers.

logic that has no timing or memory associated with it: for such systems other techniques, such as state machine analysis, are better used instead.

The simplest arrangement of gates that exhibit memory, at least while power is still applied, is the cross-coupled NAND (or NOR) gate (Figure 15.4). More complex

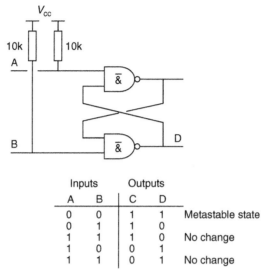

| Inputs | | Outputs | | |
|---|---|---|---|---|
| A | B | C | D | |
| 0 | 0 | 1 | 1 | Metastable state |
| 0 | 1 | 1 | 0 | |
| 1 | 1 | 1 | 0 | No change |
| 1 | 0 | 0 | 1 | |
| 1 | 1 | 0 | 1 | No change |

Convention and good practice is to use pull-up resistors
on open gate inputs.

**Figure 15.4: A simple latch. In this example, the outputs of each of two NAND gates are
cross coupled to one of the other inputs. The unused input is held high by a resistor to the
positive supply rail. The state of the gate outputs will be changed when one of the inputs is
grounded and this output state will be steady until the other input is grounded or until the
power is removed. This simple circuit, the R-S flip-flop, has often been used to debounce
mechanical contacts and as a simple memory.**

arrangements produce the wide range of flip-flop (FF) gates, including the set–reset latch,
the D-type FF, which is edge triggered by a clock pulse, the JK FF and a wide range of
counters (or dividers), and shift registers (Figure 15.5). These circuit elements and their
derivatives find their way into the circuitry of digital signal handling for a wide variety of
reasons. Early digital circuitry was based around standardized logic chips, but it is much
more common nowadays to use application-specific ICs.

## 15.4 The Significance of Bits and Bobs

Groupings of eight bits, which represent a symbol or a number, are usually referred
to as bytes, and the grouping of four bits is, somewhat obviously, sometimes called a
nibble (sometimes spelt nybble). Bytes can be assembled into larger structures, which

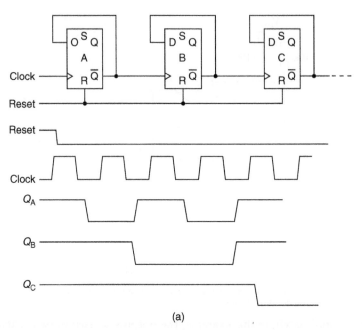

(a)

**Figure 15.5(a): The simplest counter is made up of a chain of edge-triggered D-type FFs. For a long counter, it can take a sizeable part of a clock cycle for all of the counter FFs to change state in turn. This ripple through can make decoding the state of the counter difficult and can lead to transitory glitches in the decoder output, indicated in the diagram as points where the changing edges do not exactly line up. Synchronous counters in which the clock pulse is applied to all of the counting FFs at the same time are used to reduce the overall propagation delay to that of a single stage.**

are referred to as words. Thus a three byte word will comprise 24 bits (though word is by now being used mainly to mean two bytes and DWord to mean four bytes). Blocks are the next layer of structure perhaps comprising 512 bytes (a common size for computer hard discs). Where the arrangement of a number of bytes fits a regular structure the term frame is used. We will meet other terms that describe elements of structure in due course.

Conventionally we think of bits, and their associated patterns and structures, as being represented by one of two voltage levels. This is not mandatory and there are other ways of representing the on/off nature of the binary signal. You should not forget alternatives such as the use of mechanical or solid state switches, presence or absence of a light,

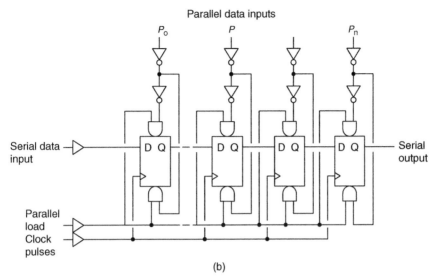

**Figure 15.5(b): This arrangement of FFs produces the shift register. In this circuit, a pattern of 1s and 0s can be loaded into a register (the load pulses) and then can be shifted out serially one bit at a time at a rate determined by the serial clock pulse. This is an example of a parallel in serial out (PISO) register. Other varieties include LIFO (last in first out), SIPO (serial in parallel out), and FILO (first in last out). The diagrams assume that unused inputs are tied to ground or to the positive supply rail as needed.**

polarity of a magnetic field, state of waveform phase, and direction of an electric current. The most common voltage levels referred to are those used in the common $74 \times 00$ logic families and are often referred to as TTL levels. A logic 0 (or low) will be any voltage that is between 0 and 0.8 V while a logic 1 (or high) will be any voltage between 2.0 V and the supply rail voltage, which will be typically 5.0 V. In the gap between 0.8 and 2.0 V the performance of a logic element or circuit is not reliably determinable as it is in this region where the threshold between low and high logic levels is located. Assuming that the logic elements are being used correctly, the worst-case output levels of the TTL families for a logic 0 is between 0 and 0.5 V and for a logic 1 is between 2.4 V and the supply voltage. The difference between the range of acceptable input voltages for a particular logic level and the range of outputs for the same level gives the noise margin. Thus for TTL families, the noise margin is typically in the region of 0.4 V for both logic low and logic high. Signals whose logic levels lie outside these margins may cause misbehavior or errors and it is part of the skill of the design and layout of such circuitry that this risk is minimized.

Logic elements made using CMOS technologies have better input noise margins because the threshold of a CMOS gate is approximately equal to half of the supply voltage. Thus, after considering the inevitable spread of production variation and the effects of temperature, the available input range for a logic low (or 0) lies in the range 0 to 1.5 V and for a logic high (or 1) in the range of 3.5 to 5.0 V (assuming a 5.0-V supply). However, the output impedance of CMOS gates is at least three times higher than that for simple TTL gates and thus in a 5.0-V supply system interconnections in CMOS systems are more susceptible to reactively coupled noise. CMOS systems produce their full benefit of high noise margin when they are operated at higher voltages but this is not possible for CMOS technologies intended to be compatible with $74 \times 00$ logic families.

## 15.5  Transmitting Digital Signals

There are two ways in which you can transport bytes of information from one circuit or piece of equipment to another. Parallel transmission requires a signal line for each bit position and at least one further signal that will indicate that the byte now present on the signal lines is valid and should be accepted. Serial transmission requires that the byte be transmitted one bit at a time and in order that the receiving logic or equipment can recognize the correct beginning of each byte of information it is necessary to incorporate some form of signaling in serial data in order to indicate (as a minimum) the start of each byte. Figure 15.6 shows an example of each type.

Parallel transmission has the advantage that, where it is possible to use a number of parallel wires, the rate at which data can be sent can be very high. However, it is not easy to maintain a very high data rate on long cables using this approach and its use for digital audio is usually restricted to the internals of equipment and for external use as an interface to peripheral devices attached to a computer.

The serial link carries its own timing with it and thus it is free from errors due to skew and it clearly has benefits when the transmission medium is not copper wire but infra-red or radio. It also uses a much simpler single circuit cable and a much simpler connector. However, the data rate will be roughly 10 times that for a single line of a

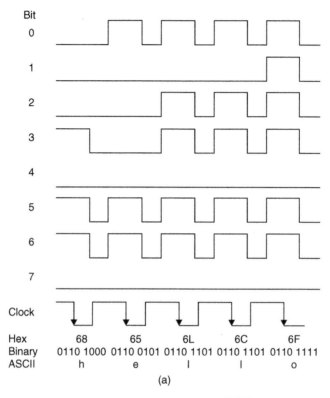

(a)

**Figure 15.6(a): Parallel transmission: a data strobe line $\overline{DST}$ (the—sign means active low)
would accompany the bit pattern to clock the logic state of each data line on its falling edge
and is timed to occur some time after the data signals have been set so that any reflections,
cross talk, or skew in the timing of the individual data lines will have had time to settle. After
the $\overline{DST}$ signal has returned to the high state, the data lines are reset to 0 (usually they
would only be changed if data in the next byte required a change).**

parallel interface. Achieving this higher data rate requires that the sending and receiving
impedances are accurately matched to the impedance of the connecting cable. Failure to
do this will result in signal reflections, which in turn will result in received data being
in error. This point is of practical importance because the primary means of conveying
digital audio signals between equipments is by the serial AES/EBU signal interface at a
data rate approximately equal to 3 Mbits per second.

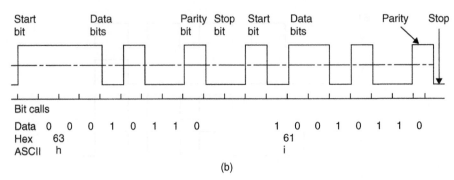

Figure 15.6(b): Serial transmission requires the sender and receiver to use and recognize the same signal format or protocol, such as RS232. For each byte, the composite signal contains a start bit, a parity bit, and a stop bit using inverted logic ($1 = -12$ V; $0 = -12$ V). The time interval between each bit of the signal (the start bit, parity bit, stop bit, and data bits) is fixed and must be kept constant.

You should refer to a text on the use of transmission lines for a full discussion of this point but for guidance here is a simple way of determining whether you will benefit by considering the transmission path as a transmission line.

- Look up the logic signal rise time, $t_R$.

- Determine the propagation velocity in the chosen cable, $v$. This will be typically about 0.6 of the speed of light.

- Determine the length of the signal path, $l$.

- Calculate the propagation delay, $\tau = l/v$.

- Calculate the ratio of $t_R/\tau$.

- If the ratio is greater than 8, then the signal path can be considered electrically short and you will need to consider the signal path's inductance or capacitance, whichever is dominant.

- If the ratio is less than 8, then consider the signal path in terms of a transmission line.

The speed at which logic transitions take place determines the maximum rate at which information can be handled by a logic system. The rise and fall times of a logic signal are

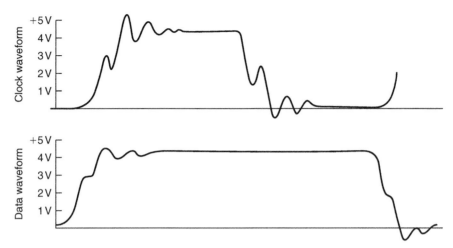

**Figure 15.7:** A practical problem arose where the data signal was intended to be clocked in using the rising edge of a separate clock line, but excessive ringing on the clock line caused the data line to be sampled twice, causing corruption. In addition, due to the loading of a large number of audio channels, the actual logic level no longer achieved the 4.5-V target required for acceptable noise performance, increasing the susceptibility to ringing. The other point to note is that the falling edge of the logic signals took the data line voltage to a negative value, and there is no guarantee that the receiving logic element would not produce an incorrect output as a consequence.

important because of the effect on integrity. The outline of the problem is shown in Figure 15.7, which has been taken from a practical problem in which serial data were being distributed around a large digitally controlled audio mixing desk. Examples such as this illustrate the paradox that digital signals must, in fact, be considered from an analogue point of view.

## 15.6 The Analogue Audio Waveform

It seems appropriate to ensure that there is agreement concerning the meaning attached to words that are freely used. Part of the reason for this is in order that a clear understanding can be obtained into the meaning of phase. The analogue audio signal that we will encounter when it is viewed on an oscilloscope is a causal signal. It is considered as having zero value for negative time and it is also continuous with time. If we observe a few milliseconds of a musical signal we are very likely to observe a waveform that can be seen to have an underlying structure (Figure 15.8). Striking a single string can produce a waveform that appears to have a relatively simple structure. The waveform resulting from striking a chord is visually more complex, although, at any one time, a snapshot of it will show the evidence of structure. From a mathematical or analytical viewpoint the complicated waveform of real sounds is impossibly complex to handle and, instead, the analytical, and indeed the descriptive, process depends on us understanding the principles

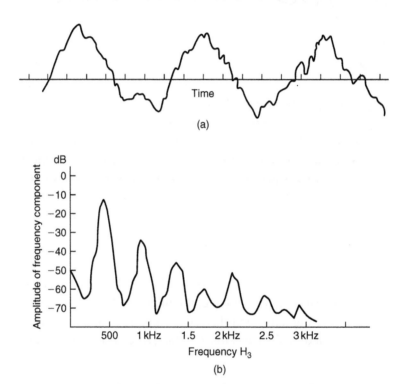

**Figure 15.8: (a) An apparently simple noise, such as a single string on a guitar, produces a complicated waveform, sensed in terms of pitch. The important part of this waveform is the basic period of the waveform, its fundamental frequency. The smaller detail is due to components of higher frequency and lower level. (b) An alternative description is analysis into the major frequency components. If processing accuracy is adequate, then the description in terms of amplitudes of harmonics (frequency domain) is identical to the description in terms of amplitude and time (time domain).**

through the analysis of much simpler waveforms. We rely on the straightforward principle of superposition of waveforms such as the simple cosine wave.

On its own an isolated cosine wave, or real signal, has no phase. However, from a mathematical point of view the apparently simple cosine wave signal, which we consider as a stimulus to an electronic system, can be considered more properly as a complex wave or function that is accompanied by a similarly shaped sine wave (Figure 15.9). It is worthwhile throwing out an equation at this point to illustrate this:

$$f(t) = \text{Re}\, f(t) + \text{j}\, \text{Im}\, f(t)$$

where $f(t)$ is a function of time, $t$, which is composed of Re $f(t)$, the real part of the function and j Im $f(t)$, the imaginary part of the function and j is $\sqrt{-1}$.

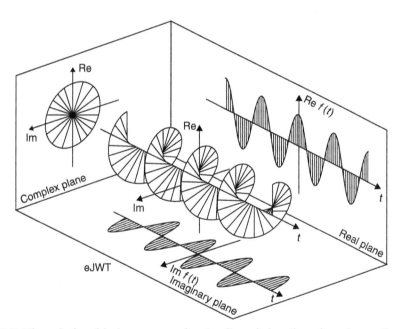

**Figure 15.9: The relationship between cosine (real) and sine (imaginary) waveforms in the complex exponential eJWT. This assists in understanding the concept of phase. Note that one property of the spiral form is that its projection onto any plane parallel to the time axis will produce a sinusoidal waveform.**

Emergence of the $\sqrt{-1}$ is the useful part here because you may recall that analysis of the simple analogue circuit (Figure 15.10) involving resistors and capacitors produces an expression for the attenuation and the phase relationship between input and output of that circuit, which is achieved with the help of $\sqrt{-1}$.

The process that we refer to glibly as the Fourier transform considers that all waveforms can be considered as constructed from a series of sinusoidal waves of the appropriate amplitude and phase added linearly. A continuous sine wave will need to exist for all time in order that its representation in the frequency domain will consist of only a single frequency. The reverse side, or dual, of this observation is that a singular event, for example, an isolated transient, must be composed of all frequencies. This trade-off between the resolution of an event in time and the resolution of its frequency components is fundamental. You could think of it as if it were an uncertainty principle.

The reason for discussing phase at this point is that the topic of digital audio uses terms such as linear phase, minimum phase, group delay, and group delay error. A rigorous treatment of these topics is outside the scope for this chapter but it is necessary to describe them. A system has linear phase if the relationship between phase and frequency is a linear one. Over the range of frequencies for which this relationship may hold the systems, output is effectively subjected to a constant time delay with respect to its input. As a simple example, consider that a linear phase system that exhibits $-180°$ of phase shift at 1 kHz will show $-360°$ of shift at 2 kHz. From an auditive point of view, a linear phase performance should preserve the waveform of the input and thus be benign to an audio signal.

Most of the common analogue audio processing systems, such as equalizers, exhibit minimum phase behavior. Individual frequency components spend the minimum necessary time being processed within the system. Thus some frequency components of a complex signal may appear at the output at a slightly different time with respect to others.

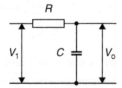

**Figure 15.10: The simple resistor and capacitor attenuator can be analyzed to provide us with an expression for the output voltage and the output phase with respect to the input signal.**

Such behavior can produce gross waveform distortion as might be imagined if a 2-kHz component were to emerge 2 ms later than a 1-kHz signal. In most simple circuits, such as mixing desk equalizers, the output phase of a signal with respect to the input signal is usually the ineluctable consequence of the equalizer action. However, for reasons which we will come to, the process of digitizing audio can require special filters whose phase response may be responsible for audible defects.

One conceptual problem remains. Up to this point we have given examples in which the output phase has been given a negative value. This is comfortable territory because such phase lag is converted readily to time delay. No causal signal can emerge from a system until it has been input, as otherwise our concept of the inviolable physical direction of time is broken. Thus all practical systems must exhibit delay. Systems that produce phase lead cannot actually produce an output that, in terms of time, is in advance of its input. Part of the problem is caused by the way we may measure the phase difference between input and output. This is commonly achieved using a dual-channel oscilloscope and observing the input and output waveforms. The phase difference is readily observed and can be readily shown to match calculations such as that given in Figure 15.10. The point is that the test signal has essentially taken on the characteristics of a signal that has existed for an infinitely long time exactly as it is required to do in order that our use of the relevant arithmetic is valid. This arithmetic tacitly invokes the concept of a complex signal, which is one which, for mathematical purposes, is considered to have real and imaginary parts (see Figure 15.9). This invocation of phase is intimately involved in the process of composing, or decomposing, a signal using the Fourier series. A more physical appreciation of the response can be obtained by observing the system response to an impulse.

Since the use of the idea of phase is much abused in audio at the present time, introducing a more useful concept may be worthwhile. We have referred to linear phase systems as exhibiting simple delay. An alternative term to use would be to describe the system as exhibiting a uniform (or constant) group delay over the relevant band of audio frequencies. Potentially audible problems start to exist when the group delay is not constant but changes with frequency. The deviation from a fixed delay value is called group delay error and can be quoted in milliseconds.

The process of building up a signal using the Fourier series produces a few useful insights [Figure 15.11(a)]. The classic example is that of the square wave and it is shown in Figure 15.11(a) as the sum of the fundamental, third and fifth harmonics. It is worth noting that

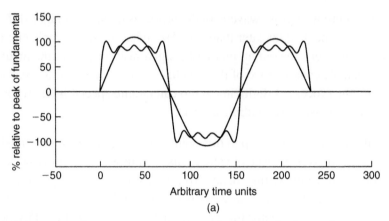

(a)

**Figure 15.11(a): Composing a square wave from the harmonics is an elementary example of a Fourier series. For the square wave of unit amplitude the series is of the form.**

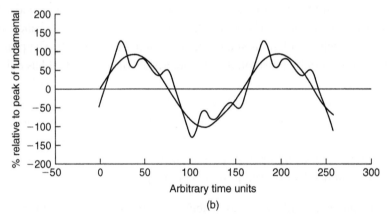

(b)

**Figure 15.11(b): In practice, a truly symmetrical shape is rare, as most practical methods of limiting the audio bandwidth do not exhibit linear phase, but delay progressively the higher frequency components. Band-limiting niters respond to excitation by a square wave by revealing the effect of the absence of higher harmonics and the so-called "ringing" is thus not necessarily the result of potentially unstable filters.**

the "ringing" on the waveform is simply the consequence of band-limiting a square wave, that simple, minimum phase systems will produce an output rather like Figure 15.11(b), and there is not much evidence to show that the effect is audible. The concept of building up complex wave shapes from simple components is used in calculating the shape of

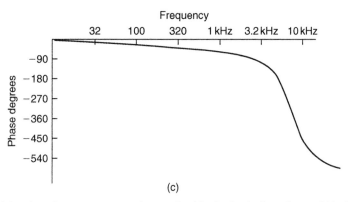

(c)

**Figure 15.11(c): The phase response shows the kind of relative phase shift that might be responsible.**

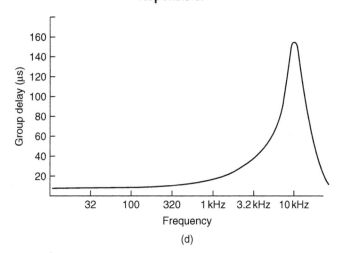

(d)

**Figure 15.11(d): The corresponding group delay curve shows a system that reaches a peak delay of around 16 μs.**

tidal heights. The accuracy of the shape is dependent on the number of components that we incorporate and the process can yield a complex wave shape with only a relatively small number of components. We see here the germ of the idea that will lead to one of the methods available for achieving data reduction for digitized analogue signals.

$$4 / \pi[\sin \omega + 1/3 \sin 3\,\omega + 1/5 \sin 5\omega + 1/7 \sin 7\omega \ldots]$$

where $\omega = 2\pi f$, the angular frequency.

The composite square wave has ripples in its shape, due to band limiting, since this example uses only the first four terms, up to the seventh harmonic. For a signal that has been limited to an audio bandwidth of approximately 21 kHz, this square wave must be considered as giving an ideal response even though the fundamental is only 3 kHz. The 9% overshoot followed by a 5% undershoot, the Gibbs phenomenon, will occur whenever a Fourier series is truncated or a bandwidth is limited.

Instead of sending a stream of numbers that describe the wave shape at each regularly spaced point in time, we first analyze the wave shape into its constituent frequency components and then send (or store) a description of the frequency components. At the receiving end these numbers are unraveled and, after some calculation, the wave shape is reconstituted. Of course this requires that both the sender and the receiver of the information know how to process it. Thus the receiver will attempt to apply the inverse, or opposite, process to that applied during coding at the sending end. In the extreme it is possible to encode a complete Beethoven symphony in a single 8-bit byte. First, we must equip both ends of our communication link with the same set of raw data, in this case a collection of CDs containing recordings of Beethoven's work. We then send the number of the disc that contains the recording which we wish to "send." At the receiving end, the decoding process uses the received byte of information, selects the disc, and plays it. A perfect reproduction using only one byte to encode 64 minutes of stereo recorded music is created … and to CD quality!

A very useful signal is the impulse. Figure 15.12 shows an isolated pulse and its attendant spectrum. Of equal value is the waveform of the signal that provides a uniform spectrum. Note how similar these wave shapes are. Indeed, if we had chosen to show in Figure 15.12(a) an isolated square-edged pulse then the pictures would be identical, save that references to the time and frequency domains would need to be swapped. You will encounter these wave shapes in diverse fields such as video and in the spectral shaping of digital data waveforms. One important advantage of shaping signals in this way is that since the spectral bandwidth is better controlled, the effect of the phase response of a band-limited transmission path on the waveform is also limited. This will result in a waveform that is much easier to restore to clean "square" waves at the receiving end.

## 15.7  Arithmetic

We have seen how the process of counting in binary is carried out. Operations using the number base of 2 are characterized by a number of useful tricks that are often used. Simple

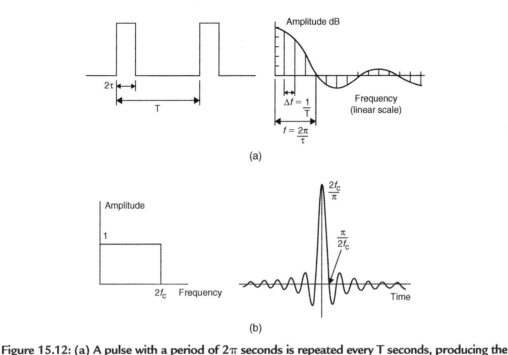

(a)

(b)

Figure 15.12: (a) A pulse with a period of $2\pi$ seconds is repeated every T seconds, producing the spectrum as shown. The spectrum appears as having negative amplitudes, as alternate "lobes" have the phase of their frequency components inverted, although it is usual to show the modulus of the amplitude as positive and to reflect the inversion by an accompanying plot of phase against frequency. The shape of the lobes is described by the simple relationship: $A = k(\sin x)/x$. (b) A further example of the duality between time and frequency showing that a widely spread spectrum will be the result of a narrow pulse. The sum of the energy must be the same for each so that we would expect a narrow pulse to be of large amplitude if it is to carry much energy. If we were to use such a pulse as a test signal we would discover that the individual amplitude of any individual frequency component would be quite small. Thus when we do use this signal for just this purpose we will usually arrange to average the results of a number of tests.

counting demonstrates the process of addition and, at first sight, the process of subtraction would need to be simply the inverse operation. However, since we need negative numbers in order to describe the amplitude of the negative polarity of a waveform, it seems sensible to use a coding scheme in which the negative number can be used directly to perform subtraction. The appropriate coding scheme is the two's complement coding scheme.

We can convert from a simple binary count to a two's complement value very simply. For positive numbers simply ensure that the MSB is a zero. To make a positive number into a negative one, first invert each bit and then add one to the LSB position thus:

$$+9_{10} > 1001_2 > 01001_{2c} \text{(using a 5 bit word length)}$$
$$-9_{10} > -01001_2 > 10110 + 1_{\text{invert and add1}} > 10111_{2c}.$$

We must recognize that since we have fixed the number of bits that we can use in each word (in this example to 5 bits) then we are naturally limited to the range of numbers we can represent (in this case from $+15$ through 0 to $-16$). Although the process of forming two's complement numbers seems lengthy, it is performed very speedily in hardware. Forming a positive number from a negative one uses the identical process. If the binary numbers represent an analogue waveform, then changing the sign of the numbers is identical to inverting the polarity of the signal in the analogue domain. Examples of simple arithmetic should make this a bit more clear:

**Table 15.2  Examples of simple arithmetic**

| Decimal, base 10 | Binary, base 2 | 2's complement |
|:---:|:---:|:---:|
| Addition | | |
| 12 | 01100 | 01100 |
| + 3 | +00011 | +00011 |
| =15 | =01111 | =01111 |
| Subtraction | | |
| 12 | 01100 | 01100 |
| −3 | −00011 | +11101 |
| =9 | | =01001 |

Since we have only a 5-bit word length any overflow into the column after the MSB needs to be handled. The rule is that if there is overflow when a positive and a negative number are added then it can be disregarded. When overflow results during the addition of two positive numbers or two negative numbers then the resulting answer will be incorrect if the overflowing bit is neglected. This requires special handling in signal processing, one approach being to set the result of an overflowing sum to the appropriate largest positive or negative number. The process of adding two sequences of numbers that represent two audio waveforms is identical to that of mixing the two waveforms in the analogue domain. Thus when the addition process results in overflow the effect is identical to the resulting mixed analogue waveform being clipped.

We see here the effect of word length on the resolution of the signal and, in general, when a binary word containing $n$ bits is added to a larger binary word comprising $m$ bits the resulting word length will require $m + 1$ bits in order to be represented without the effects of overflow. We can recognize the equivalent of this in the analogue domain where we know that the addition of a signal with a peak-peak amplitude of 3 V to one of 7 V must result in a signal whose peak-peak value is 10 V. Don't be confused about the rms value of the resulting signal, which will be

$$\frac{\sqrt{3^2 + 7^2}}{2.82} = 2.69 \ V_{rms},$$

assuming uncorrelated sinusoidal signals.

A binary adding circuit is readily constructed from the simple gates referred to earlier, and Figure 15.13 shows a 2-bit full adder. More logic is needed to be able to accommodate wider binary words and to handle the overflow (and underflow) exceptions.

If addition is the equivalent of analogue mixing, then multiplication will be the equivalent of amplitude or gain change. Binary multiplication is simplified by only having 1 and 0 available since $1 \times 1 = 1$ and $1 \times 0 = 0$.

Since each bit position represents a power of 2, then shifting the pattern of bits one place to the left (and filling in the vacant space with a 0) is identical to multiplication by 2. The opposite is, of course, true of division. The process can be appreciated by an example:

| Decimal | Binary |
|---------|--------|
| 3 | 00011 |
| ×5 | 00101 |
| =15 | 00011 |

(Continued)

| Decimal | Binary |
|---|---|
| + | 00000 |
| + | 00011 |
| + | 00000 |
| + | 00000 |
| = | 000001111 |
| and another example: | |
| 12 | 01100 |
| ×13 | 01101 |
| =156 | =010011100 |

The process of shifting and adding could be programmed in a series of program steps and executed by a microprocessor but this would take too long. Fast multipliers work by arranging that all of the available shifted combinations of one of the input numbers are made available to a large array of adders, while the other input number is used to determine which of the shifted combinations will be added to make the final sum. The resulting word width of a multiplication equals the sum of both input word widths. Further, we will need to recognize where the binary point is intended to be and arrange to shift the output word appropriately. Quite naturally the surrounding logic circuitry will have been designed to accommodate a restricted word width. Repeated multiplication must force the output word width to be limited. However, limiting the word width has a direct impact on the accuracy of the final result of the arithmetic operation. This curtailment of accuracy is cumulative since subsequent arithmetic operations can have no knowledge that the numbers being processed have been "damaged."

Two techniques are important in minimizing the "damage." The first requires us to maintain the intermediate stages of any arithmetic operation at as high an accuracy as possible for as long as possible. Thus although most conversion from analogue audio to digital (and the converse digital signal conversion to an analogue audio signal) takes place using 16 bits, the intervening arithmetic operations will usually involve a minimum of 24 bits.

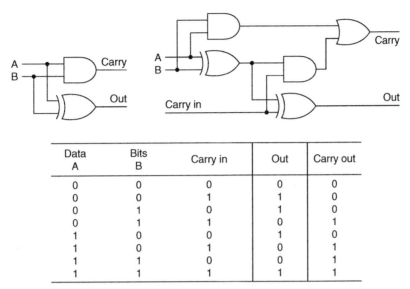

| Data<br>A | Bits<br>B | Carry in | Out | Carry out |
|:---:|:---:|:---:|:---:|:---:|
| 0 | 0 | 0 | 0 | 0 |
| 0 | 0 | 1 | 1 | 0 |
| 0 | 1 | 0 | 1 | 0 |
| 0 | 1 | 1 | 0 | 1 |
| 1 | 0 | 0 | 1 | 0 |
| 1 | 0 | 1 | 0 | 1 |
| 1 | 1 | 0 | 0 | 1 |
| 1 | 1 | 1 | 1 | 1 |

**Figure 15.13: A 2-bit full adder needs to be able to handle a carry bit from an adder handling lower order bits and similarly provide a carry bit. A large adder based on this circuit would suffer from the ripple through of the carry bit as the final sum would not be stable until this had settled. Faster adding circuitry uses look-ahead carry circuitry.**

The second technique is called *dither*, which will be covered fully later. Consider, for the present, that the output word width is simply cut (in the example given earlier such as to produce a 5-bit answer). The need to handle the large numbers that result from multiplication without overflow means that when small values are multiplied they are likely to lie outside the range of values that can be expressed by the chosen word width. In the example given earlier, if we wish to accommodate the most significant digits of the second multiplication (156) as possible in a 5-bit word, then we shall need to lose the information contained in the lower four binary places. This can be accomplished by shifting the word four places (thus effectively dividing the result by 16) to the right and losing the less significant bits. In this example the result becomes 01001, which is equivalent to decimal 9. This is clearly only approximately equal to 156/16.

When this crude process is carried out on a sequence of numbers representing an audio analogue signal, the error results in an unacceptable increase in the signal-to-noise ratio. This loss of accuracy becomes extreme when we apply the same adjustment to the lesser product of $3 \times 5 = 15$ since, after shifting four binary places, the result is zero. Truncation is thus a rather poor way of handling the restricted output word width. A slightly better approach is to round up the output by adding a fraction to each output number just prior to truncation. If we added 00000110, then shifted four places and truncated the output would become 01010 (=1010), which, although not absolutely accurate, is actually closer to the true answer of 9.75. This approach moves the statistical value of the error from 0 to −1 toward +/− 0.5 of the value of the LSB, but the error signal that this represents is still very highly correlated to the required signal. This close relationship between noise and signal produces an audibly distinct noise that is unpleasant to listen to.

An advantage is gained when the fraction that is added prior to truncation is not fixed but random. The audibility of the result is dependent on the way in which the random number is derived. At first sight it does seem daft to add a random signal (which is obviously a form of noise) to a signal that we wish to retain as clean as possible. Thus the probability and spectral density characteristics of the added noise are important. A recommended approach commonly used is to add a random signal that has a triangular probability density function (TPDF) (Figure 15.14). Where there is sufficient reserve of processing power it is possible to filter the noise before adding it in. This spectral shaping is used to modify the spectrum of the resulting noise (which you must recall is an error signal) such that it is biased to those parts of the audio spectrum where it is least audible. The mathematics of this process are beyond this text.

A special problem exists where gain controls are emulated by multiplication. A digital audio mixing desk will usually have its signal levels controlled by digitizing the position of a physical analogue fader (certainly not the only way by which to do this, incidentally). Movement of the fader results in a stepwise change of the multiplier value used.

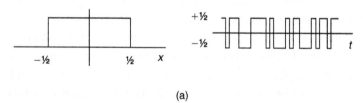

(a)

**Figure 15.14(a): The amplitude distribution characteristics of noise can be described in terms of the amplitude probability distribution characteristic. A square wave of level 0 or +5V can be described as having a rectangular probability distribution function (RPDF). In the case of the 5-bit example, which we are using, the RPDF wave form can be considered to have a value of +0.12 or −0.12 [(meaning +0.5 or −0.5), equal chances of being positive or negative].**

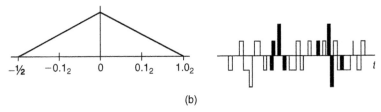

(b)

Figure 15.14(b): The addition of two uncorrelated RPDF sequences gives rise to one with triangular distribution (TPDF). When this dither signal is added to a digitized signal it will always mark the output with some noise, as there is a finite possibility that the noise will have a value greater than 0, so that as a digitized audio signal fades to zero value, the noise background remains fairly constant. This behavior should be contrasted with that of RPDF for which, when the signal fades to zero, there will come a point at which the accompanying noise also switches off. This latter effect may be audible in some circumstances and is better avoided. A wave form associated with this type of distribution will have values ranging from +1.02 through 02 to −1.02.

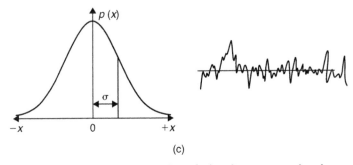

(c)

Figure 15.14(c): Noise in the analogue domain is often assumed to have a Gaussian distribution. This can be understood as the likelihood of the waveform having a particular amplitude. The probability of an amplitude x occurring can be expressed as $p(x) = [e(-(x - \mu)^2/2\sigma^2]/\sigma\sqrt{2\pi}$ where $\mu$ is the mean value, $\sigma$ is the variance, and $X$ is the sum of the squares of the deviations x from the mean. In practice, a "random" waveform, which has a ratio between the peak to mean signal levels of 3, can be taken as being sufficiently Gaussian in character. The spectral balance of such a signal is a further factor that must be taken into account if a full description of a random signal is to be described.

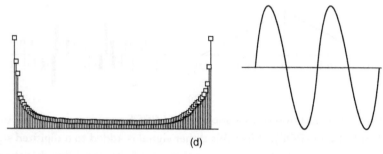

(d)

Figure 15.14(d): The sinusoidal wave form can be described by the simple equation: $x(t) = A\sin(2\pi ft)$, where $x(t)$ is the value of the sinusoidal wave at time $t$, $A$ is the peak amplitude of the waveform, $f$ is the frequency in Hz, and $t$ is the time in seconds and its probability density function is as shown here.

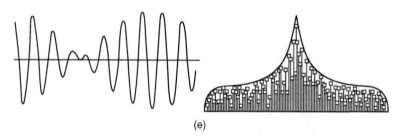

(e)

Figure 15.14(e): A useful test signal is created when two sinusoidal waveforms of the same amplitude but unrelated in frequency are added together. The resulting signal can be used to check amplifier and system nonlinearity over a wide range of frequencies. The test signal will comprise two signals to stimulate the audio system (for testing at the edge of the band, 19 and 20 kHz can be used) while the output spectrum is analyzed and the amplitude of the sum and difference frequency signals is measured. This form of test is considerably more useful than a THD test.

When such a fader is moved, any music signal being processed at the time is subjected to stepwise changes in level. Although small, the steps will result in audible interference unless the changes that they represent are themselves subjected to the addition of dither. Thus although the addition of a dither signal reduces the correlation of the error signal to the program signal, it must, naturally, add to the noise of the signal. This reinforces the need to ensure that the digitized audio signal remains within the processing circuitry with as high a precision as possible for as long as possible. Each time that the audio signal has its precision reduced it inevitably must become noisier.

## 15.8 Digital Filtering

Although it may be clear that the multiplication process controls the signal level, it is not immediately obvious that the multiplicative process is intrinsic to any form of filtering. Thus multipliers are at the heart of any significant digital signal processing, and modern digital signal processing would not be possible without the availability of suitable IC technology. You will need to accept, at this stage, that the process of representing an analogue audio signal in the form of a sequence of numbers is readily achieved and thus we are free to consider how the equivalent analogue processes of filtering and equalization may be carried out on the digitized form of the signal.

In fact, the processes required to perform digital filtering are performed daily by many people without giving the process much thought. Consider the waveform of the tidal height curve of Figure 15.15. The crude method by which we obtained this curve (Figure 15.1) contained only an approximate method for removing the effect of ripples in the water by including a simple dashpot linked to the recording mechanism. If we were to look at this trace more closely we would see that it was not perfectly smooth due to local effects such as passing boats and wind-driven waves. Of course tidal heights do not normally increase by 100 mm within a few seconds and so it is sensible to draw a line that passes through the average of these disturbances. This averaging process is filtering and, in this case, it is an example of low-pass filtering. To achieve this numerically we could measure the height indicated by the tidal plot each minute and calculate the average height for each 4-min span (and this involves measuring the height at five time points):

$$h_{\text{average}} = \frac{1}{5} \sum_{\tau=t}^{\tau=t+4} h_{\tau}.$$

Done simply, this would result in a stepped curve that still lacks the smoothness of a simple line. We could reduce the stepped appearance by using a moving average in which we calculate the average height in a 4-min span but we move the reference time forward by a single minute each time we perform the calculation. The inclusion of each of the height samples was made without weighting their contribution to the average and this is an example of rectangular windowing. We could go one step further by weighting the

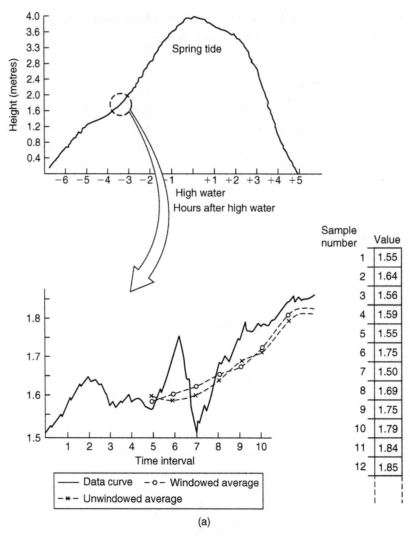

Figure 15.15(a): To explain the ideas behind digital filtering, we review the shape of the tidal height curve (Portsmouth, UK, spring tides) for its underlying detail. The pen Plotter trace would also record every passing wave, boat, and breath of wind; all are overlaid on the general shape of the curve.

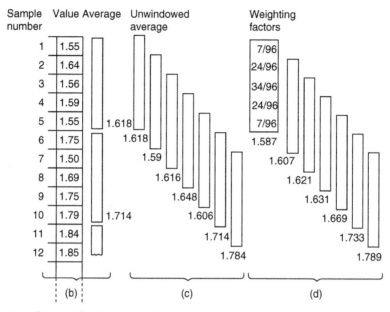

Figures 15.15(b)–15(d): For a small portion of the curve, make measurements at each interval. In the simplest averaging scheme we take a block of five values, average them, and then repeat the process with a fresh block of five values. This yields a relatively coarse stepped waveform. (c) The next approach carries out the averaging over a block of five samples but shifts the start of each block only one sample on at a time, still allowing each of the five sample values to contribute equally to the average each time. The result is a more finely structured plot that could serve our purpose. (d) The final frame in this sequence repeats the operations of (c) except that the contribution that each sample value makes to the averaging process is weighted, using a five-element weighting filter or window for this example whose weighting values are derived by a modified form of least-squares averaging. The values that it returns are naturally slightly different from those of (c).

contribution that each height makes to the average each time we calculate the average of a 4-min period. Shaped windows are common in the field of statistics and are used in digital signal processing. The choice of window does affect the result, although as it happens the effect is slight in the example given here.

One major practical problem with implementing practical finite impulse response (FIR) filters for digital audio signals is that controlling the response accurately or at low

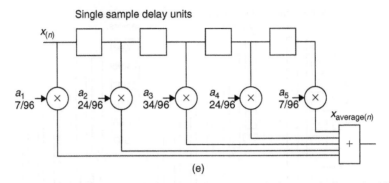

Figure 15.15(e): A useful way of showing the process being carried out in (d) is to draw a block diagram in which each time that a sample value is read it is loaded into a form of memory while the previous value is moved on to the next memory stage. We take the current value of the input sample and the output of each of these memory stages and multiply them by the weighting factor before summing them to produce the output average. The operation can also be expressed in an algebraic form in which the numerical values of the weighting coefficients have been replaced by an algebraic symbol: $x_{average}n = (a_1x_{n-1} + a_2x_{n-2} + a_3x_{n-3} + a_4x_{n-4})$. This is a simple form of a type of digital filter known as a finite impulse response or transversal filter. In the form shown here it is easy to see that the delay of the filter is constant and thus the filter will show linear phase characteristics. If the input to the filter is an impulse, the values you should obtain at the output are identical, in shape, to the profile of the weighting values used. This useful property can be used in the design of filters, as it illustrates the principle that the characteristics of a system can be determined by applying an impulse to it and observing the resultant output.

frequencies forces the number of stages to be very high. You can appreciate this through recognizing that the FIR filter response is determined by the number and value of the coefficients applied to each of the taps in the delayed signal stages. The value of these coefficients is an exact copy of the filter's impulse response. Thus an impulse response intended to be effective at low frequencies is likely to require a great many stages. This places pressure on the hardware that has to satisfy the demand to perform the necessary large number of multiplications within the time allotted for processing each sample value. In many situations a sufficiently accurate response can be obtained with less circuitry by feeding part of a filter's output back to the input (Figure 15.16).

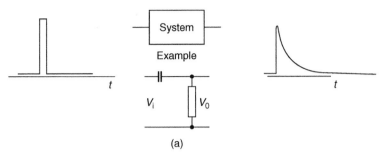

(a)

**Figure 15.16(a): An impulse is applied to a simple system whose output is a simple exponential decaying response:** $V_0 = V_i e^{-t/RC}$.

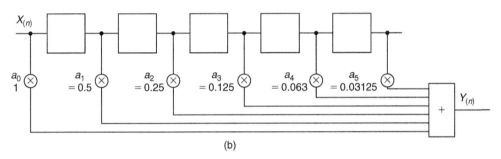

(b)

**Figure 15.16(b): A digital filter based on an FIR structure would need to be implemented as shown. The accuracy of this filter depends on just how many stages of delay and multiplication we can afford to use. For the five stages shown, the filter will cease to emulate an exponential decay after only 24 dB of decay. The response to successive n samples is**
$$Y_n = 1X_n + (1/2)X_{n-1} + (1/4)X_{n-2} + (1/8)X_{n-3} + (1/6)X_{n-4}.$$

We have drawn examples of digital filtering without explicit reference to their use in digital audio. The reason is that the principles of digital signal processing hold true no matter what the origin or use of the signal being processed. The ready accessibility of analogue audio "cook-books" and the simplicity of the signal structure have drawn a number of less than structured practitioners into the field. For whatever inaccessible reason, these practitioners have given the audio engineer the peculiar notions of directional copper cables, especially colored capacitors, and compact discs outlined with green felt tip pens. Bless them all, they have their place as entertainment and they remain free in their pursuit of the arts of the audio witch doctor. The world of digital processing

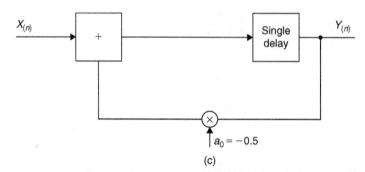

(c)

Figure 15.16(c): This simple function can be emulated by using a single multiplier and adder element if some of the output signal is fed back and subtracted from the input. Use of a multiplier in conjunction with an adder is often referred to as a multiplier-accumulator or MAC. With the correct choice of coefficient in the feedback path, the exponential decay response can be exactly emulated: $Y_n = X_n - 0.5Y_{n-1}$. This form of filter will continue to produce a response forever unless the arithmetic elements are no longer able to handle the decreasing size of the numbers involved. For this reason, it is known as an infinite impulse response (IIR) filter or, because of the feedback structure, a recursive filter. Whereas the response characteristics of FIR filters can be gleaned comparatively easily by inspecting the values of the coefficients used, the same is not true of IIR filters. A more complex algebra is needed in order to help in the design and analysis, which are not covered here.

requires more rigor in its approach and practice. Figure 15.17 shows examples of simple forms of first- and second-order filter structures. Processing an audio signal in the digital domain can provide a flexibility that analogue processing denies. You may notice from the examples how readily the characteristics of a filter can be changed simply by adjustment of the coefficients used in the multiplication process. The equivalent analogue process would require much switching and component matching. Moreover, each digital filter or process will provide exactly the same performance for a given set of coefficient values, which is a far cry from the miasma of tolerance problems that beset the analogue designer.

The complicated actions of digital audio equalization are an obvious candidate for implementation using infinite impulse response filters and the field has been heavily researched in recent years. Much research has been directed toward overcoming some of the practical problems, such as limited arithmetic resolution or precision and limited processing time. Practical hardware considerations force the resulting precision of any digital arithmetic operation to be limited. The limited precision also affects the choice of

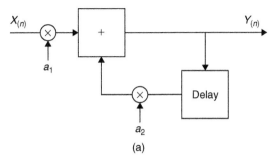

(a)

**Figure 15.17(a): The equivalent of the analogue first-order high- and low-pass filters requires a single delay element. Multipliers are used to scale the input (or output) values so that they lie within the linear range of the hardware. Digital filter characteristics are quite sensitive to the values of the coefficients used in the multipliers. The output sequence can be described as $Y_n = a_1 X_n + a_2 X_{n-1}$. If $0 > a_2 > -1$ the structure behaves as a first-order lag. If $a_2 > 0$ than the structure produces an integrator. The output can usually be kept in range by ensuring that $a_1 = 1 - a_2$.**

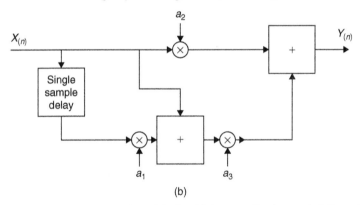

(b)

**Figure 15.17(b): The arrangement for achieving high-pass filtering and differentiation again requires a single delay element. The output sequence is given by $Y_n = a_2 X_n + a_3(a_1 X_{n-1} + X_n)$. The filter has no feedback path so it will always be stable. Note that $a_1 = -1$ and with $a_2 = 0$ and $a_3 = 1$ the structure behaves as a differentiator. These are simple examples of first-order structures and are not necessarily the most efficient in terms of their use of multiplier or adder resources. Although a second-order system would result if two first-order structures were run in tandem, full flexibility of second-order IIR structures requires recursive structures. Perhaps the most common of these emulates the analogue biquad (or twin integrator loop) filter.**

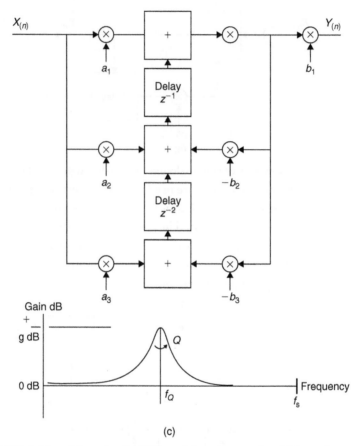

(c)

Figure 15.17(c): To achieve the flexibility of signal control, which analogue equalizers exhibit in conventional mixing desks, an IIR filter can be used. Shown here it requires two single-sample value delay elements and six multiplying operations each time it is presented with an input sample value. We have symbolized the delay elements by using the $z^{-1}$ notation, which is used when digital filter structures are formally analyzed. The output sequence can be expressed as $b_1Y_n = a_1X_n + a_2X_{n-1} - b_2Y_{n-1} + a_3X_{n-2} - b_3Y_{n-2}$. The use of $z^{-1}$ notation allows us to express this difference or recurrence equation as $b_1Y_nz^{-0} = a_1X_nz^{-0} + a_2X_nz^{-1} - b_2Y_nz^{-1} + a_3X_nz^{-2} - b_3Y_nz^{-2}$. The transfer function of the structure is the ratio of the output over the input, just as it is in the case of an analogue system. In this case the input and output happen to be sequences of numbers, and the transfer function is indicated by the notation H(z):

(Continued) ▶

values for the coefficients whose value will determine the characteristics of a filter. This limited precision is effectively a processing error, which will make its presence known through the addition of noise to the output. The limited precision also leads to the odd effects for which there is no direct analogue equivalent, such as limit cycle oscillations. The details concerning the structure of a digital filter have a very strong effect on the sensitivity of the filter to noise and accuracy, in addition to the varying processing resource requirement. The best structure thus depends a little on the processing task that is required to be carried out. The skill of the engineer is, as ever, in balancing the factors in order to optimize the necessary compromises.

While complicated filtering is undoubtedly used in digital audio signal processing, spare a thought for the simple process of averaging. In digital signal processing terms this is usually called interpolation (Figure 15.18). The process is used to conceal unrecoverable errors in a sequence of digital sample values and, for example, is used in the compact disc for just this reason.

$$H(z) = \frac{Y(z)}{X(z)} = \frac{a_1 + a_2 z^{-1} + a_3 z^{-2}}{b_1 + b_2 z^{+1} + b_3 z^{-2}}.$$

**Figure 15.17(c): continued: The value of each of the coefficients can be determined from knowledge of the rate at which samples are being made available, $F_s$, and your requirement for the amount of cut or boost and of the $Q$ required. One of the first operations is that of prewarping the value of the intended center frequency $f_c$ in order to take account of the fact that the intended equalizer center frequency is going to be comparable to the sampling frequency. The "warped" frequency is given by $f_w = F_s / \pi \tan \pi f_c / F_s$. And now for the coefficients:**

$$a_1 = 1 + \pi f_w (1 + k)/F_s Q + \left(\pi f_w/F_s\right)^2$$
$$a_2 = b_2 = -2\left(1 + \left(\pi f_w/F_s\right)^2\right)$$
$$a_3 = 1 - \pi f_w (1 + k)/F_s Q + \left(\pi f_w/F_s\right)^2$$
$$b_1 = 1 + \pi f_w/F_s Q + \left(\pi f_w/F_s\right)^2$$
$$b_3 = 1 + \pi f_w/F_s Q + \left(\pi f_w/F_s\right)^2$$

**The mathematics concerned with filter design certainly appear more complicated than that which is usually associated with analogue equalizer design. The complication does not stop here though, as a designer must take into consideration the various compromises brought on by limitations in cost and hardware performance.**

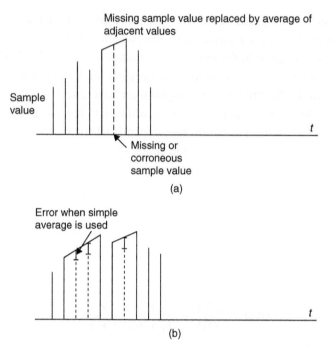

Figure 15.18: (a) Interpolation involves guessing the value of the missing sample. The fastest
guess uses the average of the two adjacent good sample values, but an average based on
many more sample values might provide a better answer. The use of a simple rectangular
window for including the values to be sampled will not lead to as accurate a replacement
value. The effect is similar to that caused by examining the spectrum of a continuous signal
that has been selected using a simple rectangular window. The sharp edges of the window
function will have frequency components that cannot be separated from the wanted signal.
(b) A more intelligent interpolation uses a shaped window that can be implemented as an
FIR, or transversal, filter with a number of delay stages, each contributing a specific fraction
of the sample value of the output sum. This kind of filter is less likely than the simple linear
average process to create audible transitory aliases as it fills in damaged sample values.

## 15.9 Other Binary Operations

One useful area of digital activity is related to filtering, and its activity can be described
by similar algebra. The technique uses a shift register whose output is fed back and

combined with the incoming logic signal. Feedback is usually arranged to come from earlier stages as well as the final output stage. The arrangement can be considered as a form of binary division. For certain combinations of feedback the output of the shift register can be considered as a statistically dependable source of random numbers. In the simplest form the random output can be formed in the analogue domain by the simple addition of a suitable low-pass filter. Such a random noise generator has the useful property that the noise waveform is repeated, which allows the results of stimulating a system with such a signal to be averaged.

When a sequence of samples with a nominally random distribution of values is correlated with itself, the result is identical to a band-limited impulse (Figure 15.19). If such a random signal is used to stimulate a system (this could be an artificial reverberation device, an equalizer, or a loudspeaker under test) and the resulting output is correlated with the input sequence, the result will be the impulse response of the system under test. The virtue of using a repeatable excitation noise is that measurements can be made in the presence of other random background noise or interference, and if further accuracy is required, the measurement is simply repeated and the results averaged. True random

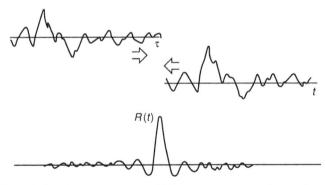

**Figure 15.19:** Correlation is a process in which one sequence of sample values is checked against another to see just how similar both sequences are. A sinusoidal wave correlated with itself (a process called auto correlation) will produce a similar sinusoidal wave. By comparison, a sequence of random sample values will have an autocorrelation function that will be zero everywhere except at the point where the samples are exactly in phase, yielding a band-limited impulse.

background noise will average out, leaving a "cleaner" sequence of sample values that describe the impulse response. This is the basis behind practical measurement systems.

Shift registers combined with feedback are also used in error detecting and correction systems.

## 15.10 Sampling and Quantizing

It is not possible to introduce each element of this broad topic without requiring the reader to have some foreknowledge of future topics. The aforementioned text has tacitly admitted that you will wish to match the description of the processes involved to a digitized audio signal, although we have pointed out that handling audio signals in the digital domain is only an example of some of the flexibility of digital signal processing.

The process of converting an analogue audio signal into a sequence of sample values requires two key operations. These are sampling and quantization. They are not the same operation, for while sampling means that we only wish to consider the value of a signal at a fixed point in time, the act of quantizing collapses a group of amplitudes to one of a set of unique values. Changes in the analogue signal between sample points are ignored. For both of these processes the practical deviations from the ideal process are reflected in different ways in the errors of conversion.

Successful sampling depends on ensuring that the signal is sampled at a frequency at least twice that of the highest frequency component. This is Nyquist's sampling theorem. Figure 15.20 shows the time domain view of the operation, whereas Figure 15.21 shows the frequency domain view.

### 15.10.1 Sampling

Practical circuitry for sampling is complicated by the need to engineer ways around the various practical difficulties. The simple form of the circuit is shown in Figure 15.22. The analogue switch is opened for a very short period, $t_{ac}$ each $1/F_s$ seconds. In this short period the capacitor must charge (or discharge) to match the value of the instantaneous input voltage. The buffer amplifier presents this voltage to the input of the quantizer or analogue-digital converter (ADC). There are several problems. The series resistance of the switch sets a limit on how large the storage capacitor can be while the input current requirements of the buffer amplifier set a limit on how low the capacitance can be. The

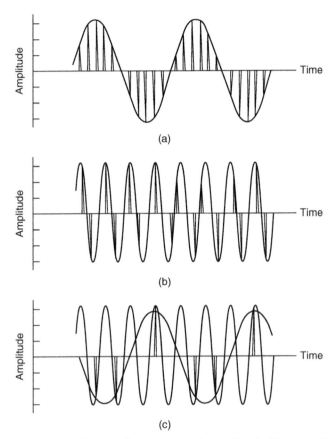

(a)

(b)

(c)

Figure 15.20: (a) In the time domain the process of sampling is like one of using a sequence of pulses, whose amplitude is either 1 or 0, and multiplying it by the value of the sinusoidal waveform. A sample and hold circuit holds the sampled signal level steady while the amplitude is measured. (b) At a higher frequency, sampling is taking place approximately three times per sinusoid input cycle. Once more it is possible to see that even by simply joining the sample spikes the frequency information is still retained. (c) This plot shows the sinusoid being under sample, and on reconstituting the original signal from the spikes the best-fit sinusoid is the one shown as the dashed line. This new signal will appear as a perfectly proper signal to any subsequent process and there is no method for abstracting such aliases from properly sampled signals. It is necessary to ensure that frequencies greater than half of the sampling frequency $F_s$ are filtered out before the input signal is presented to a sampling circuit. This filter is known as an antialiasing filter.

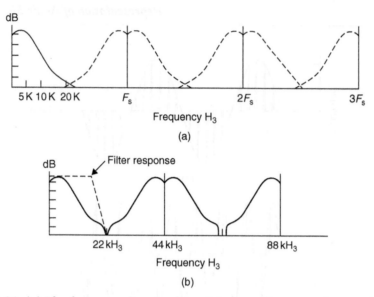

Figure 15.21: (a) The frequency domain view of the sampling operation requires us to recognize that the spectrum of a perfectly shaped sampling pulse continues forever. In practice sampling, waveforms do have finite width and practical systems do have limited bandwidth. We show here the typical spectrum of a musical signal and the repeated spectrum of the sampling pulse using an extended frequency axis. Note that even modern musical signals do not contain significant energy at high frequencies and, for example, it is exceedingly rare to find components in the 10-kHz region more than $-30\,$dB below the peak level. (b) The act of sampling can also be appreciated as a modulation process, as the incoming audio signal is being multiplied by the sampling waveform. The modulation will develop sidebands, which are reflected on either side of the carrier frequency (the sampling waveform), with a set of sidebands for each harmonic of the sampling frequency. The example shows the consequence of sampling an audio bandwidth signal that has frequency components beyond $F_2/2$, causing a small but potentially significant amount of the first lower sideband of the sampling frequency to be folded or aliased into the intended audio bandwidth. The resulting distortion is not harmonically related to the originating signal and can sound truly horrid. Use of an antialias filter before sampling restricts the leakage of the sideband into the audio signal band. The requirement is ideally for a filter with an impossibly sharp rate of cutoff, and in practice a small guard band is allowed for tolerance and finite cutoff rates. Realizing that the level of audio signal with a frequency around 20 kHz is typically 60 dB below the peak signal level, it is possible to perform practical filtering using seventh-order filters. However, even these filters are expensive to manufacture and represent a significant design problem in their own right.

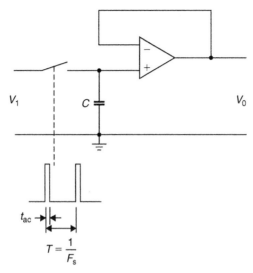

**Figure 15.22: An elementary sample and hold circuit using a fast low distortion
semiconductor switch that is closed for a short time to allow a small-valued storage capacitor
to charge up to the input voltage. The buffer amplifier presents the output to the quantizer.**

imperfections of the switch mean that there can be significant energy leaking from the
switching waveform as the switch is operated and there is the problem of cross talk from
the audio signal across the switch when it is opened. The buffer amplifier itself must be
capable of responding to a step input and settling to the required accuracy within a small
fraction of the overall sample period. The constancy or jitter of the sampling pulse must
be kept within very tight tolerances and the switch itself must open and close in exactly
the same way each time it is operated. Finally, the choice of capacitor material is itself
important because certain materials exhibit significant dielectric absorption.

The overall requirement for accuracy depends greatly on the acceptable signal-to-noise
ratio (SNR) for the process, which is much controlled by the resolution and accuracy of
the quantizer or converter. For audio purposes we may assume that suitable values for $F_s$
will be in the 44- to 48-kHz range. The jitter or aperture uncertainty will need to be in the
region of 120 pse, acquisition and settling time need to be around 1 µs, and the capacitor
discharge rate around 1 V/s for a signal that will be quantized to 16 bits if the error due to
that cause is not to exceed +/− 0.5 LSB. The jitter performance is complex to visualize
completely because of the varying amplitude and frequency component of the jitter itself.

At this stage you will need to be sure you are confident that you appreciate that the combined effect of the antialias filtering and the proper implementation of the sampling process mean that sampled data contain perfectly all of the detail up to the highest frequency component in the signal permitted by the action of the antialias filter.

### 15.10.2 Quantizing

The sampled input signal must now be measured. The dynamic range that can be expressed by an $n$-bit number is approximately proportional to $2^n$ and this is more usually expressed in dB terms. The converters used in test equipment such as DVMs are too slow for audio conversion but it is worthwhile considering the outline of their workings (Figure 15.23). A much faster approach uses a successive approximation register (SAR) and a digital-to-analogue converter (DAC) (Figure 15.24).

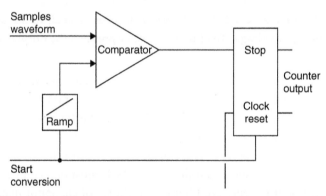

**Figure 15.23: The simplest ADC uses a ramp generator, which is started at the beginning of conversion. At the same time a counter is reset and the clock pulses are counted. The ramp generator output is compared with the signal from the sample and hold and when the ramp voltage equals the input signal the counter is stopped. Assuming that the ramp is perfectly linear (quite difficult to achieve at high repetition frequencies) the count will be a measure of the input signal. The problem for audio bandwidth conversion is the speed at which the counter must run in order to achieve a conversion within approximately 20 μs. This is around 3.2768 GHz and the comparator would need to be able to change state within 150 μs with, in the worst case, less than 150 μV of differential voltage. There have been many developments of this conversion technique for instrumentation purposes.**

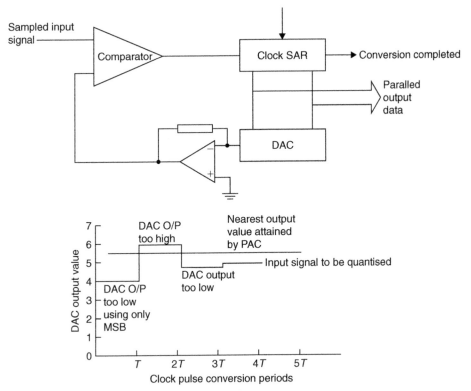

Figure 15.24: The SAR operates with a DAC and a comparator, initially reset to zero.
At the first clock period the MSB is set and the resulting output of the DAC is compared to
the input level. In the example given here the input level is greater than this and so the MSB
value is retained and, at the next clock period, the next MSB is set. In this example
the comparator output indicates that the DAC output is too high, the bit is set to 0, and
the next lower bit is set. This is carried out until all of the DAC bits have been tried.
Thus a 16-bit ADC would require only 17 clock periods (one is needed for reset)
in which to carry out a conversion.

A very simple form of DAC is based on switching currents into a virtual earth summing
point. The currents are derived from a R–2R ladder, which produces binary weighted
currents (Figure 15.25). The incoming binary data directly controls a solid state switch,
which routes a current either into the virtual earth summing point of the output amplifier
or into the local analogue ground. Since the voltage across each of the successive

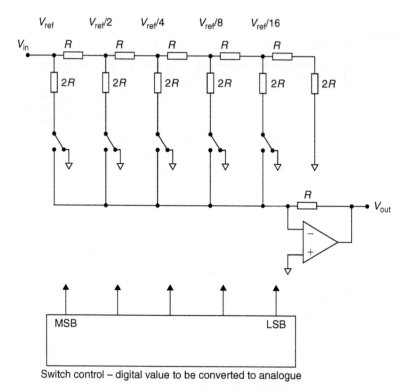

**Figure 15.25: The basic form of the R–2R digital to analogue (DAG) converter is shown here implemented by ideal current switches. The reference voltage can be an audio bandwidth signal and the DAC can be used as a true four quadrant multiplying converter to implement digitally controlled analogue level and equalization changes. Other forms of switching currents are also used and these may not offer a true multiplication facility.**

2R resistors is halved at each stage the current that is switched is also halved. The currents are summed at the input to the output buffer amplifier. The limit to the ultimate resolution and accuracy is determined partly by the accuracy of adjustment and matching of the characteristics of the resistors used and also by the care with which the converter is designed into the surrounding circuitry. Implementation of a 16-bit converter requires that all of the resistors are trimmed to within 0.0007% (half of an LSB) of the ideal value and, further, that they all maintain this ratio over the operational temperature of the converter. The buffer amplifier must be able to settle quickly and it must not contribute to the output noise significantly.

There are many variations on the use of this technique with one common approach being to split the conversion into two stages. Typically, a 16-bit converter would have the top 8 most significant bits control a separate conversion stage, which sets either the voltage or the current with which the lower 8 LSBs operate. The approach has to contend with the problem of ensuring that the changeover point between the two stages remains matched throughout the environmental range of the converter. One solution to the problem of achieving an accurate binary ratio between successive currents is to use a technique called dynamic element balancing.

Whereas sampling correctly executed loses no information, quantizing inevitably produces an error. The level of error is essentially dependent on the resolution with which the quantizing is carried out. Figure 15.26 illustrates the point by showing a sinusoid quantised to 16 quantizing levels. A comparison of the quantized output with the original has been used to create the plot of the error in the quantizing. The error waveform of this example clearly shows a high degree of structure, which is strongly related to the signal

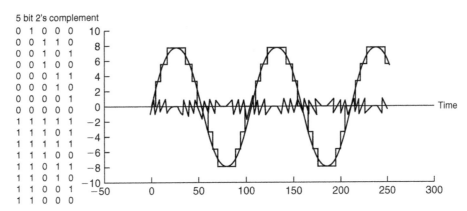

Figure 15.26: The input sinusoid is shown here prior to sampling as a dotted line superimposed on the staircase shape of the quantized input signal. The two's complement value of the level has been shown on the left-hand edge. The error signal is the difference between the quantized value and the ideal value assuming a much finer resolution. The error signal, or quantizing noise, lies in the range of $\pm I$ q. Consideration of the mean square error leads to the expression for the rms value of the quantizing noise: $V_{noise} = q/\sqrt{(12)}$ where $q$ is the size of a quantizing level. The maximum rms signal amplitude that can be described is $V_{signal} = q\, 2^{n-1}/\sqrt{2}$. Together the expression combines to give the expression for SNR(indB): $SNB_{dB} = 6.02n + 1.76.$

itself. The error can be referred to as quantizing distortion, granulation noise, or simply quantizing noise.

One obvious nonlinearity will occur when the input signal amplitude drops just below the threshold for the first quantizing level. At this stage the quantizing output will remain at zero and all knowledge of the size of the signal will be lost. The remedy is to add a small amount of noise to the signal prior to quantizing (Figure 15.27). This deliberate additional noise is known as dither noise. It does reduce the dynamic range by an amount that depends on its exact characteristics and amplitude but the damage is typically 3 dB. One virtue is that as the original input signal amplitude is reduced below the ±1 quantizing level thresholds (q) the noise is still present and therefore, by virtue of the intrinsic nonlinearity of the quantizer, so are the sidebands that contain vestiges of the original input signal. Thus the quantizer output must also contain information about the original input signal level even though it is buried in the noise. However, the noise is wideband and a spectral plot of the reconstituted waveform will show an undistorted signal component standing clear of the wideband noise.

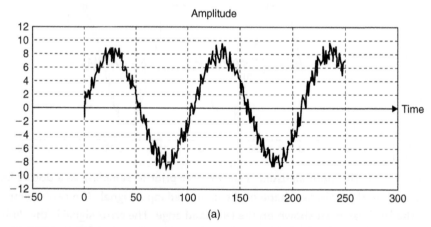

(a)

Figure 15.27(a): Adding a small amount of random noise to the signal prior to quantizing can help disguise the otherwise highly correlated quantizing noise. Aided by the binary modulation action of the quantizer, the sidebands of the noise are spread across the whole audio band width and to a very great degree their correlation with the original distortion signal is broken up. In this illustration the peak-to-peak amplitude of the noise has been set at ±1.5 q.

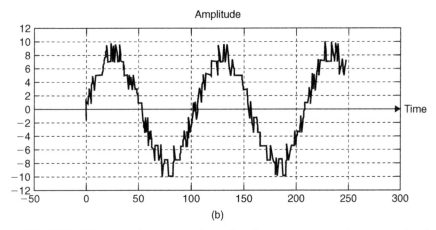

**Figure 15.27(b): The quantizer maps the noisy signal onto one of a range of unique levels as before.**

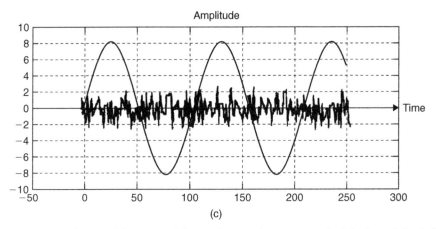

**Figure 15.27(c): The resulting quantizing noise can be compared with the original signal and this time you can see that the noise waveform has lost the highly structured relationship shown in Figure 15.26.**

Unfortunately, it is also quite a problem to design accurate and stable quantizers. The problems include linearity, missing codes, differential nonlinearity, and nonmonotonicity. Missing codes should be properly considered a fault since they arise when a converter either does not produce or respond to the relevant code. A suitable test is a simple ramp.

A powerful alternative test method uses a sinusoidal waveform and checks the amplitude probability density function. Missing, or misplaced codes, and nonmonotonic behavior can show up readily in such a plot.

Linearity can be assessed overall by using a sinusoidal test signal since the output signal will contain harmonics. The performance of an ADC can be carried out in the digital domain by direct use of the discrete fast Fourier transform. The DAC can be checked by driving it with a computer-generated sequence and observing the output in the analogue domain. The trouble with using the simple harmonic distortion test is that it is not easy to check the dynamic performance over the last decade of bandwidth and for this reason the CCIR twin tone intermodulation distortion is much preferred.

Differential nonlinearity is the random unevenness of each quantization level. This defect can be assessed by measuring the noise floor in the presence of a signal. In a good DAC the rms noise floor should be approximately 95 dB below the maximum rms level (assuming a modest margin for dither). The output buffer amplifier will contribute some noise but this should be at a fixed level and not dependent on the DAC input sample values.

The basic ADC element simply provides an output dependent on the value of the digital input. During the period while a fresh sample is settling, its output can be indeterminate. Thus the output will usually be captured by a sample and hold circuit as soon as the DAC has stabilized. The sample and hold circuit is a zero order hold circuit that imposes its own frequency response on the output signal (Figure 15.28); correction for which can be accommodated within the overall reconstruction filter. The final filter is used to remove the higher components and harmonics of the zero order hold.

### 15.10.3 Other Forms of ADC and DAC

Flash converters (Figure 15.29) function by using an array of comparators, each set to trigger at a particular quantizing threshold. The output is available directly. These days the technique is most commonly employed directly as shown in digitizing video waveforms. However, there is a use for the technique in high-quality oversampling converters for audio signals.

One great benefit of operating with digital signals is their robustness; they are, after all, construed as either 1 or 0 irrespective of the cable or optical fiber down which they travel. Their disadvantage is that the digital signals do cover a wide bandwidth. Since bandwidth is a valuable resource there has been much effort expended in devising ways in

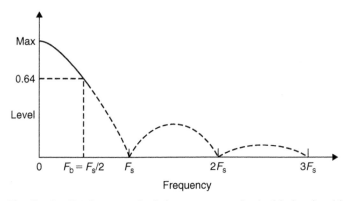

**Figure 15.28: Finally the DAC output is fed to a zero order hold circuit which performs a similar operation to the sample and hold circuit and then to a reconstruction or output antialiasing filter. The plot of the spectral shape of the zero order hold shows that there are frequency components, at decreasing levels, at harmonic intervals equal to $F_s$.**

which an apparently high-quality signal can be delivered using fewer bits. The telephone companies were among the first to employ digital compression and expansion techniques but the technology has been used for nontelephonic audio purposes. In the A and \/ law converters (Figure 15.30), the quantizing steps, $q$, do not have the same value. For low signal levels the quantizing levels are closely spaced and become more widely spaced at higher input levels. The decoder implements the matching inverse conversion.

Another approach to providing a wide coding range with the use of fewer bits than would result if a simple linear approach were to be taken is exemplified in the flying comma or floating point type of converter. In this approach a fast converter with a limited coding range is presented with a signal that has been adjusted in level such that most of the converter's coding range is used. The full output sample value is determined by the combination of the value of the gain setting and of the sample value returned by the converter. The problem here is that the change in gain in the gain stage is accompanied by a change in background noise level and this too is coded. The result is that the noise floor that accompanies the signal is modulated by the signal level, which produces a result that does not meet the performance requirement for high-quality audio. A more subtle approach is exemplified in syllabic companders. The NICAM approach manages a modest reduction from around $1\,\mathrm{Mbs}^{-1}$ to $7.04\,\mathrm{kbs}^{-1}$ and we see in it an early approach to attempts to adapt the coding process to the psychoacoustics of human hearing.

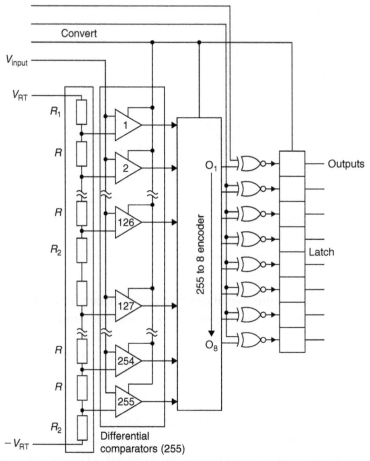

Figure 15.29: A chain of resistors provides a series of reference voltages for a set of comparators whose other input is the input signal. An 8-bit encoder will need 255 comparators. Their output will drive an encoder that maps the output state of the 255 comparators onto an 8-bit word. The NMINV control line is used to convert the output word from an offset binary count to a two's complement form. A 16-bit converter would require an impracticably large number of comparators (65536) in addition to posing serious difficulties to setting the 65536 resistor values to the required tolerance value. The technique does have one virtue in speed and in not needing a sample and hold amplifier to precede it.

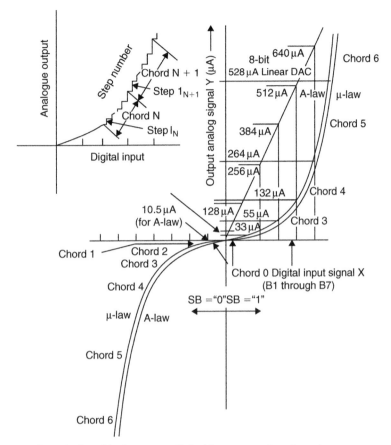

**Figure 15.30: The relationship between digital input word and analogue output current is not linear. The sign bit is the MSB and the next three bits are used to set the chord slope. The lower 4 bits set the output steps within each chord. The drawing shows the equivalent output for a linear 8-bit DAC.**

The encoder will need to have the matching inverse characteristic in order that the net transfer characteristic is unity. The dynamic range of an 8-bit m or A-law converter is around 62 dB and this can be compared to the 48 dB that a linear 8-bit converter can provide. The use of companding (compressing and then expanding the coding range) could be carried in the analogue domain prior to using a linear converter. The difficulty is then one of matching the analogue sections. This is an approach that has been taken in some consumer video equipment.

Delta sigma modulators made an early appearance in digital audio effects processors for a short while. One of the shortcomings of the plain delta modulator is the limitation in the rate at which it can track signals with high slew rates. As we have shown, each pulse is worth one unit of charge to the integrator. To make the integrator climb faster the rate of charge can be increased so that high slew rate signals can be tracked.

Oversampling techniques can be applied to both ADCs and DACs. The oversampling ratio is usually chosen as a power of 2 in order to make computation more efficient. Figure 15.31 shows the idea behind the process for either direction of conversion. The 4× oversampling shown is achieved by adding samples with zero value at each new sample point. At this stage of the process the spectrum of the signal will not have been altered and thus there are still the aliases at multiples of $F_s$. The final stage is to filter the new set

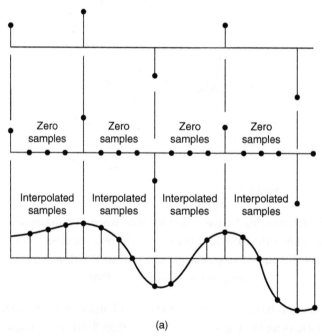

(a)

**Figure 15.31(a): The oversampling process adds zero valued dummy samples to the straight sampled signal. If oversampling is being carried out in the DAC direction, then the digital sequence of samples is treated as if these extra dummy samples had been added in. The sequence is then filtered using an interpolation filter, which creates useful values for the dummy samples.**

of samples with a low-pass filter set to remove the components between the top of the audio band and the lower sideband of the $4 \times F_s$ component.

The process effectively transfers the antialias filter from the analogue to the digital domain with the attendant advantages of digital operation. These include a near ideal transfer function, low ripple in the audio band, low group delay distortion, wide dynamic range, exactly repeatable manufacture, and freedom from a wide range of analogue component and design compromises. The four times upsampling process spreads the

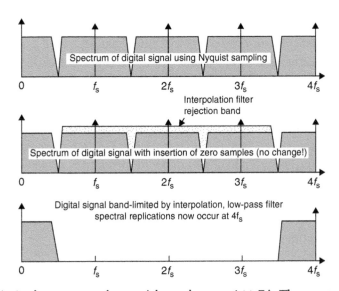

Figure 15.31(b): At the new sample rate (shown here as $4 \times F_s$). The spectrum of the signal now extends to $4 \times F_s$, although there is only audio content up to $F_s/2$. Thus when the signal is passed to the DAC element (an element that will have to be able to work at the required oversampling speed) the resulting audio spectrum can be filtered simply from the nearest interfering frequency component, which will be at $4 \times F_s$. Note that the process of interpolation does not add information. If the oversampling is being carried out in the ADC direction, the analogue audio signal itself will be sampled and quantized at the higher rate. The next stage requires the reduction of the sequence of data by a factor of four. First data are filtered in order to remove components in the band between the top of the required audio band and the lower of the $4 \times F_s$ sideband and then the data sequence can be simply subsampled (only one data word out of each four is retained).

quantization noise over four times the spectrum, thus only 1/4 of the noise power now resides in the audio band. If we assume that the noise spectrum is uniform and that dither has been judiciously applied, this is equivalent to obtaining a 1-bit enhancement in dynamic range within the audio bandwidth in either digital or analogue domains. This performance can be further improved by the process of noise shaping.

The information capacity of a communication channel is a function of the SNR and the available bandwidth. Thus there is room for trading one against the other. The higher the oversampling ratio, the wider the bandwidth in which there is no useful information. If samples were to be coarsely quantized it should be possible to place the extra quantization noise in part of the redundant spectrum. The process of relocating the noise in the redundant spectrum and out of the audio band is known as noise shaping and it is accomplished using a recursive filter structure (the signal is fed back to the filter). Practical noise shaping filters are high order structures that incorporate integrator elements in a feedback path along with necessary stabilization.

The process of oversampling and noise shaping can be taken to an extreme, and implementation of this approach is available in certain DACs for compact disc systems. The audio has been oversampled by 256× in the Philips bit stream device, 758× in the Technics' MASH device, and 1024× in Sony's device. The output is in the form of a pulse train modulated by its density (PDM), by its width (PWM), or by its length (PLM). High oversampling ratios are also used in ADCs, which are starting to appear on the market at the current time.

## 15.11  Transform and Masking Coders

We indicated very early on that there may be some advantage in terms of the reduction in data rate to taking the Fourier transform of a block of audio data and transmitting the coefficient data. The use of a technique known as the discrete cosine transform is similar in concept and is used in the AC-2 system designed by Dolby Laboratories. This system can produce a high-quality audio signal with 128 kb per channel.

The MUSICAM process also relies on a model of the human ear's masking processes. The encoder receives a stream of conventionally encoded PCM samples, which are then divided into 32 narrow bands by filtering. The allocation of the auditive significance of the contribution that each band can make to the overall program is then carried out prior to arranging the encoded data in the required format. The principle in this encoder is similar to that planned for the Philips digital compact cassette system.

The exploitation of the masking thresholds in human hearing lies behind many of the proposed methods of achieving bit rate reduction. One significant difference between them and conventional PCM converters is the delay between applying an analogue signal and the delivery of the digital sample sequence. A similar delay is involved when the digital signal is reconstituted. The minimum delay for a MUSICAM encoder is in the region of 9 to 24 ms depending on how it is used. These delays do not matter for a program that is being replayed but they are of concern when the coders are being used to provide live linking program material in a broadcast application.

A second, potentially more damaging, problem with these perceptual encoders is that there has been insufficient work carried out on the way in which concatenations of coders will affect the quality of the sound passing through. Although this problem affects the broadcaster more, the domestic user of such signals may also be affected. Be sure that perceptual coding techniques remove data from the original, as these data cannot be restored. Thus a listener who wishes to maintain the highest quality of audio reproduction may find that the use of his preamplifier's control or room equalizer provides sufficient change to an encoded signal that the original assumptions concerning masking powers of the audio signal may no longer be valid. Thus the reconstituted analogue signal may well be accompanied by unwelcome noise.

## References

There are no numbered references in the text but the reader in search of more detailed reading may first begin with some of the texts listed here. One risk exists in this multidisciplinary field of engineering and that is the rate at which the state-of-the-art of digital audio is being pushed forward. Sometimes it is simply the process of ideas that were developed for one application area (e.g., submarine holography) becoming applicable to high-quality audio processing.

A useful overall text is that of John Watkinson (*The Art of Digital Audio*, Butterworth-Heinemann, ISBN 0-240-51270-7).

No text covers every aspect of the subject and a more general approach to many topics can be found in the oft-quoted Rabiner and Gold (*Theory and Application of Digital Signal Processing*, Rabiner and Gold, ISBN 0-13-914-101-4). Although it was initially published in 1975, the principles have not changed, indeed it is salutary to read the book and to realize that so much of what we think of as being modern was developed in principle so long ago.

Undoubtedly the best book to help the reader over the tricky aspects of understanding the meaning of transforms (Laplace, Fourier, and z) is by Peter Kraniauskus (*Transforms in Signals and Systems*, Addison-Wesley, ISBN 0-201-19694-8).

Texts that describe the psychoneural, physiological, and perceptive models of human hearing can be found in Brian Moore's tome (*An Introduction to the Psychology of Hearing*, Academic Press ISBN 0-12-505624-9), and in James Pickles's *An Introduction to the Physiology of Hearing* (Academic Press, ISBN 0-12-554754-4). For both of these texts a contemporary publication date is essential as developments in our basic understanding of the hearing progress are still taking place.

Almost any undergraduate text that handles signals and modulation will cover the topic of sampling, quantizing, noise, and errors sufficiently well. A visit to a bookshop that caters for university or college requirements should prove useful.

Without doubt the dedicated reader should avail themselves of copies of the *Journal of the Audio Engineering Society* in whose pages many of the significant processes involved in digital audio have been described. The reader can achieve this readily by becoming a member. The same society also organizes two conventions each year at which a large number of papers are presented.

Additional sources of contemporary work may be found in the *Research Department Reports* of the BBC Research Department, Kingswood Warren, UK, while the American IEEE ICASSP proceedings and the related IEEE journals have also held details of useful advances.

## Other Titles of Interest

Baert et al., *Digital audio and compact disc technology*, 2nd Ed., Butterworth-Heinemann, 1992.

Pohlmann K. C., *Advanced digital audio*, Sams, 1991.

Pohlmann K. C., *Principles of digital audio*, 2nd Ed., Sams, 1989.

Sinclair R., *Introducing digital audio*, PC Publishing, 1991.

# CHAPTER 16

# Compact Disc

John Linsley Hood

## 16.1 Problems with Digital Encoding

### 16.1.1 Quantization Noise

Although a number of ways exist by which an analogue signal can be converted into its digital equivalent, the most popular, and the technique used in the CD, is the one known as "pulse code modulation," usually referred to as "PCM." In this, the incoming signal is sampled at a sufficiently high repetition rate to permit the desired audio bandwidth to be achieved. In practice, this demands a sampling frequency somewhat greater than twice the required maximum audio frequency. The measured signal voltage level, at the instant of sampling, is then represented numerically as its nearest equivalent value in binary coded form (a process which is known as "quantization").

This has the effect of converting the original analogue signal, after encoding and subsequent decoding, into a voltage "staircase" of the kind shown in Figure 16.1. Obviously, the larger the number of voltage steps in which the analogue signal can be stored in digital form (that shown in the figure is encoded at "4-bit"–$2^4$ or 16 possible voltage levels), the smaller each of these steps will be and the more closely the digitally encoded waveform will approach the smooth curve of the incoming signal.

The difference between the staircase shape of the digital version and the original analogue waveform causes a defect of the kind shown in Figure 16.2, known as "quantization error," and because this error voltage is not directly related in frequency or amplitude to the input signal, it has many of the characteristics of noise and is therefore also known as "quantization noise." This error increases in size as the number of encoding levels is reduced. It will be audible if large enough, and is the first problem with digitally encoded

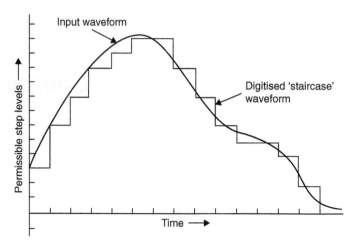

**Figure 16.1: Digitally encoded/decoded waveform.**

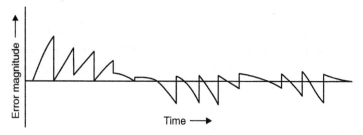

**Figure 16.2: Quantization error.**

signals. I will consider this defect, and the ways by which it can be minimized, later in this chapter.

### 16.1.2 Bandwidth

The second practical problem is that of the bandwidth necessary to store or transmit such a digitally encoded signal. In the case of the CD, the specified audio bandwidth is 20 Hz to 20 kHz, which requires a sampling frequency somewhat greater than 40 kHz. In practice, a sampling frequency of 44.1 kHz is used. In order to reduce the extent of the staircase waveform quantization error, a 16-bit sampling resolution is used in the recording of the CD, equivalent to $2^{16}$ or 65,536 possible voltage steps. If 16 bits are to be transmitted in each sampling interval, then, for a stereo signal, the required bandwidth will be $2 \times 16 \times 44100$ Hz, or 1.4112 mHz, which is already 70 times greater than the audio bandwidth of the incoming signal. However, in practice, additional digital "bits" will be added to this signal for error correction and other purposes, which will extend the required bandwidth even further.

### 16.1.3 Translation Nonlinearity

The conversion of an analogue signal both into and from its binary-coded digital equivalent carries with it the problem of ensuring that the magnitudes of the binary voltage steps are defined with adequate precision. If, for example, "16-bit" encoding is used, the size of the "most significant bit" (MSB) will be 32,768 times the size of the "least significant bit" (LSB). If it is required that the error in defining the LSB shall be not worse than $\pm 0.5\%$, then the accuracy demanded of the MSB must be at least within $\pm 0.0000152\%$ if the overall linearity of the system is not to be degraded.

The design of any switched resistor network, for encoding or decoding purposes, that demanded such a high degree of component precision would be prohibitively expensive and would suffer from great problems as a result of component aging or thermal drift. Fortunately, techniques are available that lessen the difficulty in achieving the required accuracy in the quantization steps. The latest technique, known as "low bit" or "bit-stream" decoding, side steps the problem entirely by effectively using a time-division method, since it is easier to achieve the required precision in time, rather than in voltage or current, intervals.

### 16.1.4 Detection and Correction of Transmission Errors

The very high bandwidths needed to handle or record PCM-encoded signals means that recorded data representing the signal must be very densely packed. This leads to the problem that any small blemish on the surface of the CD, such as a speck of dust, a scratch, or a thumb print, could blot out, or corrupt, a significant part of the information needed to reconstruct the original signal. Because of this, the real-life practicability of all digital record/replay systems will depend on the effectiveness of electronic techniques for the detection, correction, or, if worst comes to worst, masking of the resultant errors. Some very sophisticated systems have been devised, which are also examined later.

### 16.1.5 Filtering for Bandwidth Limitation and Signal Recovery

When an analogue signal is sampled and converted into its PCM-encoded digital equivalent, a spectrum of additional signals is created, of the kind shown in Figure 16.3(a), where

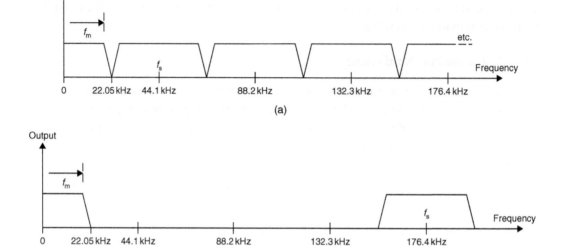

Figure 16.3: PCM frequency spectrum (a) when sampled at 44.1 kHz and (b) when four times oversampled.

$f_s$ is the sampling frequency and $f_m$ is the upper modulation frequency. Because of the way in which the sampling process operates, it is not possible to distinguish between a signal having a frequency that is somewhat lower than half the sampling frequency and one that is the same distance above it; a problem called "aliasing." In order to avoid this, it is essential to limit the bandwidth of the incoming signal to ensure that it contains no components above $f_s/2$.

If, as is the case with the CD, the sampling frequency is 44.1 kHz and the required audio bandwidth is 20 Hz to 20 kHz, +0/−1 dB, an input "antialiasing" filter must be employed to avoid this problem. This filter must allow a signal magnitude that is close to 100% at 20 kHz, but nearly zero (in practice, usually −60 dB) at frequencies above 22.05 kHz. It is possible to design a steep-cut, low-pass filter that approximates closely to this characteristic using standard linear circuit techniques, but the phase shift and group delay (the extent to which signals falling within the affected band will be delayed with respect of lower frequency signals) introduced by this filter would be too large for good audio quality or stereo image presentation.

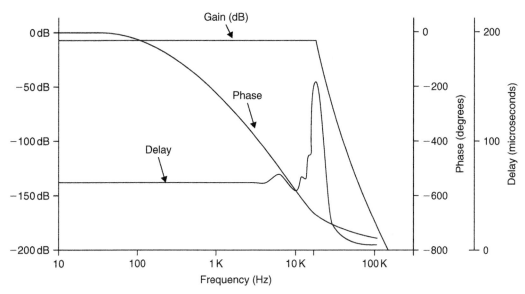

**Figure 16.4: Responses of a low-pass LC filter.**

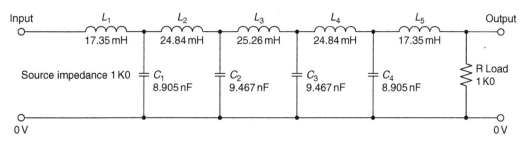

**Figure 16.5: Steep-cut LP filter circuit.**

This difficulty is illustrated by the graph of Figure 16.4, which shows the relative group delay and phase shift introduced by a conventional low-pass analogue filter circuit of the kind shown in Figure 16.5. The circuit shown gives only a modest −90-dB/octave attenuation rate, while the actual slope necessary for the required antialiasing characteristics (say, 0 dB at 20 kHz and −60 dB at 22.05 kHz) would be −426 dB/octave. If a group of filters of the kind shown in Figure 16.5 were connected in series to increase the attenuation rate from −90 to −426 dB/octave, this would cause a group delay, at 20 kHz, of about 1 ms with respect to 1 kHz and a relative phase shift of some 3000°, which would be clearly audible. (In the recording equipment it is possible to employ steep-cut filter systems in which the phase and group delay characteristics are controlled more carefully than would be practicable in a mass-produced CD replay system where both size and cost must be considered.)

Similarly, because the frequency spectrum produced by a PCM-encoded 20-kHz bandwidth audio signal will look like that shown in Figure 16.3(a), it is necessary, on replay, to introduce yet another equally steep-cut low-pass filter to prevent the generation of spurious audio signals that would result from the heterodyning of signals equally disposed on either side of $f_s/2$.

An improved performance in respect to both relative phase error and group delay in such "brick wall" filters can be obtained using so-called "digital" filters, particularly when combined with prefiltering phase correction. However, this problem was only fully solved, and then only on replay (because of the limitations imposed by the original Philips CD patents), by the use of "oversampling" techniques in which, for example, the sampling frequency is increased to 176.4 kHz ("four times oversampling"), which moves the aliasing frequency from 22.05 kHz up to 154.35 kHz, giving the spectral distribution shown in Figure 16.3(b). It is then a relatively easy matter to design a filter, such as that shown in Figure 16.14, having good phase and group delay characteristics, which has a transmission near 100% at all frequencies up to 20 kHz, but near zero at 154.35 kHz.

## 16.2  The Record-Replay System

### 16.2.1  The Recording System Layout

How the signal is handled, on its way from the microphone or other signal source to the final CD, is shown in the block diagram of Figure 16.6. Assuming that the signal has by now been reduced to a basic L–R stereo pair, this is amplitude limited to ensure that no signals greater than the possible encoding amplitude limit are passed on to the analogue-to-digital converter (ADC) stage. These input limiter stages are normally cross linked in operation to avoid disturbance of the stereo image position if the maximum permitted signal level is exceeded, and the channel gain reduced in consequence of this, in only a single channel.

The signal is then passed to a very steep-cut 20-kHz antialiasing filter (often called a "brick wall filter") to limit the bandwidth offered for encoding. This bandwidth limitation

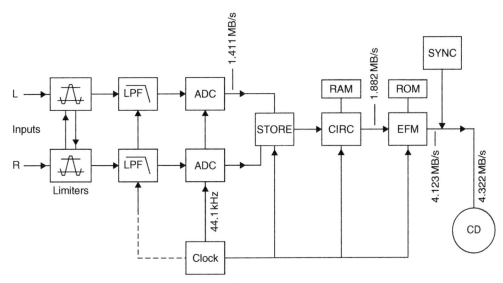

**Figure 16.6: Basic CD recording system.**

is a specific requirement of the digital encoding/decoding process, for the reasons already considered. It is necessary to carry out this filtering process after the amplitude limiting stage because it is possible that the action of peak clipping may generate additional high-frequency signal components. This would occur because "squaring off" the peaks of waveforms will generate a Fourier series of higher frequency harmonic components.

The audio signal, which is still at this stage in analogue form, is then passed to two parallel operating 16-bit ADCs and, having now been converted into a digital data stream, is fed into a temporary data-storage device—usually a "shift register"—from which the output data stream is drawn as a sequence of 8-bit blocks, with the 'L' and 'R' channel data now arranged in a consecutive but interlaced time sequence.

From the point in the chain at which the signal is converted into digitally encoded blocks of data, at a precisely controlled "clock" frequency, to the final transformation of the encoded data back into analogue form, the signal is immune to frequency or pitch errors as a result of motor speed variations in the disc recording or replay process.

The next stage in the process is the addition of data for error correction purposes. Because of the very high packing density of the digital data on the disc, it is very likely that the recovered data will have been corrupted to some extent by impulse noise or blemishes, such as dust, scratches, or thumb prints on the surface of the disc, and it is necessary to include additional information in the data code to allow any erroneous data to be corrected. A number of techniques have been evolved for this purpose, but the one used in the CD is known as the "cross-interleave Reed–Solomon code" (CIRC). This is a very powerful error correction method and allows complete correction of faulty data arising from quite large disc surface blemishes.

Because all possible '0' or '1' combinations may occur in the 8-bit encoded words, and some of these would offer bit sequences rich in consecutive '0's or '1's, which could embarrass the disc speed or spot and track location servo-mechanisms, or, by inconvenient juxtaposition, make it more difficult to read the pit sequence recorded on the disc surface, a bit-pattern transformation stage known as the "eight to fourteen modulation" (EFM) converter is interposed between the output of the error correction (CIRC) block and the final recording. This expands the recorded bit sequence into the form shown in Figure 16.7 to facilitate the operation of the recording and replay process. The functions and method of operation of all these various stages are explained in more detail later in this chapter.

### 16.2.2 Disc Recording

This follows a process similar to that used in the manufacture of vinyl EP and LP records, except that the recording head is caused to generate a spiral pattern of pits in an optically

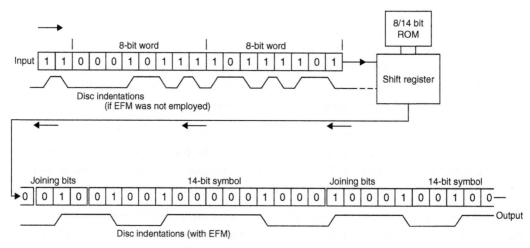

**Figure 16.7: The EFM process.**

flat glass plate, rather than a spiral groove in a metal one, and that the width of the spiral track is very much smaller (about 1/60th) than that of the vinyl groove. (Detail of the CD groove pattern is, for example, too fine to be resolved by a standard optical microscope.) When the master disc is made, "mother" and "daughter" discs are then made preparatory to the production of the stampers, which are used to press out the track pattern on a thin (1.4 mm) plastics sheet, prior to the metallization of the pit pattern for optical readout in the final disc.

## 16.3 The Replay System

### 16.3.1 Physical Characteristics

For the reasons shown earlier, the minimum bandwidth required to store the original 20-Hz to 20-kHz stereo signal in digitally encoded form has now been increased 215-fold, to some 4.3 MHz. It is, therefore, no longer feasible to use a record/replay system based on an undulating groove formed on the surface of a vinyl disc because the excursions in the groove would be impracticably close together unless the rotational speed of the disc were to be enormously increased, which would lead to other problems, such as audible replay noise, pick-up tracking difficulties, and rapid surface wear.

The technique adopted by Philips/Sony in the design of the CD replay system is therefore based on an optical pick-up mechanism, in which the binary coded '0's and '1's are read from a spiral sequence of bumps on an internal reflecting layer within a rapidly rotating (approximately 400 rpm) transparent plastic disc. Because the replay system is noncontacting, this also offers the advantage that there is no specific disc wear incurred in the replay of the records and they have, in principle, if handled carefully, an indefinitely long service life.

#### 16.3.1.1. CD Performance and Disc Statistics

Bandwidth 20 Hz to 20 kHz, ±0.5 dB

Dynamic range >90 dB

S/N ratio >90 dB

Playing time (max.) 74 min

Sampling frequency 44.1 kHz

Binary encoding accuracy 16-bit (65,536 steps)

Disc diameter 120 mm

Disc thickness 1.2 mm

Center hole diameter 15 mm

Permissible disc eccentricity (max.) $\pm 150 \mu m$

Number of tracks (max.) 20 625

Track width 0.6 $\mu m$

Track spacing 1.6 $\mu m$

Tracking accuracy $\pm 0.1 \mu m$

Accuracy of focus $\pm 0.5 \mu m$

Lead-in diameter 46 mm

Lead-out diameter 116 mm

Track length (max.) 5300 m

Linear velocity 1.2–1.4 m/s

### 16.3.1.2 Additional Data Encoded on Disc

- Error correction data.

- Control data—total and elapsed playing times, number of tracks, end of playing area, preemphasis [may be added using either 15 $\mu s$ (10,610 Hz) or 50 $\mu s$ (3183 Hz) time constants], and so on.

- Synchronization signals added to define beginning and end of each data block.

- Merging bits used with EFM.

### 16.3.1.3 Optical Readout System

This is shown, schematically, in Figure 16.8, and consists of an infra-red laser light source (GaAIAs, 0.5 mW, 780 nm), which is focused on a reflecting layer buffed about 1 mm beneath the transparent "active" surface of the disc being played. This metallic

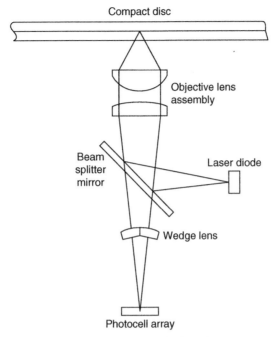

Compact disc

Objective lens
assembly

Beam
splitter
mirror

Laser diode

Wedge lens

Photocell array

**Figure 16.8: Single-beam optical readout system.**

reflecting layer is deformed in the recording process to produce a sequence of oblong humps along the spiral path of the recorded track (actually formed by making pits on the reverse side of the disc prior to metallization). Because of the shallow depth of focus of the lens, due to its large effective numerical aperture (*f*/0.5) and the characteristics of the laser light focused on the reflecting surface, these deformations of the surface greatly diminish the intensity of the incident light reflected to the receiver photocell, in comparison with that from the flat mirror-like surface of the undeformed disc. This causes the intensity of the light reaching the photocell to fluctuate as the disc rotates and causes the generation of the high-speed sequence of electrical '0's and '1's required to reproduce the digitally encoded signal.

The signals representing '1's are generated by a photocell output level transition, either up or down, while '0's are generated electronically within the system by the presence of a timing impulse that is not coincident with a received '1' signal. This confers the valuable feature that the system defaults to a '0' if a data transition is not read, and such random errors can be corrected with ease in the replay system.

It is necessary to control the position of the lens, in relation both to the disc surface and to the recorded spiral sequence of surface lumps, to a high degree of accuracy. This is done by high-speed closed-loop servo-mechanism systems, in which the vertical and lateral position of the whole optical readout assembly is precisely adjusted by electro-mechanical actuators, which are caused to operate in a manner that is very similar to the voice coil in a moving coil loudspeaker.

Two alternative arrangements are used for positioning the optical readout assembly, of which the older layout employs a sled-type arrangement that moves the whole unit in a rectilinear manner across the active face of the disc. This maintains the correct angular position of the head, in relation to the recorded track, necessary when a "three-beam" track position detector is used. Recent CD replay systems more commonly employ a single-beam lateral/vertical error detection system. Since this is insensitive to the angular relationship between the track and the head, it allows a simple pivoted arm structure to be substituted for the rectilinear-motion sled arrangement. This pivoted arm layout is less expensive to produce, is less sensitive to mechanical shocks, and allows more rapid scanning of the disc surface when searching for tracks.

Some degree of immunity from readout errors due to scratches and dust on the active surface of the disc is provided by the optical characteristics of the lens, which has a sufficiently large aperture and short focal length that the surface of the disc is out of focus when the lens is accurately focused on the plane of the buried mirror layer.

### 16.3.2 Electronic Characteristics

The electronic replay system follows a path closely similar to that used in the encoding of the original recorded signal, although in reverse order, and is shown schematically in Figure 16.9. The major differences between record and replay paths are those such as "oversampling," "digital filtering," and "noise shaping" intended to improve the accuracy of, and reduce the noise level inherent in, the digital-to-analogue transformation.

Referring to Figure 16.9, the RF electrical output of the disc replay photocell, after amplification, is fed to a simple signal detection system, which mutes the signal chain in the absence of a received signal, to ensure intertrack silence. If a signal is present, it is then fed to the EFM decoder stage where the interface and "joining" bits are removed, and the signal is passed as a group of 8-bit symbols to the CIRC error correction circuit, which permits a very high level of signal restoration.

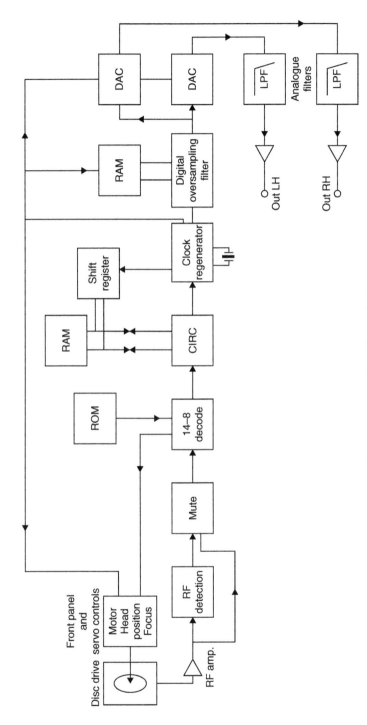

**Figure 16.9: Replay schematic layout.**

An accurate crystal-controlled clock regeneration circuit then causes the signal data blocks to be withdrawn in correct order from a sequential memory "shift register" circuit and reassembled into precisely timed and numerically accurate replicas of the original pairs of 16-bit (left and fight channel) digitally encoded signals. The timing information from this stage is also used to control the speed of the disc drive motor and ensure that signal data are recovered at the correct bit rate.

The remainder of the replay process consists of the stages in which the signal is converted back into analogue form, filtered to remove the unwanted high-frequency components, and reconstructed, as far as possible, as a quantization noise-free copy of the original input waveform. As noted earlier, the filtering and the accuracy of reconstruction of this waveform are helped greatly by the process of "oversampling" in which the original sampling rate is increased, on replay, from 44.1 kHz to some multiple of this frequency, such as 176.4 kHz or even higher. This process can be done by a circuit in which the numerical values assigned to the signal at these additional sampling points are obtained by interpolation between the original input digital levels. As a matter of convenience, the same circuit arrangement will also provide a steep-cut filter having a near-zero transmission at half the sampling frequency.

### 16.3.2.1  The "Eight to Fourteen Modulation" Technique

This is a convenient shorthand term for what should really be described as "8-bit to 14-bit encoding/decoding" and is done for considerations of mechanical convenience in the record/replay process. As noted earlier, the '1's in the digital signal flow are generated by transitions from low to high, or from high to low, in the undulations on the reflecting surface of the disc. On a statistical basis, it would clearly be possible, in an 8-bit encoded signal, for a string of eight or more '1's to occur in the bit sequence, the recording of which would require a rapid sequence of surface humps with narrow gaps between them, making this inconvenient to manufacture. Also, in the nature of things, because these pits or humps will never have absolutely square, clean-cut edges, transitions from one sloping edge to another, where there is such a sequence of closely spaced humps, would also lead to a reduction in the replay signal amplitude and might cause lost data bits.

However, a long sequence of '0's would leave the mirror surface of the disc unmarked by any signal modulation at all, and, bearing in mind the precise track and focus tolerances demanded by the replay system, this absence of signals at the receiver photocell would embarrass the control systems that seek to regulate the lateral and vertical position of the spot focused on the disc and that use errors found in the bit repetition frequency, derived from the recovered sequence of '1's and '0's, to correct inaccuracies in the disc rotation speed. All these problems would be worsened in the presence of mechanical vibration.

The method chosen to solve this problem is to translate the 256-bit sequences possible with an 8-bit encoded signal into an alternative series of 256-bit sequences found in a 14-bit code, which are then reassembled into a sequence of symbols as shown graphically in Figure 16.7. The requirements for the alternative code are that a minimum of two '0's shall separate each '1' and that no more than ten '0's shall occur in sequence. In the 14-bit code, there are 267 values that satisfy this criterion, of which 256 have been chosen and stored in a ROM-based "look-up" table. As a result of the EFM process, there are only nine different pit lengths that are cut into the disc surface during recording, varying from 3 to 11 clock periods in length.

Because the numerical magnitude of the output (EFM) digital sequence is no longer directly related to that of the incoming 8-bit word, the term "symbol" is used to describe this or other similar groups of bits.

Since the EFM encoding process cannot by itself ensure that the junction between consecutive symbols does not violate the requirements noted earlier, an "interface" or "coupling" group of three bits is also added, at this stage, from the EFM ROM store, at the junction between each of these symbols. This coupling group will take the form of a '000', '100', '010', or '001' sequence, depending on the position of the '0's or '1's at the end of the EFM symbol. As shown in Figure 16.6, this process increases the bit rate from 1.882 to 4.123 MB/s, and the further addition of uniquely styled 24-bit synchronizing words to hold the system in coherence, and to mark the beginnings of each bit sequence, increases the final signal rate at the output of the recording chain to 4.322 MB/s. These additional joining and synchronizing bits are stripped from the signal when the bit stream is decoded during the replay process.

### 16.3.2.2 Digital-to-Analogue Conversion

The transformation of the input analogue signal into, and back from, a digitally encoded bit sequence presents a number of problems. These stem from the limited time ($22.7\,\mu s$) available for the conversion of each signal sample into its digitally encoded equivalent and from the very high precision needed in allocating numerical values to each sample. For example, in a 16-bit encoded system the magnitude of the MSB will be 32,768 times as large as the LSB. Therefore, to preserve the significance of a '0' to '1' transition in the

LSB, both the initial and the long-term precision of the electronic components used to define the size of the MSB would need to be better than $\pm 0.00305\%$. (A similar need for accuracy obviously also exists in the ADC used in recording.)

Bearing in mind that even a 0.1% tolerance component is an expensive item, such an accuracy requirement would clearly present enormous manufacturing difficulties. In addition, any errors in the sizes of the steps between the LSB and the MSB would lead to waveform distortion during the encoding/decoding process: a distortion that would worsen as the signal became smaller.

Individual manufacturers have their own preferences in the choice of digital-to-analogue conversion (DAC) designs, but a Philips system is illustrated, schematically, by way of example, in Figure 16.10, is an arrangement called "dynamic element matching." In this circuit, outputs from a group of current sources, in a binary size sequence from 1 to 1/128, are summed by the amplifier A1, whose output is taken to a simple "sample and hold" arrangement to recover the analogue envelope shape from the impulse stream generated by the operation of the A1 input switches ($S_1$–$S_8$). The required precision of the ratios between the input current sources is achieved by the use of switched resistor–capacitor current dividers, each of which is only required to divide its input current into two equal streams.

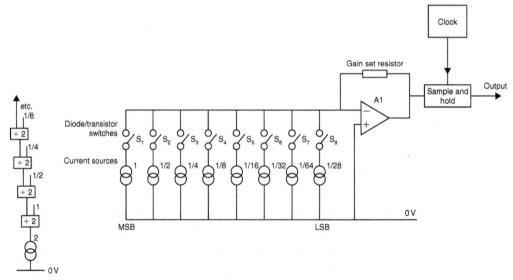

**Figure 16.10: Dynamic matching DAC.**

Since the input "16-bit" encoded signal is divided into two "8-bit" words in the CD replay process, representing the MS and LS sections from $e_1$ to $e_8$ and from $e_9$ to $e_{16}$, these two 8-bit digital words can be separately D/A converted, with the outputs added in an appropriate ratio to give the final 16-bit D/A conversion.

### 16.3.2.3 Digital Filtering and "Oversampling"

It was noted previously that Philips' original choice of sampling frequency (44.1 kHz) and of signal bandwidth (20 Hz to 20 kHz) for the CD imposed the need for steep-cut filtering both prior to the ADC and following the DAC stages. This can lead to problems caused by propagation delays and phase shifts in the filter circuitry, which can degrade the sound quality. Various techniques are available that can lessen these problems, of which the most commonly used come under the headings of "digital filtering" and "oversampling." Because these techniques are interrelated, I have lumped together the descriptions of both of these.

There are two practicable methods of filtering used with digitally encoded signals. For these signals, use can be made of the effect that if a signal is delayed by a time interval, $T_s$, and this delayed signal is then combined with the original input, signal cancellation—partial or complete—will occur at those frequencies where $T_s$ is equal to the duration of an odd number of half cycles of the signal. This gives what is known as a "comb filter" response, shown in Figure 16.11, and this characteristic can be progressively augmented to approach an ideal low-pass filter response (100% transmission up to some chosen frequency, followed by zero transmission above this frequency) by the use of a number of further signal delay and addition paths having other, carefully chosen, gain coefficients and delay times. (Although, in principle, this technique could also be used on a signal in analogue form, there would be problems in providing a nondistorting time delay mechanism for such a signal—a problem that does not arise in the digital domain.)

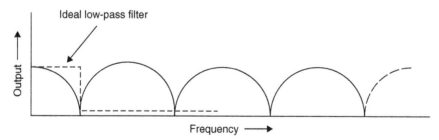

**Figure 16.11: Comb filter frequency response.**

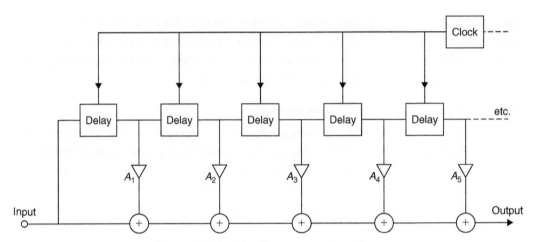

**Figure 16.12: A basic oversampling filter.**

However, this comb filter type arrangement is not very conveniently suited to a system, such as the replay path for a CD, in which all operations are synchronized at a single specific "clock" frequency or its submultiples, and an alternative digital filter layout, shown in Figure 16.12 in simplified schematic form, is normally adopted instead. This provides a very steep-cut low-pass filter characteristic by operations carried out on the signal in its binary-encoded digital form.

In this circuit, the delay blocks are "shift registers," through which the signal passes in a "first in, first out" sequence at a rate determined by the clock frequency. Filtering is achieved in this system by reconstructing the impulse response of the desired low-pass filter circuit, such as that shown in Figure 16.13. The philosophical argument is that if a circuit can be made to have the same impulse response as the desired low-pass filter, it will also have the same gain/frequency characteristics as that filter—a postulate that experiment shows to be true.

This required impulse response is built up by progressive additions to the signal as it passes along the input-to-output path, at each stage of which the successive delayed binary coded contributions are modified by a sequence of mathematical operations. These are carded out, according to appropriate algorithms, stored in "look-up" tables, by the coefficient multipliers $A_1, A_2, A_3, \ldots, A_n$. (The purpose of these mathematical manipulations is, in effect, to ensure that those components of the signal that recur more

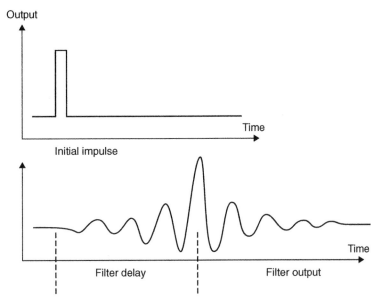

**Figure 16.13: Impulse response of low-pass FIR filter. Zeros are $l/f_s$ apart; cutoff frequency $= f_s/2$.**

frequently than would be permitted by the notional "cut-off" frequency of the filter will all have a coded equivalent to zero magnitude.) Each additional stage has the same attenuation rate as a single-pole RC filter (–6 dB/octave), but with a strictly linear phase characteristic, which leads to zero group delay.

This type of filter is known either as a "transversal filter," from the way in which the signal passes through it, or a "finite impulse response" (FIR) filter because of the deliberate omission from its synthesized impulse response characteristics of later contributions from the coefficient multipliers. (There is no point in adding further terms to the $A_1, \ldots, A_n$ series when the values of these operators tend to zero.)

Some contemporary filters of this kind use 128 sequential "taps" to the transmission chain, giving the equivalent of a –768-dB/octave low-pass filter. This demonstrates, incidentally, the advantage of handling signals in the digital domain in that a 128-stage analogue filter would be very complex and also have an unacceptably high thermal noise background.

If the FIR clock frequency is increased to 176.4 kHz, the action of the shift registers will be to generate three further signal samples and to interpolate these additional samples between those given by the original 44.1-kHz sampling intervals—a process termed "four times oversampling."

The simple sample-and-hold stage, at the output of the DAC shown in Figure 16.10, will also assist filtering, as it will attenuate any signals occurring at the clock frequency to an extent determined by the duration of the sampling operation—called the sampling "window." If the window length is near 100% of the cycle time, attenuation of the S/H circuit will be nearly total at $f_s$.

Oversampling, on its own, would have the advantage of pushing the aliasing frequency up to a higher value, which makes the design of the antialiasing and waveform reconstruction filter a much easier task to accomplish using simple analogue-mode low-pass filters whose characteristics can be tailored so that they introduce very little unwanted group delay and phase shift. A typical example of this approach is the linear phase analogue filter design, shown in Figure 16.14, used following the final 16-bit DACs in the replay chain.

However, the FIR filter shown in Figure 16.12 has the additional effect of computing intermediate numerical values for the samples interpolated between the original 44.1-kHz input data, which makes the discontinuities in the PCM step waveform

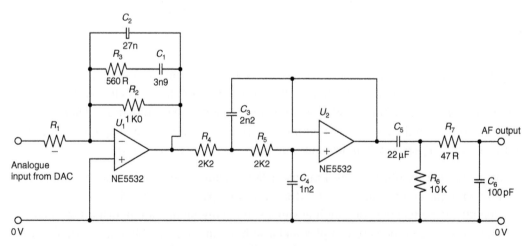

**Figure 16.14: A linear phase LP filter.**

smaller, as shown in Figure 16.15. This reduces the quantization noise and also increases the effective resolution of the DAC. As a general rule, an increase in the replay sampling rate gives an improvement in resolution equivalent to that given by a similar increase in encoding level, such that a four times oversampled 14-bit decoder would have the same resolution as a straight 16-bit decoder.

Yet another advantage of oversampling is that it increases the bandwidth over which the "quantization noise" will be spread—from 22.05 to 88.2 kHz in the case of a four times oversampling system. This reduces the proportion of the total noise that is now present within the audible (20 Hz to 20 kHz) part of the frequency spectrum—especially if "noise shaping" is also employed. This aspect is examined later in this chapter.

### 16.3.2.4 "Dither"

If a high-frequency noise signal is added to the waveform at the input to the ADC and if the peak-to-peak amplitude of this noise signal is equal to the quantization step 'Q', both the resolution and the dynamic range of the converter will be increased. The reason for this can be seen if we consider what would happen if the actual analogue signal level were to lie somewhere between two quantization levels. Suppose, for example, in the case of an ADC, that the input signal had a level of 12.4 and that the nearest quantization levels were 12 and 13. If dither had been added, and a sufficient number of samples were taken, one after another, there would be a statistical probability that 60% of these

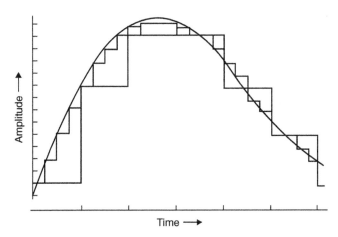

**Figure 16.15: Effect of four times oversampling and interpolation of intermediate values.**

would be attributed to level 12 and that 40% would be attributed to level 13 so that, on averaging, the final analogue output from the ADC/DAC process would have the correct value of 12.4.

A further benefit is obtained by the addition of dither at the output of the replay DACs (most simply contrived by allowing the requisite amount of noise in the following analogue low-pass filters) in that it will tend to mask the quantization "granularity" of the recovered signal at low bit levels. This defect is particularly noticeable when the signal frequency happens to have a harmonic relationship with the sampling frequency.

### 16.3.2.5 The "Bitstream" Process and "Noise Shaping"

A problem in any analogue-to-digital or digital-to-analogue converter is that of obtaining an adequate degree of precision in the magnitudes of the digitally encoded steps. It has been seen that the accuracy required, in the most significant bit in a 16-bit converter, was better than 0.00305% if '0'–'1' transitions in the LSB were to be significant. Similar, although lower, orders of accuracy are required from all the intermediate step values. Achieving this order of accuracy in a mass-produced consumer article is difficult and expensive. In fact, differences in tonal quality between CD players are likely to be due, in part, to inadequate precision in the DACs.

As a means of avoiding the need for high precision in the DAC converters, Philips took advantage of the fact that an effective improvement in resolution could be achieved merely by increasing the sampling rate, which could then be traded-off against the number of bits in the quantization level. Furthermore, whatever binary encoding system is adopted, the first bit in the received 16-bit word must always be either a '0' or a '1', and in the "two's complement" code used in the CD system, the transition in the MSB from '0' to '1' and back will occur at the midpoint of the input analogue signal waveform.

This means that if the remaining 15 bits of a 16-bit input word are stripped off and discarded, this action will have the effect that the input digital signal will have been converted—admittedly somewhat crudely—into a voltage waveform of analogue form. Now, if this '0/1' signal is 256 times oversampled, in the presence of dither, an effective 9-bit resolution will be obtained from two clearly defined and easily stabilized quantization levels: a process for which Philips coined the term "bit stream" decoding.

Unfortunately, such a low-resolution quantization process will incur severe quantization errors that manifest as a high background noise level. Philips' solution to this is to

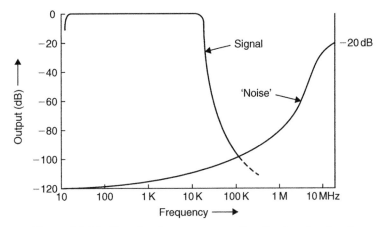

**Figure 16.16: Signal noise spectrum after "noise shaping."**

employ "noise shaping," a procedure in which, as shown in Figure 16.16, the noise components are largely shifted out of the 20-Hz to 20-kHz audible region into the inaudible upper reaches of the new 11.29-MHz bandwidth.

The proposition is, in effect, that a decoded digital signal consists of the pure signal, plus a noise component (caused by the quantization error) related to the lack of resolution of the decoding process. It is further argued that if this noise component is removed by filtering, what remains will be the pure signal—no matter how poor the actual resolution of the decoder. Although this seems an unlikely hypothesis, users of CD players employing the "bit stream" system seem to agree that the technique does indeed work in practice. It would therefore seem that the greater freedom from distortion, which could be caused by errors in the quantization levels in high bit-level DACs, compensates for the crudity of a decoding system based on so few quantization steps.

Mornington-West[1] quotes oversampling values of 758 and 1024 times, respectively, for "Technics" and "Sony" "low-bit" CD players, which would be equivalent in resolution to 10.5- and 11-bit quantization if a simple '0' or '1' choice of encoding levels was used. Since the presence of dither adds an effective 1 bit to the resolution and dynamic range, the final figures would become 10-, 11.5-, and 12-bit resolution, respectively, for the Philips, Technics, and Sony CD players.

However, such decoders need not use the single-bit resolution adopted by Philips, and if a 2- or 4-bit quantization was chosen as the base to which the oversampling process was applied—an option that would not incur significant problems with accuracy of quantization—this would provide low-bit resolution values as good as the 16-bit equivalents at a lower manufacturing cost and with greater reproducibility. Ultimately, the limit to the resolution possible with a multiple sampling decoder is set by the time "jitter" in the switching cycles and the practicable operating speeds of the digital logic elements used in the shift registers and adders. In the case of the 1024 times oversampling "Sony" system, a 44.1584-MHz clock speed is required, which is near the currently available limit.

## 16.4  Error Correction

The possibility of detecting and correcting replay errors offered by digital audio techniques is possibly the largest single benefit offered by this process because it allows the click-free, noise-free background level in which the CD differs so obviously from its vinyl predecessors. Indeed, were error correction not possible, the requirements for precision of the CD manufacturing and replay process would not be practicable.

Four possible options exist for the avoidance of audible signal errors once these have been detected. These are the replacement of the faulty word or group of words by correct ones, the substitution of the last correct word for the one found to be faulty, on the grounds that an audio signal is likely to change relatively slowly in amplitude in comparison with the 44.1-kHz sample rate, linear interpolation of intermediate sample values in the gaps caused by the deletion of incorrectly received words, and, if worst comes to worst, the muting of the signal for the duration of the error.

Of these options, the replacement of the faulty word, or group of words, by a correct equivalent is clearly the first preference, although it will, in practice, be supplemented by the other error-concealment techniques. The error correction system used in the CD replay process combines a number of error correction features and is called the cross-interleave Reed–Solomon code system. It is capable of correcting an error of 3500 bits and of concealing errors of up to 12,000 bits by linear interpolation. I will look at the CIRC system later, but, meanwhile, it will be helpful to consider some of the options that are available.

### 16.4.1  Error Detection

Errors likely to occur in a digitally encoded replay process are described as "random" when they affect single bits and "burst" when they affect whole words or groups of words. Correcting random errors is easier so the procedure used in the Reed–Solomon code endeavors to break down burst errors into groups of scattered random errors. However, it greatly facilitates remedial action if the presence and location of the error can be detected and "flagged" by some added symbol.

Although the existence of an erroneous bit in an input word can sometimes be detected merely by noting a wrong word length, the basic method of detecting an error in received words is by the use of "parity bits." In its simplest form, this would be done by adding an additional bit to the word sequence, as shown in Figure 16.17(a), so that the total (using the logic rules shown in Figure 16.17) always added up to zero (a method known as "even parity"). If this addition had been made to all incoming words, the presence of a word plus parity bit that did not add up to '0' could be detected instantly by a simple computer algorithm and it could then be rejected or modified.

### 16.4.2 Faulty Bit/Word Replacement

Although the procedure shown earlier would alert the decoder to the fact that the word was in error, the method could not distinguish between an incorrect word and an incorrect parity bit—or even detect a word containing two separate errors, although this might be a rare event. However, the addition of extra parity bits can indeed correct such errors as well as detect them, and a way by which this could be done is shown in Figure 16.17(b). If a group of four 4-bit input words, as shown in lines a–d, each has a parity bit attached to it, as shown in column q, so that each line has an even parity, and if each column has a parity bit attached to it, for the same purpose, as shown in line e, then an error, as shown in grid reference (b.n) in Figure 16.17(c), could not only be detected and localized as occurring at the intersection of row b and column n, but it could also be corrected, since if the received value '0' is wrong, the correct alternative must be '1'.

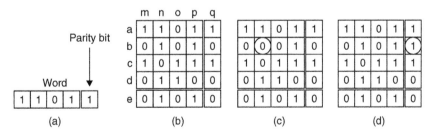

**Figure 16.17: Parity bit error correction. Logic:** $0 + 0 = 0$, $0 + 1 = 1$, $1 + 1 = 0$.

Moreover, the fact that the parity bits of column q and row e both have even parity means that, in this example, the parity bits themselves are correct. If the error had occurred instead in one of the parity bits, as in Figure 16.17(d), this would have shown up by the fact that the loss of parity occurred only in a single row—not in both a row and a column.

So far, the addition of redundant parity bit information has offered the possibility of detecting and correcting single bit "random" errors, but this would not be of assistance in correcting longer duration "burst" errors, comprising one or more words. This can be done by "interleaving," the name given to the deliberate and methodical scrambling of words, or the bits within words, by selectively delaying them and then reinserting them into the bit sequence at later points, as shown in Figure 16.18. This has the effect of converting a burst error, after deinterleaving, into a scattered group of random errors, a type of fault that is much easier to correct.

A further step toward the correction of larger duration errors can be made by the use of a technique known as "cross-interleaving." This is done by reassembling scrambled data into 8-bit groups without descrambling. (It is customary to refer to these groups of bits as "symbols" rather than words because they are unrelated to the signal.) Following this, these symbols are themselves mixed up in their order by removal and reinsertion at different delay intervals. In order to do this it is necessary to have large bit-capacity shift registers, as well as a fast microprocessor, which can manipulate the information needed to direct the final descrambling sequences and generate and insert the restored and corrected signal words.

To summarize, errors in signals in digital form can be corrected by a variety of procedures. In particular, errors in individual bits can be corrected by the appropriate addition of parity bits, and burst errors affecting words, or groups of words, can be corrected by interleaving and deinterleaving the signal before and after transmission—a process that separates and redistributes the errors as random bit faults, correctable by parity techniques.

A variety of strategies has been devised for this process, aimed at achieving the greatest degree of error removal for the lowest necessary number of added parity bits. The CIRC error correction process used for CDs is very efficient in this respect, as it only demands an increase in transmitted data of 33.3% and yet can correct burst errors up to 3500

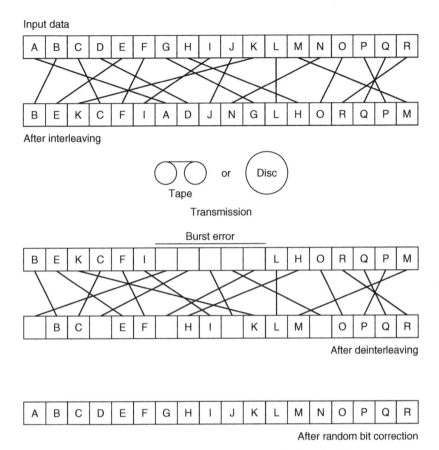

**Figure 16.18: Burst error correction by interleaving.**

bits in length. It can conceal, by interpolation, transmission errors up to 12,000 bits in duration—an ability that has contributed enormously to the sound quality of the CD player by comparison with the vinyl disc.

From the point of view of the CD manufacturer, it is convenient that the complete CIRC replay error correction and concealment package is available from several IC suppliers as part of a single large-scale integrated chip. From the point of view of the serious CD user, it is preferable that the error correction system has to do no more work than it must, since although the errors will mainly be restored quite precisely, it may be necessary, sometimes, for the system to substitute approximate, interpolated values for the signal data, and the effect of frequent corrections may be audible to the critical listener. So treat CDs with care, keep them clean, and try to avoid surface scratches.

## Reference

1. Mornington-West, A., *Newnes audio and hi-fi handbook*, 2nd Ed., 141.

# CHAPTER 17

# Digital Audio Recording Basics

Ian Sinclair

Once conversion from analogue signals into the digital domain has taken place, audio becomes data and a digital audio recorder is no more than a data recorder adapted to record samples from convertors. Provided that the original samples are reproduced with their numerical value unchanged and with their original time base, a digital recorder causes no loss of information at all. The only loss of information is due to the conversion processes unless there is a design fault or the equipment needs maintenance. In this chapter John Watkinson explains the various techniques needed to record audio data.

## 17.1 Types of Media

There is considerably more freedom of choice of digital media than was the case for analogue signals, and digital media take advantage of the research expended in computer recording.

Digital media do not need to be linear, nor do they need to be noise-free or continuous. All they need to do is allow the player to be able to distinguish some replay event, such as the generation of a pulse, from the lack of such an event with reasonable rather than perfect reliability. In a magnetic medium, the event will be a flux change from one direction of magnetization to another. In an optical medium, the event must cause the pickup to perceive a change in the intensity of the light falling on the sensor. In CD, the contrast is obtained by interference. In some discs it will be through selective absorption of light by dyes. In magneto-optical discs the recording itself is magnetic, but it is made and read using light.

### 17.1.1 Magnetic Recording

Magnetic recording relies on the hysteresis of certain magnetic materials. After an applied magnetic field is removed, the material remains magnetized in the same direction.

By definition the process is nonlinear, and analogue magnetic recorders have to use bias to linearize it. Digital recorders are not concerned with the nonlinearity, and HF bias is unnecessary.

Figure 17.1 shows the construction of a typical digital record head, which is just like an analogue record head. A magnetic circuit carries a coil through which the record current passes and generates flux. A nonmagnetic gap forces the flux to leave the magnetic circuit of the head and penetrate the medium. The current through the head must be set to suit the coercivity of the tape and is arranged to almost saturate the track. The amplitude of the current is constant, and recording is performed by reversing the direction of the current with respect to time. As the track passes the head, this is converted to the reversal of the magnetic field left on the tape with respect to distance. The recording is actually made just after the trailing pole of the record head where the flux strength from the gap is falling. The width of the gap is generally made quite large to ensure that the full thickness

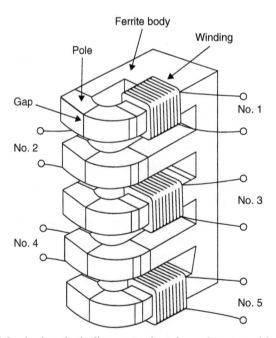

**Figure 17.1: Typical ferrite head windings are placed on alternate sides to save space, but parallel magnetic circuits have high cross talk.**

of the magnetic coating is recorded, although this cannot be done if the same head is intended to replay.

Figure 17.2 shows what happens when a conventional inductive head, that is, one having a normal winding, is used to replay the track made by reversing the record current. The head output is proportional to the rate of change of flux and so only occurs at flux reversals. The polarity of the resultant pulses alternates as the flux changes and changes back. A circuit is necessary which locates the peaks of the pulses and outputs a signal corresponding to the original record current waveform.

The head shown in Figure 17.2 has the frequency response shown in Figure 17.3. At DC there is no change of flux and no output. As a result, inductive heads are at a disadvantage at very low speeds. The output rises with frequency until the rise is halted by the onset of thickness loss. As the frequency rises, the recorded wavelength falls and flux from the shorter magnetic patterns cannot be picked up so far away. At some point, the wavelength becomes so short that flux from the back of the tape coating cannot reach the head and a decreasing thickness of tape contributes to the replay signal. In digital recorders using short wavelengths to obtain high density, there is no point in using thick coatings. As

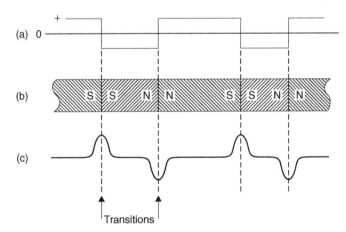

**Figure 17.2: Basic digital recording. At (a) the write current in the head is reversed from time to time, leaving a binary magnetization pattern shown at (b). When replayed, the waveform at (c) results because an output is only produced when flux in the head changes. Changes are referred to as transitions.**

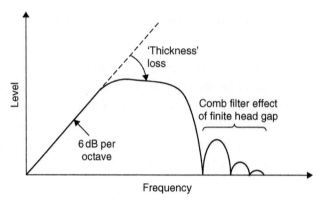

**Figure 17.3: The major mechanisms defining magnetic channel bandwidth.**

wavelength further reduces, the familiar gap loss occurs, where the head gap is too big to resolve detail on the track.

As can be seen, the frequency response is far from ideal, and steps must be taken to ensure that recorded data waveforms do not contain frequencies which suffer excessive losses.

A more recent development is the magneto-resistive (MR) head. This is a head that measures the flux on the tape rather than using it to generate a signal directly. Flux measurement works down to DC and so offers advantages at low tape speeds. Unfortunately, flux measuring heads are not polarity conscious and if used directly they sense positive and negative flux equally, as shown in Figure 17.4. This is overcome by using a small extra winding carrying a constant current. This creates a steady bias field, which adds to the flux from the tape. The flux seen by the head now changes between two levels and a better output waveform results.

Recorders that have low head-to-medium speed, such as digital compact cassette (DCC) use MR heads, whereas recorders with high speeds, such as digital audio stationary head (DASH), rotary head digital audio tape (RDAT), and magnetic disc drives, use inductive heads.

Heads designed for use with tape work in actual contact with the magnetic coating. The tape is tensioned to pull it against the head. There will be a wear mechanism and need for periodic cleaning.

In the hard disc, the rotational speed is high in order to reduce access time, and the drive must be capable of staying on line for extended periods. In this case the heads do not

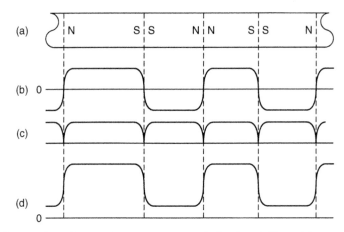

**Figure 17.4: The sensing element in a magneto-resistive head. Transitions are not sensitive to the polarity of the flux, only the magnitude. At (a) the track magnetization is shown, which causes a bidirectional flux variation in the head as at (b) resulting in the magnitude output at (c). However, if the flux in the head due to the track is biased by an additional field, it can be made unipolar as at (d) and the correct output waveform is obtained.**

contact the disc surface, but are supported on a boundary layer of air. The presence of the air film causes spacing loss, which restricts the wavelengths at which the head can replay. This is the penalty of rapid access.

Digital audio recorders must operate at high density in order to offer a reasonable playing time. This implies that the shortest possible wavelengths will be used. Figure 17.5 shows that when two flux changes, or transitions, are recorded close together, they affect each other on replay. The amplitude of the composite signal is reduced, and the position of the peaks is pushed outward. This is known as intersymbol interference, or peak-shift distortion, and occurs in all magnetic media.

The effect is primarily due to high frequency loss and it can be reduced by equalization on replay, as is done in most tapes, or by precompensation on record, as is done in hard discs.

### 17.1.2 Optical Discs

Optical recorders have the advantage that light can be focused at a distance whereas magnetism cannot. This means that there need be no physical contact between the pickup and the medium and no wear mechanism.

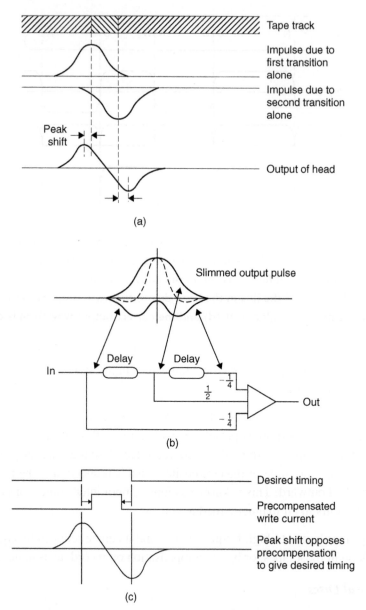

Tape track

Impulse due to
first transition
alone

Impulse due to
second transition
alone

Peak
shift

Output of head

(a)

Slimmed output pulse

Delay    Delay

In

$-\frac{1}{4}$

$\frac{1}{2}$

Out

$-\frac{1}{4}$

(b)

Desired timing

Precompensated
write current

Peak shift opposes
precompensation
to give desired timing

(c)

**Figure 17.5: (a) Peak shift distortion can be reduced by (b) equalization in replay or (c) precompensation.**

In the same way that the recorded wavelength of a magnetic recording is limited by the gap in the replay head, the density of optical recording is limited by the size of light spot which can be focused on the medium. This is controlled by the wavelength of the light used and by the aperture of the lens. When the light spot is as small as these limits allow, it is said to be diffraction limited. The recorded details on the disc are minute, and could easily be obscured by dust particles. In practice the information layer needs to be protected by a thick transparent coating. Light enters the coating well out of focus over a large area so that it can pass around dust particles, and comes to a focus within the thickness of the coating. Although the number of bits per unit area is high in optical recorders, the number of bits per unit volume is not as high as that of tape because of the thickness of the coating.

Figure 17.6 shows the principle of readout of the compact disc which is a read-only disc manufactured by pressing. The track consists of raised bumps separated by flat areas. The entire surface of the disc is metalized, and the bumps are one quarter of a wavelength in height. The player spot is arranged so that half of its light falls on top of a bump, and half on the surrounding surface. Light returning from the flat surface has traveled half a wavelength further than light returning from the top of the bump, and so there is

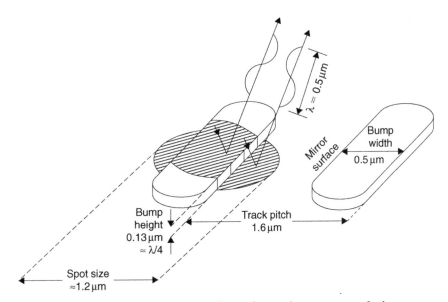

**Figure 17.6: CD readout principle and dimensions. The presence of a bump causes destructive interference in the reflected light.**

a phase reversal between the two components of the reflection. This causes destructive interference, and light cannot return to the pickup. It must reflect at angles which are outside the aperture of the lens and be lost. Conversely, when light falls on the flat surface between bumps, the majority of it is reflected back to the pickup. The pickup thus sees a disc apparently having alternately good or poor reflectivity.

Some discs can be recorded once, but not subsequently erased or rerecorded. These are known as WORM (write once read mostly) discs. One type of WORM disc uses a thin metal layer that has holes punched in it on recording by heat from a laser. Others rely on the heat raising blisters in a thin metallic layer by decomposing the plastic material beneath. Yet another alternative is a layer of photo-chemical dye that darkens when struck by the high powered recording beam. Whatever the recording principle, light from the pickup is reflected more or less, or absorbed more or less, so that the pickup once more senses a change in reflectivity. Certain WORM discs can be read by conventional CD players and are thus called recordable CDs, whereas others will only work in a particular type of drive.

### 17.1.3 Magneto-Optical Discs

When a magnetic material is heated above its Curie temperature, it becomes demagnetized, and on cooling will assume the magnetization of an applied field which would be too weak to influence it normally. This is the principle of magneto-optical recording used in the Sony MiniDisc. The heat is supplied by a finely focused laser; the field is supplied by a coil that is much larger.

Figure 17.7 assumes that the medium is initially magnetized in one direction only. In order to record, the coil is energized with the waveform to be recorded. This is too weak to influence the medium in its normal state, but when it is heated by the recording laser beam the heated area will take on the magnetism from the coil when it cools. Thus a magnetic recording with very small dimensions can be made.

Readout is obtained using the Kerr effect, which is the rotation of the plane of polarization of light by a magnetic field. The angle of rotation is very small and needs a sensitive pickup. The recording can be overwritten by reversing the current in the coil and running the laser continuously as it passes along the track.

A disadvantage of magneto-optical recording is that all materials having a Curie point low enough to be useful are highly corrodible by air and need to be kept under an effectively sealed protective layer.

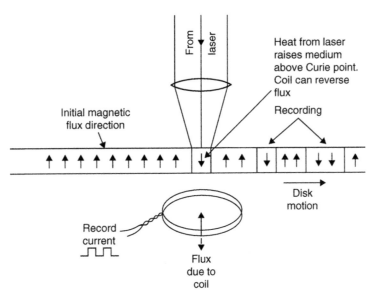

**Figure 17.7: The thermomagneto-optical disk uses the heat from a laser to allow a magnetic field to record on the disk.**

All optical discs need mechanisms to keep the pickup following the track and sharply focused on it.

The frequency response of an optical disc is shown in Figure 17.8. The response is best at DC and falls steadily to the optical cut-off frequency. Although the optics work down to DC, this cannot be used for the data recording. DC and low frequencies in data would interfere with the focus and tracking servos. In practice the signal from the pickup is split by a filter. Low frequencies go to the servos, and higher frequencies go to the data circuitry. As a result, the data channel has the same inability to handle DC as does a magnetic recorder, and the same techniques are needed to overcome it.

## 17.2  Recording Media Compared

Of the various media discussed so far, it might be thought that one would be the best and would displace all the others. This has not happened because there is no one best medium; it depends on the application.

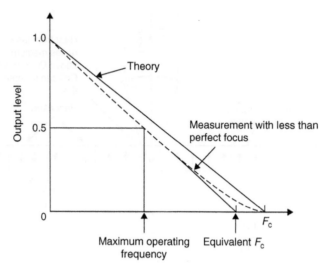

**Figure 17.8: Frequency response of laser pickup. Maximum operating frequency is about half of cutoff frequency $F_c$.**

Random access memory (RAM) offers extremely short access time, but the volume of data generated by digital audio precludes the use of RAM for anything more than a few seconds because it would be too expensive. In addition loss of power causes the recording to be lost.

Tape has the advantage that it is thin and can be held compactly on reels. However, this slows down the access time because the tape has to be wound, or shuttled, to the appropriate place. Tape is, however, inexpensive, long lasting, and is appropriate for archiving large quantities of data.

However, discs allow rapid access because their entire surface is permanently exposed and the positioner can move the heads to any location in a matter of milliseconds. The capacity is limited compared to tape because in the case of magnetic discs there is an air gap between the medium and the head. Exchangeable discs have to have a certain minimum head flying height below which the risk of contamination and a consequent head crash is too great. In Winchester technology the heads and disc are sealed inside a single assembly and contaminants can be excluded. In this case the flying height can be reduced and the packing density increased as a consequence. However, the disc is no longer exchangeable. In the case of optical discs the medium itself is extremely thick and multiple platter drives are impracticable because of the size of the optical pickup.

If the criterion is access time, discs are to be preferred. If the criterion is compact storage, tape is to be preferred. In computers, both technologies have been used in a complementary fashion for many years. In digital audio the same approach could be used, but to date the steps appear faltering.

In tape recording, the choice is between rotary and stationary heads. In a stationary head machine, the narrow tracks required by digital recordings result in heads with many parallel magnetic circuits, each of which requires its own read and write circuitry. Gaps known as guard bands must be placed between the tracks to reduce cross talk. Guard bands represent wasted tape.

In rotary head machines, the tracks are laid down by a small number of rapidly rotating heads and less read/write circuitry is required. The space between the tracks is controlled by the linear tape speed and not by head geometry and so any spacing can be used. If azimuth recording is used, no guard bands are necessary. A further advantage of rotary head recorders is that the high head to tape speed raises the frequency of the off-tape signals, and with a conventional inductive head, this results in a larger playback signal compared to the thermal noise from the head and the preamplifiers.

As a result the rotary head tape recorder offers the highest storage density yet achieved, despite the fact that available formats are not yet in sight of any fundamental performance limits.

## 17.3 Some Digital Audio Processes Outlined

While digital audio is a large subject, it is not necessarily a difficult one. Every process can be broken down into smaller steps, each of which is relatively easy to assimilate. The main difficulty with study is not following the simple step, but to appreciate where it fits in the overall picture. The next few sections illustrate various important processes in digital audio and show why they are necessary. Such processes are combined in various ways in real equipment.

### 17.3.1 The Sampler

Figure 17.9 consists of an ADC, which is joined to a DAC by way of a quantity of RAM. What the device does is determined by the way in which the RAM address is controlled. If the RAM address increases by one every time a sample from the ADC is stored in the

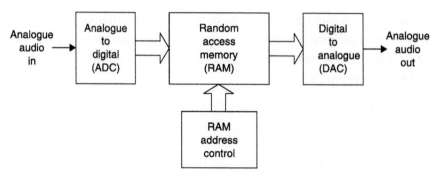

Figure 17.9: In the digital sampler, the recording medium is a RAM. Recording time available is short compared to other media, but access to the recording is immediate and flexible as it is controlled by addressing the RAM.

RAM, a recording can be made for a short period until the RAM is full. The recording can be played back by repeating the address sequence at the same clock rate but reading data from the memory into the DAC. The result is generally called a sampler. By running the replay clock at various rates, the pitch and duration of the reproduced sound can be altered. At a rate of one million bits per second, a megabyte of memory gives only 8 s worth of recording, so clearly samplers will be restricted to a fairly short playing time.

Using data reduction, the playing time of a RAM based recorder can be extended. Some telephone answering machines take messages in RAM and eliminate the cassette tape. For predetermined messages, read only memory can be used instead as it is nonvolatile. Announcements in aircraft, trains, and elevators are one application of such devices.

### 17.3.2  The Programmable Delay

If the RAM of Figure 17.9 is used in a different way, it can be written and read at the same time. The device then becomes an audio delay. Controlling the relationship between the addresses then changes the delay. The addresses are generated by counters that overflow to zero after they have reached a maximum count. As a result the memory space appears to be circular as shown in Figure 17.10. The read and write addresses are driven by a common clock and chase one another around the circle. If the read address follows close behind the write address, the delay is short. If it just stays ahead of the write address, the maximum delay is reached. Programmable delays are useful in TV studios where they allow audio to be aligned with video which has been delayed in

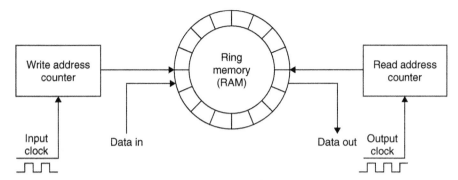

**Figure 17.10: Time base corrector (TBC) memory is addressed by a counter that overflows periodically to give a ring structure. Memory allows read side to be nonsynchronous with write side.**

various processes. They can also be used in auditoria to align the sound from various loudspeakers.

In digital audio recorders, a device with a circular memory can be used to remove irregularities from the replay data rate. The off-tape data rate can fluctuate within limits but the output data rate can be held constant. A memory used in this way is called a time base corrector (TBC). All digital recorders have TBCs to eliminate wow and flutter.

### 17.3.3  Time Compression

When samples are converted, the ADC must run at a constant clock rate and it outputs an unbroken stream of samples. Time compression allows the sample stream to be broken into blocks for convenient handling.

Figure 17.11 shows an ADC feeding a pair of RAMS. When one is being written by the ADC, the other can be read, and vice versa. As soon as the first RAM is full, the ADC output switched to the input of the other RAM so that there is no loss of samples. The first RAM can then be read at a higher clock rate than the sampling rate. As a result the RAM is read in less time than it took to write it, and the output from the system then pauses until the second RAM is full. The samples are now time compressed. Instead of being an unbroken stream, which is difficult to handle, the samples are now arranged in blocks with convenient pauses in between them. 1n these pauses numerous processes can take place. A rotary head recorder might switch heads; a hard disc might move to another track. On a tape recording, the time compression of the audio samples allows time for synchronizing patterns, subcode, and error-correction words to be recorded.

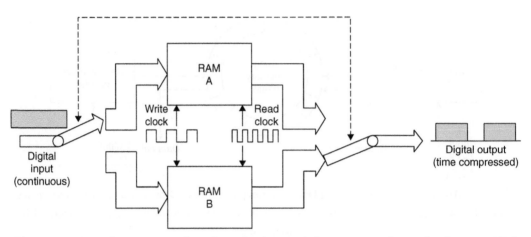

**Figure 17.11: In time compression, the unbroken real-time stream of samples from an ADC is broken up into discrete blocks. This is accomplished by the configuration shown here. Samples are written into one RAM at the sampling rate by the write clock. When the first RAM is full, the switches change over, and writing continues into the second RAM while the first is read using a higher frequency clock. The RAM is read faster than it was written and so all data will be output before the other RAM is full. This opens spaces in the data flow, which are used as described in the text.**

In digital audio recorders that use video cassette recorders (VCRs), time compression allows the continuous audio samples to be placed in blocks in the unblanked parts of the video waveform, separated by synchronizing pulses.

Subsequently, any time compression can be reversed by time expansion. Samples are written into a RAM at the incoming clock rate, but read out at the standard sampling rate. Unless there is a design fault, time compression is totally inaudible. In a recorder, the time-expansion stage can be combined with the time base-correction stage so that speed variations in the medium can be eliminated at the same time. The use of time compression is universal in digital audio recording. In general the instantaneous data rate at the medium is not the same as the rate at the convertors, although clearly the average rate must be the same.

Another application of time compression is to allow more than one channel of audio to be carried on a single cable. If, for example, audio samples are time compressed by a factor of two, it is possible to carry samples from a stereo source in one cable.

In digital video recorders, both audio and video data are time compressed so that they can share the same heads and tape tracks.

## 17.3.4 Synchronization

In addition to the analogue inputs and outputs, connected to convertors, many digital recorders have digital inputs that allow the convertors to be bypassed. This mode of connection is desirable because there is no loss of quality in a digital transfer. Transfer of samples between digital audio devices is only possible if both use a common sampling rate and they are synchronized. A digital audio recorder must be able to synchronize to the sampling rate of a digital input in order to record the samples. It is frequently necessary for such a recorder to be able to play back locked to an external sampling rate reference so that it can be connected to, for example, a digital mixer. The process is already common in video systems but now extends to digital audio.

Figure 17.12 shows how the external reference locking process works. The time base expansion is controlled by the external reference, which becomes the read clock for the RAM and so determines the rate at which the RAM address changes. In the case of a digital tape deck, the write clock for the RAM would be proportional to the tape speed. If the tape is going too fast, the write address will catch up with the read address in the memory, whereas if the tape is going too slow the read address will catch up with the write address. The tape speed is controlled by subtracting the read address from the write address. The address difference is used to control the tape speed. Thus if the tape speed is too high, the memory will fill faster than it is being emptied, and the address difference will grow larger than normal. This slows down the tape.

Thus in a digital recorder the speed of the medium is constantly changing to keep the data rate correct. Clearly this is inaudible as properly engineered time base correction totally isolates any instabilities on the medium from data fed to the convertor.

In multitrack recorders, the various tracks can be synchronized to sample accuracy so that no timing errors can exist between the tracks. In stereo recorders image shift due to phase errors is eliminated.

In order to replay without a reference, perhaps to provide an analogue output, a digital recorder generates a sampling clock locally by means of a crystal oscillator. Provision

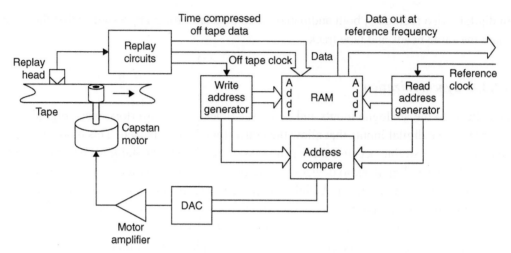

**Figure 17.12: In a recorder using time compression, the samples can be returned to a continuous stream using RAM as a TBC. The long-term data rate has to be the same on the input and output of the TBC or it will lose data. This is accomplished by comparing the read and write addresses and using the difference to control the tape speed. In this way the tape speed will automatically adjust to provide data as fast as the reference clock takes it from the TBC.**

will be made on professional machines to switch between internal and external references.

### 17.3.5 Error Correction and Concealment

As anyone familiar with analogue recording will know, magnetic tape is an imperfect medium. It suffers from noise and dropouts, which in analogue recording are audible. In a digital recording of binary data, a bit is either correct or wrong, with no intermediate stage. Small amounts of noise are rejected, but inevitably, infrequent noise impulses cause some individual bits to be in error. Dropouts cause a larger number of bits in one place to be in error. An error of this kind is called a burst error. Whatever the medium and whatever the nature of the mechanism responsible, data are either recovered correctly, or suffer some combination of bit errors and burst errors. In compact disc, random errors can be caused by imperfections in the moulding process, whereas burst errors are due to contamination or scratching of the disc surface.

The audibility of a bit error depends on which bit of the sample is involved. If the LSB of one sample was in error in a loud passage of music, the effect would be totally masked and no one could detect it. Conversely, if the MSB of one sample was in error in a quiet passage, no one could fail to notice the resulting loud transient. Clearly a means is needed to render errors from the medium inaudible. This is the purpose of error correction.

In binary, a bit has only two states. If it is wrong, it is only necessary to reverse the state and it must be right. Thus the correction process is trivial and perfect. The main difficulty is in identifying the bits that are in error. This is done by coding the data by adding redundant bits. Adding redundancy is not confined to digital technology, airliners have several engines and cars have twin braking systems. Clearly the more failures that have to be handled, the more redundancy is needed. If a four-engined airliner is designed to fly normally with one engine failed, three of the engines have enough power to reach cruise speed, and the fourth one is redundant. The amount of redundancy is equal to the amount of failure that can be handled. In the case of the failure of two engines, the plane can still fly, but it must slow down; this is graceful degradation. Clearly the chances of a two-engine failure on the same flight are remote.

In digital audio, the amount of error that can be corrected is proportional to the amount of redundancy and within this limit the samples are returned to exactly their original value. Consequently *corrected* samples are inaudible. If the amount of error exceeds the amount of redundancy, correction is not possible, and, in order to allow graceful degradation, concealment will be used. Concealment is a process where the value of a missing sample is estimated from those nearby. The estimated sample value is not necessarily exactly the same as the original, and so under some circumstances concealment can be audible, especially if it is frequent. However, in a well-designed system, concealments occur with negligible frequency unless there is an actual fault or problem.

Concealment is made possible by rearranging or shuffling the sample sequence prior to recording. This is shown in Figure 17.13 where odd-numbered samples are separated from even-numbered samples prior to recording. The odd and even sets of samples may be recorded in different places, so that an uncorrectable burst error only affects one set. On replay, the samples are recombined into their natural sequence, and the error is now split up so that it results in every other sample being lost. The waveform is now described half as often, but can still be reproduced with some loss of accuracy. This is better than not being reproduced at all even if it is not perfect. Almost all digital recorders use such

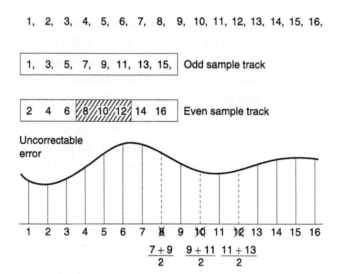

**Figure 17.13:** In cases where the error correction is inadequate, concealment can be used provided that the samples have been ordered appropriately in the recording. Odd and even samples are recorded in different places as shown here. As a result an uncorrectable error causes incorrect samples to occur singly, between correct samples. In the example shown, sample 8 is incorrect, but samples 7 and 9 are unaffected and an approximation to the value of sample 8 can be had by taking the average value of the two. This interpolated value is substituted for the incorrect value.

an odd-even shuffle for concealment. Clearly if any errors are fully correctable, the shuffle is a waste of time; it is only needed if correction is not possible.

In high-density recorders, more data are lost in a given sized dropout. Adding redundancy equal to the size of a dropout to every code is inefficient. Figure 17.14(a) shows that the efficiency of the system can be raised using interleaving. Sequential samples from the ADC are assembled into codes, but these are not recorded in their natural sequence. A number of sequential codes are assembled along rows in a memory. When the memory is full, it is copied to the medium by reading down columns. On replay, the samples need to be deinterleaved to return them to their natural sequence. This is done by writing samples from tape into a memory in columns, and when it is full, the memory is read in rows. Samples read from the memory are now in their original sequence so there is no effect on the recording. However, if a burst error occurs on the medium, it will damage sequential

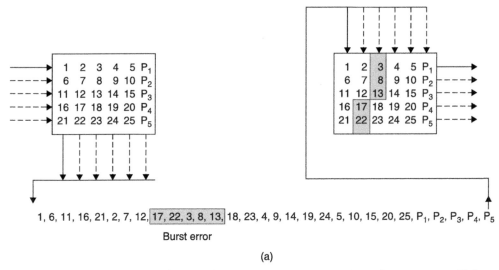

1, 6, 11, 16, 21, 2, 7, 12, $\boxed{17, 22, 3, 8, 13,}$ 18, 23, 4, 9, 14, 19, 24, 5, 10, 15, 20, 25, $P_1$, $P_2$, $P_3$, $P_4$, $P_5$

Burst error

(a)

**Figure 17.14(a): Interleaving is essential to make error correction schemes more efficient. Samples written sequentially in rows into a memory have redundancy P added to each row. The memory is then read in columns and data are sent to the recording medium. On replay, the nonsequential samples from the medium are deinterleaved to return them to their normal sequence. This breaks up the burst error (shaded) into one error symbol per row in the memory, which can be corrected by the redundancy P.**

samples in a vertical direction in the deinterleave memory. When the memory is read, a single large error is broken down into a number of small errors whose size is exactly equal to the correcting power of the codes and the correction is performed with maximum efficiency.

An extension of the process of interleave is where the memory array has not only rows made into code words, but also columns made into code words by the addition of vertical redundancy. This is known as a product code. Figure 17.14(b) shows that in a product code the redundancy calculated first and checked last is called the outer code, and the redundancy calculated second and checked first is called the inner code. The inner code is formed along tracks on the medium. Random errors due to noise are corrected by the inner code and do not impair the burst correcting power of the outer code. Burst errors are declared uncorrectable by the inner code, which flags the bad samples on the way into the deinterleave memory. The outer code reads the error flags in order to locate erroneous

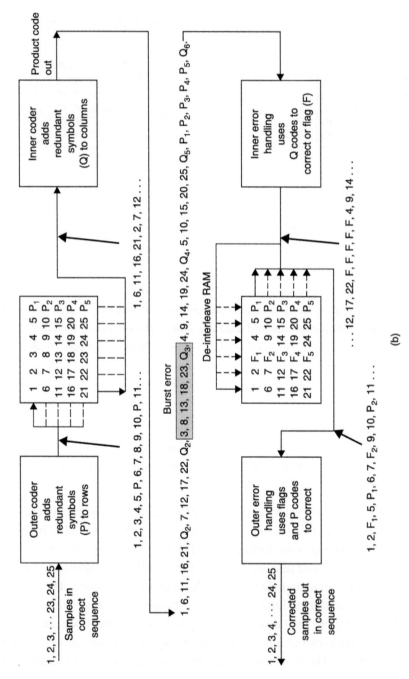

Figure 17.14(b): In addition to the redundancy P on rows, inner redundancy Q is also generated on columns. On replay, the Q code checker will pass on flags F if it finds an error too large to handle itself. The flags pass through the deinterleave process and are used by the outer error correction to identify which symbol in the row needs correcting with P redundancy. The concept of crossing two codes in this way is called a product code.

data. As it does not have to compute the error locations, the outer code can correct more errors.

An alternative to the product block code is the convolutional cross interleave, shown in Figure 17.14(c). In this system, data are formed into an endless array and the code words are produced on columns and diagonals. The compact disc and DASH formats use such a system because it needs less memory than a product code.

The interleave, deinterleave, time-compression, and time base-correction processes cause delay and this is evident in the time taken before audio emerges after starting a digital machine. Confidence replay takes place later than the distance between record and replay heads would indicate. In DASH format recorders, confidence replay is about one-tenth of a second behind the input. Synchronous recording requires new techniques to overcome the effect of the delays.

The presence of an error-correction system means that the audio quality is independent of the tape/head quality within limits. There is no point in trying to assess the health of a

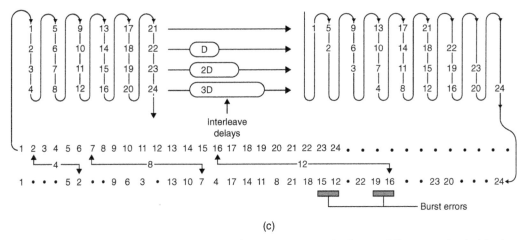

(c)

**Figure 17.14(c): Convolutional interleave is shown. Instead of assembling samples in blocks, the process is continuous and uses RAM delays. Samples are formed into columns in an endless array. Each row of the array is subject to a different delay so that after the delays, samples in a column are available simultaneously which were previously on a diagonal. Code words which cross one another at an angle can be obtained by generating redundancy before and after the delays.**

machine by listening to it, as this will not reveal whether the error rate is normal or within a whisker of failure. The only useful procedure is to monitor the frequency with which errors are being corrected and to compare it with normal figures. Professional digital audio equipment should have an error rate display.

Some people claim to be able to hear error correction and misguidedly conclude that the aforementioned theory is flawed. Not all digital audio machines are properly engineered, however, and if the DAC shares a common power supply with the error correction logic, a burst of errors will raise the current taken by the logic, which loads the power supply and can interfere with the operation of the DAC. The effect is harder to eliminate in small battery-powered machines where space for screening and decoupling components is hard to find, but it is only a matter of design: there is no flaw in the theory.

### 17.3.6 Channel Coding

In most recorders used for storing digital information, the medium carries a track that reproduces a single waveform. Clearly data words representing audio samples contain many bits and so they have to be recorded serially, a bit at a time. Some media, such as CD, only have one track, so it must be totally self-contained. Other media, such as DCC, have many parallel tracks. At high recording densities, physical tolerances cause phase shifts, or timing errors, between parallel tracks and so it is not possible to read them in parallel. Each track must still be self-contained until the replayed signal has been time base corrected.

Recording data serially is not as simple as connecting the serial output of a shift register to the head. In digital audio, a common sample value is all zeros, as this corresponds to silence. If a shift register is loaded with all zeros and shifted out serially, the output stays at a constant low level, and nothing is recorded on the track. On replay there is nothing to indicate how many zeros were present or even how fast to move the medium. Clearly serialized raw data cannot be recorded directly, it has to be modulated into a waveform that contains an embedded clock irrespective of the values of the bits in the samples. On replay a circuit called a data separator can lock to the embedded clock and use it to separate strings of identical bits.

The process of modulating serial data to make it self-clocking is called channel coding. Channel coding also shapes the spectrum of the serialized waveform to make it more efficient. With a good channel code, more data can be stored on a given medium.

Spectrum shaping is used in CD to prevent data from interfering with the focus and tracking servos and in RDAT to allow rerecording without erase heads.

A self-clocking code contains a guaranteed minimum number of transitions per unit time, and these transitions must occur at multiples of some basic time period so that they can be used to synchronize a phase locked loop. Figure 17.15 shows a phase-locked loop that contains an oscillator whose frequency is controlled by the phase error between input transitions and the output of a divider. If transitions on the medium are constrained to occur at multiples of a basic time period, they will have a constant phase relationship with the oscillator, which can stay in lock with them even if they are intermittent. As the damping of the loop is a low-pass filter, jitter in the incoming transitions, caused by peak-shift distortion or by speed variations in the medium, will be rejected and the oscillator will run at the average frequency of the off-tape signal. The phase-locked loop must be locked before data can be recovered, and to enable this, every data block is preceded by a constant frequency recording known as a preamble. The beginning of data is identified by a unique pattern known as a sync pattern.

Irrespective of the channel code used, transitions always occur separated by a range of time periods which are all multiples of the basic clock period. If such a replay signal is viewed on an oscilloscope, a characteristic display called an eye pattern is obtained. Figure 17.16 shows an eye pattern, and in particular the regular openings in the trace. A decision point is in the center of each opening, and the phase-locked loop acts to keep it centered

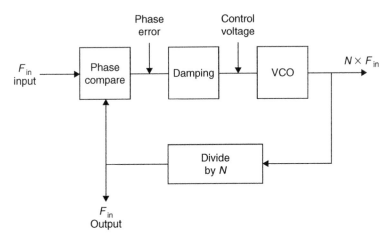

Figure 17.15: A typical phase-locked loop where the VCO is forced to run at a multiple of the input frequency. If the input ceases, the output will continue at the same frequency until it drifts.

laterally, in order to reject the maximum amount of jitter. At each decision point along the time axis, the waveform is above or below the point, and can be returned to a binary signal.

Occasionally, noise or jitter will cause the waveform to pass the wrong side of a decision point, and this will result in an error that will require correction.

Figure 17.17 shows an extremely simple channel code known as frequency modulation (FM), which is used for the AES/EBU digital interface and for recording time code on tape.

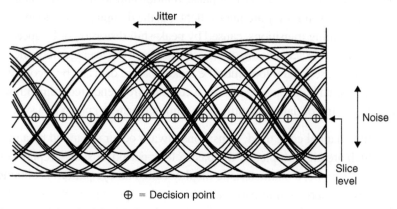

Figure 17.16: At the decision points, the receiver must make binary decisions about the voltage of the signal, whether it is above or below the slicing level. If the eyes remain open, this will be possible in the presence of noise and jitter.

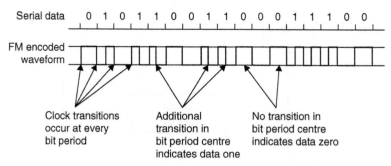

Figure 17.17: FM channel code, also known as Manchester code or biphase mark, is used in AESEBU interface and for time code recording. The waveform is encoded as shown here. See text for details.

Every bit period begins with a transition, irrespective of the value of the bit. If the bit is a one, an additional transition is placed in the center of the bit period. If the bit is a zero, this transition is absent. As a result, the waveform is always self-clocking irrespective of the values of the data bits. Additionally, the waveform spends as much time in the low state as it does in the high state. This means that the signal has no DC component and will pass through capacitors, magnetic heads, and transformers equally well. However simple FM may be, it is not very efficient because it requires two transitions for every bit and jitter of more than half a bit cannot be rejected.

More recent products use a family of channel codes known as group codes. In group codes, groups of bits, commonly eight, are associated together into a symbol for recording purposes. Eight-bit symbols are common in digital audio because two of them can represent a 16-bit sample. Eight-bit data have 256 possible combinations, but if the waveforms obtained by serializing them are examined, it will be seen that many combinations are unrecordable. For example, all ones or all zeros cannot be recorded because they contain no transitions to lock the clock and they have excessive DC content. If a larger number of bits is considered, a greater number of combinations is available. After the unrecordable combinations have been rejected, there will still be 256 left which can each represent a different combination of eight bits. The larger number of bits are channel bits; they are not data because all combinations are not recordable. Channel bits are simply a convenient way of generating recordable waveforms. Combinations of channel bits are selected or rejected according to limits on the maximum and minimum periods between transitions. These periods are called run-length limits and as a result group codes are often called run-length-limited codes.

In RDAT, an 8/10 code is used where 8 data bits are represented by 10 channel bits. Figure 17.18 shows that this results in jitter rejection of 80% of a data bit period: rather better than FM. Jitter rejection is important in RDAT because short wavelengths are used and peak shift will occur. The maximum wavelength is also restricted in RDAT so that low frequencies do not occur.

In CD, an 8/14 code is used where 8 data bits are represented by 14 channel bits. This only has a jitter rejection of 8/14 of a data bit, but this is not an issue because the rigid CD has low jitter. However, in 14 bits there are 16K combinations, and this is enough to impose a minimum run length limit of 3 channel bits. In other words, transitions on the disc cannot occur closer than 3 channel bits apart. This corresponds to 24/14 data bits.

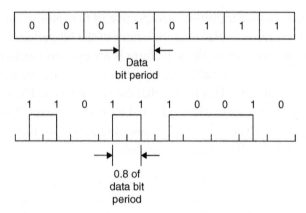

**Figure 17.18: In RDAT an 8/10 code is used for recording. Each 8 data bits are represented by a unique waveform generated by 10 channel bits. A channel bit one causes a transition to be recorded. The transitions cannot be closer than 0.8 of a data bit, and this is the jitter resistance. This is rather better than FM, which has a jitter window of only 0.5 bits.**

Thus the frequency generated is less than the bit rate and a result is that more data can be recorded on the disc than would be possible with a simple code.

## 17.4  Hard Disc Recorders

The hard disc recorder stores data on concentric tracks, which it accesses by moving the head radially. Rapid access drives move the heads with a moving coil actuator, whereas lower cost units will use stepping motors, which work more slowly. The radial position of the head is called the cylinder address, and as the disc rotates, data blocks, often called sectors, pass under the head. To increase storage capacity, many discs can be mounted on a common spindle, each with its own head. All the heads move on a common positioner. The operating surface can be selected by switching on only one of the heads. When one track is full, the drive must select another head. When every track at that cylinder is full, the drive must move to another cylinder. The drive is not forced to operate in this way; it is equally capable of obtaining data blocks in any physical sequence from the disc.

Clearly while the head is moving it cannot transfer data. Using time compression to smooth out the irregular data transfer, a hard disc drive can be made into an audio recorder with the addition of a certain amount of memory.

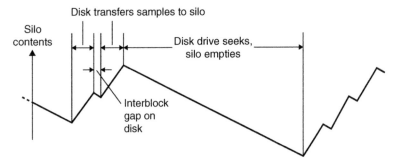

**Figure 17.19: During an audio replay sequence, the silo is constantly emptied to provide samples and is refilled in blocks by the drive.**

Figure 17.19 shows the principle. The instantaneous data rate of the disc drive is far in excess of the sampling rate at the convertor, and so a large time-compression factor can be used. The disc drive can read a block of data from disc and place it in the TBC in a fraction of the real time it represents in the audio waveform. As the TBC steadily advances through the memory, the disc drive has time to move the heads to another track before the memory runs out of data. When there is sufficient space in the memory for another block, the drive is commanded to read and fills up the space. Although the data transfer at the medium is highly discontinuous, the buffer memory provides an unbroken stream of samples to the DAC and so continuous audio is obtained.

Recording is performed using the memory to assemble samples until the contents of one disc block are available. These are then transferred to disc at high data rate. The drive can then reposition the head before the next block is available in memory.

An advantage of hard discs is that access to the audio is much quicker than with tape, as all of the data are available within the time taken to move the head. This speeds up editing considerably.

After a disc has been in use for some time, the free blocks will be scattered all over the disc surface. The random access ability of the disc drive means that a continuous audio recording can be made on physically discontinuous blocks. Each block has a physical address, known as the block address, which the drive controller can convert into cylinder and head selection codes to locate a given physical place on the medium. The size of each block on the disc is arranged to hold the number of samples that arrive during a whole number of time code frames. It is then possible to link each disc block address used during a recording with the time code at which it occurred. The time codes and the corresponding blocks are stored in a table. The table is also recorded on the disc when the recording is completed.

In order to replay the recording, the table is retrieved from the disc, and a time code generator is started at the first code. As the generator runs, each code is generated in sequence, and the appropriate data block is read from the disc and placed in memory, where it can be fed to the convertor.

If it is desired to replay the recording from elsewhere than the beginning, the time code generator can be forced to any appropriate setting, and the recording will play from there. If an external device, such as a video recorder, provides a time code signal, this can be used instead of the internal time code generator, and the machine will automatically synchronize to it.

The transfer rate and access time of the disc drive are such that if sufficient memory and another convertor are available, two completely independent playback processes can be supplied with data by the same drive. For the purpose of editing, two playback processes can be controlled by one time code generator. The time code generator output can be offset differently for each process, so that they can play back with any time relationship. If it is required to join the beginning of one recording to the end of another, the operator specifies the in point on the second recording and the out point on the second recording. By changing the time code offsets, the machine can cause both points to occur simultaneously in data accessed from the disc and played from memory. In the vicinity of the edit points, both processes are providing samples simultaneously and a cross fade of any desired length can be made between them.

The arrangement of data on the disc surface has a bearing on the edit process. In the worst case, if all the blocks of the first recording were located at the outside of the disc and all of the blocks of the second recording were located at the inside, the positioner would spend a lot of time moving. If the blocks for all recordings are scattered over the entire disc surface, the average distance the positioner needs to move is reduced.

The edit can be repeated with different settings as often as necessary without changing the original recordings. Once an edit is found to be correct, it is only necessary to store the handful of instructions which caused it to happen, and it can be executed at any time in the future in the same way. The operator has the choice of archiving the whole disc contents on tape, so different edits can be made in the future, or simply recording the output of the current edit so that the disc can be freed for another job.

The rapid access and editing accuracy of hard disc systems make them ideal for assembling sound effects to make the sound tracks of motion pictures.

The use of data reduction allows the recording time of a disc to be extended considerably. This technique is often used in plug-in circuit boards, which are used to convert a personal computer into a digital audio recorder.

## 17.5 The PCM Adaptor

The PCM adaptor was an early solution to recording the wide bandwidth of PCM audio before high-density digital recording developed. The video recorder offered sufficient bandwidth at moderate tape consumption. While they were a breakthrough at the time of their introduction, by modern standards PCM adaptors are crude and obsolescent, offering limited editing ability and slow operation.

Figure 17.20 shows the essential components of a digital audio recorder using this technique. Input analogue audio is converted to digital and time compressed to fit into the parts of the video waveform which are not blanked. Time-compressed samples are then odd-even shuffled to allow concealment. Next, redundancy is added and data are interleaved for recording. Data are serialized and set on the active line of the video signal as black and white levels shown in Figure 17.21. The video is sent to the recorder, where the analogue FM modulator switches between two frequencies representing the black and white levels, a system called frequency shift keying (FSK). This takes the place of the channel coder in a conventional digital recorder.

On replay, the FM demodulator of the video recorder acts to return the FSK recording to the black/white video waveform, which is sent to the PCM adaptor. The PCM adaptor extracts a clock from the video sync pulses and uses it to separate the serially recorded bits. Error correction is performed after deinterleaving, unless the errors are too great, in which case concealment is used after the deshuffle. The samples are then returned to the standard sampling rate by the time base expansion process, which also eliminates any speed variations from the recorder. They can then be converted back to the analogue domain.

In order to synchronize playback to a reference and to simplify the circuitry, a whole number of samples is recorded on each unblanked line. The common sampling rate of

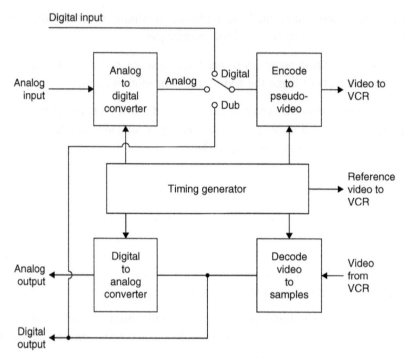

**Figure 17.20: Block diagram of PCM adaptor. Note the dub connection needed for producing a digital copy between two VCRs.**

44.1 kHz is obtained by recording three samples per line on 245 active lines at 60 Hz. The sampling rate is thus locked to the video sync frequencies and the tape is made to move at the correct speed by sending the video recorder syncs which are generated in the PCM adaptor.

## 17.6 An Open Reel Digital Recorder

Figure 17.22 shows the block diagram of a machine of this type. Analogue inputs are converted to the digital domain by converters. Clearly there will be one convertor for every audio channel to be recorded. Unlike an analogue machine, there is not necessarily one tape track per audio channel. In stereo machines the two channels of audio samples may be distributed over a number of tracks each in order to reduce the tape speed and extend the playing time.

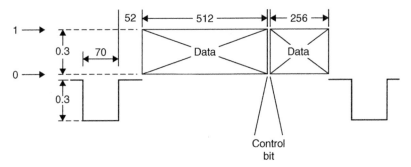

**Figure 17.21: Typical line of video from PCM-1610. The control bit conveys the setting of the preemphasis switch or the sampling rate, depending on the frame. The bits are separated using only the timing information in the sync pulses.**

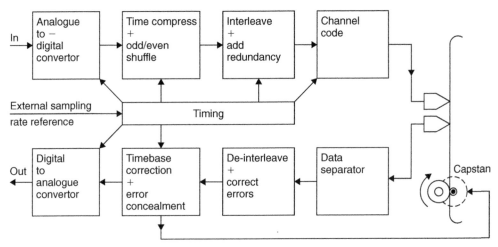

**Figure 17.22: Block diagram of one channel of a stationary head digital audio recorder. See text for details of the function of each block. Note the connection from the TBC to the capstan motor so that the tape is played at such a speed that the TBC memory neither underflows nor overflows.**

The samples from the convertor will be separated into odd and even for concealment purposes, and usually one set of samples will be delayed with respect to the other before recording. The continuous stream of samples from the convertor will be broken into blocks by time compression prior to recording. Time compression allows the insertion of edit gaps, addresses, and redundancy into the data stream. An interleaving process is also necessary to reorder the samples prior to recording. As explained earlier, the subsequent deinterleaving breaks up the effects of burst errors on replay.

The result of the processes so far is still raw data, which will need to be channel coded before they can be recorded on the medium. On replay a data separator reverses the channel coding to give the original raw data with the addition of some errors. Following deinterleave, the errors are reduced in size and are more readily correctable. The memory required for deinterleave may double as the TBC memory, so that variations in the speed of the tape are rendered undetectable. Any errors that are beyond the power of the correction system will be concealed after the odd-even shift is reversed. Following conversion in the DAC an analogue output emerges.

On replay a digital recorder works rather differently to an analogue recorder, which simply drives the tape at constant speed. In contrast, a digital recorder drives the tape at constant sampling rate. The TBC works by reading samples out to the convertor at constant frequency. This reference frequency comes typically from a crystal oscillator. If the tape goes too fast, the memory will be written faster than it is being read and will eventually overflow. Conversely, if the tape goes too slow, the memory will become exhausted of data. In order to avoid these problems, the speed of the tape is controlled by the quantity of data in the memory. If the memory is filling up, the tape slows down; if the memory is becoming empty, the tape speeds up. As a result, the tape will be driven at whatever speed is necessary to obtain the correct sampling rate.

## 17.7 Rotary Head Digital Recorders

The rotary head recorder borrows technology from video recorders. Rotary heads have a number of advantages over stationary heads. One of these is extremely high packing density: the number of data bits that can be recorded in a given space. In a digital audio recorder, packing density directly translates into the playing time available for a given size of the medium.

In a rotary head recorder, the heads are mounted in a revolving drum and the tape is wrapped around the surface of the drum in a helix, as can be seen in Figure 17.23. The helical tape path results in the heads traversing the tape in a series of diagonal or slanting tracks. The space between the tracks is controlled not by head design but by the speed of the tape, and in modern recorders this space is reduced to zero with a corresponding improvement in packing density.

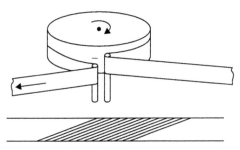

**Figure 17.23: Rotary head recorder. Helical scan records long diagonal tracks.**

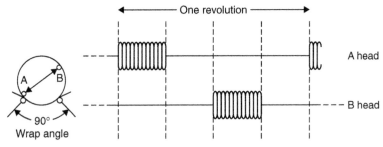

**Figure 17.24: The use of time compression reduces the wrap angle necessary, at the expense of raising the frequencies in the channel.**

The added complexity of the rotating heads and the circuitry necessary to control them are offset by the improvement in density. The discontinuous tracks of the rotary head recorder are naturally compatible with time-compressed data. As Figure 17.24 illustrates, the audio samples are time compressed into blocks, each of which can be contained in one slant track.

In a machine such as RDAT, there are two heads mounted on opposite sides of the drum. One rotation of the drum lays down two tracks. Effective concealment can be had by recording odd-numbered samples on one track of the pair and even-numbered samples on the other. Samples from the two audio channels are multiplexed into one data stream, which is shared between the two heads.

As can be seen from the block diagram shown in Figure 17.25, a rotary head recorder contains the same basic steps as any digital audio recorder. The record side needs ADCs, time compression, the addition of redundancy for error correction, and channel coding. On replay the channel coding is reversed by the data separator, errors are broken up by

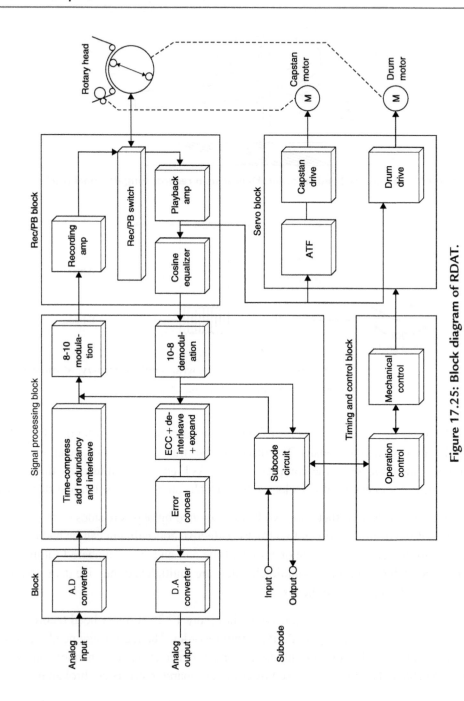

**Figure 17.25: Block diagram of RDAT.**

the deinterleave process and corrected or concealed, and the time compression and any fluctuations from the transport are removed by time base correction. The corrected, time-stable samples are then fed to the DAC.

One of the reasons for the phenomenal recording density at which RDAT operates is the use of azimuth recording. In this technique, alternate tracks on the tape are laid down with heads having different azimuth angles. In a two-headed machine this is easily accommodated by having one head set at each angle. If the correct azimuth head subsequently reads the track there is no difficulty, but as Figure 17.26 shows, the wrong head suffers a gross azimuth error.

Azimuth error causes phase shifts to occur across the width of the track and, at some wavelengths, this will result in cancellation except at very long wavelengths where the process is no longer effective. The use of 8110 channel coding in RDAT ensures that no low frequencies are present in the recorded signal and so this characteristic of azimuth recording is not a problem. As a result the pickup of signals from the adjacent track is effectively prevented, and the tracks can be physically touching with no guard bands being necessary.

As the azimuth system effectively isolates the tracks from one another, the replay head can usually be made wider than the track. A typical figure is 50% wider. A tracking error of up to $+/- 25\%$ of the track width then causes no loss of signal quality.

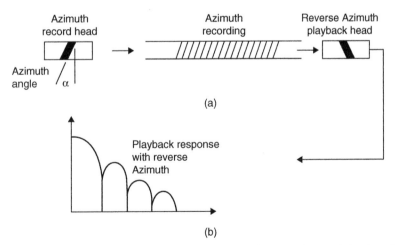

Figure 17.26: In azimuth recording (a), the head gap is tilted. If the track is played with the same head, playback is normal, but the response of the reverse azimuth head is attenuated (b).

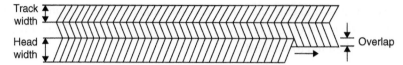

**Figure 17.27: In azimuth recording, the tracks can be made narrower than the head pole by overwriting the previous track.**

In practice the same heads can also be used for recording, even though they are too wide. As can be seen in Figure 17.27, the excess track width is simply overwritten during the next head sweep. Erase heads are unnecessary, as the overlapping of the recorded tracks guarantees that the whole area of a previous recording is overwritten. A further advantage of the system is that more than one track width can be supported by the same mechanism simply by changing the linear tape speed. Prerecorded tapes made by contact duplication have lower coercivity coatings, and to maintain the signal level the tracks are simply made wider by raising the tape speed. Any RDAT machine can play such a recording without adjustment.

In any rotary head recorder, some mechanism is necessary to synchronize the linear position of the tape to the rotation of the heads, otherwise the recorded tracks cannot be replayed. In a conventional video recorder, this is the function of the control track, which requires an additional, stationary head. In RDAT the control track is dispensed with, and tracking is obtained by reading patterns in the slant tracks with the normal playback heads.

Figure 17.28 shows how the system works. The tracks are divided into five areas. PCM audio data are in the center and subcode data are at the ends. Audio and subcode data are separated by tracking patterns. The tracking patterns are recorded and played back along with data. The tracking is measured by comparing the level of a pilot signal picked up from the tracks on each side of the wanted track. If the replay head drifts toward one side, it will overlap the next track on that side by a greater amount and cause a larger pilot signal to be picked up. Pilot pickup from the track on the opposite side will be reduced. The difference between the pilot levels is used to change the speed of the capstan, which has the effect of correcting the tracking.

Ordinarily, azimuth effect prevents the adjacent tracks being read, but the pilot tones are recorded with a wavelength much longer than that of data. They can then be picked up by a head of the wrong azimuth.

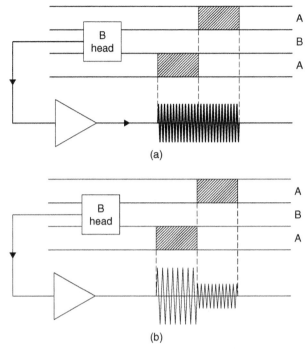

(a)

(b)

**Figure 17.28: (a) A correctly tracking head produces pilot-tone bursts of identical amplitude. (b) The head is off-track, and the first pilot burst becomes larger, whereas the second becomes smaller. This produces the tracking error.**

The combination of azimuth recording, an active tracking system, and high coercivity tape (1500 Oersteds compared to 200 Oersteds for analogue audio tape) allows the tracks to be incredibly narrow. Heads of 20 μm width write tracks 13 μm wide. About 10 such tracks will fit in the groove of a vinyl disc. Although the head drum spins at 2000 rpm, the tape speed needed is only 8.15 mm per second.

The subcode of RDAT functions in a variety of ways. In consumer devices, the subcode works in the same way as in CD, having a table of contents and flags allowing rapid access to the beginning of tracks and carrying signals to give a playing time readout.

In professional RDAT machines, the subcode is used to record time code. A time code format based on hours, minutes, seconds, and DAT frames (where a DAT frame is one drum revolution) is recorded on the tape, but suitable machines can convert the tape code to any video, audio, or film time code and operate synchronized to a time code reference. As the heads are wider than the tracks, a useful proportion of the data can be read even when the tape is being shuttled. Subcode data are repeated many times so that they can be read at any speed. In this way an RDAT machine can chase any other machine and remain synchronized to it.

While there is nothing wrong with the performance of RDAT, it ran into serious political problems because its ability to copy without loss of quality was seen as a threat by copyright organizations. The launch of RDAT as a consumer product was effectively blocked until a system called serial copying management system was incorporated. This allows a single generation of RDAT copying of copyright material. If an attempt is made to copy a copy, a special flag on the copy tape defeats recording on the second machine.

In the meantime, RDAT found favor in the professional audio community where it offered exceptional sound quality at a fraction of the price of professional equipment. Between them, the rapid access of hard disc-based recorders and the low cost of RDAT have effectively rendered %inch analogue recorders and stereo open reel digital recorders obsolete.

## 17.8  Digital Compact Cassette

DCC is a consumer stationary head digital audio recorder using data reduction. Although the convertors at either end of the machine work with PCM data, these data are not directly recorded, but are reduced to one-quarter of their normal rate by processing. This allows a reasonable tape consumption similar to that achieved by a rotary head recorder. In a sense, the complexity of the rotary head transport has been exchanged for the electronic complexity of the data reduction and subsequent expansion circuitry.

Figure 17.29 shows that DCC uses stationary heads in a conventional tape transport that can also play analogue cassettes. Data are distributed over nine parallel tracks, which occupy half the width of the tape. At the end of the tape the head rotates about an axis perpendicular to the tape and plays the other nine tracks in reverse. The advantage of the conventional approach with linear tracks is that tape duplication can be carried out at high speed. This makes DCC attractive to record companies.

However, reducing the data rate to one-quarter and then distributing it over nine tracks means that the frequency recorded on each track is only about 1/32 that of a PCM

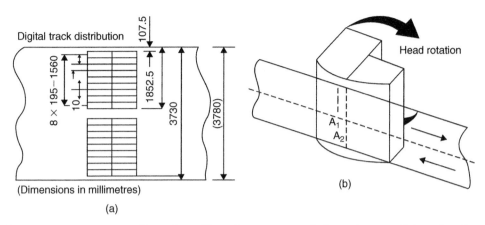

**Figure 17.29: In DCC audio and auxiliary data are recorded on nine parallel tracks along each side of the tape as shown at (a) The replay head shown at (b) carries magnetic poles, which register with one set of nine tracks. At the end of the tape, the replay head rotates 180° and plays a further nine tracks on the other side of the tape. The replay head also contains a pair of analogue audio magnetic circuits that will be swung into place if an analogue cassette is to be played.**

machine with a single head. At such a low frequency, conventional inductive heads that generate a voltage from flux changes cannot be used, and DCC has to use active heads that actually measure the flux on the tape at any speed. These magneto-resistive heads are more complex than conventional inductive heads and have only recently become economic as manufacturing techniques have been developed.

Data reduction relies on the phenomenon of auditory masking and this effectively restricts DCC to being a consumer format. It will be seen from Figure 17.30 that the data reduction unit adjacent to the input is complemented by the expansion unit or decoder prior to the DAC. The sound quality of a DCC machine is not a function of the tape, but depends on the convertors and on the sophistication of the data reduction and expansion units.

## 17.9  Editing Digital Audio Tape

Digital recordings are simply data files, and editing digital audio should be performed in the same way that a word processor edits text. No word processor attempts to edit on the

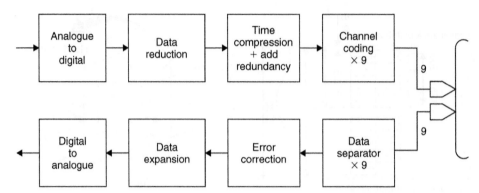

**Figure 17.30: In DCC, PCM data from the convertors are reduced to one-quarter of the original rate prior to distribution over eight tape tracks (plus an auxiliary data track). This allows a slow linear tape speed that can only be read with an MR head. The data reduction unit is mirrored by the expansion unit on replay.**

medium, but brings blocks of data to a computer memory where it is edited before being sent back for storage.

In fact, this is the only way that digital audio recordings can be edited because of the use of interleave and error correction.

Interleave reorders the samples on the medium, and so it is not possible to find a physical location on the medium that corresponds linearly to the time through the recording. Error correction relies on blocks of samples being coded together. If part of a block is changed, the coding will no longer operate.

Figure 17.31 shows how an audio edit is performed. Samples are played back, deinterleaved, and errors are corrected. Samples are now available in their natural real-time sequence and can be sent to a cross-fader where external material can be inserted. The edited samples are then recoded and interleaved before they can be rerecorded. Deinterleave and interleave cause delay, and by the time these processes have been performed, the tape will have moved further through the machine. In simple machines, the tape will have to be reversed, and new data recorded in a second pass. In more sophisticated machines, an edit can be made in a single pass because additional record heads are placed further down the tape path.

In a stationary head machine, these are physically displaced along the head block. In a rotary head machine, the extra heads are displaced along the axis of the drum.

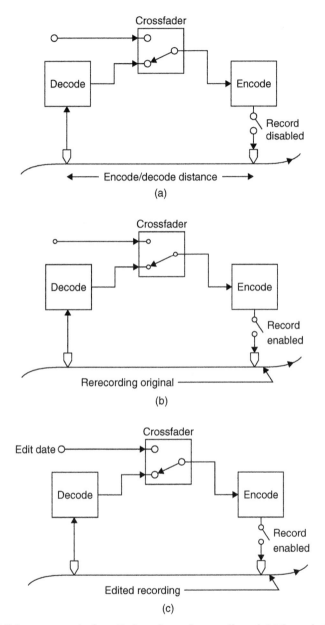

**Figure 17.31:** Editing a convolutionally interleaved recording. (a) The existing recording is decoded and re-encoded. After some time, record can be enabled at (b) when the existing tape pattern is being rerecorded. The crossfader can then be operated, resulting (c) in an interleaved edit on the tape.

Displaced heads also allow synchronous recording to be performed on multitrack digital audio recorders.

Some stationary head digital formats allow editing by tape cutting. This requires use of an odd-even sample shift and concealment to prevent the damaged area of the tape being audible. With electronic editing, now widely available, tape-cut editing is obsolete as it does not offer the ability to preview or trim the result and causes damage to the medium. The glue on the splicing tape tends to migrate in storage and cause errors.

## References

1. Baert, L., Theunissen., L., and Vergult G., *Digital Audio and Compact Disc Technology*, 2nd Ed, Buttenvorth-Heinemann, 1992.
2. Pohlmann, K., *The Compact Disc*, Oxford University Press, 1989.
3. Pohlmann, K. C., *Advanced Digital Audio*, Sams, 1991.
4. Rumsey, F., *Tapeless Sound Recording*, Focal Press, 1990.
5. Rumsey, F., *Digital Audio Operations*, Focal Press, 1991.
6. Sinclair, R., *Introducing Digital Audio*, PC Publishing, 1991.
7. Watkinson, J., *Art of Digital Audio*, Focal Press, Butterworth-Heinemann, 1988.
8. Watkinson, J., *Coding for Digital Recording*, Focal Press, 1990.

# CHAPTER 18

# Digital Audio Interfaces

Richard Brice

## 18.1 Digital Audio Interfaces

Many of the advantages of digital signal processing are lost if signals are repeatedly converted back and forth between the digital and analogue domain. So that the number of conversions could be kept to a minimum, as early as the 1970s, manufacturers started to introduce proprietary digital interface standards enabling various pieces of digital audio hardware to pass digital audio information directly without recourse to standard analogue connections. Unfortunately, each manufacturer adopted its own standard, and the Sony digital interface (SDIF) and the Mitsubishi interface both bear witness to this early epoch in digital audio technology when compatibility was very poor between different pieces of equipment. It wasn't long before customers were demanding an industry-standard interface so that they could "mix and match" equipment from different manufacturers to suit their own particular requirements. This pressure led to the introduction of widespread, standard interfaces for the connection of both consumer and professional digital audio equipment.

The requirements for standardizing a digital interface go beyond those for an analogue interface in that, as well as defining the voltage levels and connector style, it is necessary to define the data format the interface will employ. The two digital audio interface standards described here are:

(1) The two-channel, serial, balanced, professional interface (the so-called AES/EBU or IEC958 type 1 interface).

(2) The two-channel, serial, unbalanced, consumer interface (the so-called SPDIF or IEC958 type 2 interface).

In fact, both these interfaces are very similar, with variations being more due to electrical differences than between differences in data format.

### 18.1.1 AES/EBU or IEC958 Type 1 Interface

This electrically balanced version of the standard digital interface was originally defined in documents produced by the Audio Engineering Society (AES) and the European Broadcasting Union (EBU) and is, consequently, usually referred to as the AES/EBU standard. This is the standard adopted mainly by professional and broadcast installations. Mechanically, this interface employs the ubiquitous XLR connector and adopts normal convention for female and male versions for inputs and outputs, respectively. Electrically, pin 1 is specified as shield and pins 2 and 3 for balanced signal. One of the advantages of the digital audio interface over its analogue predecessor is that polarity is not important, so it is not necessary to specify which pin of 2 and 3 is "hot." The balanced signal is intended to be carried by balanced, twisted-pair, and screen microphone-style cable and voltage levels are allowed to be between 3 and 8 V pk-pk (EMF, measured differentially). Both inputs and outputs are specified as transformer coupled and earth free. The output impedance of the interface is defined as 110 ohms, and a standard input must always terminate in 110 ohms. A drawing for the electrical standard for this interface is given in Figure 18.1.

### 18.1.2 The SPDIF or IEC985 Type 2 Interface

This consumer version of the two-channel, serial digital interface is very different electrically from the AES/EBU interface described earlier. It is a 75-ohm, matched termination interface intended for use with coaxial cable. It therefore has more in common with an analogue video signal interface than with any analogue audio counterpart. Mechanically the connector style recommended for the SPDIF interface is RCA style phono with sockets always being of the isolated type. Voltage levels are

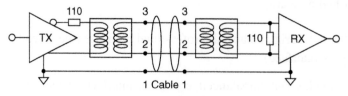

**Figure 18.1: AES/EBU interface.**

defined as 1 V pk-pk when unterminated. Transformer coupling is by no means always used with this interface but it is recommended on at least one end. Figure 18.2 is a drawing of a common implementation of the SPDIF interface.

### 18.1.3 Data

Despite the very considerable electrical differences between the AES/EBU interface and the SPDIF interface, their data formats are very similar. Both interfaces have capacity for the real-time communication of 20 bits of stereo audio information at sampling rates between 32 and 48 kHz, as well as provision for extra information, which may indicate to the receiving device various important parameters about the data being transferred (such as whether preemphasis was used on the original analogue signal prior to digitization). There is also a small overhead for limited error checking and for synchronization.

Some of the earlier digital–audio interfaces such as Sony's SDIF and the Mitsubishi interface sent digital audio data and synchronizing data clocks on separate wires. Such standards obviously require multicore cable and multiway connectors, which looked completely different from any analogue interface that had gone before. The intention of the designers of the AES/EBU and SPDIF interfaces was to create standards that created as little "culture shock" as possible in both the professional and the consumer markets and therefore they chose connector styles that were both readily available and operationally convenient. This obviously ruled out the use of multicore and multiway connectors and resulted in the use of a digital coding scheme that buries the digital synchronizing signals in with the data signal. Such a code is known as "serial and self-clocking." The type of code adopted for AES/EBU and SPDIF is biphase mark coding. This scheme is sometimes known as Manchester code and it is the same type of self-clocking, serial code used for SMPTE and EBU time code. Put at its simplest, such a code represents the "ones and noughts" of a digital signal by two different frequencies where frequency $F_n$ represents a zero and $2F_n$ represents a one. Such a signal eliminates almost all DC content, enabling it to be transformer coupled, and also allows for phase

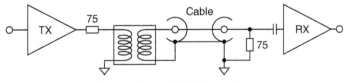

**Figure 18.2: SPDIF interface.**

inversion since it is only a frequency (and not its phase) that needs to be detected. The resulting signal has much in common with an analogue FM signal and since the two frequencies are harmonically related (an octave apart), it is a simple matter to extract the bit clock from the composite incoming data stream.

In data format terms the digital audio signal is divided into frames. Each digital audio frame contains a complete digital audio sample for both left and right channels. If 48-kHz sampling is used, it is obvious that the 48 thousand frames pass over the link in every second, leading to a final baud rate of 3.072 Mbit/s. If 44.1-kHz sampling is employed, 44 thousand one-hundred frames are transmitted every second, leading to a final baud rate of 2.8224 Mbit/s. The lowest allowable transfer rate is 2.084 Mbit/s when 32 kHz is used. Just as each complete frame contains a left and right channel sample, so each frame may be further divided into individual audio samples known as subframes. A diagram of a complete frame consisting of two subframes is given in Figure 18.3.

It is extremely important that any piece of equipment receiving the digital audio signal, as shown in Figure 18.3, must know where the boundaries between frames and subframes lie. That is the purpose of the "sync preamble" section of each frame and subframe. The sync preamble section of the digital audio signal differs from all other data sent over the digital interface in that it violates the rules of a biphase mark encoded signal. In terms of the FM analogy given earlier you can think of the sync preamble as containing a third nonharmonically related frequency, which, when detected, establishes the start of each subframe. There exists a family of three slightly different sync preambles, one to mark the beginning of a left sample subframe and another to mark the start of the right channel subframe. The third sync preamble pattern is used only once every 192 frames (or once every 4 ms in the case of 48-kHz sampling) and is used to establish a 192 bit repeating pattern to the channel-status bit labeled C in Figure 18.3.

The 192 bit repeat pattern of the C bit builds up into a table of 24 bytes of channel-status information for the transmitted signal. It is in this one bit of data every subframe that the

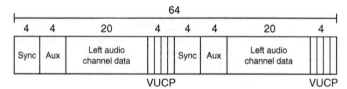

**Figure 18.3: Digital audio data format.**

difference between the AES/EBU interface data format and the SPDIF data format is at its most significant. The channel status bits in both the AES/EBU format and SPDIF format communicate to the receiving device such important parameters as sample rate, whether frequency preemphasis was used on the recording, and so on. Channel-status data are normally the most troublesome aspect of practical interfacing using the SPDIF and AES/EBU interface, especially where users attempt to mix the two interface standards. This is because the usage of channel status in consumer and professional equipment is almost entirely different. It must be understood that the AES/EBU interface and the SPDIF interface are thus strictly incompatible in data format terms and the only correct way to transfer data from SPDIF to AES/EBU and AES/EBU to SPDIF is through a properly designed format converter that will decode and recode digital audio data to the appropriate standard.

Other features of the data format remain pretty constant across the two interface standards. The validity bit, labeled V in Figure 18.3, is set to 0 every subframes if the signal over the link is suitable for conversion to an analogue signal. The user bit, labeled U, has a multiplicity of uses defined by particular users and manufacturers. It is used most often over the domestic SPDIF interface. The parity bit, labeled P, is set such that the number of ones in a subframe is always even. It may be used to detect individual bit errors but not conceal them.

It's important to point out that both the AES/EBU interface and its SPDIF brother are designed to be used in an error-free environment. Errors are not expected over digital links and there is no way of correcting for them.

### 18.1.4 Practical Digital Audio Interface

There are many ways of constructing a digital audio interface, and variations abound from different manufacturers. Probably the simplest consists of an HC family inverter IC, biased at its midpoint with a feedback resistor and protected with diodes across the input to prevent damage from static or overvoltage conditions. (About the only real merit of this circuit is simplicity!) Transformer coupling is infinitely preferred. Happily, while analogue audio transformers are complex and expensive items, digital audio—containing no DC component and very little low-frequency component—can be coupled via transformers, which are tiny and inexpensive! So, it represents a false economy indeed to omit them in the design of digital interfaces. Data-bus isolators manufactured by Newport are very suitable. Two or four transformers are contained within one IC-style package. Each transformer costs about $2—a long way from the $20 or so required for analogue transformers. Remember too that "in digits" only, one transformer is required to couple both channels of the stereo signal. You'll notice, looking at the circuit diagrams (Figure 18.4), RS422 (RS485) receiver-chips buffer and reslice digital audio data. The SN75173J is a quad receiver in a single 16 pin package costing a few dollars. The part has the added advantage that, to adapt the interface between SPDIF and AES, all that is required is to change the value of the terminating resistor on the secondary side of the input transformer. SPDIF digital output can be derived by inverters driving in tandem. If AES/EBU output is required, it is best performed by an RS422 driver IC.

### 18.1.5  TOSlink Optical Interface

In many ways an optical link seems to be the ideal solution for joining two pieces of digital audio equipment together. Obviously a link that has no electrical contact cannot introduce ground-loop, hum problems. Also, because the bandwidth of an optical link is so high, it would appear from a superficial inspection that an optical link would provide the very fastest (and therefore "cleanest") signal path possible. However, the optical TOSLink is widely regarded as sounding a little less crisp than its coaxial, electrical counterpart. There are a number of possible reasons for this. In the first place, the speed of the link is compromised by the relatively slow light-emitting diode transmitter and phototransistor receiver housed within the connector shells. Second, inexpensive optical fibers, which allow the optical signal more than one direct path between transmitter and receiver (the technical term is multimodes), cause a temporal smearing of the audio pulses, resulting in an effect known as modal dispersion. This can cause a degree of timing instability in digital audio circuits (jitter) and can affect sound quality. The only advantage the optical link confers, therefore, is its inherent freedom from ground path-induced interference signals such as hum and RF noise. Yet at digital audio frequencies, ground isolation—if it is required—is obtained much better by means of a transformer. If you want to modify a piece of equipment with an optical interface to include SPDIF coaxial output, a modification is shown in Figure 18.5.

### 18.1.6  Transmission of AES3-Formatted Data by Unbalanced Coaxial Cable

In October 1995, the AES produced an information document (AES-3id-1995) relating to the transmission of digital audio information (utilizing the professional data format)

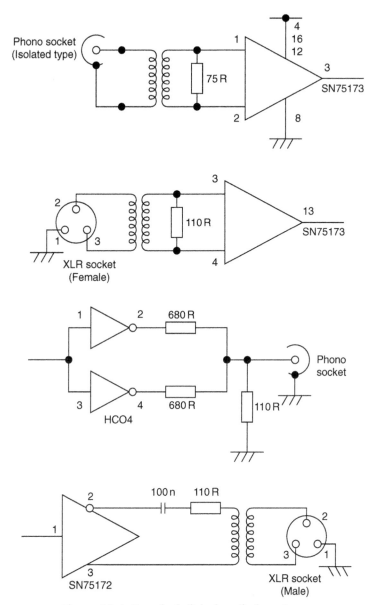

**Figure 18.4: Practical digital audio interfaces.**

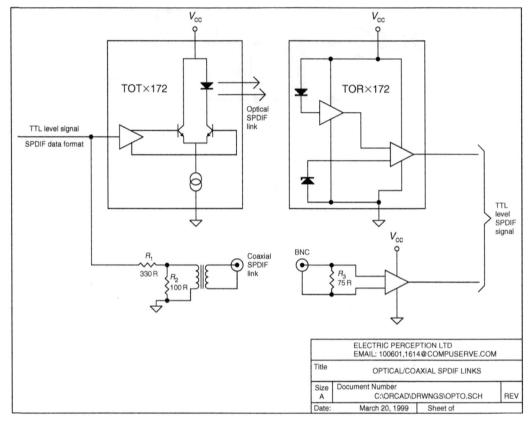

**Figure 18.5: Optical digital audio interface and adaptation to coaxial SPDIF.**

over an electrical interface that has much in common with the interconnection standards employed in analogue video. Limitations of AES data traveling on twisted pairs and terminated in XLRs include poor RF radiation performance and a limitation of maximum transmission distance to 100 m. The proposed unbalanced interface is suitable for transmission distances of up to 1000 m. Furthermore, by a prudent choice of impedance and voltage operating level, coupled with a sensible specification of minimum rise time, the signal is suitable for routing through existing analogue video cables, switchers, and distribution amplifiers.

The salient parts of the signal and interface specification are given in Table 18.1.

**Table 18.1 Salient parts of the signal and interface specification**

| General | |
|---|---|
| Transmission data format | Electrically equivalent to AES |
| Impedance | 75 ohms |
| Mechanical | BNC connector |
| Signal characteristics | 1 V, measured when terminated |
| Output voltage | in 75 ohms |
| DC offset | <50 mV |
| Rise/fall time | 30 to 44 nS |
| Bit width (at 48 kHz) | 162.8 nS |

## 18.2 MADI (AES10–1991) Serial Multichannel Audio Digital Interface

The MADI standard is a serial transmission format for multichannel linearly represented PCM audio data. The specification covers transmission of 56, mono, 24-bit resolution channels of audio data with a common sampling frequency in the range of 32 to 48 kHz. Perhaps this is conceived more easily of in terms of 28 stereo "AES" audio channels (i.e., of AES3–1985 data) traveling on a common bearer, as illustrated in Figure 18.6. The MADI standard is not a "networking" standard; in other words, it only supports point-to-point interconnections.

### 18.2.1 Data Format

The MADI serial data stream is organized into frames that consist of 56 channels (numbered 0–55). These channels are consecutive within the frame, and audio data remain, just as it is in the original digital audio interface, in linearly coded, 2's-complement form, although this is scrambled as described later. The frame format is illustrated in Figure 18.6. Each channel "packet" consists of 32 bits (as also shown in Figure 18.6), in which 24 are allocated to audio data (or possibly nonaudio data if the nonvalid flag is invoked) and 4 bits for the validity (V), user (U), channel status (C), and parity (P) bits as they are used in the AES3–1985 standard audio interface. In this manner the structure and data within contributing dual-channel AES bit streams can be preserved intact when traveling in the MADI multichannel bit stream. The remaining 4 bits

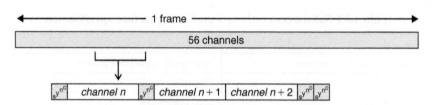

**Figure 18.6: Data structure of MADI, multichannel audio interface.**

per channel (called, confusingly, mode bits) are used for frame synchronization on the MADI interface and for preserving information concerning A/B preambles and start of channel-status block within each of the contributing audio channels.

### 18.2.2  Scrambling and Synchronization

Serial data are transmitted over the MADI link in a polarity-insensitive (NRZI) form. However, before data are sent, they are subjected to a 4- to 5-bit encoding, as defined in Table 18.2. MADI has a rather unusual synchronization scheme in order to keep transmitter and receiver in step. The standard specifies that the transmitter inserts a special synchronizing sequence (1100010001) at least once per frame. Note that this sequence cannot be derived from data, as specified in Table 18.2. Unusually, this sync signal need not appear between every frame, as Figure 18.6 illustrates. This sync signal is simply repeated wherever required in order to regulate the final data rate of 100 megabits/second specified in the standard.

### 18.2.3  Electrical Format

MADI travels on a coaxial cable interface with a characteristic impedance of 75 ohms. Video-style BNC connectors are specified. Because the signal output is practically DC free, it may be AC coupled and must sit around $0\,V \pm 100\,mV$. This signal is specified to have a peak-to-peak amplitude of 300–600 mV when terminated—this choice of amplitude being determined by the practical consideration that the signal could be directly derived from the output of an ECL gate.

### 18.2.4  Fiber-Optic Format

Oddly, the MADI standard did not define a fiber implementation, despite the fact that the copper implementation was based on a widely used fiber interface known as FDDI

Table 18.2 Input and output data sequence

| Input data sequence | Output data sequence |
|---|---|
| 0000 | 11110 |
| 0001 | 01001 |
| 0010 | 10100 |
| 0011 | 10101 |
| 0100 | 01010 |
| 0101 | 01011 |
| 0110 | 01110 |
| 0111 | 01111 |
| 1000 | 10010 |
| 1001 | 10011 |
| 1010 | 10110 |
| 1011 | 10111 |
| 1100 | 11010 |
| 1101 | 11011 |
| 1110 | 11100 |
| 1111 | 11101 |

(ISO 9314). It is this standard, which predates the MADI, that specified the 4- to 5-bit mapping defined in Table 18.2! This lack of standardization has resulted in a rather disorganized situation regarding MADI over fiber. The AES's own admission is simply that "any fiber-system could be used for MADI as long as the basic bandwidth and data-rate can be supported … However, adoption of a common implementation would be advantageous."

# CHAPTER 19

# Data Compression

Richard Brice

Data reduction or compression techniques are important because universal laws put a premium on information. You couldn't read all the books in the world, neither could you store them. You might make a start on reading every book by making it a team effort. In other words, you might tackle the problem with more than one brain and one pair of eyes. In communication theory terms, this approach is known as increasing the channel capacity by broadening the bandwidth. But you wouldn't have an infinite number of people at your disposal unless you had an infinite amount of money to pay them! Likewise no one has an infinite channel capacity or an infinite bandwidth at their disposal. The similar argument applies to storage. Stated axiomatically: information, in all its forms, is using up valuable resources, so the more efficiently we can send it and store it the better. That's where compression comes in.

If I say to you, "Wow, I had a bad night, the baby cried from three 'til six!" You understand perfectly what I mean because you know what a baby crying sounds like. I might alternatively have said, "Wow, I had a bad night, the baby did this; wah, bwah, bwah, wah ..." and continue doing it for 3 h. Try it. You'll find you lose a lot of friends because nobody needs to have it demonstrated. Most of the 3-h impersonation is superfluous. The second message is said to have a high level of redundancy in the terms of communication theory. The trick performed by any compression system is sorting out the necessary information content—sometimes called the entropy—from the redundancy. (If, like me, you find it difficult to comprehend the use of entropy in this context consider this: entropy refers here to a lack of pattern; to disorder. Everything in a signal that has a pattern is, by definition, predictable and therefore redundant. Only those parts of the signal that possess no pattern are unpredictable and therefore represent necessary information.)

All compression techniques may be divided between lossless systems and lossy systems. Lossless compression makes use of efficiency gains in the manner in which data are coded. All that is required to recover original data exactly is a decoder that implements the reverse process performed by the coder. Such a system does not confuse entropy for redundancy and hence dispense with important information. However, neither does the lossless coder perfectly divide entropy from redundancy. A good deal of redundancy remains and a lossless system is therefore only capable of relatively small compression gains. Lossy compression techniques attempt a more complete distinction between entropy and redundancy by relying on knowledge of the predictive powers of the human perceptual systems. This explains why these systems are referred to as implementing perceptual coding techniques. Unfortunately, not only are these systems inherently more complicated, they are also more likely to get things wrong and produce artifacts.

## 19.1  Lossless Compression

Consider the following contiguous stream of luminance bytes taken from a bit-map graphic:

```
00101011
00101011
00101011
00101011
00101011
00101011
00101100
00101100
00101100
00101100
00101100
```

There must be a more efficient way of coding this! "Six lots of 00101011 followed by five lots of 00101100" springs to mind. Like this:

```
00000110
00101011
00000101
00101100
```

This is the essence of a compression technique known as run-length encoding (RLE). RLE works really well but it has a problem. If a data file is composed of data that are predominantly nonrepetitive, RLE actually makes the file bigger! So RLE must be made adaptive so that it is only applied to strings of similar data (where redundancy is high); when the coder detects continuously changing data (where entropy is high), it simply reverts back to sending the bytes in an uncompressed form. Evidently it also has to insert a small information overhead to instruct the decoder when it is (and isn't) applying the compression algorithm.

Another lossless compression technique is known as Huffman coding and is suitable for use with signals in which sample values appear with a known statistical frequency. The analogy with Morse code is frequently drawn, in which letters that appear frequently are allocated simple patterns and letters that appear rarely are allocated more complex patterns. A similar technique, known by the splendid name of the Lempel-Ziv-Welch (LZW) algorithm, is based on the coding of repeated data chains or patterns. A bit like Huffman's coding, LZW sets up a table of common patterns and codes specific instances of patterns in terms of "pointers," which refer to much longer sequences in the table. The algorithm doesn't use a predefined set of patterns but instead builds up a table of patterns that it "sees" from incoming data. LZW is a very effective technique—even better than RLE. However, for the really high compression ratios, made necessary by the transmission and storage of high-quality audio down low bandwidth links, different approaches are required, based on an understanding of human perception processes.

In principle, the engineering problem presented by low-data rates, and therefore in reduced digital resolution, is no different to the age-old analogue problems of reduced dynamic range. In analogue systems, noise reduction systems (either complementary, i.e., involving encoding and complementary decoding such as Dolby B and dbx1, or single ended) have been used for many years to enhance the dynamic range of inherently noisy transmission systems such as analogue tape. All of these analogue systems rely on a method called "compansion," a word derived from the contraction of compression and expansion. The dynamic range is deliberately reduced (compressed) in the recording stage processing and recovered (expanded) in the playback electronics. In some systems this compansion acts over the whole frequency range (dbx is one such type). Others work over a selected frequency range (Dolby A, B, C, and SR). We shall see that the principle of compansion applies in just the same way to digital systems of data reduction. Furthermore, the distinction made between systems that act across the whole audio frequency spectrum and those that act selectively on ranges of frequencies (subbands) is true too of digital implementations. However, digital systems have carried the principle of subband working to a sophistication undreamed of in analogue implementations.

## 19.2 Intermediate Compression Systems

Consider the 8-bit digital values: 00001101, 00011010, 00110100, 0110100, and 11010000. (Eight bit examples are used because the process is easier to follow but the following principles apply in just the same way to digital audio samples of 16 bits or, indeed, any word length.) We might just as correctly write these values thus:

| |
|---|
| 00001101 = 1101 * 1 |
| 00011010 = 1101 * 10 |
| 00110100 = 1101 * 100 |
| 01101000 = 1101 * 1000 |
| 11010000 = 1101 * 10000 |

If you think of the multipliers 1, 10, 100, and so on as powers of two then it's pretty easy to appreciate that the representation given earlier is a logarithmic description (to the log of base two) with a 4-bit mantissa and a 3-bit exponent. So already we've saved 1 bit in 8 (a 20% data reduction). We've paid a price of course because we've sacrificed accuracy in the larger values by truncating the mantissas to 4 bits. However, this is possible in any case with audio because of the principle of masking, which underlies the operation of all noise reduction systems. Put at its simplest, masking is the reason we strain to listen to a conversation on a busy street and why we cannot hear the clock ticking when the television set is on: loud sounds mask quiet ones. So the logarithmic representation makes sense because resolution is maintained at low levels but sacrificed at high levels where the program signal will mask the resulting, relatively small, quantization errors.

### 19.2.1 NICAM

Further reductions may be made because real audio signals do not change instantaneously from very large to very small values, so the exponent value may be sent less often than the mantissas. This is the principle behind the stereo television technique of NICAM, which stands for near instantaneous companded audio multiplex. In NICAM 782, 14-bit samples are converted to 10-bit mantissas in blocks of 32 samples with a common 3-bit exponent. This is an excellent and straightforward technique but it is only possible to secure relatively small reductions in data throughput of around 30%.

## 19.3  Psychoacoustic Masking Systems

Wideband compansion systems view the phenomenon of masking very simply and rely on the fact that program material will mask system noise. But, actually masking is a more complex phenomenon. Essentially it operates in frequency bands and is related to the way in which the human ear performs a mechanical Fourier analysis of the incoming acoustic signal. It turns out that a loud sound only masks a quieter one when the louder sound is lower in frequency than the quieter, and only then, when both signals are relatively close in frequency. It is due to this effect that all wideband compansion systems can only achieve relatively small gains. The more data we want to discard the more subtle must our data reduction algorithm be in its appreciation of the human masking phenomena. These compression systems are termed psychoacoustic systems and, as you will see, some systems are very subtle indeed.

## 19.4  MPEG Layer 1 Compression (PASC)

It's not stretching the truth too much to say that the failed Philips' digital compact cassette (DCC) system was the first nonprofessional digital audio tape format. As we have seen, other digital audio developments had ridden on the back of video technology. The CD rose from the ashes of Philips Laserdisc, and DAT machines use the spinning-head tape recording technique originally developed for B and C-Format 1-inch video machines, later exploited in U-Matic and domestic videotape recorders. To their credit then, that, in developing the DCC, Philips chose not to follow so many other manufacturers down the route of modified video technology. Inside a DCC machine, there's no head wrap, no spinning head, and few moving precision parts. Until DCC, it had taken a medium suitable for recording the complex signal of a color television picture to store the sheer amount of information needed for a high-quality digital audio signal. Philips' remarkable technological breakthrough in squeezing two high-quality, stereo digital audio channels into a final data rate of 384 kBaud was accomplished by, quite simply, dispensing with the majority (75%) of the digital audio data! Philips named their technique of bit-rate reduction or data-rate compression precision adaptive sub-band coding (PASC). PASC was adopted as the original audio compression scheme for MPEG video/audio coding (layer 1).

In MPEG layer 1 or PASC audio coding, the whole audio band is divided up into 32 frequency subbands by means of a digital wave filter. At first sight, it might seem that this process will increase the amount of data to be handled tremendously—or by 32 times anyway! This, in fact, is not the case because the output of the filter bank, for any one frequency band, is at 1/32nd of the original sampling rate. If this sounds counterintuitive, take a look at the Fourier transform and note that a very similar process is performed here. Observe that when a periodic waveform is sampled $n$ times and transformed, the result is composed of $n$ frequency components. Imagine computing the transform over a 32- sample period: 32 separate calculations will yield 32 values. In other words, the data rate is the same in the frequency domain as it is in the time domain. Actually, considering that both describe exactly the same thing with exactly the same degree of accuracy, this shouldn't be surprising. Once split into subbands, sample values are expressed in terms of a mantissa and exponent exactly as explained earlier. Audio is then grouped into discrete time periods, and the maximum magnitude in each block is used to establish the masking "profile" at any one moment and thus predict the mantissa accuracy to which the samples in that subband can be reduced, without their quantization errors becoming perceivable (see Figure 19.1).

Despite the commercial failure of DCC, the techniques employed in PASC are indicative of techniques now widely used in the digital audio industry. All bit-rate reduction coders have the same basic architecture, pioneered in PASC: however, details differ. All

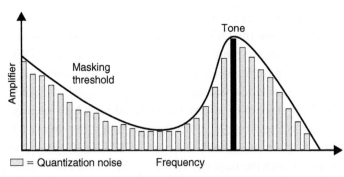

**Figure 19.1: Subband quantization and how it relates to the masking profile.**

systems accept PCM dual channel, digital audio (in the form of one or more AES pairs) is windowed over small time periods and transformed into the frequency domain by means of subband filters or via a transform filter bank. Masking effects are then computed based on a psychoacoustic model of the ear. Note that blocks of sample values are used in the calculation of masking. Because of the temporal, as well as frequency dependent, effects of masking, it's not necessary to compute masking on a sample-by-sample basis. However, the time period over which the transform is performed and the masking effects computed are often made variable so that quasi-steady-state signals are treated rather differently to transients. If coders do not include this modification, masking can be predicted incorrectly, resulting in a rush of quantization noise just prior to a transient sound. Subjectively this sounds like a type of pre-echo. Once the effects of masking are known, the bit allocation routine apportions the available bit rate so that quantization noise is acceptably low in each frequency region. Finally, ancillary data are sometimes added and the bit stream is formatted and encoded.

### 19.4.1 Intensity Stereo Coding

Because of the ear's insensitivity to phase response above about 2 kHz, further coding gains can be achieved by sending by coding the derived signals $(L + R)$ and $(L - R)$ rather than the original left and right channel signals. Once these signals have been transformed into the frequency domain, only spectral amplitude data are coded in the HF region; the phase component is simply ignored.

### 19.4.2 The Discrete Cosine Transform

The encoded data's similarity to a Fourier transform representation has already been noted. Indeed, in a process developed for a very similar application, Sony's compression scheme for MiniDisc actually uses a frequency domain representation utilizing a variation of the discrete fourier transform (DFT) method known as the discrete cosine transform (DCT). The DCT takes advantage of a distinguishing feature of the cosine function, which is illustrated in Figure 19.2, that the cosine curve is symmetrical about the time origin. In fact, it's true to say that any waveform that is symmetrical about an arbitrary "origin" is made up of solely cosine functions. This is difficult to believe, but consider adding other cosine functions to the curve illustrated in Figure 19.2. It doesn't matter what size or what period waves you add, the curve will always be symmetrical about the origin. Now, it would obviously be a great help, when we come to perform a Fourier transform, if we

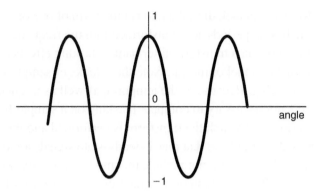

**Figure 19.2: Cosine function.**

knew the function to be transformed was only made up of cosines because that would cut down the maths by half. This is exactly what is done in the DCT. A sequence of samples from the incoming waveform is stored and reflected about an origin. Then one-half of the Fourier transform performed. When the waveform is inverse transformed, the front half of the waveform is simple ignored, revealing the original structure.

## 19.5  MPEG Layer 2 Audio Coding (MUSICAM)

The MPEG layer 2 algorithm is the preferred algorithm for European DTV and includes a number of simple enhancements of layer 1 (or PASC). Layer 2 was originally adopted as the transmission coding standard for the European digital radio project (Digital Audio Broadcasting or DAB) where it was termed MUSICAM. The full range of bit rates for each layer is supported, as are all three sampling frequencies, 32, 44.1, and 48 kHz. Note that MPEG decoders are always backward compatible, that is, a layer 2 decoder can decode layer 1 or layer 2 bit streams; however, a layer 2 decoder cannot decode a layer 3-encoded stream.

MPEG layer 2 coding improves compression performance by coding data in larger groups. The layer 2 encoder forms frames of 3 by 12 by 32 = 1152 samples per audio channel. Whereas layer 1 codes data in single groups of 12 samples for each subband, layer 2 codes data in three groups of 12 samples for each subband. The encoder encodes with a unique scale factor for each group of 12 samples only if necessary to avoid audible distortion. The encoder shares scale factor values between two or all three groups when the values of the scale factors are sufficiently close or when the encoder anticipates that temporal noise masking will hide the consequent distortion. The layer 2 algorithm also improves performance over layer 1 by representing the bit allocation, the scale factor values, and the quantized samples with a more efficient code. Layer 2 coding also added 5.1 multichannel capability. This was done in a scaleable way so as to be compatible with layer 1 audio.

MPEG layers 1 and 2 contain a number of engineering compromises. The most severe concerns the 32 constant-width subbands which do not reflect accurately the equivalent filters in the human hearing system (the critical bands). Specifically, the bandwidth is too wide for the lower frequencies so the number of quantizer bits cannot be specifically tuned for the noise sensitivity within each critical band. Furthermore, the filters have insufficient Q so that signal at a single frequency can affect two adjacent filter bank outputs. Another limitation concerns the time frequency–time domain transformations achieved with the wave filter. These are not transparent so, even without quantization, the inverse transformation would not perfectly recover the original input signal.

## 19.6 MPEG Layer 3

The layer 3 algorithm is a much more refined approach. Layer 3 is finding its application on the Internet where the ability to compress audio files by a large factor is important in download times. In layer 3, time to frequency mapping is performed by a hybrid filter bank composed of the 512-tap polyphase quadrature mirror filter (used in layers 1 and 2) followed by an 18-point modified cosine transform filter. This produces a signal in 576 bands (or 192 bands during a transient). Masking is computed using a 1024-point FFT: once again more refined than the 512-point FFT used in layers 1 and 2. This extra complexity accounts for the increased coding gains achieved with layer 3, but increases the time delay of the coding process considerably. Of course, this is of no account at all when the result is an encoded .mp3 file.

### 19.6.1 Dolby AC-3

The analogy between data compression systems and noise reduction has already been drawn. It should therefore come as no surprise that one of the leading players in audio data compression should be Dolby, with that company's unrivaled track record in noise reduction systems for analogue magnetic tape. Dolby AC-3 is the adopted coding standard for terrestrial digital television in the United States; however, it was actually implemented for the cinema first, where it was called Dolby Digital. It was developed to provide multichannel digital sound with 35-mm prints. In order to retain an analogue track so that these prints could play in any cinema, it was decided to place the new digital optical track between the sprocket holes, as illustrated in Figure 19.3. The physical space limitation (rather than crude bandwidth) was thereby a key factor in defining its maximum practical bit rate. Dolby Labs did a great deal of work to find a channel format that would best satisfy the requirements of theatrical film presentation. They discovered that five discrete channels—left (L), right (R), center (C), left surround (LS), and right surround (RS)—set the right balance between realism and profligacy! To this they added a limited (1/10th) bandwidth subwoofer channel; the resulting system being termed 5.1 channels. Dolby Digital provided Dolby Labs a unique springboard for consumer formats for the new DTV (ATSC) systems.

Like MPEG, AC-3 divides the audio spectrum of each channel into narrow frequency bands of different sizes optimized with respect to the frequency selectivity of human hearing. This makes it possible to sharply filter coding noise so that it is forced to stay very close in frequency to the frequency components of the audio signal being coded. By reducing or eliminating coding noise wherever there are no audio signals to mask it, the sound quality of the original signal can be preserved subjectively. In this key respect, a perceptual coding system such as AC-3 is essentially a form of very selective and powerful Dolby-A type noise reduction! Typical final data-rate applications include 384 kb/s for 5.1-channel Dolby Surround Digital consumer formats and 192 kb/s for two-channel audio distribution.

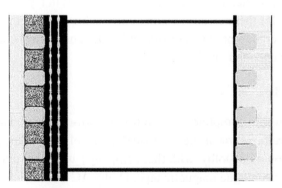

**Figure 19.3: Dolby Digital as originally coded on film stock.**

### 19.6.2 Dolby E

Dolby E is a digital audio compression technology designed for use by TV broadcast and production professionals which allows an AES/EBU audio pair to carry up to eight channels of digital audio. Because the coded audio frame is arranged to be synchronous with the video frame, encoded audio can be edited without mutes or clicks. Dolby E can be recorded on a studio-level digital VTR and switched or edited just like any other AES digital audio signal, as long as some basic precautions are observed. Data must not be altered by any part of the system it passes through. That's to say, the gain must not be changed, data must not be truncated or dithered, and neither must the sample rate be converted. Dolby E technology is designed to work with most popular international video standards. In its first implementation, Dolby E supported 29.97 fps, 20-bit word size, and 48-kHz audio. Newer versions will support 25 fps, 24 fps, and 16-bit and 24-bit audio.

### 19.6.3 DTS

DTS, the rival to Dolby Digital in the cinema, uses an entirely different approach to AC-3-coded data on film stock. In DTS, the digital sound (up to 10 channels and a subwoofer channel) are recorded on CDs, which are synchronized to the film by means of a time code. Because of the higher data rate available (CD against optical film track), DTS uses a relatively low 4:1 compression ratio.

### 19.6.4 MPEG AAC

MPEG-2 advanced audio coding (AAC) was finalized as a standard in 1997 (ISO/IEC 13818–7). AAC constitutes the coding algorithms of the new MPEG-4 standard.

## 19.7 MPEG-4

MPEG-4 will define a method of describing objects (both visual and audible) and how they are "composited" and interact together to form "scenes." The scene description part of the MPEG-4 standard describes a format for transmitting the spatiotemporal positioning information that describes how individual audiovisual objects are composed within a scene. A "real world" audio object is defined as an audible semantic entity recorded with one microphone—in case of a mono recording—or with more microphones, at different positions, in case of a multichannel recording. Audio objects can be grouped or mixed together, but objects cannot be split easily into subobjects.

Applications for MPEG-4 audio might include "mix minus 1" applications in which an orchestra is recorded minus the concerto instrument, allowing a musician to play along with his or her instrument at home, or where all effects and music tracks in a feature film are "mix minus the dialogue," allowing very flexible multilingual applications because each language is a separate audio object and can be selected as required in the decoder.

In principle, none of these applications is anything but straightforward; they could be handled by existing digital (or analogue) systems. The problem, once again, is bandwidth. MPEG-4 is designed for very low bit rates and this should suggest that MPEG have designed (or integrated) a number of very powerful audio tools to reduce necessary data throughput. These tools include the MPEG-4 Structured Audio format, which uses low bit-rate algorithmic sound models to code sounds. Furthermore, MPEG-4 includes the functionality to use and control postproduction panning and reverberation effects at the decoder, as well as the use of a SAOL signal-processing language enabling music synthesis and sound effects to be generated, once again, at the terminal rather than prior to transmission.

### 19.7.1 Structured Audio

We have already seen how MPEG (and Dolby) coding aims to remove perceptual redundancy from an audio signal, as well as removing other simpler representational redundancy by means of efficient bit-coding schemes. Structured audio (SA) compression schemes compress sound by, first, exploiting another type of redundancy in signals—structural redundancy.

Structural redundancy is a natural result of the way sound is created in human situations. The same sounds, or sounds which are very similar, occur over and over again. For example, a performance of a work for solo piano consists of many piano notes. Each time the performer strikes the "middle C" key on the piano, a very similar sound is created by the piano's mechanism. To a first approximation, we could view the sound as exactly the same upon each strike; to a closer one, we could view it as the same except for the velocity with which the key is struck and so on. In a PCM representation of the piano performance, each note is treated as a completely independent entity; each time the "middle C" is struck, the sound of that note is independently represented in the data sequence. This is even true in a perceptual coding of the sound. The representation has been compressed, but the structural redundancy present in rerepresenting the same note as different events has not been removed.

In structured coding, we assume that each occurrence of a particular note is the same, except for a difference described by an algorithm with a few parameters. In the model-transmission stage we transmit the basic sound (either a sound sample or another algorithm) and the algorithm which describes the differences. Then, for sound transmission, we need only code the note desired, the time of occurrence, and the parameters controlling the differentiating algorithm.

## 19.7.2 SAOL

SAOL (pronounced "sail") stands for "Structured Audio Orchestra Language" and falls into the music-synthesis category of "Music V" languages. Its fundamental processing model is based on the interaction of oscillators running at various rates. Note that this approach is different from the idea (used in the multimedia world) of using MIDI information to drive synthesis chips on sound cards. This latter approach has the disadvantage that, depending on IC technology, music will sound different depending on which sound card is realized. Using SAOL (a much "lower-level" language than MIDI) realizations will always sound the same.

At the beginning of an MPEG-4 session involving SA, the server transmits to the client a stream information header, which contains a number of data elements. The most important of these is the orchestra chunk, which contains a tokenized representation of a program written in SAOL. The orchestra chunk consists of the description of a number of instruments. Each instrument is a single parametric signal-processing element that maps a set of parametric controls to a sound. For example, a SAOL instrument might describe a physical model of a plucked string. The model is transmitted through code, which implements it, using the repertoire of delay lines, digital filters, fractional-delay interpolators, and so forth that are the basic building blocks of SAOL.

The bit stream data itself, which follows the header, is made up mainly of time-stamped parametric events. Each event refers to an instrument described in the orchestra chunk in the header and provides the parameters required for that instrument. Other sorts of data may also be conveyed in the bit stream; tempo and pitch changes, for example.

Unfortunately, at the time of writing (and probably for some time beyond!) the techniques required for automatically producing a structured audio bit stream from an arbitrary, prerecorded sound are beyond today's state of the art, although they are an active research topic. These techniques are often called "automatic source separation" or "automatic transcription." In the meantime, composers and sound designers will use special content creation tools to directly create SA bit streams. This is not considered to be a fundamental obstacle to the use of MPEG-4 structured audio because these tools are very similar to the ones that contemporary composers and editors use already; all that is required is to make their tools capable of producing MPEG-4 output bit streams. There is an interesting corollary here with MPEG-4 for video. For, while we are not yet capable of integrating and coding real-world images and sounds, there are immediate applications for directly synthesized programs. MPEG-4 audio also foresees the use of text-to-speech conversion systems.

### 19.7.3 Audio Scenes

Just as video scenes are made from visual objects, audio scenes may be usefully described as the spatiotemporal combination of audio objects. An "audio object" is a single audio stream coded using one of the MPEG-4 coding tools, such as structured audio. Audio objects are related to each other by mixing, effects processing, switching, and delaying them, and may be panned to a particular three-dimensional location. The effects processing is described abstractly in terms of a signal-processing language—the same language used for SA.

## 19.8 Digital Audio Production

We've already looked at the technical advantages of digital signal processing and recording over its older analogue counterpart. We now come to consider the operational impact of this technology, where it has brought with it a raft of new tools and some new problems.

# CHAPTER 20

# Digital Audio Production

Richard Brice

## 20.1 Digital Audio Workstations (DAWs)

When applied to a digital audio application, a computer hardware platform is termed a digital audio workstation (DAW). The two ubiquitous standards in the audio arena are the Apple Macintosh computer family (or "Macs"), which use Motorola processors, and the IBM PC and compatibles (PCs), which use Intel-based processors. The Macintosh was always "audio ready" because it was designed with an audio capacity beyond the PC's dumb "beep." Other small personal computers (especially Atari ST) became very popular in music applications. The Atari ST computer (Figure 20.1) was another 68000 based computer (like the Apple Mac). Including a powerful ROM- based operating system and a desktop metaphor very like that now commonplace in Windows, the Atari was pretty much ahead of its time in the early 1980s. However, the Atari ST owes its tremendous success, and continuing long-overdue existence in many recording studios, to the decision

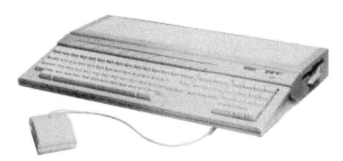

Figure 20.1: Atari ST computer.

of the designers to include MIDI IN and OUT sockets on the side of the machine. This made the Atari the only ready-to-go, "plug and play" MIDI sequencer platform, a fact reflected in the number of software products designed for it.

PowerPCs and PowerMacs are machines built around a reduced instruction set computing (RISC) processor developed jointly by IBM, Apple, and Motorola. They are designed to run operating system software that supports both PC and Mac applications and are designed to be especially good at handling large data files typical of media components. RISC technology is especially noted in workstation computers. Three computers designed and manufactured by the American high-end computer company Silicon Graphics Inc. (SGI) make extensive use of RISC technology. SGI's subsidiary, MIPS Technologies, Inc., designs the RISC processor technology inside SGI machines. MIPS' new R5000 MIPS RISC processor delivers a peak of 480 million floating point operations per second (MFLOPS)—up to twice as fast as Intel's 200-MHz Pentium Pro and over seven times as fast as a 133-MHz Pentium! Workstations from SGI are also finding their way into high-end audio production.

SGI is the leading manufacturer of high-performance visual computing systems. The company delivers interactive three-dimensional graphics, digital media, and multiprocessing super-computing technologies to technical, scientific, and creative professionals. SGI manufactures some of the best tools for multimedia creation, as well as white-heat video, audio, and graphics stand-alone packages. They also provide specialist tools for HTML, hypermedia page creation, and serving for the creation of multimedia creations on the Internet/World Wide Web (WWW). SGI has its headquarters in Mountain View, California. SGI's products include the Indy, which is a "value" RISC workstation utilizing a 64-bit system architecture and MIPS processors. On the audio side it has digital audio I/O as well as analogue ports. The Indigo 2 is aimed as a cost-effective desktop alternative to older style dedicated video production hardware. The Onyx is a super-computer with a graphics bias! SGI also manufactures the CHALLENGE Media Server for the broadcast television environment. Table 20.1 is the audio subsystem specification for Onyx and CHALLENGE and represents the typical digital audio performance from desktop audio. The option also provides for microphone input and headphone output but these figures are not quoted here.

Note that the input ports for audio (both analogue and digital) conform to consumer levels and practice even though SGI themselves refer to the digital inputs as AES/EBU.

**Table 20.1 Audio subsystem specification for onyx and CHALLENGE**

| | |
|---|---|
| Number of channels | 4 analogue (16 bit), 2 digital (24 bit) |
| Input analogue route | |
| Input Z | 20 k ohms |
| Input amplitude | From 0.63 V pp to 8.4 V pp for full-scale modulation (this level is adjustable under software control) |
| Frequency response | ±0.81 dB 20 Hz to 20 kHz |
| THD + noise | Less than 0.007% 20 Hz to 20 kHz |
| Residual noise | –86 dB unweighted |
| Cross talk | –82 dB at 1 kHz, –67 dB at 20 kHz |
| ADC type | 16-bit Delta-Sigma |
| Output analogue route | |
| Output Z | 600 ohms |
| Output level | 4.7 V pp (4.4 dBV) for 0 dBFS |
| Sampling rates | 32, 44.1, 48 kHz, or divisors selectable |
| Frequency response | ±1.2 dB 20 Hz to 20 kHz |
| THD + noise | Less than 0.02% 20 Hz to 20 kHz |
| Residual noise | –81 dB unweighted |
| Cross talk | –80 dB at 1 kHz, –71 dB at 20 kHz |
| Digital serial I/O | |
| Type | Coaxial only |
| Input Z | 75 ohms, transformer coupled |
| Input level | 0.5 V pp into 75 ohms |
| Sample rates | 30 kHz to 50 kHz |
| Output Z | 75 ohms, transformer coupled |
| Output level | 0.5 V when terminated in 75 ohms |
| Coding | AES-3, IEC 958 |

## 20.1.1  Hard-Disk Editing

Not long ago, most recordings were mastered on two-track analogue tape. Whether the performance was by a soloist, small classical or rock ensemble, or full orchestra, the good "takes" of a recording session were separated from the bad and joined together using razor-blade editing. Using this technique, the tape was physically cut with a razor blade and joined using a special sticky tape. With the high tape speeds employed in professional recording, accurate editing was possible using this technique and many fine recordings were assembled this way. Any engineer who has been involved with razor-blade editing will know that it is a satisfying skill to acquire but it is tricky and it is always fraught with difficulties. The reason being that a mistimed or misjudged edit is difficult to put right once the first "incision" has been made! So much so that a dub or copy of the original master tapes was sometimes made for editing lest the original master should be irreparably damaged by a poor edit. Unfortunately, because analogue recording is never perfect, this meant that editing inevitably meant one tape-generation of quality loss before the master tape had left the studio for production. The advent of digital audio has brought about a new vista of possibility in sound editing. Apart from the obvious advantages of digital storage, that once the audio signal is in robust digital form it can be copied an infinite number of times, thus providing an identical tape "clone" for editing purposes, the arrival of the ability to process digital audio on desktop PCs has revolutionized the way we think about audio editing, providing a flexibility undreamed of in the days of analogue mastering machines.

Editing digital audio on a desktop microcomputer has two major advantages.

(1) An edit may be made with sample accuracy, by which is meant, a cut may be made with a precision of 1/40,000th of a second!

(2) An edit may be made nondestructively, meaning that when the computer is instructed to join two separate takes together, it doesn't create a new file with a join at the specified point, but instead records two pointers that instruct on subsequent playback to vector or jump to another data location and play from the new file at that point.

In other words, it "lists" the edits in a new file of stored vector instructions. Indeed this file is known as an edit decision list. (Remember that the hard-disk doesn't have to jump instantaneously to another location because the computer holds a few seconds of audio data in a RAM cache memory.) This opens the possibility of almost limitless editing in order to assemble a "perfect" performance. Edits may be rehearsed and auditioned many times without ever "molesting" the original sound files.

## 20.1.2 Low-Cost Audio Editing

Most audio capture cards like the Creative Labs Soundblaster come bundled with a primitive sound-file editing software. Usually this permits a "butt-join" edit (the exact analogy of a razor blade splice) between different sound files or between data loaded onto the clipboard in rather the same way as a word processor works on text files. In fact, Creative Labs' Wave Studio utility is quite powerful and affords some manipulations (such as the ability to operate on left and right channels of a stereo signal separately) that exceed the capabilities of some low-end professional editing software. However, the big disadvantage with Wave Studio is that it does not allow for nondestructive editing. An inexpensive and truly excellent package is authored by Minnetonka Software Inc. and is known as FastEddie. This "value" package permits nondestructive editing to sample accuracy, preaudition of edit points, the ability to mix files, time "stretch" or compress the WAV file, and normalize gain so that a WAV file may be increased in gain just enough so that the largest excursion on the file is brought almost to within clipping level, thus maximizing dynamic range. The utility can also be used to generate audio effects such as echo and reverb and reversal of the sound file so that it plays backward. In order to facilitate the judgment of edit points, most editing software provides an on-screen waveform display. This may be zoomed at will: IN to examine individual samples and OUT to reveal musical sections or the whole file.

An example is shown in Figure 20.2, which is an off-screen capture of a FastEddie window. The highlighted sections in the lower, read-only part of the window are ready for editing in the top editing part of the window. Given its price, FastEddie has a number of very professional features, among these is the ability to cross-fade at edit points rather than to produce a butt join (Figure 20.3). The advantage of this feature is due to the complex nature of musical sounds. Even if an edit is made at a precise point in a musical score, a discontinuity is produced in the audio waveform. This discontinuity usually manifests itself as an audible click or "bump." The use of cross-fading ameliorates these effects. FastEddie also provides facilities for producing fade-outs and fade-ins on complete sound-file data.

## 20.1.3 Professional Audio Editing

For professional music editing, most sophisticated editing systems still exist, ranging from the reasonable to the very expensive. At the higher end, the platforms

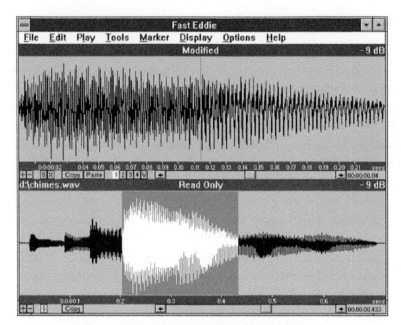

Figure 20.2: FastEddie digital audio editor.

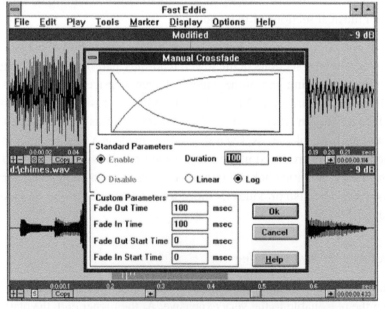

Figure 20.3: Audio editing with cross-fades for edit points.

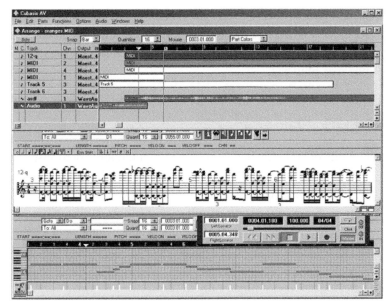

**Figure 20.4: Multitrack audio recording combined with MIDI sequencing.**

are predominantly Mac based. The crème-de-la-crème in this field being the Sonic Solutions—the choice of music editing system for most classical music producers. The high-end systems mostly show up with their own proprietary hardware and it is here, as much as in software, that these systems score over desktop PC systems using a 16-bit sound card. The professional units typically offer better quality A to D and D to A conversion and more transparent signal processing.

### 20.1.4 Multitrack Hard-Disk Recording

Hard-disk capacity and computer speeds are not so high that multitrack recording on a DAW is commonplace. A typical home-studio application (Cubase AV) is illustrated in Figure 20.4, which shows the combination of multitrack audio and MIDI data, all within one screen environment.

### 20.1.5 Plug-ins

As more and more audio production is gradually taking place on computer platforms, or on the desktop, hardware manufacturers are now producing software programs that imitate their outboard hardware in a digital signal processor (DSP) program. This software interfaces with mixing and editing programs (via mutually agreed data-exchange standards) so as to provide the metaphor of "outboard" facilities on the desktop. These programs are known as plug-ins.

## 20.2 Audio Data Files

Digital representations of sound, when stored on computer, are stored just like any other kind of data, as files. A number of different file formats exist in common usage. Most sound files begin with a header consisting of information describing the format of that file. Characteristics such as word length, number of channels, and sampling frequency are specified so that audio applications can read the file properly. One very common type of file format is the WAV (or Wave) format. This is a good example because it demonstrates all the typical features of a typical audio file.

### 20.2.1 WAV Files

WAV files are a version of the generic RIFF file format. This was codeveloped by Microsoft and IBM. RIFF represents information in predefined blocks, preceded by a header that identifies exactly what the data are. This format is very similar to the audio interchange file format (AIFF) format developed by Apple (see later) in that it supports monaural and multichannel samples and a variety of sample rates. Like AIFF, WAVE files are big and require approximately 10 Mbytes per minute of 16-bit stereo samples with a sampling rate of 44.1 kHz.Here is a hexadecimal representation of the first 128 bytes of a WAV file.

```
26B7:0100 52 49 46 46 28 3E 00 00-57 41 56 45 66 6D 74 20 RIFF(>..WAVEfmt
26B7:0110 10 00 00 00 01 00 01 00-22 56 00 00 22 56 00 00 ....."V.. "V..
26B7:0120 01 00 08 00 64 61 74 61-04 3E 00 00 80 80 80 80 ...data.> .........
26B7:0130 80 80 80 80 80 80 80 80-80 80 80 80 80 80 80 80 ............
26B7:0140 80 80 80 80 80 80 80 80-80 80 80 80 80 80 80 80 ............
26B7:0150 80 80 80 80 80 80 80 80-80 80 80 80 80 80 80 80 ............
26B7:0160 80 80 80 80 80 80 80 80-80 80 80 80 80 80 80 80 ............
26B7:0170 80 80 80 80 80 80 80 80-80 80 80 80 80 80 80 80 ............
```

The header provides Windows with all the information it needs. First off, it defines the type of RIFF file, in this case, WAVEfmt. Notice the bytes, which are shown underlined. The first two, 22 and 56, relate to the audio sampling frequency. Their order needs reversing to read; 5622 hexadecimal, which is equivalent to 22 050 in decimal—in other words, 22-kHz sampling. The next two inform the Media Player that the sound file is 1 byte per sample (mono) and 8 bits per sample.

### 20.2.2 AU Files

AU (or μ-law—pronounced mu-law) files utilize an international standard for compressing audio data. It has a compression ratio of 2:1. The compression technique is optimized for speech (in the United States it is a standard compression technique for telephone systems; in Europe, a-law is used). This file format is found most frequently on the Internet where it is used for '.au' file formats, alternately know as 'Sun audio' or 'NeXT' format. Even though it's not the highest quality audio file format available, its nonlinear, logarithmic coding scheme results in a relatively small file size, which is ideal for applications where download time is a problem.

### 20.2.3 AIFF and AIFC

The Audio Interchange File Format (AIFF) allows for the storage of monaural and multichannel sample sounds at a variety of sample rates. AIFF format is frequently found in high-end audio recording applications. Originally developed by Apple, this format is used predominantly by SGI and Macintosh applications. Like WAV, AIFF files can be quite large; 1 min of 16-bit stereo audio sampled at 44.1 kHz usually takes up about 10 megabytes. To allow for compressed audio data, Apple introduced the new AIFF-C, or AIFC, format, which allows for the storage of compressed and uncompressed audio data. AIFC supports compression ratios as high as 6:1. Most of the applications that support AIFF playback also support AIFC.

### 20.2.4 MPEG

The International Standard Organisation's Moving Picture Expert Group (MPEG) is responsible for one of the most popular compression standards in use on the Internet today. Designed for both audio and video file compression, MPEG-I audio compression specifies three layers, and each layer specifies its own format. The more complex layers take longer to encode but produce higher compression ratios while keeping much of an audio file's original fidelity. Layer 1 takes the least amount of time to compress, but layer 3 yields higher compression ratios for comparable quality files.

### 20.2.5 VOC

Creative Voice (.voc) is the proprietary sound file format that is recorded with Creative Lab's Sound Blaster and Sound Blaster Pro audio cards. This format supports only 8-bit mono audio files up to sampling rates of 44.1 kHz and stereo files up to 22 kHz.

### 20.2.6 Raw PCM Data

Raw Pulse Code Modulated (PCM) data are sometimes identified with the .pcm, but it sometimes has no extension at all. Since no header information is provided in the file, you must specify the waveform's sample rate, resolution, and number of channels to the application to which it is loaded.

## 20.3 Sound Cards

Available in a bewildering array of different guises, for serious audio work, 16-bit cards only are suitable, and even then beware of very poor noise levels. Computers differ widely in their suitability as devices for high-quality audio. The Creative Technology Ltd. Sound Blaster card family has some of the most widespread sound cards used in the PC world. Supplied standard with a four operator FM sound generator chip for sound synthesis, Creative Labs offer a wavetable-based upgrade. Sound Blaster ships with sound file editing software. The card comes with a utility program for controlling the analogue mixer on the card where the various sound sources are combined and routed. This utility is called Creative Mixer and it's illustrated in Figure 20.5. Note that fader style controls are implemented in software so as to provide a familiar user interface. Control over CD, MIDI synthesizer, and WAV file replay, as well as line and microphone inputs, are provided. Global (all sources) equalization is also provided.

## 20.4 PCI Bus Versus ISA Bus

Most PCs, until the arrival of the Pentium, were provided with a PC/AT bus (or ISA bus) for connecting peripherals (such as sound cards, frame grabbers, and so on). The ISA bus operates with 16-bit data bus and a 16-bit address bus and operates with a divided clock. The ISA bus limits real-world transfer rates to around 1–2 Mbytes/s, which is just enough

**Figure 20.5: Creative Labs' Creative Mixer utility.**

for high-quality, dual-channel audio. The Peripheral Component Interconnect (PCI) bus is incorporated in newer Pentium-based IBM PCs. PCI is a local bus, so named because it is a bus which is much "closer" to the CPU. Local buses run at a much higher rate and PCI offers considerable performance advantages over the traditional ISA bus, allowing data to be transferred at between 5 and 70 Mbytes/s; allowing the possibility of real-time, multitrack audio applications. The PCI bus is a processor-independent bus specification that allows peripheral boards to access system memory directly (under the aegis of a local bus controller) without directly using the CPU, employing a 32-bit data bus and a 64-bit address bus at full clock speed. Installation and configuration of PCI bus plug-in cards are much simpler than the equivalent installation on the ISA bus. Commonly referred to as the "plug-and-play" feature of the PCI bus, this user transparency is achieved by having the PCs BIOS configure the plug-in card's base address and interrupt level at power-up. Because all cards are automatically configured, conflicts between them are eliminated. It is a process that can only be done manually with cards on the ISA bus. The PCI Bus is not limited to PCs; it is the primary peripheral bus in the PowerPC and PowerMacs from Apple. Incorporation of the PCI bus is planned for other RISC-based processor platforms.

## 20.5 Disks and Other Peripheral Hardware

Read/write compact disk (CD-R) drives are now available at a price within the reach of the small recording studio, and recordable media are less than $2. CD-R drives are usually SCSI based so PCs usually have to have an extra expansion card fitted to provide this interface (see later). Recordable CDs rely on a laser-based magneto-optical system to "burn" data into the recorded medium. Once written, data cannot be erased. Software exists (and usually comes bundled with the drive) which enables the drive to be used as a data medium or an audio carrier (or sometimes as both). There exist a number of different variations of the standard ISO-9600 CD-ROM. The two most important are the (HFS/ISO) Hybrid disk, which provides support for CD-ROM on Mac and PC using separate partitions, and the Mixed mode disk, which allows one track of either HFS (Mac) or ISO-9600 information and subsequent tracks of audio.

A number of alternative removable media are available and suitable for audio use; some are based on magnetic storage (like a floppy disk or a Winchester hard-drive) and some on magneto-optical techniques—nearer to CD technology: Bernoulli cartridges are based on floppy disk, magnetic storage technology. Disks up to 150 MByte are available. Access times are fast; around 20 ms. SyQuest are similar. Modern SyQuest cartridges and drives are now available in up to 1.3 GByte capacity and 11-ms access times, making SyQuest the nearest thing to a portable hard drive. Magneto-optical drives use similar technology to CD, they are written and read using a laser (Sony is a major manufacturer of optical drives). Sizes up to 1.3 GBytes are available with access times between 20 and 30 ms.

## 20.6 Hard Drive Interface Standards

There are several interface standards for passing data between a hard disk and a computer. The most common are the SCSI or Small Computer System Interface, the standard interface for Apple Macs; the IDE or Integrated Drive Interface, which is not as fast as SCSI; and the Enhanced IDE interface, which is a new version of the IDE interface that supports data transfer rates comparable to SCSI.

### 20.6.1 IDE Drives

The Integrated Drive Electronics interface was designed for mass storage devices, in which the controller is integrated into the disk or CD-ROM drive. It is thereby a lower cost alternative to SCSI interfaces in which the interface handling is separate from the drive electronics. The original IDE interface supports data transfer rates of about 3.3 Mbytes per second and has a limit of 538 Mbytes per device. However, a recent version of IDE, called enhanced IDE (EIDE) or Fast IDE, supports data transfer rates of about

12 Mbytes per second and storage devices of up to 8.4 Gbytes. These numbers are comparable to what SCSI offers. However, because the interface handling is handled by the disk drive, IDE is a very simple interface and does not exist as an interequipment standard, that is, you cannot connect an external drive using IDE. Due to demands for easily upgradable storage capacity, and for connection with external devices such as recordable CD players, SCSI has become the preferred bus standard in audio applications.

### 20.6.2 SCSI

An abbreviation of Small Computer System Interface and pronounced "scuzzy," SCSI is a parallel interface standard used by Apple Macintosh computers (and some PCs) for attaching peripheral devices to computers. All Apple Macintosh computers starting with the Macintosh Plus come with a SCSI port for attaching devices such as disk drives and printers. SCSI interfaces provide for fast data transmission rates,; up to 40 Mbytes per second. In addition, SCSI is a multidrop interface, which means that you can attach many devices to a single SCSI port.

Although SCSI is an ANSI standard, unfortunately, due to ever higher demands on throughput, SCSI comes in a variety of "flavors!" Each is used in various studio and mastering applications and, as a musician engineer, you will need to be aware of the differences. The following varieties of SCSI are currently implemented:

SCSI-1: Uses an 8-bit bus and supports data rates of 4 Mbytes/s.

SCSI-2: Same as SCSI-1, but uses a 50-pin connector instead of a 25-pin connector. This is what most people mean when they refer to plain SCSI.

Fast SCSI: Uses an 8-bit bus and supports data rates of 10 Mbytes/s.

Ultra SCSI: Uses an 8-bit bus and supports data rates of 20 Mbytes/s.

Fast Wide SCSI: Uses a 16-bit bus and supports data rates of 20 Mbytes/s.

Ultra Wide SCSI: Uses a 16-bit bus and supports data rates of 40 Mbytes/s; this is also called SCSI-3.

### 20.6.3 Fiber Channel

Fiber channel is a data transfer architecture developed by a consortium of computer and mass storage device manufacturers. The most prominent Fibre Channel standard is Fibre

Channel Arbitrated Loop (FC-AL), which was designed for new mass storage devices and other peripheral devices that require very high bandwidth. Using an optical fiber to connect devices, FC-AL supports full-duplex data transfer rates of 100 Mbit/s. This is far too high a transfer rate to be relevant as an audio-only standard. However, in multichannel applications and in multimedia applications (with video, for example) Fibre Channel may well find its way into the modern studio, so much so that FC-AL is expected to eventually replace SCSI for high-performance storage systems.

### 20.6.4  Firewire (IEEE 1394) Interface

The "Firewire" (IEEE 1394 interface) is an international standard, low-cost digital interface intended to integrate entertainment, communication, and computing electronics into consumer multimedia. Originated by Apple Computer as a desktop LAN, Firewire has been developed by the IEEE 1394 working group. Firewire supports 63 devices on a single bus (SCSI supports 7, SCSI Wide supports 15) and allows buses to be bridged (joined together) to give a theoretical maximum of thousands of devices. It uses a thin, easy to handle cable that can stretch further between devices than SCSI, which only supports a maximum "chain" length of 7 meters (20 feet). Firewire supports 64-bit addressing with automatic address selection and has been designed from the ground up as a "plug and play" interface. Firewire can handle 10 Mbytes per second of continuous data with improvements in the design promising continuous throughput of 20–40 Mbytes per second in the very near future and a long-term potential of over 100 Mbytes/s. Much like LANs and WANs, IEEE 1394 is defined by the high-level application interfaces that use it, not a single physical implementation. Therefore, as new silicon technologies allow high higher speeds and longer distances, IEEE 1394 will scale to enable new applications.

## 20.7  Digital Noise Generation—Chain Code Generators

The binary sequence generated by a chain code generator appears to have no logical pattern; it is, to all intents and purposes, a random sequence of binary numbers. The code is generated by a shift register that is clocked at a predetermined frequency and whose input is derived from a network that develops a function of the outputs from the register.

A basic form of chain code generator is illustrated in Figure 20.6, which consists of a 4-bit shift register whose input is derived from the output of an exclusive-OR gate, itself fed from the penultimate and last output of the shift register. The output from the chain

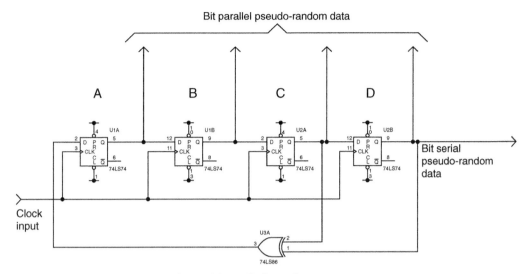

**Figure 20.6: Chain-code generator.**

code generator may be taken in serial form (from any of the data-latch outputs) or in parallel form (from all the data latch outputs). In operation, imagine that the output of stage B starts with a 1 as power is applied, but that all the other outputs start with a 0. Note that the output of the XOR gate will only equal 1 when its inputs are of a different state (i.e., nonidentical). We can now predict the ensuing pattern that results as the shift register is clocked:

| State | Output (A,B,C,D) |
|---|---|
| 0 (initial) | 0,1,0,0 |
| 1 | 0,0,1,0 |
| 2 | 1,0,0,1 |
| 3 | 1,1,0,0 |
| 4 | 0,1,1,0 |
| 5 | 1,0,1,1 |
| 6 | 0,1,0,1 |
| 7 | 1,01,0 |

(Continued)

| State | Output (A,B,C,D) |
|-------|------------------|
| 8 | 1,1,0,1 |
| 9 | 1,1,1,0 |
| 10 | 1,1,1,1 |
| 11 | 0,1,1,1 |
| 12 | 0,0,1,1 |
| 13 | 0,0,0,1 |
| 14 | 1,0,0,0 |
| 15 | 0,1,0,0 |
| 16 | 0,0,1,0 |
| 17 | 1,0,0,1 |
| 18 | 1,1,0,0 |
| etc. | |

Note that, at state 15, the pattern starts to repeat. This sequence is known as the maximum-length sequence. The fact that the outputs states are predictable illustrates that the output of the code generator is not really random at all but is a pseudo-random binary sequence (PRBS). The sequence does, however, have some very "random" qualities—like a very nearly equal number of 1 s and 0 s (think of it a coin-tossing machine)! Practically, this lack of true randomness does not matter provided that the sequence is long enough to appear random in any particular application. In every case of an $n$-stage, chain code generator, the longest (maximal length) sequence of 1 s and 0 s repeats after $(2 e\, n - 1)$ states. Note that, as illustrated in Figure 20.6, a pathological condition can occur if all outputs power up in an identical 0 state—in which case 0 s will propagate indefinitely around the chain code generator, resulting in no output. Practical circuits have to include provision to prevent this situation from ever occurring. Indeed it is precisely because of the necessity to avoid this "forbidden" all zeros state that the output of the chain code generator illustrated in Figure 20.6 consists of a cycle of 15 (rather than the more intuitively expected 16) states.

It can be shown mathematically that the output binary sequence from the chain code generator has a frequency spectrum extending from the repeat frequency of the entire sequence up to the clock frequency and beyond. The noise is effectively flat (within 0.1 dB) to about 0.12 of the clock frequency ($F_c$). The noise source is −3 dB at $0.44 F_c$

and falls off rapidly after that. For most applications (audio included), simple low-pass filtering of the digital maximal-length sequence results in white Gaussian noise, provided that the breakpoint of the low-pass filter is well below the clock frequency of the register (say $0.05 F_c$ to $0.1 F_c$), that is, in the region where the spectrum of the digital noise is constant. The analogue filter may be a simple 6-dB/octave RC circuit, but usually a sharper active filter is desirable.

## Reference

1.    Brice, R., *Newnes Guide to Digital Television*, Newnes, 1999.

# CHAPTER 21

# *Other Digital Audio Devices*

Ian Sinclair

Digital radio, using Eureka-3 digital audio broadcasting (DAB), is now up and running, although receiver prices are high, and the emphasis so far has been on in-car units rather than home units. There are, however, several other digital options that have now opened out for audiophiles, particularly for those with computing interests. The problem might be that, with so many options either available or promised, no one can really decide what to buy until the situation settles.

## 21.1 Video Recorders

At the time when Beta and VHS video recorders where competing for the United Kingdom market, several makes of Beta recorders, notably Sanyo, offered the option of sound recording by digitizing an audio input and recording it as if it were a video signal. These recorders have become prized possessions of some audiophiles because of their good sound recording quality and low-cost media. That's assuming they can still get hold of Beta tapes, which, although now rare in the United Kingdom, are still easily available in other parts of the world and are still manufactured for the professional grade of Betamax camcorders.

Looking at more modern equipment, manufacturers such as Hitachi have incorporated audio facilities (including audio dubbing) into Nicam recorders. The input audio signals are converted to Nicam stereo digital format, which implies some compression and recorded. This offers at least 3 h of good-quality music on a standard El 80 tape. One drawback is that automatic gain control settings often result in rather low-level recording so that you need to adjust your volume control settings on replay.

## 21.2 High Definition Compatible Digital (HDCD)

High Definition Compatible Digital (HDCD) is one of several recent improvements in the coding of the familiar compact disc. HDCD discs, developed by Pacific Microsomes, are created using a faster sampling rate of 96 kHz, as compared to the conventional 44.1 kHz used on present CDs, and with 20-bit data units. If this were coded directly on to the CD it would not be compatible with existing CDs, so data are compressed to 16-bit units and a 44.1-kHz pulse rate.

The result is that the HDCD discs can be played on a normal CD deck, but these discs will deliver more dynamic range and overall better sound on a player that uses HDCD decoding. Discs prepared in this way can be recognized by the use of a distinctive HDCD logo.

The first firms to offer players with HDCD capability were Denon, Harman Kardon, Rotel, and Toshiba. Although more than 4000 titles are available in the United States at the time of this writing, these and HDCD players are not easy to find in the United Kingdom (one of the recent offerings is from the respected firm of Linn). The players contain interpolation circuitry that can also enhance conventional CDs, and HDCD has, as you would expect, arrived mainly on the players in the £1000 upwards price bracket. Players indicate the presence of an HDCD disc by lighting an indicator. Note that the most recent Philips CD recorder caters to copying HDCD discs.

## 21.3 CD Writers

CD writer drives have become commonplace on PC computers where they are used primarily for making backup copies of valuable data and programs, but they are still a fairly rare sight on hi-fi installations. The CD writer drives as used in computers have several advantages:

1. They are considerably less expensive.

2. They allow you to store computer data, including still or moving images, as well as sound.

3. They are compact, fitting into a $5\frac{1}{4}$-inch drive space.

4. You can decide for yourself what software to use with them.

5. They can use either write-once (CD-R) or read-rewrite (CD-RW) discs.

6. They can make recordings at a higher speed than music can be played.

If you do not have a suitable computer, of course, this option is not open to you, and you will need to look at one of the CD writer units intended to be used along with a hi-fi system. Such units are more expensive because they need to incorporate several items of circuitry and software that would be available within the computer. In this chapter, the term *drive* refers to a unit incorporated into, or connected to, a computer, and *deck* means a unit that is part of a hi-fi stack or assembly.

Until 1990, the idea of creating your own CDs would have been considered ridiculous because the creation of a CD involved many processes that called for elaborate and expensive equipment. The availability of compact disc writing equipment that is well within a normal domestic budget is due to evolution of CD technology, dispensing with the need to burn into the disc material. The system that is mainly used for home sound recording, or for computer use, is CD-R, meaning CD recordable. This system allows you to write once to a disc and read it as many times as you like. Early versions also allowed this, but later technology allows you to add more tracks to a disc if you did not fill it on earlier sessions. A disc that permits this type of use is described as *multisession*. At the time of writing, a blank CD-R disc costs around £0.75, making this the least expensive method of recording that has ever been devised. A CD-R disc will hold up to 74 min of full CD-quality music, or the equivalent in computer data, about 650 Mbytes.

Computers and some more recent hi-fi CD recording decks can also use a different form of technology, CD-RW, which allows a disc to be recorded, played, wiped, and recorded again, much like as you reuse a tape or a floppy disc. This technology is, at present, not so well suited to audio use, and although the blank discs that once cost around £10 each are down to less than £3 each, they are not so popular for computing use either. Many of the better computer CD writer drives can use either type of disc, and prices are remarkably low, typically £125 if you shop around. Most CD-R drives can write at 2× or 4×, or even 6×, depending on the model, which means that they can make a recording of existing digital files faster than a tape. A 2× recorder will record at twice the speed at which the music can be played. This is an advantage for the drive in a computer, because data files of music can be processed as fast as the CD writer allows, but for the CD-writer drive in a hi-fi installation you cannot speed up the music at the input and high recording speed is pointless. A drive or deck that allows both CD-R and CD-RW discs to be recorded and replayed is known as a CD-R/RW drive or deck.

Unlike DAT, there are no copyright barriers to CD recording. DAT developed as a medium for sound recording, and the record industry worked overtime to make sure that the system was not released until it incorporated safeguards that prevented serial copying. This so hindered the acceptance of DAT that it never became widely used, certainly not in Europe. In contrast, the writeable CD was developed as a computer peripheral, and the record industry did not realize what was happening until it was too late to stop it. Compared to DAT, CD recording is fast, inexpensive, and easy, with no hindrance to making copies. Copyright protection has, however, been developed for DVD (see 21.9 DVD).

To understand how the change in technology has come about, think back to how the early CDs were manufactured, and are still manufactured. The CDs that you buy are made by burning indentations with a powerful laser into the track surfaces of a master disc, using the presence or absence of a pit to indicate a 1 or 0 digital bit. The player also uses a laser, operating at a much lower power level and aimed at the track. The amount of light that is reflected from the laser beam depends on whether the beam hits a pit or an unpitted piece of track. As the disc spins, these changes in intensity are detected and converted into electrical signals, duplicating perfectly the digital signals that were used to create the original. The advantage of this system is that it permits record pressing analogous to the old vinyl disc method. The CD that is burned by the recording laser is used as a master to make copies that can be used for stamping out plastic discs with the information intact.

This process, incidentally, is much less expensive than the method of recording tapes, which need to be recorded from one end to the other, albeit at a faster speed than they are played. It follows then that a CD is much less expensive to produce than a tape, and some bargain CDs, even in the United Kingdom, are sold at prices that reflect the lower cost. The majority of issues, however, maintain the "CD premium" in prices, in the belief that buyers will pay more for them even if they have cost less to make. You may have noticed the low prices of magazines that have CDs attached, pointing out the low price that the magazine has paid for the CD.

The more modern CD-R drives are recorded using a low-power laser that does not burn pits into the plastic of the disc. Instead, the discs are coated with a dye that is affected by the intense light from the laser. The effect is to change the dye color and, although the change is not a vast one, it can be seen by the eye. If you look at a partly recorded CD-R disc, you can see that the recorded portion (the inner part) is a quite distinctly lighter shade of blue (usually) than the outer unrecorded portion. Because this change is irreversible, the disc tracks can be written only once. This type of process is called *dye sublimation*, and the surface appearance of the disc is also due to a thin metallic coating, silver or gold, to make the surface more reflective. CDs created with CD-Rs are compatible with all other computer CD-ROM drives.

Oddly enough, a "premier price" situation has developed with blank CD-R discs. The requirements for recording computer data are more onerous than for sound recordings; after all, your sound system does not shut down if there is a mistake in a tiny fraction of a musical note. This should mean that any blank CD good enough for data recording should certainly be good enough for audio. Some shops, however, will try to sell audio-grade CD-R blanks at a very substantial premium.

The rapidly growing use of CD-R/RW has spawned a whole set of new terms that are probably better known to computer users than to audio enthusiasts. Some of the more important terms that have not been explained so far are summarized here.

**Disc at once** (DAO): A CD-R/RW writing mode that requires the whole of the data to be written in one uninterrupted session. Compare **track at once, incremental writing**.

**Finalized disc**: A CD-R disc that has had its overall lead-in and lead-out information written so that no further sessions can be recorded.

**Fixation**: The set of actions used at the end of a writing session on a CD-R drive. Fixation writes lead-in and lead-out information and creates a table of contents for the disc so that the disc can be read on a normal CD-ROM drive or audio CD player. If the option *of fixation for append* is used, further sessions can be added to the disc until it is full. See also **finalized disc**.

**Incremental writing** or **packet writing**: A method of writing data to a CD-R or CD-RW disc in which several sets of data can be written in each track. This reduces the effect of the overhead of 150 recorded blocks that are used for *run-in, run-out* and linking.

**Lead-in**: A section of all CD ROM or music discs, prerecorded, CD-R or CD-RW, that contains information on the data or music contents. The lead-in area immediately precedes the recorded area. For a fully recorded disc, the lead-in contains the **table of contents.**

**Lead-out**: A section of all pr-recorded compact discs that follows the recorded area (on the outer rim of a fully recorded disc). On the CD-R or CD-RW discs, the lead-out is not created until the disc is declared as fully recorded (preventing further recording). With no lead-out, the disc cannot be replayed on music players, and some older CD-ROM drives on computers may not accept it.

**Multisession**: Refers to a CD-ROM that can be recorded more than once, adding new material on the subsequent recording, until the disc is full. All computer CD-ROM drives and most hi-fi CD decks should be capable of playing CDs recorded in this way.

**PCA**: Program calibration area, the portion of a CD-R disc used for making a trial recording to calibrate the laser intensity needed for the disc that is being used. This allows for differences in disc materials, particularly between CD-R and CD-RW discs.

**PMA**: Program memory area, the portion of a CD-R or CD-RW disc that contains a table of track numbers along with start and stop data positions for each track.

**Session**: A recording made on CD-R or CD-RW that can consist of between 1 and 99 tracks. A session is preceded by a lead-in and ended by a lead-out, and a multisession disc is one that can be recorded at different times, writing a complete session on each occasion, with all data readable.

**Table of contents**: A table of track locations and extents prepared by the CD-R/RW software so that the player can locate each track and data it contains.

**Track at once**: A system for writing a CD-R or CD-RW disc that writes the session as a set of complete tracks. Compare **disc at once**.

### 21.3.1 Uses

The hi-fi version of the CD recorder is used much as you would use a cassette recorder to record music from any other sections of the equipment, such as tuner, cassette deck, vinyl-disc deck, and DAT deck. You may also, subject to the restrictions of equipment and copyright, be able to record from an existing CD player, and this type of transfer is much better if the CD player allows a direct digital output that can be connected to the recorder.

The computer type of CD-R/RW drive must be used along with a sound card that allows line and microphone analogue audio inputs. The quality of recording that you can obtain depends very much on the quality of the analogue-digital conversion in the sound card, and few provide anything like what we accept as CD sound quality. If you are using the system to copy sound tracks from a cassette recorder to a CD, however, the quality level of most cards is acceptable. The line input level of most sound cards is lower than we are accustomed to in hi-fi equipment, and you may need to use the microphone input. This, however, may be too sensitive, causing distortion at high sound levels, and an attenuator may be needed. Some cassette decks allow you to vary the output, which is an ideal way of tackling the problem.

The computer type of drive is well suited to CD copying and to making compilations from a variety of discs. This is not to say that these actions cannot be carried out on the hi-fi type of deck, but you can be certain that the computer type is using direct digital transfers, not converting the CD output into analogue and then converting back to digital in the recorder. The main advantage of using the computer drive is that you can add images, text, and other data into the same CD if you wish (and if you can cope with the mixture). This is particularly useful if you want to make multimedia shows of sound and images.

We shall look at audio and other file transfers for the computer CD-R/RW drives in more detail in the following sections, as many of the steps are almost identical. For the moment, a description of a popular hi-fi CD-recording deck will give you an idea of what is currently "state of the art" in this field.

The Philips CDR 770 (Figure 21.1) was launched in September 1999 and was initially marketed mainly in Germany, where the main demand for CD recorders seems to be at present (as an Internet search will confirm). The initial price in Germany was DM 699, roughly £233, which compares well with earlier models from other manufacturers. The CDR 770 uses the 43.5-cm width that is now standard for hi-fi components. Like any other recorders, the CDR 770 allows consumers to make their own recordings from digital sources, as well as from any analogue sources connected to their audio system.

The CDR 770 performs analogue to digital conversion using the Philips system called DLR (Direct Line Recording). This uses the normal CD 44.1-kHz sampling frequency for bit-by-bit conversion, and for CD copying actions it ensures highly accurate recordings by matching the speed of the recording disc to that of the playing (source) disc. For work with other digital sources, different sampling rates are automatically detected, allowing the CDR 770 to deal with any sampling rate from 11 to 56 kHz.

**Figure 21.1: The Philips CDR 770 CD recorder deck.**

DLR also allows full bit-by-bit recordings to be made of the new HDCD-encoded discs (see earlier, this chapter). The entire encoding of these discs is therefore reproduced on the copy and is available for playback on compatible CD players with a HDCD decoder.

Audio conversions from other analogue sources are also of high quality using the analogue inputs provided. This makes it easy for the user to transfer LP or older disc collections to CD, as well as to record from other analogue sources such as tapes, radio, or even live music (given suitable microphones).

The CDR 770 also incorporates a CD text function, allowing the consumer to put in text information such as album, artist, or track name. When you make a CD recording using the CDR 770 you can enter your own personal text for each disc, for each track, or for each artist. Each of these text items can contain up to 60 characters. The text is then shown on the display during playback. CD text that is present on prerecorded discs will also be displayed when playing the disc back in a Philips Audio CD-Recorder.

Conscious that the hi-fi user is less accustomed to setting up digital equipment than a computer user, Philips has redesigned the user interface of the CDR 770 so as to make the recording action easier and more intuitive. This uses clear messages at every stage to prevent errors and shows the user exactly what to do next. For example, the new *Make CD* function allows discs to be recorded and finalized quickly and conveniently, using a single command rather than a set of operations in sequence. Another useful feature is multitrack erase, allowing multiple tracks to be selected and erased at the same time. In addition, the CDR 770 features 99-track programming, easy recording start, an FTD display that gives a clear, at-a-glance indication of the set status, and a music calendar with track bar.

One common problem that users have with hi-fi CD recording is mistaken starts, starting the wrong piece for recording or starting in the wrong place. Because CD-R is a medium that does not allow erasing, this action either makes a set of tracks that you do not want to play or makes the whole disc unusable. This is no problem for the computer user who makes use of a CD-R/RW drive, because the digital files are stored and can be edited before recording, but this is not the way that the hi-fi type of CD recording deck works. Philips has included a buffer memory into the CDR 770, allowing storage of up to 3 s of music.

### CDR 770 Technical Specifications

| Number of channels: 2 (stereo) | Applicable supply: AC 230 V (50/60 Hz) |
|---|---|
| Power consumption: 15 w | Operating temperature: 5–35°C |
| Weight: 4 kg | Dimensions: 435 × 305 × 88 mm (w × d × h) |

## Audio, General

| | |
|---|---|
| Frequency response (digital in): 20 Hz–22.05 kHz | Playback S/N: 100 dB |
| Playback dynamic range: 95 dB | Playback total harmonic distortion: 85 dB (0.0056%) |
| Recording S/N (analogue): 90 dB | Recording S/N (digital): recording quality equal to source |
| Recording dynamic range: 92 dB | Recording total harmonic distortion: 85 dB (0.0056%) |
| Headphones 0–5 V rms / 8–2000 Ω | |

## Recording Values for Line Input/Output

| |
|---|
| Digital coaxial input (direct recording): 12–56 kHz ± 100 ppm |
| Digital optical input (direct recording): 12–56 kHz ± 100 ppm |
| Analogue input (level potentiometer): 700 mV rms/50 kΩ = 0 dB |
| Line output voltage: 2 V rms ± 2 dB |
| Digital coaxial output: 0.5 Vpp/75 Ω |

## Recording Functions, CD-R and CD-RW Discs

| | |
|---|---|
| Auto start recording per disc | Erase last track (CD-RW disc) |
| Erase disc (CD-RW disc) | Erase table of contents (for rerecording on finalized RW-disc) |
| Manual/auto track increment | Remaining recording time display |
| Autofinalize (make disc compatible to CD player) | SCMS (serial copy management system) |
| RID code (recorder unique identifier) | |

## Playback Functions

| | | | | |
|---|---|---|---|---|
| Play | Pause | Stop | Direct track selection | Next/Previous track selection |

**Accessories**

| | |
|---|---|
| Search forward/reverse | Remote control (+batteries) |
| Repeat (all/per track) | Audio cable (×2) |
| Program play (30 tracks) | Digital coaxial cable (×1) |
| Time display switching | AC mains cord |

The buffer allows a mistakenly started recording to be stopped within the first 3s, before the start of any disc writing actions, and it also ensures that recordings can be made without loss of music at the start of a track when using synchronized CD recording. Buffering also permits the use of synchronized starting from analogue sources. This is done by monitoring the incoming audio signal for the rise in level that indicates the start of play, after which the first few seconds of music are recovered from the buffer for recording.

The recorder incorporates a digital recording level and balance control that allows manual adjustments. This can be used to correct variations between individual discs, a valuable feature if you are making compilations, as it allows you to adjust the volume levels of all tracks. Digital recording level adjustments can be made easily using the *Easy Jog* control.

The CDR 770 has three sets of input sockets, allowing easy connection to a wide variety of audio sources. Both optical and coaxial digital inputs are provided for the highest quality connection to digital sources, along with standard stereo sockets for connection to virtually all analogue sources, as well as coaxial digital and analogue outputs.

Like its computer drive counterpart, the CDR 770 can make use of both CD-R and CD-RW blanks, and the CD-R discs can be played on any other CD player, either on audio equipment or in a computer system. The CD-RW discs can be replayed on the CDR 770 and on many of the most recent CD playing decks or drives. If your CD player is not of recent design, however, it will not be able to read the CD-RW discs.

## 21.4 MPEG Systems

When the CD system was launched, following the commercial failure (in the United Kingdom, but not elsewhere) of the earlier laser-disc moving picture system, there was no form of compression of data used. The whole system was designed with a view to recording an hour of music on a disc of reasonable size, and the laser scanning system that was developed from the earlier "silver disc" was quite capable of achieving tight packing of data, sufficient for the needs of audio.

Data compression was, by that time, fairly well developed, but only for computer data, and by the start of the 1980s several systems were in use. Any form of compression for audio use had to be standardized so that it would be as universal as the compact cassette and the CD, and in 1987 the standardizing institutes started to work on a project known as EUREKA, with the aim of developing an algorithm (a procedure for manipulating data) for video and audio compression. This has become the standard known as ISO MPEG Audio Layer-3. The letters MPEG stand for Moving Picture Expert Group, because the main aim of the project was to find a way of tightly compressing digital data that could eventually allow a moving picture to be contained in a CD, even though the CD as used for audio was not of adequate capacity (see 21.9 DVD).

As far as audio signals are concerned, the standard CD system uses 16-bit samples that are recorded at a sampling rate of more than twice the actual audio bandwidth, typically 44 kHz. Without any compression, this requires about 8.8 Mbytes of data per minute of playing time. The MPEG coding system for audio allows this to be compressed by a factor of 12, without losing perceptible sound quality. If a small reduction in quality is allowable, then factors of 24 or more can be used. Even with such high compression ratios the sound quality is still better than can be achieved by reducing either the sampling rate or the number of bits per sample. This is because MPEG operates by what are termed perceptual coding techniques, meaning that the system is based on how the human ear perceives sound.

The MPEG-1 Layer III algorithm is based on removing data relating to frequencies that the human ear cannot cope with. Taking away sounds that you cannot hear will greatly reduce the amount of data required, but the system is lossy, in the sense that the removed data cannot be reinstated. The compression systems used for computer programs, by contrast, cannot be lossy because every data bit is important; there is no unperceived data. Compressing other computer data, notably pictures, can be very lossy, so that the JPEG (Joint Photographic Expert Group) form of compression can achieve even higher compression ratios.

The two features of human hearing that MPEG exploits are its nonlinearity and the adaptive threshold of hearing. The threshold of hearing is defined as the level below which a sound is not heard. This is not a fixed level; it depends on the frequency of the sound and varies even more from one person to another. Maximum sensitivity occurs in the frequency range 2–5 kHz. Whether or not you hear a sound therefore depends on the frequency of the sound and the amplitude of the sound relative to the threshold level for that frequency.

The threshold of hearing adapts to the sounds that are heard, so that the threshold increases greatly, for example, when loud noises accompany soft music. The louder sound masks the softer, and the term masking is used of this effect. Note that this is in direct contradiction of the "cocktail-party effect," which postulates the ability of the ear to focus on a wanted sound in the presence of a louder unwanted sound.

The masking effect is particularly important in orchestral music recording. When a full orchestra plays fortissimo then the instruments that contribute least to the sound are, according to many sources, not heard. A CD recording will contain all of this information, even if a large part of it is redundant because it cannot be perceived. By recording only what can be perceived, the amount of music that can be recorded on a medium such as a CD is increased greatly, which can be done without any perceptible loss of audio quality.

Musicians will feel uneasy about this argument because they and many others feel that every instrument makes a contribution. Can you imagine what an orchestra would sound like if the softer instruments were not played in any fortissimo passage? Would it still be fortissimo? Would we end up with a brass band, without strings or woodwinds? My own view is that the masking theory is not applicable to live music, but it may well apply to sound that we hear through the restricted channels of loudspeakers. In addition, how will a compressed recording sound when compared to a version using HCDC technology?

MPEG coding starts with circuitry described as a *perceptual subband audio encoder*. The action of this section is to analyze continually the input audio signal and, from this information, prepare data (the masking curve) that define the threshold level below which nothing will be heard. The input is then divided off in frequency bands, called subbands. Each subband is quantized separately, controlling the quantization so that the quantization noise will be below the masking curve level for that subband. Data on the quantization used for a subband are held along with the coded audio for that subband so that the decoder can reverse the process. Figure 21.2 shows the block diagram for the encoding process.

### 21.4.1 Layers

MPEG1, as applied to audio signals, can be used in three modes, called layers I, II, and III. An ascending layer number means more compression and more complex encoding.

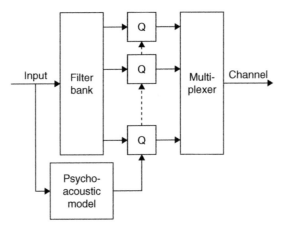

**Figure 21.2: Block diagram for MPEG encoding.**

Layer I is used in home recording systems and for solid-state audio (sound that has been recorded on chip memory, used for automated voices, etc.).

Layer II offers more compression than layer I and is used for digital audio broadcasting, television, telecommunications, and multimedia work. The bit rates that can be used range from 32 to 192 Kbit/s for mono and from 64 to 384 Kbit/s for stereo. The highest quality, approaching CD levels, is obtained using about 192–256 Kbit/s per stereo pair of channels. The precise figure depends on how complex an encoder is used. In general, the encoder is from two to four times more complex than the level I encoder, but the decoder need be only about 25% more complex. MPEG level II is used in applications such as CD-i full-motion video, video CD, solid state audio, disc storage and editing, DAB, DVD, cable and satellite radio, cable and satellite TV, ISDN links, and film sound tracks.

Layer III offers even more compression and is used for the most demanding applications for narrow band telecommunications and other specialized professional audio areas of audio work. It has found much more use as a compression system for MP3 files (see 21.5 MP3).

MPEG-1 is intended to be flexible in use, so that a wide range of bit rates from 32 to 320 Kbit/s can be used, with a low sampling frequency (LSF) of 8 Kbit/s added later. Layer III allows the use of a variable bit rate, with the figure in the header taken as the average. Decoders for layers I and II need not support this feature, but most do.

Table 21.1 shows the relative complexity of encoding and decoding for the three levels of MPEG-1. The encoding process is always more complex, but the relative complexity of the decoder is less.

MPEG-1 coding can be applied to mono or stereo signals, and the stereo system makes use of joint stereo coding, a system that achieves further compression by seeking out redundancy between the two channels of a stereo signal. The system supports four modes:

**Table 21.1 Comparing Complexity of Circuitry for MPEG-1 Levels**

| Complexity | | |
|---|---|---|
| **Layer** | **Encoder** | **Decoder** |
| I | 1.5 to 3 | 1.0 |
| II | 2 to 4 | 1.25 |
| III | 7.5 or more | 2.5 |

| | |
|---|---|
| mono stereo | joint stereo |
| (intensity stereo or mid/side stereo) | dual channel (two independent channels, e.g., for two languages) |

When the digital signal has been encoded, it is divided into blocks of 384 samples (layer I) or 1152 samples (layers II and III) to form the unit MPEG-1 frame. A complete MPEG-1 audio stream consists of a set of consecutive frames, with each frame consisting of a header and encoded sound data. The header of a frame contains general information such as the MPEG layer, the sampling frequency, the number of channels, whether the frame is CRC protected, whether the sound is an original, and so on. Each audio frame uses a separate header so as to simplify synchronization and bit stream editing, even if much of the information is repeated and hence redundant. A layer III frame can achieve further compression by distributing its encoded sound data over several other consecutive frames if those frames do not require all of their bits.

One important point about all digital audio systems is that the analogue concept of S/N ratio is no longer relevant, and so far no replacement has been found. If we try to measure S/N in any of the ways that work perfectly well for analogue signals, the results are widely variable and have no correspondence with the signal as heard by the listener.

- Note that the MUSICAM algorithm is no longer used, it was developed into MPEG-1 Audio Layers I and II. The name MUSICAM is a trademark used by several companies.

- MPEG-1 is one of several (seven at the last count) MPEG standards, and we seem to be in danger of being buried under the weight of standards at a time when development is so rapid that each standard becomes out of date almost as soon as it has been adopted. Think, for example, how soon NICAM has become upstaged by digital TV sound.

## 21.5 MP3

MP3 is a high-compression coding and decoding system that is now used for transmitting audio signals over Internet links and for storing audio signals in compact computer file form. MP3 allows the construction of small players that store, typically, 40 min of music, but contain no moving parts. Because MP3 is a lossy form of compression, the MP3 deck for hi-fi systems has not emerged so far, but we should remember that the compact cassette was also considered unfit for hi-fi uses in its initial days. The Minidisk uses similar compression methods.

The name MP3 began as an extension to a filename, devised to distinguish sound files created using MPEG-1 Layer III encoding and decoding software. The PC type of computer makes use of these extension letters, up to three of them, placed following a dot and used to distinguish file types. For example, *thoughts.txt* would be a file called thoughts, consisting purely of text, and *thoughts.doc* would be a document called thoughts, which could contain illustrations and formatted text, even sounds. A file called *thoughts.jpg* would be a compressed image file, and *thoughts.bmp* would be an uncompressed image file. There are many such extensions, each used to identify a specific type of file.

The same MP3 extension is used for sound files that have used MPEG-2 Layer III with a reduced sampling rate, but there is no connection between MP3 and MPEG-3. MP3 files use a compression ratio of around 12:1, so that MP3 files stored on a recordable CD will provide about 12 h of sound. See later for a description of DAM-CD.

The main use of MP3, however, has been the portable MP3 player, which allows MP3 files to be recorded from downloads over the Internet. This has made MP3 very much of an audio system for the computer buff, but like all matters pertaining to computing, this use is likely to spread. MP3 is unlikely to appeal to those who seek perfection in orchestral music (let's face it, what system does?) but for many other applications it offers a sound quality that is at least as good as anything that can be transmitted by FM radio or obtained from a high-quality cassette.

- The advantages are many. You can load the memory up with music that you like, deleting anything you don't want to hear again. You can play tracks in any order, select tracks at random, and store other music on your PC until you want it on your MP3 player.

One other attraction is rather an illusion, that music is free. Many Internet sites offer MP3 files at no direct cost, but you have to pay for the large amount of telephone time you need for downloading them. Unless you want only fairly small-scale works you will need a fast Internet connection, such as you get with cable TV firms. The alternative is to spend a lot of your income paying for BT phone calls, although alternatives are appearing almost daily. If you really want to download a lot of music it makes sense to take out a fast line or use one of the offers of a fixed charge for unlimited Internet use. Either way, your music is not exactly free.

In addition, "free" music is often made by unknown artists in strange places. Sometimes you will find an excellent recording made by a Russian orchestra that is unable to raise the money to make CDs or to go on tour. Other recordings may have quite awful quality, and there is always the suspicion about some of the worst recordings that some tracks may even be acquired illegally, using miniature recorders taken to live concerts. Some may even be copied from existing CDs. However, the MP3 system is an excellent way for any person or group to make and record their own music and distribute it worldwide without the costs involved in making a CD.

- There is nothing illegal about possessing MP3 files, no matter how they were obtained originally, on your computer. If you make copies and distribute them, that's another matter, and the usual laws of copyright apply. It is certainly illegal to convert CD tracks and distribute the music in MP3 format without the permission of the copyright holder.

## 21.6 Transcribing a Recording by Computer

The standard hi-fi methods of copying music for your own use include cassette recording, DAT, and now CD-R or CD-RW. With the help of a computer you can go considerably further by editing the music (cutting out scratches, for example, in old vinyl-disc recordings) or by recording to MP3. The following paragraphs summarize the methods used for computer manipulation of sound for any digital form of recording, mainly CD-R and MP3. The computer must contain a sound card with a A-D converter that is up to CD quality standards, and if you want to record your own CDs you will also need a CD writing drive with appropriate software such as Adaptec *Easy CD Creator.* For MP3 files you will need software such as *Winamp.*

This is not intended to be an exhaustive guide to using a computer for manipulating audio files, as space does not permit a thorough treatment of such a large topic. If you are an experienced user of a PC computer, this is a guide to its use for audio work, and if you do not use a computer, it is a guide to what you are missing.

The first step to the creation of either an MP3 file or a CD-R disc is to extract music tracks and digitize them in an uncompressed format using a type of file distinguished by the extension letters WAV, hence called a WAV file. Some software will carry out this action automatically, reading in the audio tracks and converting to MP3 or to CD-R without leaving a WAV file behind on the computer's hard disc. As applied to a CD as source, this action is often termed *CD ripping.* Whether you are aware of it or not, WAV files are always created as an intermediary, and it's an advantage if you can store them in the computer, check them, and possibly edit them before you save them in MP3 or CD-R format and delete the WAV versions.

You are not obliged to use a CD as a source, though, and many users of MP3 or CD-R are more concerned with taking tracks from old 78s, from LPs, or from cassettes, even from radio or private recordings. Remember, however, that no matter what source you use, working at CD quality will require disc space on your computer of around 700 Mbytes for a full CD.

If you are using a CD as your source, you must use the digital output from the CD drive or deck. It is certainly possible to connect the audio output of a CD deck to the line input of the sound card on your computer and to create WAV files in this way, but this sacrifices quality. Most computers fitted with a CD writer will also have a fast CD reader, allowing you to read digital data at 36 times (or more) the normal recording speed. This also ensures that the digital output of the CD is used.

- The normal setting on most CD copying software gives you a 2-s gap between tracks when you are working in "normal mode," which is *track-at-once.* If you specify *disc-at-once,* you will not get any added gaps between the tracks, so if you want extra time between the tracks you have to edit the WAV files so as to include silent intervals. If your list of tracks shows separate files, the recording will always place track markers so that you can move to any track in the usual way.

You can also create WAV files using any other audio source, such as 78s, LPs, cassettes, and DAT tapes. The conversion quality will be lower, because these sources all provide analogue signals of varying quality and signal level. You will need to do a few experiments with connections and signal levels, and this is why it is such an advantage to make a separate conversion to WAV, because you can play back a WAV file that is stored in the computer without the need to waste space on CD-R with an unsatisfactory recording or to make a useless MP3 file. You must, incidentally, use a modern 16-bit sound card—do not try to work with analogue to digital conversions using the older 8-bit type of card. Any computer that is fast enough to cope with audio work will almost certainly be fitted with a 16-bit card. Remember, however, that the quality of A-D conversion may not be as good as you would like.

The usual advice is to connect the audio output from the source device to the line- in connector, usually a 3.5-mm stereo jack socket, on the sound card of the computer. Depending on the sound card that you are using, you may find that the line- in is much too insensitive and that you hear virtually nothing when you replay the WAV file. The only option, unless you have a spare preamp to connect between the signal source and the sound card, is to use the MIC input. This, by contrast, may be too sensitive, leading to overloading.

The important thing is to try this out with a short piece of music before you start making any recordings to CD-R or MP3. The typical software that you will be using for creating the WAV file is Creative *Wave Studio,* which permits you to make a short recording to test sound levels. Using the software control panel illustrated in Figure 21.3, you can adjust

**Figure 21.3: A typical software mixer panel as seen on the computer screen.**

the level of the signal on its way to the WAV file, but this will not help if the input stage of the sound card is overloading.

You can then play this back, either through the loudspeakers of the computer or by way of a connection to a hi-fi system, using the output jack of the sound card. The same software, incidentally, allows you to edit a WAV file to remove gaps and, after some practice, unwanted sounds, such as scratches and thumps.

You may need to make some setting-up steps, and although some software will do this almost automatically you should check the following:

- Record options must be set to stereo, 8 or 16 bits, 44.1 kHz sampling rate. Use 16-bit data for CD or other high-quality sources.

- Type in the name that you want to use for the WAV file and a folder (directory) on the computer's hard disc where you want to store the file. The usual pattern is to select a name that will describe the music, such as Beethoven 5 Symphony, and store it where you can accommodate a large file of up to 700 Mbytes.

- Check that you have set the recording levels correctly.

- Click the Record icon on the screen and start the source playing. The screen display will probably show the progress of the recording.

You can make your recording one track at a time, making a separate WAV file from each track, or you can make a single file of the whole input. A single file uses less space (because it eliminates the "overhead" involved in making a separate file for each track), but it makes finding individual tracks (if you need to) more difficult. Software products such as *LP Ripper* or Adaptek *Spin Doctor* will work on the WAV file and separate out the tracks for you, whereas others will edit the WAV files manually.

The topic of editing a WAV file is too specialized for this book, but the principles are not difficult, and some practice with a short file is more useful than any amount of text instruction.

## 21.7 WAV Onward

The WAV file is, however, an intermediary. It takes up a large amount of space on your hard disc because it is totally uncompressed, and its main purposes are to allow you to edit the sound and to provide a source for conversion to MP3 or CD-R. Conversion to CD-R is a critical process because the recording will be ruined if digital data are not available when the CD writer software needs it. By using the WAV intermediate you are assured of this, because this is a file that is already in digital form, and there is no need to wait for audio signals to be converted at the risk of not keeping up with the demand from the CD writer.

To record the WAV files in CD form, place a blank CD-R disc into the recorder and use a good piece of software such as Adaptek's *Easy CD Creator.* This allows you to choose to make a Data or a Music CD, to select WAV files to record, and put them into the order you want. Once you have files ready, the software will test the files and then make the recording. The disc will be ejected once the recording has been made. You can use another option of the software to print front and back covers for the CD jewel case. A CD made in this way will play in any reasonably modern CD player. On test, a CD I prepared in this way worked even on a very old Philips CD deck (the first model sold in the United Kingdom) with no problems.

If you are preparing MP3 files you need no hardware, only software. All MP3 software is not equal, and some are concerned much more with tricks than with quality or speed of conversion. The software will usually allow a choice of bit rate, and the usual rates are 128 Kbits/s for files sent over the Internet and 198 Kbits/s for files to be stored on CD-R. At 128 Kbits/s, a 4-min piece of sound will need 3.8 Mbytes of storage space, as distinct from closer to 40 Mbytes for an uncompressed CD-quality file.

Recommended software includes MusicMatch *JukeBox 4.1*, AudioSoft *Virtuosa Gold 3.1,* and *XingMP3*.

## 21.8 DAM CD

DAM is an acronym for digital automatic music, and a DAM CD is one that can contain music both in MP3 format and in normal uncompressed CD format. The MP3 tracks can be copied into any MP3 player, and the normal CD tracks can be played using any normal CD player. It is equally possible to make a DAM CD that contains only MP3 tracks, so packing about 10 times more music on to the CD than would be possible using uncompressed CD methods. If you use the computer extensively as a music player, you can transfer your favorite music into this format for easy access and compact storage.

You can also buy DAM CDs over the Internet. They often feature unknown artists and are priced accordingly, although you should not expect a large selection of classical music to be available. The CD is usually offered on the Web page for the artist, and costs are kept low by using CD-R so that the music you want is transcribed to a CD when you place your order.

- Some older CD decks cannot cope with DAM CDs that inevitably use multisession methods.

## 21.9 DVD and Audio

The CD format was standardized at a time when digital recording of sound on disc was still an uncharted realm, full of possibilities and surprises, and CD technology strained at the limits of what was possible, particularly A-D and D-A conversion methods. The use of lasers to write the master discs, although not new because of the Philips "silver discs" used for video recordings, was unfamiliar to many recording companies, and the extent of the packing of bits on the CD stretched the pressing capabilities of all but a few users. Now, at least two decades on, we can see that the potential of the little CD is much greater than we could have hoped for.

DVD, originally the acronym of digital video disc, is now taken to mean *digital versatile disc* and refers to a more recent development of CD technology. This was originally directed to recording full-length films on CD, hence the "video" in the original title, but the idea has been extended to a universal type of disc that can be used for films, audio, or computer data interchangeably. The main difference, at present, is that there are very few DVD writing drives available, and these few are expensive by computing (although not by hi-fi) standards.

- An important feature of a modern DVD computer or TV drive unit is that it will accept conventional CDs as well as DVD discs.

The DVD holds much more data, can transfer it faster, but is as easy to reproduce by stamping processes as the older CDs (which, alas, does not mean that it will be sold at reasonable prices in the United Kingdom, even if a DVD costs so much less to produce than a videotape). Eventually, DVD will be the one uniform recording format, replacing cassettes, DAT, videotape, and CD-ROM. A DVD drive is already virtually a standard item on computers, and the manufacturers claim that in a time of 3 years it has become the most successful electronics product of all time for home use (Figure 21.4).

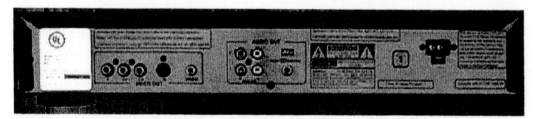

**Figure 21.4: The rear of an APEX AD 600A DVD player that also plays CDs and MP3 files, priced in the United States at about $150.**

Computers are the main end use of DVD at present, but DVD drives to replace videocassette players are already widely available. The spread of DVD as a replacement for VCR, however, is not likely to spread widely until the recording version reaches an acceptable price level. Surveys have shown repeatedly that the most common use for VCRs in the United Kingdom is to record TV programs either when the viewer is not at home or when two interesting programs are being broadcast at the same time. Use of DVD simply to play prerecorded discs is very restrictive—I cannot think of more than a handful of films I would ever want to see again, and some of my own videotapes have not been played since the day I recorded them. This is mainly a United Kingdom attitude, and the laser disc that was rejected in the United Kingdom has survived up until now in other countries.

- With the primary markets of computers and film viewing now being supplied, we are waiting for a standardized DVD format for audio that reportedly will allow up to 17 h of CD quality to be stored on a single disc.

DVD offers so much more storage space than CD that the options it allows are more than most users can cope with at first. A single-layer disc can store just over 2 h of digital video signals at a higher quality than is possible using VCR (which relies on considerable bandwidth reduction). More than one layer of CD recording can be placed on a disc, however, because the layers are transparent, and by altering the focus of the reading laser, it is not technically difficult to read either of two superimposed layers that are only a fraction of a millimeter apart.

By making two-layer DVDs the recording time can be doubled, and by adding double-sided recording it can be doubled again to 8 h of video. The discs can contain up to eight audio tracks, each using up to eight channels, so that films can contain soundtracks in more than one language and cater for surround sound systems.

The DVD can also end the concept of a film as a single story because, unlike tape, it can switch from one set of tracks to another very quickly, allowing films to be recorded with several options endings, for example. Different camera angles can also be selected by the viewer from the set recorded on the disc, and displays of text, in more than one language, can be used for audio and video tracks. Like CD and so unlike VCR, winding and rewinding are obsolete concepts, and a DVD can be searched at a very high speed that seems instantaneous compared to VCR. The disc is also smaller than a videocassette, does not wear out from being played many times, and resists damage from magnets or heat.

- DVD for video uses MPEG-2 coding and decoding, but there is nothing to prevent cut-price producers from coding with MPEG-1, producing the same video quality as a VCR. Even MPEG-2, however, is a lossy compression method, which sometimes shows in video quality as shimmering, fuzzy detail, and other effects.

- In contrast, DVD audio quality is excellent. DVD audio can optionally use CD methods (PCM) with higher sampling rates for even better quality than CD. Other options, used mainly in connections with films, are Dolby Digital or DTS-compressed audio.

### 21.9.1  Regionalization

Unlike audio CD, DVD is more regionalized than we would wish. Taking a cynical view, this is done to prevent European users from flying over to the United States to stock up with DVDs at bargain prices. Film studios have taken the same attitude to DVD as the record companies did to DAT—if you can't ban it, cripple it. The official reasons are that regionalization prevents premature release of a file in another country and protects the distribution rights of suppliers in different countries. Regionalization does *not* apply to DVDs that consist purely of audio signals.

Apart from regional codes, DVDs must be designed to work with the type of color TV coding that different countries use, so that DVDs have to be manufactured in NTSC, PAL, and SECAM versions.

The DVD standard includes regional codes, and each DVD drive or deck is allocated a code for the region in which it is marketed. A disc bought in one region will not play on a deck/drive bought in another region, because the codes will not match. Several DVD users in the United Kingdom have countered this by buying their DVD equipment in the United States and then also buying the discs in the United States.

The established regions are:

1.  United States, Canada, U.S. territories
2.  Japan, Europe, South Africa, Middle East (including Egypt)
3.  Southeast Asia, East Asia (including Hong Kong)
4.  Australia, New Zealand, Pacific Islands, Central America, Mexico, South America, Caribbean
5.  Eastern Europe (Former Soviet Union), Indian Subcontinent, Africa (also North Korea, Mongolia)
6.  China
7.  Reserved
8.  Special international venues, such as airliners and cruise ships

The manufacturers of discs are not obliged to use these codes, and if they do not do so the discs can be used on any drive/deck anywhere in the world. Some types of drives/decks can be modified so that they will play DVDs irrespective of regional coding.

- DVD-ROM discs that are used for computer software are not subject to region codes, nor are audio DVDs.

### 21.9.2  Copy Protection

DVDs can use four different methods of copy protection systems. The *Macrovision* system includes signals that will cause a VCR to record incorrectly by feeding incorrect information to the synchronization and automatic level control circuits. *CGMS* is designed to prevent serial copying (making copies of copies). *CSS* (Content Scrambling System) is a form of data coding supported by film studios, but the coding algorithm has been cracked and posted in the Internet (along with methods for defeating other protection systems), casting doubt on the future of this method. Finally, the *DCPS* (Digital Copy Protection System) is designed to prevent perfect digital copying between devices that incorporate this coding system.

### 21.9.3  DVD-Audio

The first DVD drives started to appear around 1996, but at that time there was no agreed format for DVD-Audio, despite the obvious advantages of DVD for audio recording.

Considerable effort has gone into defining standards, but the final specification was approved in February 1999. Any delays were caused by the introduction of copy-protection codes as demanded by the music industry.

The situation now is that it is possible to design universal DVD players that will deal with both DVD-Video and DVD-Audio, but decks intended for DVD-Audio only will not play DVD-Video. As a further complication, because DVD-Audio is a rather different format, some DVD-Audio discs will not be fully usable in any DVD-Video player other than a universal type, which at the moment is not in production or even planned. With some cooperation from manufacturers it would be possible to turn out DVD-Audio discs that would operate on all DVD decks or drives. As usual, it is unwise to be a pioneer consumer, just as it was in the Beta/VHS days.

The protection system that has been adopted uses what the manufacturers call a digital watermark. This adds signals that appear as low-level noise, and the recording companies claim that this is completely inaudible. If enough audiophiles can hear the difference, then it is a distinct possibility that two separate audio markets could develop, one using the older CD format for music acceptable to enthusiasts, with DVD used for all other recordings. However, we may feel that the golden-eared brigade can always detect the inaudible, even on discs that have not had the coding added, but most users will not be affected.

- Sony and Philips, who developed the CD standards, have joined forces again to make their Super Audio CD format that competes directly with DVD-Audio. This takes us all back to the VCR battles of VHS and Beta, but manufacturers are likely to respond by making playing decks that will allow the use of either type of disc. At the moment, neither players nor discs are in plentiful supply.

Although DVD drives will read CDs, they will not in general read CD-R discs, and since recordable DVD is still rather distant, this might be a stumbling block to anyone who is contemplating transferring a treasured collection of tracks to CD-R. However, most DVD players can read CD-RW discs. This difference arises because DVDs use a laser whose light color is not the same as is used for CD players, and this light does not match that used for CD-R, although it is better adapted to CD-RW.

At present, I would not urge anyone to rush out and buy DVD-Audio, even if equipment becomes available on the United Kingdom market. The list of incompatibilities already suffered by DVD-Video (films that will not play on specific players) is very long, and we simply can't guess how many problems of the same type we might see with DVD-Audio. For the long term, however, the medium must be the future of audio and video distribution. In the United Kingdom, much depends on making recordable DVD at a price that is not too much out of line with VCR, which is a tall order even when all TV is digital. As to DVD-Audio, its day will come when all record companies start to distribute on DVD rather than on CD.

# PART 5

# Microphone and Loudspeaker Technology

# CHAPTER 22

# Microphone Technology

Don Davis and Eugene Patronis

## 22.1 Microphone Sensitivity

In order to determine the electrical input level to a sound system, we need to measure the electrical output generated by the system microphone when it is subjected to a known sound pressure (SP). In making such measurements an $L_P$ of 94 dB (1 Pa) is recommended as this value is well above the normally encountered ambient noise levels.

Everyone seriously interested in the field of professional sound should own or have easy access to a precision sound level meter (SLM). Among other uses, an SLM is required to measure ambient noise, to calibrate sources, and, on occasion, to serve as input for frequency response, reverberation time, signal delay, distortion, and acoustic gain measurements.

Setting up the microphone measurement system shown in Figure 22.1 requires a pink noise generator, a micro-voltmeter, a high-pass and low-pass filter set such as the one illustrated in Figure 22.2, a power amplifier, and a well-constructed test loudspeaker, in addition to the SLM.

Select a measuring point (about 5 to 6 ft) in front of the loudspeaker and place the SLM there. Adjust the system until the SLM reads an $L_P$ of 94 dB (a band of pink noise from 250 to 5000 Hz is excellent for this purpose). Now substitute the microphone to be tested for the SLM. Take the microphone open circuit voltage reading on the micro-voltmeter. The voltage sensitivity of the microphone can then be defined as

$$S_V = 20 \text{ dB} \log(E_o) \qquad (22.1)$$

where $S_V$ is the voltage sensitivity expressed in decibels referenced to 1 V for a 1-Pa acoustic input to the microphone and $E_o$ is the open circuit output of the microphone in volts.

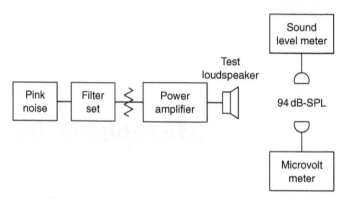

**Figure 22.1: Measuring microphone sensitivity.**

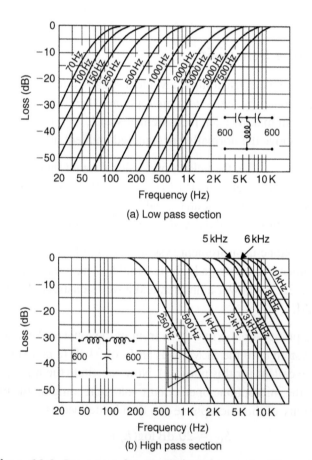

(a) Low pass section

(b) High pass section

**Figure 22.2: Response characteristics of a passive filter set.
(Courtesy of United Recording Electronics Industries.)**

The open circuit voltage output of the microphone when exposed to some other arbitrary acoustic level $L_P$ is calculated from

$$E_o = 10^{\left(\frac{S_V + L_P - 94}{20}\right)}$$     (22.2)

where $E_o$ is now the open circuit voltage output of the microphone for an arbitrary acoustic input of level $L_P$.

For example, suppose a sample microphone is tested by the conditions of Figure 22.1 with the result that the open circuit voltage is found to be 0.001 V. The voltage sensitivity of this microphone as calculated from Equation (22.1) is then

$$S_V = 20 \text{ dB} \log(0.001)$$
$$= -60 \text{ dB}.$$

This result would be read as .60 dB referenced to 0 dB being 1 V per pascal (1 V/Pa). If this same microphone were exposed to an acoustic input level of 100 dB rather than the test value of 94 dB, then its open circuit output voltage from Equation (22.2) would become

$$E_o = 10^{\left(\frac{-60 + 100 - 94}{20}\right)}$$
$$= 0.002 \text{ V}.$$

Many current microphone preamplifiers have input impedances that are at least an order of magnitude or larger than the output impedances of commonly encountered microphones. In such instances, Equation (22.2) can be employed to determine the maximum voltage that a given microphone and sound field will supply to the preamplifier input. The voltage sensitivity of Equation (22.1) is the one currently employed by most microphone manufacturers.

Another useful sensitivity rating for a microphone is that of power sensitivity. In this instance the focus is placed upon the maximum power that the microphone can deliver to a successive device such as a microphone preamplifier when the microphone is exposed to a reference sound field. In this instance the reference power is 1 mW or 0 dBm and the reference sound field pressure is 1 Pa or 94 dB. This rating is more complicated as it involves the microphone output impedance. All microphones, regardless of whether the construction is moving coil, capacitor, ribbon, etc., have intrinsic output impedance that in general is complex and frequency dependent. Strictly speaking, in order for such a device to deliver maximum power, it must work into a load that is matched on a conjugate basis with the reactance of the load being the negative of the reactance of the source and the resistance of the load being equal to the resistance of the source.

Suppose then that the real part of the microphone's output impedance is $R_o$. This being the case, the available input power in watts that the microphone can deliver to the input of a successive device, AIP, is given by

$$AIP = \left(\frac{1}{4}\right)(E_o{}^2)\left(\frac{1}{R_o}\right). \tag{22.3}$$

If AIP is referenced to 1 mW and the microphone is exposed to a sound field of 1 Pa, then

$$\frac{AIP}{0.001} = \left(\frac{1}{4}\right)(10^3)\left(10^{\frac{S_V}{10}}\right)\left(\frac{1}{R_o}\right). \tag{22.4}$$

This can be converted to a power level by taking the logarithm to the base 10 of Equation (22.4) and then multiplying by 10 dBm to yield

$$L_{AIP} = (-6 + 30 + S_V - 10 \log R_o) \text{ dBm.} \tag{22.5}$$

$L_{AIP}$ expresses the power sensitivity of a microphone in terms of dBm/Pa. If our example microphone has an $R_o$ of 200 Ω along with its voltage sensitivity of $-60$, then its power sensitivity would be

$$-6 + 30 - 60 - 23 = -59 \text{ dBm/Pa.}$$

Another useful way to express the power sensitivity of a microphone would be to reference the available input power to a sound field of 0.00002 Pa. This would produce a result 94 dBm lower than that of Equation (22.5). If we symbolize this rating by $G_{AIP}$, then

$$G_{AIP} = (S_V - 10 \log R_o - 70) \text{ dBm.} \qquad (22.6)$$

In this rating system the example microphone would produce $-53$ dBm at the threshold of hearing. The advantage of this system is that the power level supplied by a given talker's microphone is obtained by simply adding $G_{AIP}$ to the pressure level of the talker's voice at the microphone's position. $G_{AIP}$ as defined here is very similar to the EIA rating for microphones. The EIA rating system differs in that rather than employing the actual output resistance of the microphone, a nominal microphone impedance rating is employed instead.

## 22.2 Microphone Selection

Microphones are usually selected on the basis of mechanism, sensitivity, nature of response, polar response pattern, and handling characteristics. Mechanism refers to the physical nature of the transducing element of the microphone. Sensitivity in current practice refers to the voltage sensitivity, $S_V$. The nature of response refers to whether the microphone output is proportional to acoustic pressure, acoustic pressure gradient, or acoustic particle velocity. Polar response patterns summarize a microphone's directional characteristics. Handling characteristics are a result of whether the structure of the microphone housing is mechanically isolated from the transducing structure of the microphone. The following is a list of popular microphones according to the transducing mechanism:

1. Carbon.

2. Capacitor.

3. Moving coil.

4. Ribbon.

5. Piezoelectric.

### 22.2.1 Carbon

Carbon microphones made their advent as transmitters in early telephones. Pressure variations on a metallic diaphragm actuated a metallic button contact to either increase or decrease the compaction of carbon granules contained in a brass cup so as to decrease or increase the resistance of the assembly. The impinging sound thus modulated the direct current in a circuit containing a battery and the microphone element. Carbon microphones are quite sensitive and inexpensive to construct. In addition to the normal thermal noise, such microphones suffer from fluctuations in contact resistance between carbon granules even in the absence of acoustic excitation. The high noise floor and restricted frequency response limit the application of such microphones in sound reinforcement systems.

### 22.2.2 Capacitor

Capacitor microphones exist in two basic forms. In one form a capacitor has a front plate formed by a flexible low-mass, metallic, or metal film diaphragm separated by an air gap from an insulated, rigid metallic perforated back plate. Air motion through the perforations in the back plate serves to damp the mechanical resonance of the diaphragm. This resonance occurs at a high frequency as a result of a stiff, low mass diaphragm. The diaphragm is operated at ground potential while the back plate is charged through a very high resistance by a DC voltage source ranging up to 200 V.

In a second form, a permanently polarized dielectric or electret is positioned on the surface of the back plate removing the necessity for an external polarizing voltage source. In both instances the capacitor circuit is completed through a resistance of the order of $10^9\,\Omega$ and the charge on the capacitor remains approximately constant. Pressure variations on the flexible diaphragm produce changes in the air gap dimension, thus raising or lowering the capacitance by a small amount depending on the degree of diaphragm displacement. With a constant charge on the variable capacitor, the voltage variations track the diaphragm displacement variations.

The capacitor circuitry itself is of high impedance and requires that a field effect transistor (FET) source follower be contained within the microphone housing. The source follower may be energized by a local battery in the case of the electret form or may derive its power from the polarizing voltage source in the pure air capacitor form. These microphones, although not the most rugged, can be of extremely high quality with regard to frequency response. As discussed later, the construction details of the microphone capsule may be varied to make the microphone capsule sensitive to either acoustic pressure or acoustic pressure gradient.

### 22.2.3 Moving Coil

The moving coil microphone and the ribbon microphone are collectively referred to as being dynamic microphones. Much discussion has been given previously with regard to some of the features of the moving coil microphone. The mechanical resonance of the moving coil structure is usually made to occur at the geometric mean of the low frequency and high frequency limits describing the microphone's pass band. In a typical case this resonance occurs at about 630 Hz. In the pressure responsive version of such a microphone the back chamber to the rear of the diaphragm contains an acoustic resistance that highly damps the diaphragm mechanical resonance. This damping greatly broadens the resonance, forcing the response to be uniform except at the frequency extremes.

Oftentimes a small resonant tube tuned to a low frequency and vented to the outside is incorporated in the rear cavity. In addition to extending the response at low frequencies, this tube allows the static air pressure in the rear chamber to track slow changes in atmospheric pressure. Even in microphone structures featuring an otherwise sealed rear cavity, a slow leak must always be provided for static pressure equalization. A small air chamber that is resonant at a high frequency may also be located in the rear cavity in order to enhance the response at high frequencies. Moving coil microphone structures are usually quite rugged.

### 22.2.4 Ribbon

The ribbon microphone employs a conductor in a magnetic field, as does a moving coil microphone. Unlike the moving coil, which is located in a radially directed magnetic field, the conductor in a ribbon microphone is a narrow, corrugated metal ribbon located in a linearly directed magnetic field that is perpendicular to the length of the ribbon. The ribbon itself constitutes the diaphragm, both faces of which are exposed to external sound fields.

The driving force on the ribbon is directly proportional to the pressure difference acting on the two faces of the ribbon and hence is proportional to the space rate of change of acoustic pressure. The space rate of change of pressure is called the pressure gradient. The ribbon responds to the acoustic particle velocity with maximum response occurring when the incident sound is normal to a face of the ribbon. This microphone is inherently directional with a figure eight polar pattern. Although featuring excellent performance over a wide frequency range, the structure is inherently fragile and is not suitable for exterior use under windy conditions.

## 22.2.5 Piezoelectric

Piezoelectric microphones depend on a structural property possessed by certain dielectric crystals and especially prepared ceramics. The nature of this property is that if the crystal or ceramic is subjected to a mechanical stress, its shape will be distorted. When this occurs, an electric field appears in the substance as a result of shifted ion positions within the structure. A capacitor can be formed employing such a dielectric that will generate a voltage that is proportional to the mechanical stress. The mechanical stress can be made to result from the motion of a diaphragm exposed to acoustic pressure. In this fashion it is possible to construct a relatively simple, inexpensive pressure-sensitive microphone. Piezoelectric microphones have very high capacitive output impedances. In the past the high voltage sensitivity of such microphones made them popular for recorders and simple public address applications where quite short connecting cables were possible. They are still employed in some sound level meters but other professional application is quite restricted.

## 22.2.6 Matching Talker to Microphone

Distant or bashful talkers require microphones of higher voltage sensitivity in order to produce voltage levels matching those required by microphone input amplifiers. Nearby and professional talkers require microphones of less sensitivity in order to match amplifier input requirements without the use of pads in the input circuitry. Rock singers are an extreme case requiring the least input sensitivity and further requiring both breath blast and pop filters particularly when pressure gradient microphones are employed. Table 22.1 lists representative voltage sensitivity ranges typical of microphones classified according to the mechanism.

### Table 22.1: Microphone Sensitivity Comparison

| Microphone mechanism | $S_V$ in dBV/Pa range |
|---|---|
| Carbon | −20 to 0 |
| Capacitor | −50 to −25 |
| Dynamic | −60 to −50 |
| Piezoelectric | −40 to −20 |

## 22.3 Nature of Response and Directional Characteristics

Pressure microphones are those where only one side of the diaphragm is exposed to the actuating sound field. Such devices are basically insensitive to the direction of the arriving sound as long as the wavelength is large compared with the diaphragm circumference. At high frequencies when the wavelength becomes comparable to or even less than the diaphragm circumference, two directional effects become evident. For sound directly incident on the exposed face of the diaphragm, the partial reflection of the pressure waveform at the diaphragm surface increases the acoustic pressure amplitude over that which would exist in an undisturbed sound field. For sound incident from the rear of the exposed face of the diaphragm, the active face of the diaphragm is in the shadow of the microphone's housing structure and experiences a pressure less than that of the undisturbed sound field. This front-to-back discrimination can only be avoided by employing physically small microphone structures. This is the reason why measurement microphones often have capsules of ¼ inch diameter or even less.

A controlled directional response can be obtained by employing a sensing diaphragm, both faces of which are exposed to the sound field of interest. Such diaphragms experience a driving force that depends on the spatial rate of change of pressure rather than on the pressure itself. Consider the situation shown in Figure 22.3.

Figure 22.3 is a bare bones illustration of a diaphragm stripped of details of the transducing mechanism. Both sides of the diaphragm are exposed to a sound wave that

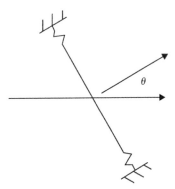

**Figure 22.3: Compliantly mounted diaphragm with both sides exposed to a sound field.**

is propagating along the horizontal axis. The diaphragm may be circular as in a capacitor or moving coil microphone or rectangular as in a ribbon microphone. The principal axis of the microphone is directed perpendicular to the plane containing the diaphragm and, as illustrated, forms an angle $\theta$ with the direction of the incident sound. When $\theta$ has the value $\pi/2$, both faces of the diaphragm experience identical pressures and the net driving force on the diaphragm is zero. Now when $\theta$ is 0, the sound wave is incident normally on the diaphragm and the driving force on the left face of the diaphragm will be the pressure in the sound wave at the left face's location multiplied by the area of the left face.

The diaphragm material, however, is not porous so sound must follow an extended path around the diaphragm along which the sound pressure can undergo a change before reaching the right face. The net driving force on the diaphragm will be the difference in the pressures on the two faces multiplied by the common diaphragm surface area. The pressure difference can be calculated by taking the product of the space rate of change of acoustic pressure, known as the pressure gradient, with the effective acoustical distance separating the two sides of diaphragm. The least value of this distance is the diaphragm diameter in the case of a circular diaphragm.

For a ribbon diaphragm the appropriate value would approximate the geometric mean of the diaphragm's length and width. Details of a particular microphone housing structure that provide a baffle-like mounting will tend to increase the effective separation. If $\theta$ is not zero, the microphone axis is inclined to the direction of the incident sound and the pressure difference is lowered according to the cosine of the angle.

As a first case, consider that the sound source is quite distant from the microphone location so that that the incident sound can be described by a plane wave. The mathematical description of such a wave where the direction of propagation is that of the $x$ axis is

$$p(x,t) = p_m e^{j(\omega t - kx)}, \tag{22.7}$$

where $p_m$ is the acoustic pressure amplitude, $\omega$ is angular frequency $= 2\pi f$, $k$ is propagation constant $= \omega/c = 2\pi/\lambda$, $c$ is phase velocity, and $\lambda$ is wavelength.

Under this circumstance, the net driving force acting on the diaphragm in the direction of increasing $x$ is given by evaluating the following expressions with $x$ set equal to the coordinate of the diaphragm's center.

$$F(t) = \left[ p(x,t) - \left\{ p(x,t) + \frac{\partial}{\partial_x} p(x,t) d \cos\theta \right\} \right] S$$

$$= \left[ -\frac{\partial}{\partial_x} p(x,t) d \cos\theta \right] S \qquad (22.8)$$

where $S$ is the surface area of one side of the diaphragm and $(\partial/\partial_x)p(x,t)$ is the gradient of the acoustic pressure in the direction of increasing $x$.

The pressure gradient is calculated by taking the partial derivative with respect to $x$ of Equation (22.7) as follows:

$$\frac{\partial}{\partial_x} p(x,t) = -jkp_m e^{(\omega t - kx)}$$

$$= -jkp(x,t). \qquad (22.9)$$

Upon substituting the result of Equation (22.9) into Equation (22.8), the driving force becomes

$$F(t) = jkp(x,t) Sd \cos\theta. \qquad (22.10)$$

In a given sound wave of normally encountered intensities, a relationship exists between the acoustic pressure and the acoustic particle velocity. The ratio of the acoustic pressure to the particle velocity is called the specific acoustic impedance of air for the wave type in question. This ratio for plane waves is a real number equal to the normal density of air multiplied by the phase velocity of sound. One can then substitute for the acoustic pressure in Equation (22.10) in terms of the particle velocity to obtain an alternative expression for the driving force:

$$F(t) = jk\rho_0 cu(x,t) Sd \cos\theta \qquad (22.11)$$
$$= j\omega\rho_0 u(x,t) Sd \cos\theta$$

The significance of the imaginary operator $j$ in this equation simply means that the phase angle of the driving force leads that of the particle velocity by $\pi/2$ radians or 90°. The amplitude of the driving force would be

$$F_m = \omega\rho_0 u_m Sd \cos\theta, \qquad (22.12)$$

where $u_m$ is the particle velocity amplitude.

The more often encountered case is where the source is nearby to the microphone location. In such an instance the appropriate wave description is that of a spherical wave propagating along a radial line from the sound source. Mathematically, such a wave is described by

$$p(r,t) = \frac{A}{r} e^{j(\omega t - kr)},$$
(22.13)

where $P_m = \frac{A}{r}$ is the pressure amplitude that is now position dependent and $A$ is a constant determined by the sound source.

The pressure gradient is now more complicated as the space variable $r$ appears in both the denominator and the exponent of the expression for the acoustic pressure.

$$\frac{\partial}{\partial_r} p(r,t) = -\left(\frac{1}{r} + jk\right) p(r,t).$$
(22.14)

If the center of the diaphragm is located at a distance $r$ from the sound source, then the driving force on the diaphragm for the spherical wave becomes

$$F(t) = \left(\frac{1}{r} + jk\right) p(r,t)\, Sd \cos\theta.$$
(22.15)

The driving force now has two components, one of which is in phase with the acoustic pressure while the other leads the acoustic pressure by 90°. The specific acoustic impedance of air for spherical waves is not as simple as was the plane wave case. The ratio of the acoustic pressure to the particle velocity is now

$$\frac{p(r,t)}{u(r,t)} = \frac{j\omega\rho_0}{\frac{1}{r} + jk}.$$
(22.16)

Upon solving Equation (22.16) for the acoustic pressure in terms of the particle velocity and substituting into Equation (22.15), one obtains the very important result

$$F(t) = j\omega\rho_0 u(r,t)\,Sd\cos\theta. \tag{22.17}$$

The importance of this result is apparent when Equation (22.17) is compared with Equation (22.8). With the exception of the identity of the space variable, the two equations are identical, implying that pressure gradient microphones respond to the particle velocity in exactly the same fashion whether the incident sound wave is plane, spherical, or a combination of the two. In contrast, pressure-sensitive microphones respond to acoustic pressure whether the source is nearby (spherical case) or distant (plane case). In fact, for a pressure-sensitive microphone the driving force depends only on the acoustic pressure and is given by the direction independent expression

$$F(t) = p(r,t)\,S. \tag{22.18}$$

Another very important aspect of pressure gradient microphones is the proximity effect. This phenomenon becomes apparent by a rearrangement of Equation (22.16). This equation is solved for the particle velocity in terms of the pressure and the terms and then multiplied in both numerator and denominator by the radial distance while making use of the fact that $k\,\omega/c =$ to obtain

$$u(r,t) = \frac{1 + jkr}{jkr}\left(\frac{p(r,t)}{\rho_0 c}\right). \tag{22.19}$$

The significance of this result is more pronounced when one examines the magnitude of the particle velocity:

$$\left|u(r,t)\right| = \frac{\sqrt{1 + k^2 r^2}}{kr}\left(\frac{\left|p(r,t)\right|}{\rho_0 c}\right). \tag{22.20}$$

When the radial distance is large or the wavelength is short or of course both of these are true, then Equation (22.20) reduces to

$$\left|u(r,t)\right| = \frac{\left|p(r,t)\right|}{\rho_0 c},$$

with the significance that the particle velocity is directly proportional to the acoustic pressure. However, when $r$ is small or the wavelength is large or a combination is true, the reduction becomes

$$|u(r,t)| = \frac{1}{\omega r}\left(\frac{|p(r,t)|}{\rho_0}\right),$$

with the significance that the particle velocity varies inversely with frequency. As a consequence, when a sound source is in close proximity to a pressure gradient microphone the lower frequencies of the source produce a larger response than the higher frequencies. This is the basis for the proximity effect.

One final observation regards the directional characteristics of pressure gradient microphones. From Equation (22.15), when $\theta$ is in the range $\pi/2 < \theta < 3\pi/2$, the cosine of $\theta$ is itself a negative quantity and the polarity of the driving force, as well as the electrical output signal of the microphone, is reversed. A use will now be made of this fact in discussing a microphone structure that possesses a variety of several different directional patterns.

A structure consisting of both a pressure gradient microphone element and a pressure microphone element makes possible a microphone possessing adjustable directional characteristics. The elements should individually be small and located close together with the diaphragms of the two elements located in the same plane. A single signal based on a linear sum of the signals from the individual elements is generated by the combination. The root mean square open circuit electrical output of the assembly can be written as

$$E_o = \alpha(\beta + \gamma\cos\theta), \tag{22.21}$$

where $\alpha$ is a dimensional constant, $\beta$ is the fraction of the pressure microphone electrical signal, $\gamma$ is the fraction of the pressure gradient microphone electrical signal, and $\theta$ is the angle of incidence of the acoustic signal.

The fractional signals can be formed and summed through the employment of passive circuitry contained within the microphone housing. The polar response curve of the microphone for a given choice of coefficients is obtained by allowing $\theta$ to range continuously from 0 to $2\pi$ while plotting the curve

$$r = \left| \beta + \gamma \cos \theta \right|, \tag{22.22}$$

where $r$ is the radial distance from the origin and has a maximum value of 1, $\beta$ and $\gamma$ are fractional coefficients with $\beta + \gamma = 1$, and $\theta$ is the angle of incident sound relative to principal axis of microphone.

Although $\beta$ and $\gamma$ are arbitrary within the constraint that they sum to unity, there are particular values that have proven to be quite useful. This information is listed in Table 22.2.

Some practitioners prefer to employ directional microphones because such microphones respond to reverberant acoustical power arriving from all directions with reduced sensitivity as compared with the same acoustical power arriving along the principal axis of the microphone. This property is expressed by the entry labeled *RE* in Table 22.2. *RE* stands for random efficiency. The hypercardioid pattern, for example, has a random efficiency of ¼. The response to power distributed uniformly over all possible directions is thus only ¼ that for the same total power arriving on axis.

The entry labeled *DF* in Table 22.2 compares the working distance of a directional microphone to that of an omnidirectional microphone. The *DF* for a hypercardioid microphone is 2, meaning that the working distance for a source on axis for this microphone can be twice as large as that for an omni in order to achieve the same direct to reverberant sound ratio in the output signal.

These factors when considered alone would lead one to believe that higher gain before acoustic feedback instability would be achievable through the employment of directional

**Table 22.2: Polar Pattern Parameters for Microphone Directional Characteristics**

| Polar pattern | $\beta$ | $\gamma$ | RE[a] | DF[a] |
|---|---|---|---|---|
| Omni | 1 | 0 | 1 | 1.0 |
| Gradient | 0 | 1 | 1/3 | 1.7 |
| Subcardioid | 0.7 | 0.3 | 0.55 | 1.3 |
| Cardioid | 0.5 | 0.5 | 1/3 | 1.7 |
| Supercardioid | 0.37 | 0.63 | 0.268 | 1.9 |
| Hypercardioid | 0.25 | 0.75 | 1/4 | 2.0 |

[a] Based on data from Shure, Inc.

microphones. This is not necessarily the case. As a class, omnidirectional microphones exhibit smoother frequency responses than directional microphones. The frequencies of oscillation triggered by acoustic feedback, the ring frequencies, depend on a number of factors.

Prominent causative agents are peaks in microphone response and peaks in loudspeaker response coupled with antinodes in the normal modes of the room. Room modes at even moderate frequencies can be quite dense. As a consequence, a single peak in either microphone or loudspeaker response may trigger an entire chorus of slightly different ring frequencies. This set of facts would tend to favor omnidirectional microphones over directional ones. The deciding factor is usually not immunity to feedback from the reverberant field but rather the necessity to reject a nearby source of objectionable sound, including possible strong discrete reflections.

A microphone consisting of a separate pressure and pressure gradient element is quite versatile in that it offers all of the polar response patterns listed in Table 22.2, assuming that it contains the appropriate switch selectable passive circuitry necessary to properly combine the signals from the individual elements. Such a microphone, however, inherently has a shortcoming in that the centers of the two elements are physically offset.

Sound waves incident on the device in other than the principal plane arrive at the two elements at slightly different times. The difference in arrival times introduces a phase difference between the electrical signals generated by the two elements. This phase difference can be significant at high frequencies and can distort the directional response pattern in the high-frequency region. Fortunately, it is possible to avoid the offset problem through the design of a single diaphragm device that also has useful directional characteristics.

Figure 22.4 is a bare bones illustration of a compliantly mounted diaphragm and a back enclosure that is vented through a porous screen to the external environment. The diaphragm may be part of either a capacitor or moving coil type of transducer, the details of which are not shown for simplicity. A sound wave is incident on the left face of the diaphragm. The direction of the incident wave makes an angle $\theta$ with the principal axis of the system. The principal axis is perpendicular to the plane that contains the diaphragm. The acoustic pressure on the left face of the diaphragm assuming a spherical wave is given by

$$p_1 = \frac{A}{r} e^{j(\omega t - kr)}. \tag{22.23}$$

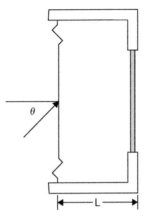

**Figure 22.4: Simplified illustration of a single diaphragm that is sensitive to a combination of pressure and pressure gradient.**

The center of the porous screen to the right of the diaphragm is separated from the corresponding point at the center of the diaphragm by an acoustical distance that amounts to $(d + L)$, where $d$ is the diameter of the diaphragm. We need now to calculate the acoustic pressure at a point just to the right of the center of the porous screen. The acoustic pressure in the incident wave on the diaphragm is a known quantity, $p_1$. As was done in the case of the pressure gradient microphone, we first calculate the rate of pressure change with distance along the direction of propagation. Next, we find the component of this change in the direction of interest. Finally, we multiply this component by the acoustical distance between the points of interest. This last step yields the pressure change. What is desired of course is the pressure at the second point. This is the pressure at the initial point plus the change in pressure. Upon letting $p_2$ represent the acoustical pressure at a point immediately to the right of the center of the porous screen, then

$$p_2 = p_1 + (d + L)\cos\theta\,\frac{\partial}{\partial_r}\,p_1. \tag{22.24}$$

The driving force that actuates the diaphragm, however, is the pressure difference between $p_1$ and the pressure in the cavity to the rear of the diaphragm multiplied by the surface area of one side of the diaphragm. A detailed analysis would show that the pressure in the cavity, $p_e$, depends on both $p_1$ and $p_2$. Recall that for a pressure-sensitive microphone, the diaphragm driving force is directly proportional to the acoustic pressure;

whereas for a pressure gradient microphone, it is directly proportional to the gradient of the acoustic pressure.

In the capsule described earlier, the driving force on the diaphragm is proportional to a linear combination of the pressure and pressure gradient terms. The sizes of the coefficients in the linear combination and consequently the particular directional polar pattern hinge on the volume of the cavity, the areas occupied by the diaphragm and the porous screen, the mechanical properties of the diaphragm, and the porosity of the screen. Such microphones are usually constructed having a dedicated directional pattern. The majority of the cardioid family of directional microphones is constructed in this fashion.

Most microphones have cylindrical symmetry and basically circular diaphragms. The principal axis of such a microphone is centered on the diaphragm, perpendicular to the plane of the diaphragm, and directed along the cylindrical axis, as illustrated in Figure 22.5. The directional polar pattern in a plane is obtained by varying the angle of incident sound relative to the principal axis of the microphone.

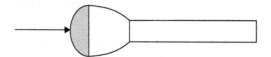

Figure 22.5: Illustration of the principal axis of a cylindrically symmetric microphone.

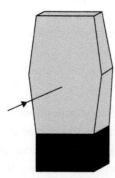

Figure 22.6: Position of the principal axis of a classic ribbon microphone.

The three-dimensional directional response of such a microphone is obtained by revolving the directional polar pattern about the cylindrical axis of the microphone. Ribbon microphones, however, don't follow the aforementioned rules, as their diaphragms do not possess cylindrical symmetry. Such microphones are usually designed to be addressed from the side, as illustrated in Figure 22.6.

The directional response in the horizontal plane of the depicted ribbon microphone is a figure eight. Revolving this pattern about the principal axis generates two spheres that describe the microphone's response in three dimensions. The polar directional patterns listed in Table 22.2 are displayed in Figure 22.7A, while the three-dimensional directional response is sketched in Figure 22.7B.

The polar patterns of Figure 22.7A are theoretical ideals and have a linear radial axis consistent with the form of the describing equations. Real microphones fall short of the theoretical ideal in two ways. They never display complete nulls in response and the polar response curves are frequency dependent. Compare the measured polar response curves of a cardioid microphone presented in Figure 22.8 with its counterpart in Figure 22.7A.

Manufacturer's polar response data are usually presented employing a logarithmic polar axis while excluding a small region in the vicinity of the origin. Such a presentation for yet again a different cardioid microphone is given in Figure 22.9.

In examining Figure 22.9, note that the reference axis has a different orientation and that the radial coordinate represents attenuation expressed in decibels relative to the on-axis value.

## 22.4 Wireless Microphones

Modern wireless microphones allowing untethered motion of the user have proven themselves to be indispensable in concerts, religious services, dramatic arts, and motion picture or video production.

Wireless microphones for use in the performing arts and sound reinforcement first made their appearance about 1960. The first transmitter units were designed to operate in the broadcast FM band between 88 and 108 MHz. The receivers were conventional FM broadcast units. The transmitters did not have to be licensed as the low radiated powers involved complied with Part 15 of the FCC rules. Frequency modulation was accomplished in the transmitter by allowing the audio voltage signal to vary the junction

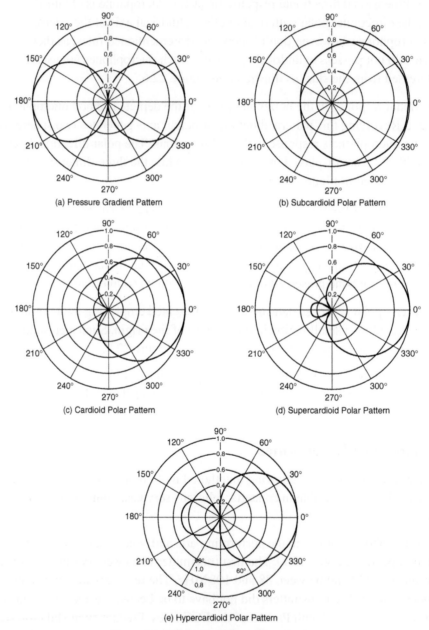

(a) Pressure Gradient Pattern

(b) Subcardioid Polar Pattern

(c) Cardioid Polar Pattern

(d) Supercardioid Polar Pattern

(e) Hypercardioid Polar Pattern

**Figure 22.7A: Standard polar patterns.**

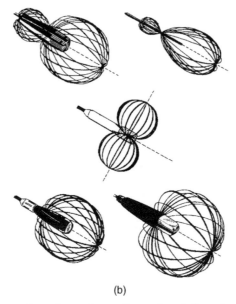

(b)

Figure 22.7B: Microphone three-dimensional directional response.
(Courtesy of Shure Brothers Incorporated)

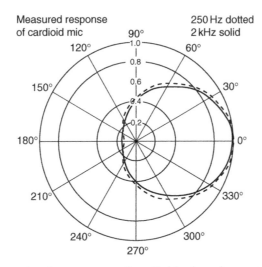

Figure 22.8: Measured polar response of cardioid microphone at 250 Hz and 2 kHz.

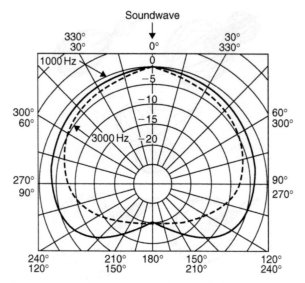

Figure 22.9: Polar response of a cardioid microphone.

capacitance of a bipolar transistor connected as a Hartley or other simple oscillator tuned to the desired carrier frequency in the FM band.

Such oscillators were prone to drift in operating frequency as the transistor characteristics were sensitive to both temperature and supply voltage variations. This required periodic retuning of the receiver to compensate for transmitter frequency drift. This was particularly true of the very early units that employed germanium transistors. Significant improvement in this regard was made possible with the availability of suitable silicon transistors.

One of the authors well remembers hand crafting several body pack transmitters in 1965 for use by lecturers at Georgia Tech. The receivers employed were H. H. Scott units that had been modified to incorporate automatic frequency control circuitry to compensate for the transmitter drift within reasonable limits. These early units had acceptable audio bandwidths but the simple modulation technique employed did not produce large frequency deviations, resulting in a small dynamic range of the recovered audio signal.

Those of us who have experienced the entire history of wireless microphones consider the present-day versions to be truly remarkable. Not only have the early shortcomings been addressed but also features not even envisioned by the early practitioners have been added. Frequency space has been made available in both the VHF and UHF frequency bands with UHF units currently being more popular. The UHF band offers more flexibility with regard to the number of different frequencies that may be employed simultaneously as well as a higher probability of finding unused frequency space in a given locale. Additionally, required receiving antenna lengths are much more manageable in the UHF band. For example, with a carrier frequency of 900 MHz and a wave speed of $3 \times 10^8$ m/s, the wavelength becomes one-third of a meter or about 13 inches. The required receiving antennas range between ¼ and ½ wavelength and thus have lengths falling between about 3 and 6 inches.

There are several significant technical innovations incorporated in current wireless microphone systems that are worthy of note. Each of these will be discussed in turn.

1. Receiver assisted setup.

2. Space diversity reception.

3. Transmitter preemphasis—receiver deemphasis.

4. Transmitter compression—receiver expansion.

A difficult problem associated with setting up wireless microphone systems in the past has been that associated with determining interference-free operating frequencies. This was particularly true when the application required the simultaneous operation of a large number of separate audio channels, each of which required an individual radio frequency assignment. Receivers having assisted setup facilities have built in protocols for scanning the entire operating band and identifying those potential operating frequencies that are free of any radio frequency carrier at the time of scan. Several such scans performed over a period of time usually are quite successful in defining interference-free operating frequencies.

Space diversity reception solves a problem depicted in Figure 22.10(a) by means of an arrangement suggested by Figure 22.10(b).

In Figure 22.10(a), a single receiving antenna is employed. This antenna receives a signal via a direct path to the transmitter as well as a transmitter signal that has been reflected by a nearby object and thus follows a longer more indirect path along its way to the receiving antenna. The phases of these two signals having the same frequency are different and hence they can interfere with each other. The interference may be

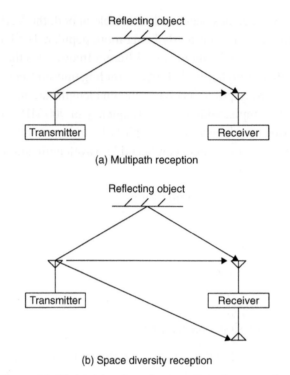

(a) Multipath reception

(b) Space diversity reception

**Figure 22.10: Multipath and space diversity reception.**

either constructive or destructive according to the degree of phase difference. When the interference is destructive, the resultant signal may be so weak that the receiver will not be able to recover the program material.

The arrangement shown in Figure 22.10(b) greatly reduces the probability that there will be a complete loss of program material. In this arrangement, two antennas located somewhat less than a wavelength apart are employed. In this arrangement, the reflected signal may not even arrive at the second antenna as shown. Even when this is not the case or when there are other reflecting objects, the chances that both antennas are subjected to destructive interference simultaneously are reduced greatly.

There are several techniques for handling the signals that appear in the space diversity antennas. In one technique the space diversity receiver is fitted with separate radio frequency amplifiers for each antenna. The signals from each of these amplifiers are compared as to strength with the stronger signal at any instant being switched to the remainder of the single receiver circuitry.

In a variation on this technique, the signals from both radio frequency amplifiers are summed and then fed to the rest of the circuitry of a single receiver with no switching being involved. Finally, two receivers set to receive the same carrier frequency are employed, one for each receiving antenna. The automatic gain control voltages that are developed at each receiver's detection stage are compared with the audio output circuitry being switched to that of the receiver having the larger control voltage. This last technique is the most expensive and, even though it involves switching, has perhaps the best performance overall.

Wireless microphone transmitters employ a relatively small frequency deviation in the frequency modulation process. The modulation index is thus small. This restricts the dynamic range that is available for program material and weak signals may be lost in the noise floor. A long-term average of the spectral density associated with both voice and music programs exhibits a broad maximum in the vicinity of 500 Hz accompanied by a roll off in density beyond about 2 kHz. The spectral density is the average power per unit frequency interval. This being the case, it is necessary to pre emphasize the higher frequencies in the audio material prior to further signal processing.

The normal range of the audio material to be transmitted may well be as large as 80 dB while the available range in the small deviation FM transmitter may be only 40 dB. The 80-dB range of the audio material is squeezed into the 40-dB range available by 2 into 1 compression prior to the modulation process. After transmission and reception at the receiver, the recovered audio material occupying a 40-dB range is first subjected to a 1 into 2 expansion in order to restore the full dynamic range of 80 dB.

This is then followed by a de emphasis of the audio material above 2 kHz in order to restore the natural spectral balance of the audio material. Figure 22.11 displays typical preemphasis and complementary deemphasis curves with the upper curve being that of preemphasis. The combination of the two yields a flat response across the audio band.

The process of compressing the audio dynamic range prior to transmission and expanding the range of the audio material following reception has been termed compansion. A typical compression curve employed in the audio circuitry of the transmitter, followed by the complementary expansion curve employed in the audio circuitry of the receiver, is displayed in Figure 22.12.

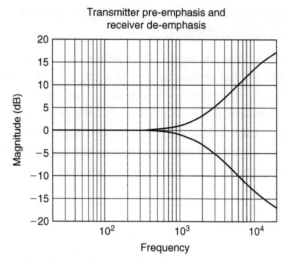

**Figure 22.11: Typical preemphasis and deemphasis curves.**

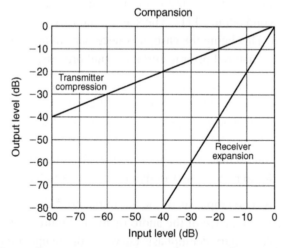

**Figure 22.12: Dynamic range compression at the transmitter and complementary expansion at the receiver.**

Transmitter units may be hand-held with a built-in microphone element or a body pack unit provided with a minireceptacle for a microphone connection. The microphones employed with body pack units are usually miniature dynamic or electret capacitor microphones attached to short cables fitted with mating connectors to that of the transmitter.

The microphone elements are fitted with clips for attachment to the user's clothing. Occasionally, the microphone element may be part of a head microphone boom structure.

Typical transmitter features are:

1. Power on–off switch.

2. Carrier frequency selection and indicator.

3. Battery level indicator.

4. Audio gain control.

5. Audio overload indicator.

6. Audio mute switch on body pack units.

7. Nine-volt battery.

Receiver units may be stand-alone or rack mounted and are usually powered from conventional power mains. Audio outputs are provided at both line and microphone levels.

A typical space diversity receiver providing assisted setup has the following features:

1. Power on–off switch.

2. Scan or operate control.

3. Carrier frequency indicator.

4. Squelch control.

5. Active receive antenna indicator.

6. Radio frequency level indicator.

7. Transmitted audio level indicator.

8. Transmitter battery life indicator.

9. Audio output level control.

A photograph of a space diversity wireless microphone system is presented in Figure 22.13.

**Figure 22.13: A wireless microphone system.**
**(Photo courtesy of Michael Pettersen of Shure, Inc.)**

One final note with regard to wireless microphone systems distilled from years of sad personal experience. The first three rules for dealing with wireless microphone systems are:

1. Batteries.

2. Batteries.

3. Batteries!

Wireless transmitters are usually powered by 9-V batteries that may be composed from primary or nonrechargeable cells or secondary cells that are rechargeable. Even if one ordinarily uses rechargeable batteries, it is well to keep a fresh supply of nonrechargeable units on hand. The histories of rechargeable batteries must be managed carefully in order to assure their proper performance. Many practitioners prefer to employ only fresh nonrechargeable batteries along with frequent replacement because of sad experiences with rechargeable units. Battery failure at a critical moment can lead to years of bad dreams.

## 22.5 Microphone Connectors, Cables, and Phantom Power

It is almost universal practice in professional audio to provide signal sources with male connectors and signal receivers with female connectors. Additionally, it is common

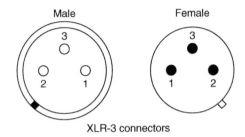

XLR-3 connectors

**Figure 22.14: Pin arrangements of XLR-3 connectors.**

practice to employ balanced circuits for both input and output in those instances where the signal levels are low and susceptible to electrical noise or cross talk interference. Indeed, many systems maintain balanced linking circuits throughout regardless of the signal levels.

If one excludes miniature microphones that constitute a special case, the de facto standard microphone connector is the XLR-3. The male and female versions of this connector are illustrated in Figure 22.14.

Through the years the assignment of functions to the various pins has varied. The present standard assignment of the male connector at the microphone has pin 1 connected to the microphone case. Pin 2 is connected to the microphone circuitry such that a positive pressure on the microphone diaphragm drives the voltage at pin 2 in the positive sense. Pin 3 is connected to the microphone circuitry such that a positive pressure on the microphone diaphragm drives the voltage at pin 3 in the negative sense. Pins 2 and 3 are balanced with respect to pin 1.

Quality microphone cables consist of a twisted pair of insulated, color-coded inner conductors formed from stranded copper wire covered by a tightly woven copper braided shield with the combination encased in an insulating jacket. The conductors may be tinned, although this is not always the case. The cable is fitted with a female connector at one end and a male connector at the other. The connector pin assignments in this instance have pin 1 connected to the shield with the option of also strapping the connector shell to pin 1. Pin 2 is connected to the positive signal conductor while pin 3 is connected to the negative signal conductor. Microphone cable is also often used as the connecting cable in link circuits between mixers, subsequent signal processing units, and power amplifiers.

Shielded, twisted pairs in balanced circuits are an absolute necessity in handling low-level signals in order to avoid electromagnetic interference. The braided shield alone offers protection from electrostatic fields but offers very little protection from changing magnetic fields. The practice of employing twisted pair conductors stems from experience gleaned from the early days of the telephone industry.

In former times, long distance circuits between cities and local circuits in rural areas employed open bare wire pairs affixed to separate glass insulators attached to multiple cross arms, which were in turn elevated by poles. It was learned early on that open-air electrical power lines that often followed parallel paths caused interference. It was found that by periodically transposing the positions occupied by the two conductors of a given circuit pair, that the interference could be reduced greatly if not eliminated altogether. This transposition amounted to periodically twisting without touching one conductor of a circuit pair over the other, in effect forming an insulated, twisted pair even though the distance between twists was relatively large. The explanation for this annulment of the interference appears in Figure 22.15.

In Figure 22.15 imagine that the twisted pair of conductors is replicated to the right as well as to the left to form an extended circuit. Imagine also that in the vicinity a magnetic field is instantaneously directed into the figure as indicated by the X's and that the field strength is increasing with time. Examine the two closed paths as indicated by the circles. According to Lenz's law, the induced voltage acting in the loops has the sense indicated by the arrows. Now look at the white conductor in the upper left, the induced voltage in this portion of conductor is in the same direction as is the arrow adjacent to it. Compare

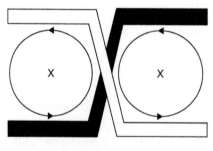

Twisted pair

**Figure 22.15: Twisted pair exposed to a time-changing magnetic field.**

that with the induced voltage in the white conductor in the lower right in which the induced voltage is oppositely directed.

The same analysis applied to the two similar segments of the black conductor yields identical results. There is no voltage induced in either conductor in the transposition region as the arrows in the adjacent circles are oppositely directed. In practice, the magnetic field alternates but as it changes its direction of growth, the induction in the loops reverses direction also while the net voltage induced in the transposed conductors remains at zero. Static magnetic fields are of no consequence unless a conductor is moving through them. Even so, a twisted pair translated through a magnetic field that is static in time will experience a net-induced voltage only if the magnetic field varies rapidly with position in space.

Air capacitor microphones require a source of polarization voltage, as well as a DC power source for operating the source follower that handles the microphone signal. Electret capacitor microphones are self-polarized but still require power for the source follower signal circuitry. This power is usually supplied by the microphone mixer via the cable connecting the microphone to the mixer. The circuitry employed for accomplishing this must maintain balance of the microphone signal circuitry. DC circuits that perform this task are called phantom power supplies. One such arrangement is depicted in Figure 22.16.

The arrangement of Figure 22.16 features an output transformer internal to the microphone housing, as well as an input transformer internal to the mixer. The DC voltage is applied equally to the microphone signal conductors at pins 2 and 3. The DC

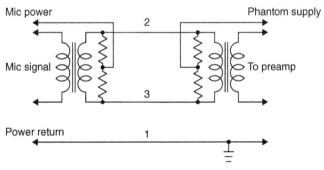

**Figure 22.16: Phantom power arrangement for capacitor microphones.**

return circuit is through the shield on the microphone cable at pin 1. Conductors 2 and 3 have the same DC potential and hence there is no direct current in the transformer windings. In order to accomplish this, the resistors denoted as $R$ must be carefully matched to be equal to within $\pm 0.1\%$. This precision is required not only for DC balance but also to maintain a large common mode rejection ratio. Commonly encountered voltage and resistor values are listed here.

| Supply voltage | Resistor value |
|---|---|
| 12 V | 680 Ω ± 0.1% |
| 24 V | 1200 Ω ± 0.1% |
| 48 V | 6800 Ω ± 0.1% |

There is a trend by some designers to employ electronically balanced inputs in the mixer input microphone circuitry. In such instances, blocking capacitors must be employed to isolate the differential mixer input from DC while maintaining continuity for the microphone signal. Such an arrangement appears in Figure 22.17.

The phantom power circuits of Figures 22.16 and 22.17 work well but both have an undesirable feature. The necessity of the employment of matched balancing resistors in both instances limits the current that may be supplied to power the microphone circuitry. This limitation can be removed through the employment of transformers that are center tapped on the appropriate windings. Such transformers would be quite expensive because of the necessity of very accurately having both an equal number of turns on either side

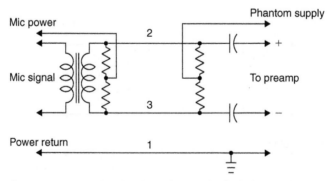

**Figure 22.17: Phantom power circuit when electronically balanced inputs are employed.**

of the center tap as well as exact resistance of the turns on either side of the center tap. If this is not accomplished, direct current will exist in the transformer winding and the signal circuit will no longer be exactly balanced.

Finally, a word of caution is in order. Sound systems may employ just a few or a very large number of microphone cables not only for microphones but also for link circuits. It is important to maintain correct signal polarity in all microphones, microphone cables, link circuits, processing electronics, loudspeaker wiring, and loudspeakers. There are convenient commercial devices called polarity checkers that can be employed to check individual microphones, cables, and overall system polarity. An investment in such devices is modest, time saving, and will earn its keep many times over.

## 22.6 Measurement Microphones

A collection of measurement microphones, whether residing in sound level meters or stand-alone devices, is an absolute necessity for sound system installers as well as acoustical consultants. Such a collection must also be supported by an appropriate microphone calibrator system that consists of both the calibrator itself and a set of adapters to accommodate the various individual sizes of the microphones in the collection.

For many years there were only two suppliers of quality measurement microphones: Brüel and Kjaer, a Danish firm, and GenRad, a domestic firm. Brüel and Kjaer still exists, although not under the original ownership, while GenRad no longer exists. Fortunately, there are now several new domestic suppliers of quality measurement microphones.

Measurement microphones are dominantly air capacitor or electret capacitor microphones while ceramic piezoelectric units may still be encountered. The standard sizes in terms of capsule diameter are 1 inch, ½ inch, ¼ inch, and ⅛ inch. The larger units have higher sensitivity and lower noise floors. The 1-inch unit is favored for making measurements in quiet environments at frequencies below about 8 kHz. The ½-inch unit is a general purpose one but has high frequency limitations.

Broad frequency band measurements usually require the ¼- or ⅛-inch variety, particularly if high sound levels are to be encountered. All sizes can have low-frequency responses that extend almost to 0 Hz, with 3 to 5 Hz being typical with even lower values being possible. A slow leak for allowing the capsule's rear chamber pressure to follow weather-induced atmospheric pressure variations determines the low-frequency limit.

The geometry of a measurement microphone's physical structure is that of a cylinder with the central axis of the cylinder being perpendicular to the plane that contains the microphone capsule's circular diaphragm. This central axis serves as a reference direction for sound incident on the microphone. Direct sound arrives at 0° relative to this axis while grazing incidence occurs at 90°, as illustrated in Figure 22.18.

Any measurement microphone should be encased in such a fashion that the microphone's physical structure disturbs the sound field in which it is immersed to a minimum degree. When the microphone capsules are smaller than ½ inch in diameter it is impossible to incorporate the necessary circuitry and connector in a uniform cylinder having a diameter equal to that of the capsule. In such instances it is necessary to enclose the circuitry and connector in a larger cylinder that is joined to the capsule by a smoothly tapered section matching the larger diameter to the smaller diameter. A notable example of this is displayed in Figure 22.19.

### 22.6.1 Measurement Microphone Types

Despite the smoothness of the microphone enclosure, one cannot escape the fact that at high frequencies the microphone capsule diameter, $d$, is comparable to the sound

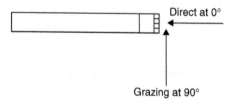

**Figure 22.18: Illustration of direct and grazing sound incidences.**

**Figure 22.19: An example of a well-engineered tapered microphone structure. (Photo courtesy of Alex Khenkin of Earthworks, Inc.)**

wavelength, $\lambda$. When this occurs, the sound field is disturbed by reflection from the capsule's diaphragm as well as diffraction by the capsule's protective grid and the microphone housing. The degree of this disturbance depends on the angle of incidence of the sound and is greatest for direct incidence.

The acoustic pressure at the diaphragm for directly incident sound at high frequencies can, in fact, exceed by several decibels that which would have existed in the free field. The free field pressure is that which would have existed if the obstacle

$$p = \gamma P_0 \frac{A x_m \cos(2\pi ft)}{V_0}.$$ (22.25)

## Further Reading

Ballou, G. M., *Handbook for sound engineers*, 3rd ed., Boston: Focal Press, 2002.

Beranek, L. L., *Acoustics*, New York: Mc-Graw-Hill, 1954.

Brüel and Kjaer, *Measuring microphones*, Technical review: Naerum, 1972.

Eargle, J., *The microphone book*, 2nd Ed., Boston: Focal Press, 2004.

Morse, P. M., *Vibration and sound*, 2nd ed., New York: Mc-Graw-Hill, 1948.

# CHAPTER 23

# *Loudspeakers*

Ian Sinclair

The conversion from electronic signals to sound is the formidable task of the loudspeaker. In this chapter, Ian Sinclair examines principles and practice of modern loudspeaker design.

A loudspeaker is a device that is actuated by electrical signal energy and radiates acoustic energy into a room or open air. The selection and installation of a speaker, as well as its design, should be guided by the problem of coupling an electrical signal source as efficiently as possible to an acoustical load. This involves the determination of the acoustical load or radiation impedances and selection of a diaphragm, motor, and means for coupling the loaded loudspeaker to an electrical signal source. The performance of the speaker is intimately connected with the nature of its acoustic load and should not be considered apart from it.

## 23.1 Radiation of Sound

Sound travels through the air with a constant velocity depending on the density of the air; this is determined by its temperature and the static air pressure. At a normal room temperature of 22°C and static pressure $p_0$ of 751 mm Hg ($10^5 N/m^2$), the density of the ambient air is 1.18 Kg/m³. Under these conditions, the velocity of sound is 344.8 m/s, but 340 m/s is a practical value. The wavelength of a sound ($X$) is equal to the velocity of propagation described by its frequency:

$$\lambda = \frac{340 \text{ m/s}}{f}. \tag{23.1}$$

The sensation of sound is caused by compressions and rarefactions of the air in the form of a longitudinal oscillatory motion. The energy transmitted per unit area varies as the square of the distance from the source. The rate with which this energy is transmitted expresses the *intensity* of the sound, which directly relates to the sensation of *loudness*. This incremental variation of the air pressure is known as *sound pressure*, which, for practical purposes, is what is measured in determining the loudness of sound.

Sound pressure level (*SPL*) is defined as 20 times the logarithm to base 10 of the ratio of the effective sound pressure (*P*) to the reference sound pressure ($P_{ref}$):

$$SPL = 20 \log \frac{P}{P_{ref}} \text{ dB.} \tag{23.2}$$

$P_{ref}$ approximates to the threshold of hearing and numerically is 0.0002 microbar ($2 \times 10^{-5}$ N/m$^2$).

The intensity (*I*) of a sound wave in the direction of propagation is

$$I = \frac{p^2}{P_0 C} W/m^2$$
$$P_0 = 1.18 Kg/m^2 \tag{23.3}$$
$$C = 340 \, m/s.$$

The intensity level (*IL*) of a sound in decibels is

$$IL = 10 \log \frac{I}{IL_{ref}} \text{ dB}$$
$$I_{ref} = 10^{-12} \, w/m^2 = 2 \times 10^{-5} N/m^2. \tag{23.4}$$

The relationship between *IL* and *SPL* is found by substituting Equation (23.2) for intensity (*I*) in Equation (23.4). Inserting values for $P_{ref}$ and $7_{ref}$ gives

$$IL = SPL + 10 \log \frac{400}{P_0 C} \text{ dB.} \tag{23.5}$$

It is apparent that the intensity level *IL* will equal the sound pressure level *SPL* only if $p_oC = 400$ Rayls. For particular combinations of temperature and static pressure this will be true, but under "standard measuring conditions" of

$$T = 22°C \text{ and } P_0 = 751 \text{ mm Hg}, P_0C = 407 \text{Rayls}. \qquad (23.6)$$

The error of –0.1 dB can be neglected for practical purposes.

## 23.2 Characteristic Impedance

The characteristic impedance is the ratio of the effective sound pressure to the particle velocity at that point in a free, plane, progressive sound wave. It is equal to the product of the density of the medium times the speed of sound in the medium ($p_0C$). It is analogous to the characteristic impedance of an infinitely long, dissipation-less, transmission line. The unit is the Rayl, or Newton s/m$^3$.

## 23.3 Radiation Impedance

When a vibrating diaphragm is placed in contact with air, its impedance to motion is altered; the added impedance seen by the surfaces that emit useful sound energy is termed "radiation impedance." The radiation reactance is usually positive, corresponding to an apparent mass. Both reflective mass and resistance as seen by the diaphragm depend on its size, shape, frequency, and the acoustical environment in which it radiates.

## 23.4 Radiation From a Piston

Many radiating sources can be represented by the simple concept of a vibrating piston located in an infinitely large rigid wall. The piston is assumed to be rigid so that all parts of its surface vibrate in phase and its velocity amplitude is independent of the mechanical or acoustic loading on its radiating surface.

Figure 23.1 shows the problem: we wish to know the sound pressure at a point A located distance *r* and angle $\theta$ from the center of the piston. To do this, we divide the surface of the piston into a number of small elements, each of which is a simple source vibrating in phase with all the other elements. The pressure *A* is, then, the sum in magnitude and

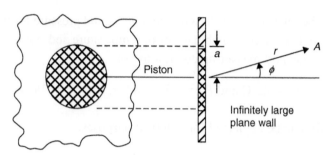

**Figure 23.1: Piston in infinitely plane wall.**

phase of the pressures from these elementary elements. For $r$ large compared with the radius of the piston $a$, the equation will be

$$P(\text{sound pressure N/m}^2) = \frac{\sqrt{2}\, jfp_0 u_0 \pi a^2}{v} \left[ \frac{2J_1(K_a\phi)}{K_a \sin\phi} \right] e^{j\bar\omega(t-r)}. \qquad (23.7)$$

where $u_0 = $ RMS velocity of the piston and $J_1()$ is Bessel function of the first order. Note that the portion of Equation (23.7) in square brackets yields the directivity pattern.

## 23.5 Directivity

At frequencies where the wavelength of sound $(X)$ is large compared with the diameter of the piston, the radiation is spherical. As the frequency is increased, the wavelength becomes comparable or less than the piston diameter and the radiation becomes concentrated into a progressively narrowed angle.

The ratio of pressure $P_0$ at a point set at an angle $\theta$ off the axis, to the on axis pressure $P_A$ at the same radial distance, is given by

$$\frac{P_\phi}{P_A} = \frac{2J_1 \left[ \dfrac{2\pi a}{\lambda} K_a \sin\phi \right]}{\dfrac{2\pi a}{\lambda} K_a \sin\phi.} \qquad (23.8)$$

Figure 23.2 shows radiation patterns for different ratios of $X/D$. The radiation from a piston is directly related to its velocity, and we can compute the acoustic power radiated and the sound pressure produced at any given distance in the far field.

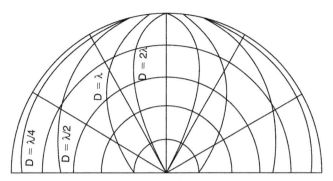

**Figure 23.2: Directivity of piston as a function of piston diameter and wavelength.**

## 23.6 Sound Pressure Produced at Distance *r*

*Low frequencies.* When the piston diameter is less than one-third wavelength ($AT_a < 1.0$) it is essentially nondirectional and can be approximated by a hemisphere whose RMS volume velocity $u_x$ equals:

$$u_1 \text{(diaphragm velocity)} = Sd \times u_c = \pi a^2 u_c \tag{23.9}$$

and the RMS sound pressure at distance *r* is:

$$p(r) = \frac{sdfp_0}{r} = \frac{\pi a^2 fp_0}{r} \text{ N/m}^2 \tag{23.10}$$

and total power radiated $W_t$ is:

$$W_t = \frac{4\pi p_0}{c}(Sd \times f \times u_c)^2 W. \tag{23.11}$$

*Medium frequencies.* At frequencies where the radiation from the piston becomes directional but still vibrates as one unit, the pressure produced at a distance *r* depends on the power radiated and the directivity factor *Q:*

$$p(r) = \sqrt{\left(\frac{W_r Q P_0 C}{a\pi r^2}\right)} \tag{23.12}$$

where

$$Q = \frac{4\pi Pax^2}{2\pi \int_0^x p^2 \phi \sin\phi \, d\phi}.$$  (23.13)

The mechanical impedance in MKS mechanical ohms (Newton-seconds/meter) of the air load upon one side of a plane piston mounted in an infinite baffle and vibrating sinusoidally is

$$Z_m = R_{mR} + jX_m = a^2\pi c\rho_0 \left[1 - \frac{J_1(2K_a)}{K_a}\right] + \frac{j\pi\rho_0 c}{2k^2} K_1(2K_a),$$  (23.14)

where $Z_m$ is mechanical impedance in Newton seconds/meter, $\alpha$ is radius of piston in meters, $\rho_0$ is density of gas in kg/cubic meter, $c$ is velocity of sound in meters/second, $R_{mR}$ is mechanical resistance in Newton seconds/meter (this component varies with frequency), $X$ is mechanical reactance in Newton seconds/meter, $K$ is $\infty/c = 2\pi/\lambda$ 5 wave number, and $J_1 K_1$ is two types of Bessel function given by the series:

$$J_1(W) = \frac{W}{2} - \frac{W^3}{2^2 \times 4} + \frac{W^5}{2^2 \times 4^2 \times 6} - \frac{W^7}{2^2 \times 4^2 \times 6^2 \times 8}$$  (23.15)

$$K_1(W) = \frac{2}{\pi}\left(\frac{W^3}{3} - \frac{W^5}{3^2 \times 5} - \frac{W^7}{3^2 \times 5^2 \times 7}\right),$$  (23.16)

where $W$ is $2K_a$.

Figure 23.3 shows graphs of the real and imaginary parts of this equation:

$$Z_m = R_{mR} + jx_m \text{ as a function of } ka$$

It will be seen that for values of $K_a < 1$, the reactance $X$ varies as the first power of frequency, while the resistive component varies as the second power of frequency. At high frequencies (i.e., $K_a > 5$) the reactance becomes small compared with resistance, which approaches a constant value. The graph can be closely approximated by the analogue (Figure 23.4), where

$$R_{m1} = 1.386a^2\rho_0 c \text{ MKS mechanical ohms}$$
$$R_{m2} = \pi^2\rho_0 c \quad\quad \text{MKS mechanical ohms}$$
$$C_{m1} = 0.6/a\rho_0 c^2 \quad \text{metrea/Newton}$$
$$M_{m1} = 8a^3\rho_0/3 \quad\quad Kg.$$

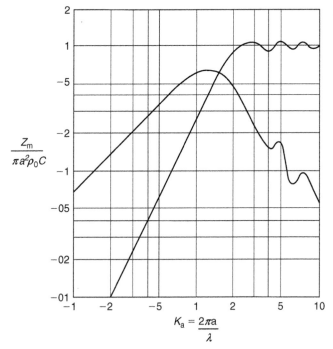

**Figure 23.3: Air load on a plane piston; mechanical impedance ref.; driving point.**

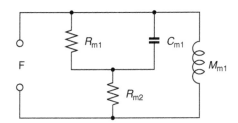

**Figure 23.4: Impedance analogue of Figure 23.3.**

Numerically, for $K_a$ , $l$:

$$R_{mR} = 1.5a^4 \rho_0/c \; ohms$$
$$M_m = 2.67a^3 \rho_0 \; kg.$$

It will be seen that the reactive component behaves as a mass loading on the diaphragm and is a function of diaphragm area only.

The term $K_a$ has special significance: it relates the diaphragm radius to the wavelength of sound at any particular frequency. It is numerically equal to

$$K_a = \frac{2\pi a}{\lambda},$$

(23.17)

where $a$ is radius of the diaphragm and $\lambda$ is wavelength.

When the wavelength $\lambda$ is greater than the circumference of the diaphragm, the loudspeaker behaves substantially as a point source and the sound field pattern is essentially omnidirectional. At the same time the radiation resistance increases with frequency. Thus at frequencies below $K_a = 1$, the increase in radiation resistance with frequency is exactly balanced by the reduction in velocity of the diaphragm with frequency due to its mass reactance (assuming there are no resonances in the diaphragm) and the sound pressure will be constant. At values above $K_a = 1$, the radiation resistance (neglecting the minor "wiggles") becomes constant, but because of focusing due to the diaphragm dimensions being greater than $\lambda$, the sound pressure on the axis remains more or less constant. The velocity of sound in air is approximately 340 m/s, therefore a 150-mm (6 inch)-diameter diaphragm will behave as a point source to a limiting frequency of about 720 Hz; thereafter it begins to focus. Various artifices (such as corrugations) are used with paper cones to extend this range, with more or less success. This was the classic premise that Rice and Kelogg postulated in 1925 when they reinvented the moving coil loudspeaker, and it is still fundamental today.

To summarize, the loudspeaker should operate under mass-controlled conditions and (neglecting directional effects due to focusing of the diaphragm) sound pressure will be constant and independent of frequency; for a given magnet and coil system it will be inversely proportional to the total mass of the diaphragm and moving coil system.

## 23.7  Electrical Analogue

The analysis of mechanical and acoustical circuits is made very much easier by the application of analogues in which mass is equivalent to inductance, compliance to capacitance, and friction to resistance. Using SI units, direct conversion among acoustical, mechanical, and electrical elements can be performed.

The three basic elements (RCL) of acoustical, electrical, and mechanical, systems are shown schematically in Figure 23.5. The inertance $M$ of an acoustic system is represented by the mass of gas contained in a constriction that is short enough so that all particles are assumed to move in phase when actuated by a sound pressure. The compliance $C$ of the system is represented by an enclosed volume, with its associated stiffness. It should be noted that the mechanical analogue of acoustic compliance is not mechanical stiffness, but rather its reciprocal, mechanical compliance $C_m = V_s$. Although resistance of an acoustic system may be due to a combination of a number of different factors, irrespective of its origin, it is conveniently represented by narrow slits in a pipe, for the viscous forces that arise when gas is forced to flow through these slits always results in the dissipation of energy.

The Helmholtz resonator may be graphically represented by Figure 23.6, but converting it to its electrical analogue shows it to be a simple resonant circuit, which can be analyzed easily using general circuit theorems.

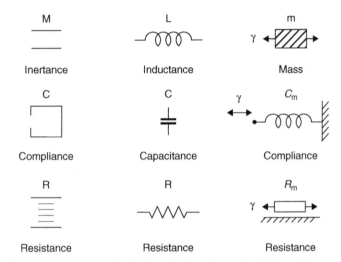

Figure 23.5: Acoustical, electrical, and mechanical analogues.

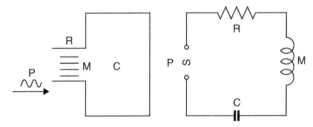

Figure 23.6: Schematic representations of a Helmholz resonator.

The beauty of the analogue method of analysis is that it is possible, by using various transformation equations, to refer the acoustic and electrical parameters to the mechanical side or, conversely, the mechanical and acoustic parameters to the electrical side, etc. For the purpose of this analysis the electrical and acoustical parameters are referred to the mechanical side. The diaphragm can be thought of as an acoustic/mechanical transducer, that is, a device for transforming acoustic energy to mechanical energy, and vice versa. Under these circumstances it will also act as an impedance transformer, that is, it will convert acoustic inertance into mechanical mass and acoustic compliance into mechanical compliance and acoustic resistance into mechanical resistance. The equivalent mechanical values of the acoustical quantities may be obtained from the following relationships:

Mechanical    Acoustic

Force     = pressure × Area

$$F_m \qquad = p \times sd \text{ Newtons} \qquad (23.18)$$

Velocity    $= \dfrac{\text{Volume velocity}}{\text{Area}}$

$$u = \frac{U}{Sd} \qquad (23.19)$$

Displacement $= \dfrac{\text{Volume displacement}}{\text{Area}}$

$$x = \frac{x_v}{Sd} \qquad (23.20)$$

Resistance = Acoustical resistance × Area squared

$$R_m = R_a \times Sd^2 \qquad (23.21)$$

Mass = Inertance × Area squared

$$m = M \times Sd^2 \qquad (23.22)$$

Compliance $= \dfrac{\text{Acoustical capacitance}}{\text{Area squared}}$

$$C_m = \frac{C_a}{Sd^2} \qquad (23.23)$$

## 23.8 Diaphragm/Suspension Assembly

Assuming that the diaphragm behaves as a rigid piston and is mass controlled, the power response is shown in Figure 23.7 where $f_0$ is the system fundamental resonant frequency. Above this, the system is mass controlled and provides a level response up to $f_t$; this corresponds to $K_a = 2$ (see Figure 23.3). Above this frequency the radiation resistance is independent of frequency, and the response would fall at 12 dB/octave, but because of the "directivity" effect, the sound field is concentrated into a progressively narrower beam. The maximum theoretical rate of rise due to this effect is 12 dB/octave, thus the on-axis HF response should be flat. In real life this is only approximated.

## 23.9 Diaphragm Size

It has been found experimentally that the effective area of the cone is its projected or base area. This should not be confused with the advertised diameter of the loudspeaker, which is anything from 25 to 50 mm greater than the effective cone diameter. In direct radiator loudspeakers and at low frequencies, radiation resistance is proportional to the fourth power of the radius (square of the area) and the mass reactance to the cube of the radius. The resistance/reactance ratio (or power factor) of the radiation impedance is therefore proportional to piston radius. Thus the electroacoustic efficiency, other factors being constant, increases with diaphragm area at low frequencies. For constant radiated power, the piston displacement varies inversely with area, hence "long throw" type of small diaphragm area loudspeakers. With fixed amplitude, the radiated power is proportional to the square of the area at a given frequency, or a frequency one octave lower may be reproduced if the area is increased by a factor of four. The upper limit of diaphragm size is set by increased weight per unit area required to give a sufficiently rigid structure.

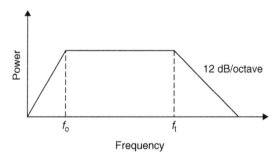

**Figure 23.7: Power response of an infinitely rigid piston.**

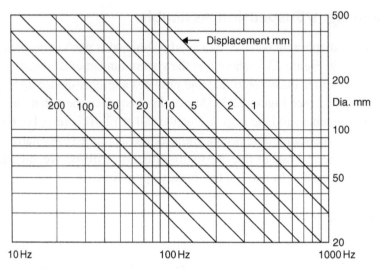

**Figure 23.8: Peak amplitude of a piston to radiate 1 W.**

Figure 23.8 shows the necessary peak amplitude of a piston mounted in an infinite baffle to radiate one acoustic watt of sound power at various frequencies (one side only of the piston radiating). Peak amplitudes in millimeters are marked on the family of curves. For any other value of acoustic power output ($P$), multiply peak amplitude by VP. With an average room of 2000 ft³, a reverberation time of 1 s, and a sound pressure level of +94 dB, the total sound output power is of the order of 30 mW. To radiate this power at 50 Hz, the peak amplitude of a 250-mm radiator will be about 2 mm, while a 100-mm piston to radiate the same power would require a peak displacement of just over 13 mm. Even with "long throw" loudspeakers, it is not possible to obtain a peak-to-peak displacement of 26 mm, thus the sound power capabilities must be severely limited at low frequencies. One will often see response curves of these small speakers taken to apparently extraordinarily low frequency limits, but these are always undertaken at low power input levels.

The directional radiation characteristics of a diaphragm are determined by the ratio of the wavelength of the emitted sound to the diaphragm diameter. Increasing the ratio of diaphragm diameter to wavelength decreases the angle of radiation. At frequencies in which the wavelength is greater than four times the diaphragm diameter, the radiation can be considered substantially hemispherical, but as this ratio decreases, the radiation pattern narrows. Figure 23.9 shows the polar response of a piston in terms of the ratio of diameter

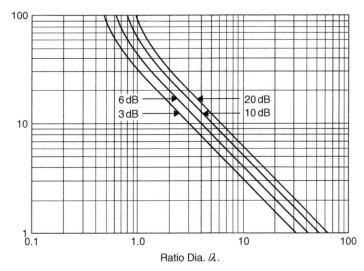

**Figure 23.9: Directional radiation pattern with a circular piston in infinite baffle.**

over wavelength. This shows the degrees off the normal axis at which the attenuation is 3, 6, 10, and 20 dB (as marked on the curves) as a function of the ratio of the piston diameter over the wavelength of the generated sound wave.

## 23.10  Diaphragm Profile

A practically flat disc is far removed from the theoretical "rigid piston." With the exception of foamed plastic, the mass, for a given rigidity, will be excessive, resulting in very low efficiency, and if the cross section is reduced, the system becomes very flexible and inefficient.

Decreasing the angle from 180° increases the stiffness enormously; concomitantly the thickness can be reduced, resulting in a lighter cone for the same degree of self support. The flexure amplitudes will be reduced, but the bell modes will make an appearance. As the angle is reduced, it reaches an optimum value for level response at the transition frequency. There will be another angle for a maximum high-frequency response, resulting ultimately in peaking in the upper treble region. Continuing the reduction in angle, the high-frequency peak will be reduced, but the response above the peak will fall rapidly.

If instead of a straight-sided cone, the profile is curved, the "smoothness" of the overall response can be improved considerably: bell modes are discouraged and the on-axis high-frequency response is improved. The price charged for this facility is reduced low-frequency power handling capacity because, for a given weight, the curved cone is just not as stiff (and as strong) as the straight-sided version.

The most efficient shape at low frequencies is circular. Theoretical and experimental investigations have shown that an ellipse with a major–minor axis of 2 has an average of 7% lower radiation resistance in the useful low-frequency range than a circle of the same area; the loss becomes progressively greater as the shape departs still further from circular. The shape of the cross section or profile of the cone depends on the power handling and response desired.

For domestic loudspeaker systems, which must be cost-conscious, the loudspeaker size is limited to 150 to 200 mm and a frequency response of 100 Hz to about 7 kHz with, possibly, a 25-mm soft dome to accommodate the high frequencies. Straight-sided cones are usually employed when a good 2- to 5-kHz response is required and when reproduction above, say, 7 kHz may be undesirable. Curved cones improve the response above 6–7 kHz by providing an impedance viewed from the voice coil, which has a more uniformly high negative reactance and therefore absorbs more power from the high positive reactance due to voice coil mass seen looking back into the voice coil. This improvement is obtained at the expense of response in the 2- to 5-kHz region, a weaker cone structure, and reduced power handling in the extreme bass.

## 23.11  Straight-Sided Cones

The most important parameter affecting the performance of a loudspeaker is "cone flexure." Because real materials are not infinitely rigid and have mass, the velocity of propagation through the material is finite. The cone is driven at the apex and the impulse travels outward toward the periphery where it is reflected back to the source. At particular frequencies when the distance to the edge are odd quarter wavelengths, the returning impulse will be 180° out of phase and tend to cancel; conversely, when the distance is multiples of half wavelengths they will augment—under these conditions the system can be considered as a transmission line, and theoretically (and to some extent, practically), if the outer annulus were made resistive and of the correct value, the line would be terminated and no reflections would occur [see Figure 23.10(a)].

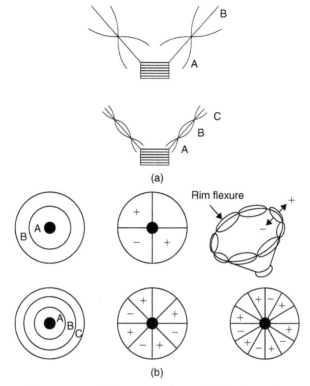

Figure 23.10: (a) Concentric and (b) bell modes.

The conical diaphragm also has radial or "bell" modes of flexure. These are similar to the resonances in a bell and occur when the circumference is an integral number of wavelengths [see Figure 23.10(b)].

Obviously, both modes occur simultaneously, and at some frequencies reinforce each other and at others tend to cancel. Their main effects on performance are the "wiggles" on the response curve and transient and delay distortions. It is instructive to apply a short "tone burst"; it will be seen that at particular frequencies during the duration of the input signal the diaphragm is stationary and on cessation of the pulse it will burst into oscillation at some frequency unrelated to the driving current.

The art of diaphragm design is to minimize these deleterious effects. One method is to introduce concentric corrugations; the effect is to increase the stiffness seen by the bell modes and decrease stiffness for the concentric modes. By correctly proportioning the number, width, and wall thickness of these corrugations, the outer edges of the cone are progressively decoupled as frequency increases. This results in the "working" diameter of the diaphragm being reduced at high frequencies, thus improving the high-frequency performance.

## 23.12  Material

Hard impregnated or filled pressed calendered papers are used when loudness efficiency and apparent high-frequency response are important. The impregnant is usually a hard thermo-setting resin. The radiation response provides very little dissipation in direct radiator cones, hence by using paper having low internal flexural losses, the transmission line is made to have strong resonances. The transient response of this type is necessarily poor, as noncenter moving modes of the cone are unappreciably damped by the motor unit. Soft, loosely packed, felted cones are used when some loss in the high-frequency response can be tolerated and a smooth response curve with reduced transient distortion is required. The apparent loudness efficiency of high loss cones of this type is anything up to 6 dB lower than that of low loss cones of similar weight.

In an effort to overcome the intransigencies of paper cones, resort has been made to other materials. Light-weight metal (aluminum alloys, etc.) immediately springs to mind because of its stability, homogeneity, and repeatability but, because of the very low internal frictional losses, strong multiple resonances occur in the upper frequencies. A diaphragm of, say, 250 mm in diameter made from a 0.1-mm-thick aluminum alloy with a total mass of 40 g will show a "ruler" level response up to approximately 2 kHz when multiple resonances occur. These are extremely narrow band (in some cases only 1 or 2 Hz wide) with an amplitude of anything up to 40 dB and an effective $Q$ of several hundred. Putting a low-pass filter cutting very sharply at, say, 1 kHz does not eliminate shock excitation of these resonances at low frequencies and the result is a "tinny" sound. Reducing the cone diameter and making the flare exponential reduce this effect and also places the resonant frequency a few octaves higher, but does not entirely eliminate the problem. Using foamed plastic materials (and sometimes coating the surfaces with a metal to form an effective girder structure) has met with some success. There are problems associated with the solid diaphragm in that the different finite times taken for the sound wave to travel directly from the voice coil through the material to the front and along the back edge of the diaphragm to the anulus and then across the front cause interference patterns that result in some cancellation of the emitted sound in the mid upper frequencies, say 800–1100 Hz. This effect can be mitigated by using a highly damped anulus, with the object of absorbing as much as possible of the "back wave." Expanded polystyrene is the favorite material for these diaphragms, although expanded polyurethane has met with some success. An extension of this principle is exemplified where the diaphragm is almost the full size of the front of the cabinet (say 24 × 18 inches). In this case the diaphragm, even at low frequencies, does not behave as a rigid piston. The overall performance is impossible of any mathematical solution and must be largely determined experimentally, but the lower bass (because of multiple resonances) is, in the opinion of its advocates, "fruity" and "full!" It has been developed to use two or even three voice coils at strategic places on the diaphragm. For synthesized noise it is possible, but in the writer's opinion, for "serious" music listening, it adds nuances to the music never envisaged or intended by the composer.

Vacuum-formed sheet thermoplastic resins have become very popular. Their mechanical stability is excellent, they are nonhygroscopic, and repeatability (a very important facet when mass producing units in hundreds of thousands) is several orders of magnitude better than paper cones. However, there is a price to pay: most of them contain a plasticizer, which increases the internal mechanical losses in the structure, and hence the magnitude of diaphragm resonances is reduced. However, under user conditions, dependent on electrical power input and thus operating temperature, they tend to migrate. This results in a changed cone (or dome) shape, and because the internal mechanical loss is reduced, the frequency response is changed. In extreme cases, especially with small thin diaphragms, cracking has occurred, but it must be emphasized that with a correctly designed unit operating within its specified power and frequency limits, these "plastic" diaphragms (especially those using specified grades of polypropylene) give a cost-effective efficient system.

## 23.13 Soft Domes

For use at medium and high frequencies, the "soft dome" system has found favor. It consists of a preformed fabric dome with integral surround and usual voice coil assembly. It is very light and its rigidity can be controlled by the amount of impregnant, but the beauty of the concept is that the damping can be adjusted by the quantity and viscosity of the "dope" applied to the dome. Responses flat ±1 dB to 20 kHz are standard, even on inexpensive mass-produced units!

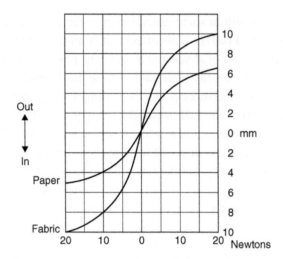

**Figure 23.11: Displacement relative applied force.**

## 23.14 Suspensions

The purpose of suspensions is to provide a known restoring force to the diaphragm/voice coil assembly and at the same time to have sufficient lateral rigidity to prevent any side-to-side movement of the system. This latter requisite is most important when it is remembered that the clearance between the voice coil and the magnet pole pieces is of the order of 0.15 mm for tweeters and 0.4 mm for 150-W woofers. The average domestic 200-mm (8-inch) loudspeaker is about 0.25 mm.

The combined stiffness of the front and rear suspensions is formulated to resonate with the total moving mass of the diaphragm/voice coil assembly and air load to the designed LF resonance. The front suspension radial width is usually about half that of the rear (in order to maximize cone diameter for a given cradle size) and it is this factor that limits the peak-to-peak displacement. Figure 23.11 shows displacement/force for a roll surround. It will be seen that the maximum linear movement is limited: it follows the familiar hysteresis curve of nonlinear dissipative systems.

The annulus of the diaphragm can either be an extension of the cone material itself or, as is more usual with high-fidelity loudspeakers, be a highly compliant surround produced from cotton or man-made fibers, neoprene, or plasticized PVC. In the case of woven materials, these must be sealed and the sealant is usually used to provide some mechanical termination of the cone.

The front suspension represents a discontinuity in the diaphragm system and because it has its own mass and compliance, it is capable of a separate resonance. When this takes place it presents a very high impedance to the edge of the cone, reducing its output and causing a dip in the response. Because of its nonlinearity, it radiates considerable distortion, especially at low frequencies where the amplitude is greatest. The requirements are high flexibility and high interval losses. Probably the most successful material is plasticized PVC, using a very stable nonmigrant plasticizer such as dibutyl sebacate.

The rear suspension is the major restoring force, the radial width is usually at least twice that of the front suspension, and is a multiroll concentrically corrugated fabric disc, impregnated with a phenolic resin. The weave of the material, number of corrugations, diameter, and amount of impregnant determine the stiffness. It should provide a substantially linear restoring force over the designed maximum amplitude of displacement. The whole structure behaves mechanically as a series resonant circuit. The mass is determined by the weight of the cone, voice coil, and former, and the stiffness by the combined effects of the rear suspension and the annulus, the $Q$ of the circuit being determined almost wholly by the losses of the restoring force.

## 23.15 Voice Coil

The dimensions of a voice coil are determined primarily by the rated power handling of the loudspeaker. It must be emphasized that with direct radiators, 95–99% of the input electrical power is dissipated in the form of heat; even with the most efficient horn loaded units a minimum of 50% is used for heating purposes only. Figure 23.12 shows voice coil temperature versus input power. The limiting temperature is set by:

1. Maximum temperature rating of the former: 100°C for paper-based materials; 150°C for "Nomex," which is an aromatic polyamide; 250°C for polyimide;

2. Temperature rating of the wire enamel: maximum 220°C for ML insulation, down to 110°C for self-bonding and self-fluxing wires;

3. Adhesive; from 110°C for cyanoacrylic to 250°C for those with polyimide base; and

4. Mechanical expansion of the voice coil diameter at elevated temperatures.

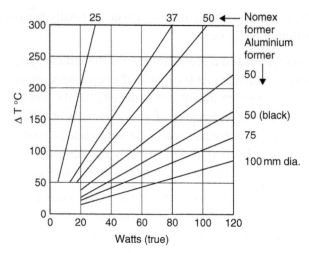

Figure 23.12: Voice coil temperature versus input power (300 Hz).

Fortunately, there is in-built semiprotection for the assembly, namely temperature coefficient of resistance of the wire, which is +0.4% for a 1°C rise in temperature! Thus at a temperature of +250°C above ambient, the voice coil resistance has doubled, and for a constant voltage input (which is the norm for modern amplifiers), the indicated power $E^2/R_{nom}$ is twice the actual power. Note that $R_{nom}$ is the manufacturer's specified resistance and is, or should be, the value at the series resonant frequency $R_{co}$ where the input impedance is minimum and resistive.

## 23.16 Moving Coil Loudspeaker

Figure 23.13 shows the structural features of a moving coil direct radiator loudspeaker, and for purposes of analysis it will be convenient to divide it into two parts: the "motor" or "drive" unit and the acoustic radiator, or diaphragm, described earlier.

The driving system consists of a solenoid situated in a radial magnetic field. It is free (within certain restraints) to move axially (see Figure 23.14). When a current is passed through the coil, a magnetic field will be generated, with the magnitude being directly proportional to //, and this will react with the steady field of the permanent magnet and a mechanical force will be developed that will tend to move the coil axially,

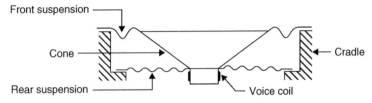

Figure 23.13: Moving coil loudspeaker cone, suspensions, and voice coil assembly.

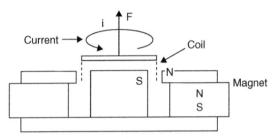

Figure 23.14: Moving coil motor element.

either inward or outward depending on the direction of the current. The magnitude of this force is

$$F = Bli \text{ (Newtons)},\qquad(23.24)$$

where $B$ is flux density in Webers/m², $l$ is conductor length in meters, and $i$ is current in amperes.

It should be noted that $l$ refers only to that portion of the coil situated in the magnetic field. As shown later, to reduce distortion the coil is often longer than the working magnetic field defined by the gap dimensions.

If the coil is free to move, its velocity will be determined by the applied force $F$ and the mechanical impedance $Z_m$. The mechanical impedance $Z_m$ will be a function of total mass ($L_m$) of the system, that is, voice coil and former, diaphragm, air loading, etc., resistance $R_m$ due to losses in the suspension and radiation, and total stiffness ($1/C_m$) due to the restoring force.

Using normal circuit theory, the impedance will be

$$Z_m = R_m + j\left[\omega L_m - \left[\frac{1}{\omega C_m}\right]\right].\qquad(23.25)$$

This represents a simple series resonant circuit shown in Figure 23.15. While one can predict resonant frequency from lumped mechanical constants, the analysis must be carried several stages further to arrive at the correct transfer characteristic from electrical input to sound output in practical loudspeaker design.

Figure 23.16 takes the analogue a stage further:

- $L_{md}$, $C_{md}$, and $R_{md}$ are the mechanical components of diaphragm, voice coil, and suspension.

- $L_{ma}$, $R_{ma}$: mechanical impedance components of air load.

- $Z_{me}$: mechanical impedance due to electrical system.

- $Z_{ma}$: mechanical impedance due to air load on rear of cone.

- $Z_{mx}$: normally zero, but see motional impedance.

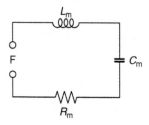

**Figure 23.15: Analogue of lumped mechanical constants.**

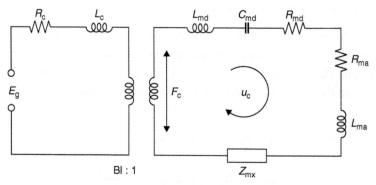

**Figure 23.16: Analogue referred to mechanical side.**

## 23.17 Motional Impedance

The moving coil system is a reversible transducer, a current through the coil will produce a force, and the resultant velocity of the coil will produce an EMF. This voltage will be a function of velocity, conductor length, and magnetic field strength; thus if an external EMF is applied to the coil the resultant motion will generate a back EMF (180° out of phase), which will tend to counteract the forward current flow, thus increasing the electrical impedance. This is "motional impedance."

If the motion of the system can be prevented by applying an infinite mechanical impedance ($Z_{mx}$ in Figure 23.16), there will be no back EMF and the electrical impedance will be only the voice coil resistance and inductance (blocked impedance). Reducing the mechanical impedance (reducing mass and resistance and increasing compliance) will result in an increase in velocity and the motional impedance will be increased. Intuitively, this indicates that motional impedance will be proportional to $Bl$ and an inverse function of the mechanical impedance. The common factor is the velocity of motion.

$$\text{Back EMF } E_b = Blv \text{ V} \tag{23.26}$$

$$\text{from (15.26) velocity } v = \frac{\text{Force}}{Z_m} = \frac{Bli}{Z_m} \text{ m/s.} \tag{23.27}$$

$$\text{Thus motional impedance } Z_{cm} = \frac{E_b}{i} = \frac{B^2 l^2}{Z_{ml}}$$

$$Z_m = \text{Kg/s } Z_{em} \text{ ohms} \tag{23.28}$$

from Equations (23.27) and (23.28)

$$Z_{em} = \frac{B^2 l^2}{R_m + j\left(\omega L_m - \dfrac{1}{\omega C_m}\right)} \tag{23.29}$$

$$Z_{em} = \frac{1}{Y_{em}} = \frac{1}{G_{em} + jB_{em}}, \tag{23.30}$$

where $Y_{em}$ is electrical admittance due to mechanical circuit

$$G_{em} = \text{electrical conductance} \left( \frac{1}{R_{em}} \right)$$

$$B_{em} = \text{electrical subceptance} \left( \frac{1}{X_{em}} = \omega C_{em} - \frac{1}{\omega L_{em}} \right)$$

$$\text{thus } ZL_{em} = \frac{B^2 l^2}{R_m + j \left( \frac{1}{\omega C_m} \right)} \qquad (23.31)$$

$$\text{from } R_{em} = \frac{B^2 l^2}{R_m} \text{ ohms} \qquad (23.32)$$

$$C_{em} = \frac{Lm}{B^2 l^2} \text{ Farads} \qquad (23.33)$$

$$L_{em} = C_m B^2 l^2 \text{ Hendries,} \qquad (23.34)$$

where $R_m$ is Kg/s (mechanical ohms), $L_m$ is Kg, and $C_m$ is meters/Newton and, by inversion,

$$R_{em} = \frac{B^2 l^2}{R_{me}} \text{ Mechanical ohms} \qquad (23.35)$$

$$C_{em} = \frac{L_e}{B^2 l^2} \text{ Metres/newton} \qquad (23.36)$$

$$L_{me} = \frac{L_e}{B^2 l^2} \text{ Kg.} \qquad (23.37)$$

It will be noted that in the analogue inductance (mass) in the mechanical circuit will become a capacitance in the electrical circuit, etc.

At mechanical resonance

$$\left( \omega L_m - \frac{1}{\omega C_m} \right) = 0$$

$$(23.38)$$

and the mechanical impedance $Z_m$ will have a minimum value $= R_m$. The velocity will be maximum, thus the back EMF and motional impedance will be maximum, indicating a parallel resonance. It will be seen that series components in the mechanical circuit appear as parallel components in the electrical side and vice versa.

### 23.17.1 Analogue Models

We can now assemble the various parameters to produce a basic analogue for a loudspeaker. Figure 23.17 shows the low-frequency analogue referred to the mechanical side. The quantity $f_c$ represents the total force acting in the equivalent circuit to produce the voice coil velocity $u_c$:

$$u_c = \frac{E_g Bl}{R_c + (R_m + jX_m)}.$$

(23.39)

Let us divide the frequency region into five parts and treat each part separately by simplifying the circuit in Figure 23.18 to correspond to that part alone. In region A, where the loudspeaker is stiffness controlled, the power output increases as the fourth power of frequency, or 12 dB/octave. In region B, at resonance frequency $0)_o$ the power output is determined by the total resistance because $X_m$ passes through zero. For large values of $Bl$ and small values of $R_c$, the total circuit resistance becomes sufficiently large so that the resonance is more than critically damped. The sound pressure will increase linearly with frequency (+6 dB/octave). In region C, the power output (and sound pressure) approaches a constant value, provided that the circuit impedance approximates a pure mass reactance. That is to say, $RMR$ and $X_m^2$ both increase as the square of the frequency.

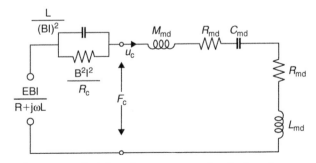

**Figure 23.17: A low-frequency mechanical analogue.**

Figure 23.17 shows that the inductance of the voice coil is reflected into the mechanical circuit as a compliance [very much smaller than $C_{ms}$, which will resonate with the total mass at a midfrequency (usually 150–700 Hz); see Figure 23.18]. At this frequency, the total electrical impedance is resistive, has the lowest absolute value, is the value that is (or should be) quoted as the rated impedance in the manufacturer's specification, and corresponds to (d) in Figure 23.18.

Instead of referring all the parameters to mechanical mesh, it is sometimes more convenient to refer to the electrical input (see Figure 23.19) in which

$$R_c = \text{voice coil resistance}$$
$$C_c = \text{voice coil inductance}$$

$$C_{em} = \frac{m_{md}}{B^2 l^2} \text{ Farads} \tag{23.40}$$

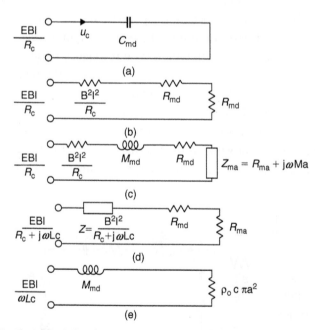

Figure 23.18: Simplified circuit, valid over restricted frequency ranges: (a) very low frequencies, (b) at principal resonance frequency $\omega_0$, (c) above principal resonance frequency, (d) at second resonance frequency, and (e) at high frequencies.

$$L_{cm} = C_{md}B^2l^2 \text{ Henries} \tag{23.41}$$

$$R_{em} = \frac{B^2l^2}{R_{em}} \text{ Ohms} \tag{23.42}$$

$R_c$ and $L_c$ are the "blocked" impedance values and not the DC resistance and inductance measured in "air."

$$C_A = \frac{\rho}{B^2l^2} \times \frac{8a}{3\pi} \times Sd \text{ Farads} \tag{23.43}$$

$$R_A = \frac{B^2l^2}{\rho_0 CSd} \text{ Ohms,} \tag{23.44}$$

where $a$ is diaphragm radius and $S_d$ is diaphragm area.

It should be noted that the factor $8a/\pi$ in Equation (23.43) is actually the "end correction" used to describe the accession to inertia acting on one side only of a rigid piston of radius 'a' vibrating in an infinite baffle. The air loading on the back side of the diaphragm is determined by the loading presented by the enclosure.

Figure 23.20 shows the impedance of a 300-mm (12-inch) loudspeaker in an 85-liter enclosure (3 ft$^3$). It will be seen that the modulus of impedance rises to 125 ohms at the mechanical resonant frequency of 55 Hz, drops to 8 ohms at the series resonance, and rises to 40 ohms at 10 kHz. Of equal interest is the reactive component: below the first resonance an inductive reactance is presented to generator (rising to infinity at resonance), while between the two resonances a capacitive reactance is presented. At 100 Hz the effective capacitance is about 90 μF, and at that frequency the phase angle

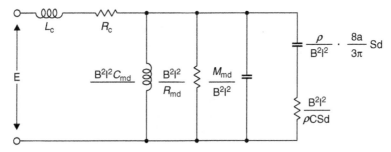

**Figure 23.19: Analogue referred to electrical input.**

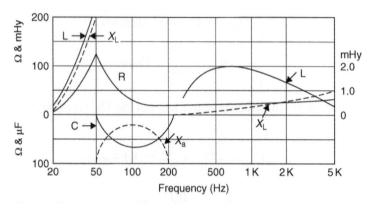

Figure 23.20: Complex impedance of moving coil loudspeaker.

Figure 23.21: A 25-mm OEM soft-dome tweeter.

Figure 23.22: A 200-mm OEM bass midrange loudspeaker drive unit.

is 45°. Above the second resonance the impedance rises slowly. It will be seen that the design of a successful moving coil loudspeaker owes as much to art as science (Figures 23.21 and 23.22). CAD can and does simplify much of the detail work, but after the basic design parameters (diaphragm size, magnetic field strength, conductor length, etc.) have been calculated, the nub of the problem is what diaphragm material, adhesives, cradle material and shape, and so on to use—the art of loudspeaker design is 5% inspiration, 95% perspiration, plus the essential compromise.

## Further Reading

Beranek, L. L., *Acoustics* McGraw-Hill.

Beranek, L. L., *Acoustic measurements*. McGraw-Hill.

Borwick, J., *Loudspeaker and headphone handbook*, Butterworth-Heinemann, 1988.

Colloms, M., *High performance loudspeakers*, 4th Ed., Wiley, 1991.

Olson, H. F., *Acoustical engineering*, D. Van Nostrand.

Olson, H. F., *Dynamical analogies*, D. Van Nostrand.

Walker, P. J., 'New developments in electrostatic loudspeakers', *J.A.E.S.*, Vol 28, No. 11, 1980.

Weens, D. B., *Designing, building and testing your own speaker system*, 3rd Ed, TAB, 1990.

Weens, D. B., *Great sound stereo speaker manual*, New York: NY: TAB, 1989.

# CHAPTER 24

# Loudspeaker Enclosures

Ben Duncan

## 24.1 Loudspeakers

Knowing a little about loudspeakers, the load that is audio amplifiers' raison d'etre is a prerequisite to understanding amplifiers. In the following sections—indeed most of the rest of this chapter—those features of loudspeakers that most define or affect the design and specification of power amplifiers are introduced.

### 24.1.1 Loudspeaker Drive-Unit Basics

There are six main types of speaker drive-units or *drivers* used for quality audio reproduction. Another name for a driver is a *transducer*, a reminder that they *transduce* electric energy into acoustic energy via mechanical energy.

#### 24.1.1.1 Cone Drivers

The most universal, everyday form of the "moving coil" or "electrodynamic" type of drive-unit (Figure 24.1) has a moving cone, with a neatly wound coil of wire (the "voice coil") attached to its rear. The coil has to be connected to and driven by an amplifier. The coil sits in a powerful magnetic field and can move back and forth without rubbing against anything. When driven, a signal-varying, counteractive magnetic field is set up, causing the coil and the attached cone to vibrate in sympathy with (as an analogue of) the driving signal. The principles are akin to an electric motor, except that the vibration is linear ("in and out") rather than rotational.

Moving coil drive-units can be made in many ways. Most have to be mounted in some kind of enclosure before they can be used. Drive-unit size (strictly, the piston diameter) broadly defines frequency range. Most cone drivers range from 1″ (25 mm) up to 24″

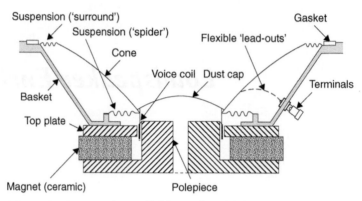

**Figure 24.1: A moving coil drive-unit, with key parts identified.**
**(Courtesy Funktion One Research)**

(0.6 m) in diameter for use at high treble down to low bass. There are at least 15,000 different types of cone materials, textures, and weights available. Most are made of paper pulp, but plastics, metals, composite materials, and laminated combinations are also used. Every one sounds different and measures differently.

There are as many permutations again for the voice coil's diameter, height, and wire gauge; the type of dust cap, magnet, the chassis, and the flexible jointing, called the *surround* (at the front), and centering device at the rear, the suspension or *spider*. With all the moving parts, ruggedness and stiffness are pitted against the need for agility, hence levity. This is the main reason why the radiating part is cone shaped. This shape can stiffen the most limp paper against the axial force applied to it by movements of the voice coil.

### 24.1.1.2 Compression Drivers

The second most common type of driver, at least in professional sound, is the compression driver (Figure 24.2). This is simply a specialized form of moving coil drive-unit. The depth of the cone is replaced by a much shallower and usually opposite-facing and dome-shaped radiating surface, called the diaphragm. The voice coil is attached peripherally between the edge of the dome and the suspension. This type is made for some midrange but mainly hf speakers, which are *horn loaded*. All bass bins and most midrange horns employ specially adapted but ordinary-looking cone drivers; these alone can handle the larger excursions required. A compression driver cannot handle more than very small excursions. To avoid large excursions and potential ripping of the

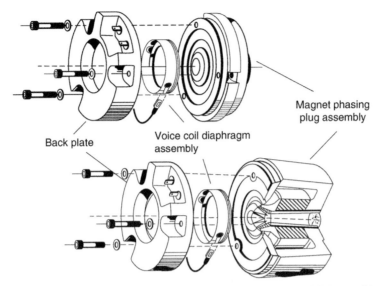

**Figure 24.2: Two types of hf compression drivers made by Emilar, widely used in PA systems from the mid-1970s.**

diaphragm, a compression driver must *never* be driven with a program having frequencies below its rated range and not driven without being attached to a suitable horn. Usually, the diaphragm is pressed out of a plastic film, or a phenolic resin-impregnated cloth or other composite, or from very light, but stiff metal, usually aluminum or else titanium or beryllium. Size ranges from about 6″ (150 mm) for midrange down to 1″ (25 mm) for high treble and above.

### 24.1.1.3 Soft and Hard Dome Drivers

The equal-second most common type of speaker drive-unit is familiar enough. It has an almost hemispherical diaphragm shaped like some compression drivers, but the dome is forward (like a fried egg) and nearly always working into free air. This is used on its own, instead of a small diameter cone, as a tweeter (hf drive-unit). The material can be any of those used in cone or compression drivers.

### 24.1.1.4 Common Voice Coil

The three types of drive-units discussed so far all have similar voice coils. They may range widely in weight, diameter (from 0.75719 up to 67150 mm), and power handling (commonly from 3 to 1000 W), but they will all mostly have a DC resistance of 5 to 10 ohms and a nominal (AC, 400 Hz) impedance of 8, 15, or 16 ohms.

### 24.1.1.5  The Ribbon Driver

The ribbon speaker is a fourth kind of electrodynamic drive-unit. Instead of a voice coil attached to the radiating part, the amplifier signal is connected across a length of flat (planar) conductor foil or "ribbon," which is again placed in a magnetic field like a voice coil, but also radiates sound like a cone, diaphragm, or dome. Compared to ordinary voice coils, this arrangement can be lighter and certainly presents a much purer ("resistive") impedance to the amplifier. The classic ribbon had a very low DC resistance and was transformer coupled. Modern ribbon speakers have longer strips, amounting to 3 or 5 ohms of near pure resistance, benign to most audio amplifiers connected to it. When "built big" as a *panel loudspeaker*, a ribbon drive-unit forms a wide-range loudspeaker in its own right, that is, no cabinet required. There is little breakup in the ribbons' surface to mar the sonic quality. And, unlike other drive-units, absence of a cabinet means the sound source radiates as a dipole, that is, from both sides. This can be important to the amplifier, in far as room interaction can change the impedance seen by *reflection*. Small ribbon drive-units are used as tweeters. They may be horn loaded to magnify their rather low output. Sonic quality can be very high, although naturally favoring the reproduction of stringed instruments.

### 24.1.1.6  The Electrostatic Source

The electrostatic loudspeaker (ESL) employs the inverse or dual principle of the electrodynamic or "motor" types of drive-units that we have just looked at. The movement is provided by electrostatic (electric field) force rather than magnetic attraction and repulsion. The vibrating part is a thin, critically stretched sheet called the diaphragm. The fixed part, after the capacitor it mimics, is called a plate. Electrostatic drivers are commonly made in the form of panels, like ribbon speakers. A power source (usually from the AC mains) provides the high EHT DC voltage of over 1000 V that is needed to *polarize* the plates. A high signal voltage swing is also required. This, together with isolation from the EHT, is attained by interposing a transformer. In practically sized and costed electrostatic speakers, the transformer and the diaphragm have a surprisingly limited capacity for handling high levels at low frequencies. In primitive designs, overdrive in the bass can cause the diaphragm to short against the opposite plate. In modern ESLs, the diaphragm is insulated. A well-known electrostatic employs an aggressive *crowbar* circuit for protection. If the ESL is subjected to potentially damaging high levels at low enough frequencies, this shorts the speaker's electrical input, possibly blowing up the amplifier, or at least blowing a fuse or shutting down the music. Under most other conditions, the ESL appears as an almost purely capacitive load, with resistive damping across it.

### 24.1.1.7 The Piezo Driver

The two fundamental types of drive-unit "motor" looked at so far all date back (in principle) to the early years of this century, or even to the beginnings of the modern harnessing of electricity, 200 or more years ago.

The *piezo* drive-units' principle is the dual of the familiar household act of creating large voltages by squeezing crystals. Although piezoelectricity precedes humankind, as it can occur naturally, it has only been widely harnessed in the past 50 or so years, first in crystal microphones and pickups and more recently in fuel-less "push button" gas fire lighting. The dual, or reverse process, that of making a crystal vibrate by applying electricity to it, was first harnessed by Motorola, who have been producing hf drive-units employing this principle since at least 1977. This type of drive-unit looks capacitive, rather like an electrostatic, but has a higher DC resistance so that it can draw no long-term power. Despite potentially useful high hf performance, since there is still a limited range of piezo drive-units, most being fitted to integral, out-dated horn designs, piezo tweeters are not used much in high-performance systems, but they are occasionally used in PA systems and may be found optimally applied in refined custom speaker systems.

The Motorola piezo element cannot be "burned out" by too much "power" as it presents a high impedance. However, it is rated at about 25 V rms, and excess voltage will quickly destroy the crystal. The crystal can even be harmed by room heating. For use with amplifiers having headroom above 25 V rms, operating two or three in series is suggested. This should not degrade damping as it would with a low impedance speaker.

### 24.1.1.8 Inductive Coupling

Eli Boaz at Goodmans, part of the TGI Group (comprising speaker manufacturers Tannoy, Goodmans, and Martin Audio in the United Kingdom) spearheaded the development of two-way drive-units where the hf driver is inductively coupled (*inductive coupling technology* or ICT). It comprises a radiator with a conductive collar (Figure 24.3). Placed *within* the bass/midrange voice coil, it acts as a single turn transformer, picking up magnetic field most efficiently at hf. This arrangement is limited to use at hf, but there is no need for a crossover, and it is highly rugged—a tweeter that cannot readily burn out.

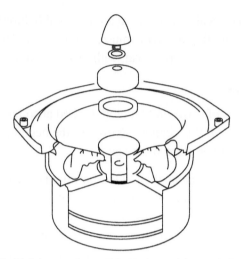

**Figure 24.3: The exploded hf dome above this Tannoy drive-unit has no ohmic connections
and cannot be burnt out. It employs inductive coupling technology, the first completely new
type of drive-unit to enter mass production for many years. Each new drive-unit type has its
loading peculiarities, which add a new layer of variables to the considerations of amplifier
users and designers alike.**
**(Courtesy of Tannoy Ltd.)**

Also, without having the capacitative loading region of a conventional tweeter (and the
load dip of any passive crossover), an ICT driver's load impedance at hf is benign.

### 24.1.2  Loudspeaker Sensitivity versus Efficiency

Loudspeaker drive-units have to be "packaged" to be usable in the real world. Together,
enclosures and drive-units define the efficiency of the resultant loudspeaker. Efficiency
(or its derivative, sensitivity) then decides the scale of amplifier power needed. With
different high-performance loudspeaker types, efficiency varies over an unusually wide
range of at least a hundredfold, from about 20% down to 0.2%.

Efficiency is not often cited, but can be inferred from the vertical and horizontal polar
radiation patterns, the impedance plot, and the sensitivity. Sensitivity is the derivative
of efficiency that makers use to specify "how much SPL for a given excitation." In part,
sensitivity is universally specified because it's easier to measure. It is given as an SPL
with a given input (nearly always 1 W) at a given distance at close range (1 m normally).
So the spec is the one that reads: "Sensitivity 96 dB @ 1w @ 1 m."

For most domestic speakers, 96 dB is a high sensitivity but low for professional types. The sensitivity is but a broad measure of efficiency differences, since two factors are missing.

One is how the sound energy is spread in space. If it is all focused forward, sensitivity (dB SPL @ 1 W @ 1 m) is raised as the sound "density" at the measuring position increases. At low frequencies, rated sensitivity commonly falls as the sound radiation becomes more nearly spherical, while efficiency is unaffected.

Factor two is the impedance. Where mainly resistive, efficiency is about the norm, as computed by integrating the SPL over all the solid angles. But around the resonant frequency where the impedance changes rapidly from capacitative to inductive, efficiency is high, as little energy is dissipated.

With these four dimensions of variables (3D space + ID impedance), converting efficiency into sensitivity figures and vice versa is not straightforward. However, as a rough idea, an 86-dB @ 1-W @ 1-m rated domestic speaker is about 0.5% efficient. With a two-sided (planar) speaker, the efficiency might be the same 0.5% but sensitivity would ideally halve toward 80 dB.

### 24.1.3 Loudspeaker Enclosure Types and Efficiencies

Horn loading is by far the most efficient technique. It is between 10 and over 100 times more efficient than any others. "Efficiency" means it gives the most acoustic intensity for a given power input, from the amplifier. Only when a horn (or "flare") is coupled to a transducer with a low output (e.g., a ribbon driver) is the overall efficiency *not* "streets ahead" of all the other driver-t-enclosure combinations.

The most efficient drivers are the familiar electrodynamically driven cone, dome, and compression types, particularly those with an optimum balance between the strength of magnetic and electric coupling, the levity of the moving parts, and the *compliance* of the suspension. In the midrange, some ESLs can be as efficient as the cone driver, both in the context of a refined domestic speaker.

The least efficient enclosures are:

1. **None** (this holds true at low frequencies only),

2. The **sealed box** (SB) or "infinite baffle" (IB), and

3. The **transmission line** (TL)—used to extend bass response.

Of these, the latter two are important, practical forms that have to be lived with. They can in any event be made relatively quite efficient by making the enclosure *big*. To some extent, *Colloms' law* holds here: "*Loudness (per watt or volt) is inversely proportional to bandwidth and smoothness*."

This is fine until we come to consider the refined horn speakers, which do not attempt 50% efficiency, and where a minimum of three types are needed to cover the audio band. While at least 10 times more efficient, there is little or no bandwidth narrowing over ordinary speakers.

Compression and piezo drivers are those usually coupled to horns (flares) and may need no other boxing or at least not any specific enclosure, as their rear chamber is usually already sealed. Sound radiation is then mainly defined by the horn, subject to mounting.

Ribbon drive-units may be also horn mounted or, if "planar" types, then along with ESLs, they may be simply mounted in a frame that has little effect on the sound radiation, which is *dipolic*, that is, two sided, like a harp's.

The other two types of drive-units—the cone and soft dome—are usually mounted in closed ("infinite baffle" or "sealed box") enclosures or, in the case of cone drivers alone, in ported ("Thiele-small," "vented," or "reflex") enclosures.

Cone drivers are also used "coupled to" horns—either midrange, or bass ("bins"). In practice, as the rear of the cone's basket mounting frame is open, and the fragile magnet is also unprotected, cone drivers in bins and horns are almost always mounted inside the overall enclosure.

Horns, transmission lines, sealed and vented boxes, and other loading types may form complete loudspeakers in free permutation. Of these, only combinations of horns or sealed boxes can cover the *full* audio frequency and dynamic range within their own family, that is, without involving each other or the other types.

### 24.1.4 Loudspeaker Configurations: A Résumé

Few or no *single* loudspeaker drive-units offer overall, high-performance audio reproduction. The nearest contender is an ESL or ribbon type panel, which can work as the sole drive-unit for the kinds of music that have no loud, low bass content. But to listen without restriction and risk of damage to every other kind of music, two, three, or more drive-units must be used to cover low, medium, and high frequencies (which from 10 Hz to 20 kHz span a wavelength range of some 2000-fold!), and over the 120-dB+ dynamic range required for high-performance sound reproduction.

### 24.1.4.1 Matching Levels

Often, the sensitivities (loudness) of the individual units that are optimum for each frequency band differ. Commonly, the tweeter is more sensitive than the driver covering bass/midfrequencies. If efficiency is unimportant, the mismatched sensitivities (which would otherwise cause an uneven, "toppy" frequency response) may be overcome by adding a series "padding" resistor in line with the hf drive-unit (tweeter).

### 24.1.4.2 Parallel Connection

Alternatively, in touring PA and wherever else efficiency matters, or wherever high SPL capability is sought, an overall flat response may be attained by using two, parallel-connected bass/mid drive-units. Applicability is *always subject to coherence in the acoustic result*, hence suitable mutual positioning of the paralleled drivers, so they work together. If "ordinary" drivers (i.e., electro-dynamic types) are used, a 15 or 16 Q rating will be likely chosen, so the resultant load is about 8 Q rather than 4 Q if the paralleled drivers were each the usual 8 Q.

Again, *always subject to coherence in the acoustic result* (and thus suitable mutual positioning, generally closer than a quarter of the shortest wavelength), drive-units or complete speakers can be paralleled across either a given amplifier or, when this runs out of drive capability, across amplifiers ad infinitum that are driven with an identical signal and have either identical or acoustically justifiable different gains. Despite this, the fewer drive-units or speakers reproducing a given program in a given frequency range, the better the sonic results. *As is so often the case in high-performance sound reproduction, least is best—if it is usable.*

### 24.1.4.3 Why Crossovers?

When two or more drive-units are used to cover the audio range, it is usually *very* important that they only receive program over their respective, intended frequency ranges. Program at other frequencies must be omitted, as it will usually degrade a drive-unit's sonics by exciting resonances and subharmonics. At moderate to high drive levels, physical damage is also likely. The "frequency division" or "frequency conscious routing" that ensures different drivers receive their intended range, and not other frequencies, is the *crossover* (no relation to crossover *distortion*).

The crossover may also perform phase shifting or signal delay in one or more bands. Principally this is to synchronize (UREI use the phrase "time align") the signals delivered by the drive-units, for example, because the sound from some has slightly further to travel.

A variety of crossover types exist, with the more sophisticated designs aiming to have the bands meld neatly in the crossover regions, without boosting or cutting any frequencies.

### 24.1.4.4 Crossover Point (XOF)

The crossover frequency or "point" is chosen by speaker designers on the basis of:

1. **Power handling.** Drops dramatically if the XOF is too low for the higher driver. Caused by the exponential increase in excursion as LF limits are reached.

2. **Distortion.** For the same reason, rises rapidly for the higher driver below a comfortable XOF.

3. **Frequency response**, both on- and off-axis. If either of these changes abruptly, near the proposed XOF, the XOF had better be moved. But with some drive-unit combinations, there is no ideal XOF.

### 24.1.4.5 Passive Crossovers

In the majority of domestic speaker systems, but particularly where low cost or simplicity is paramount, the crossover is passive (unpowered) and operates at a "high level," being placed within and supplied as part of the speaker cabinet (Figure 24.4).

In this form, the crossover comprises physically large and heavy, high voltage-rated capacitors, high current-rated inductors (coils), and high power-rated resistors.

### 24.1.4.6 Passive Low Level

It is possible (but not very common) to have a passive crossover operating at line level installed before the signal enters the power amplifier, or otherwise before the power stage (Figure 24.5). The crossover parts can then be smaller and lighter, as the voltage and current ratings (that in part determine size) can then be 30 to over 100 times lower. This is a fine arrangement for all-integrated active cabinets.

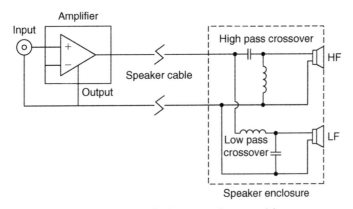

**Figure 24.4: Passively crossed over cabinet.**

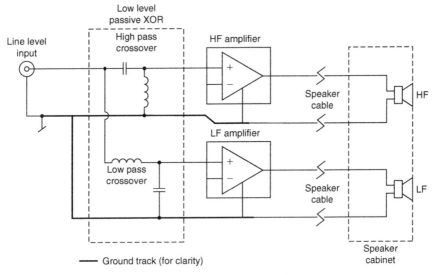

**Figure 24.5: Low-level passive.**

Otherwise, in ordinary "mix and match" sound systems, one problem arising is that because the crossover is physically divorced from the speaker, careful connection is needed. Another, less daunting, is that any existing, passive high-level crossover component values cannot be simply transferred, as they will have been "tweaked" to best suit the vagaries of the drive-units' impedances.

### 24.1.4.7 Active Crossovers

The active crossover (first suggested by Norman Crowhurst in the 1950s) takes the preceding concept a stage further. Frequency division is accomplished *actively*. This means using active devices—and a DC power source—to provide filtering in a highly predictable manner. For example, the filters are able to work in an ideal environment, having well-defined and resistive loading. This and the filter function are defined potentially very precisely by active electronics, usually employing high NFB (Figure 24.6).

The main disadvantage of active crossover systems is cost, not just of the active crossover, but of the added amplification and cabling. In DIY domestic setups there is also the *bulk* of equipment (if using, say, three stereo amplifiers, placed centrally, or six mono block amps and two mono crossovers, half to be placed by each speaker) and their cabling. Such inconvenience is irrelevant in concert sound systems, and even in recording studios. It is

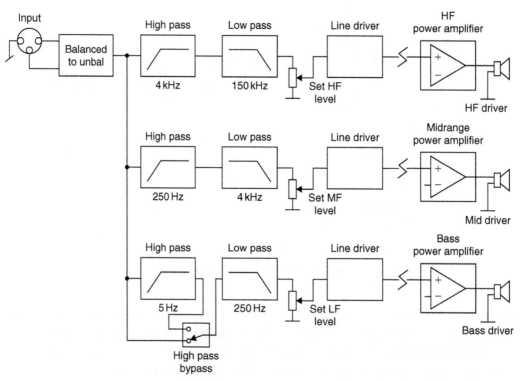

**Figure 24.6: Classic three-way active crossover.**

also absent in active powered enclosures and is the rather the reverse—1 integrated active cabs are "plug in and go" systems. For everyone else, more care is needed when connecting up an active system, taking care that the drive units are going to be fed their appropriate band. In professional systems, multipin speaker and line connectors are commonly used so that the two or more frequency bands' connections are always routed correctly.

Notable advantages compared to the common, high-level passive crossover are:

1. Reduced "congestion" and similar intermodulation distortion symptoms as each power amplifier handles only a section of the audio range.

2. Differences in driver sensitivities (considered under "matching levels," see earlier discussion) are ironed out and without compromise, except the requirement for, or use of amplifier headroom, by simply adjusting gain controls.

3. Higher dynamic headroom *by diversity*, as the program peaks occurring in the respective out-of-band frequency ranges do not steal any headroom. Also, brief clipping (overdrive) of the bass band (etc.) has little effect on clarity when the other bands are not driven into clipping at the same time.

4. Amplifiers are connected directly to their respective drive-unit(s). There are no reactive components (i.e., the passive crossover) to steal current, but the drive-unit's impedance dips may still demand significant current headroom from the amplifier.

5. The ease and capability of creating highly conjugate, highly matched, and closely toleranced crossover functions.

Of these, the low-level *passive* crossover potentially has all the same advantages—except possibly this last item.

### 24.1.4.8 Active Manifestations

Active crossovers commonly take one of three forms. In pro-audio, "the crossover" is commonly used packaged in its own box, like other signal processors, as a *stand alone*. As a power amplifier presents a (literally) well-placed opportunity to share a box and a power supply, a few power amp makers offer a pluggable option or "octal socket," where a crossover module or card can be retrofitted. Often, the performance of the card's filtering (e.g., the frequency and slope rate) may be decided upon by the user. This is often a low-budget option limited to installation work, but while it can save money, it need not be shoddy. If carried out with as much care as any crossover is due, turning an amplifier into a "speaker-driving filter" has the advantage of minimizing superfluous hardware and signal path complexity.

Otherwise, the active crossover, along with the power amplifiers, is placed in the loudspeaker enclosure. This creates an "active enclosure," "active speaker," or "Tri-amp cab" (if three way; else "bi-amped-cab," "quad-amped-cab," etc.) that has the advantage of hiding the complexity and offers a fait accompli, but takes away the flexibility that touring PA users often need.

*Active systems* are widely used in pro-audio, for installed and touring PA systems, and studio control room monitors. In nearly all cases, stand-alone active crossovers, discrete power amplifiers, and individual enclosures are brought together. In high-end hi-fi, discrete active systems are comparatively rare, except in DIY circles. Active speakers are becoming somewhat more common, at least in the United Kingdom and Europe.

### 24.1.4.9 Bi-wiring

Bi-wiring is a "part-way house" to having a low level active (or passive) crossover. A separate speaker connecting wire is provided for each drive unit, or for each frequency band (Figure 24.7). This lessens interaction and intermodulation that is otherwise caused by communal speaker cabling.

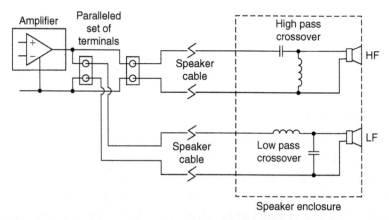

Figure 24.7: Bi-wiring improves sonic quality by avoiding superimposition voltage drops over the greater length of the output stage to speaker connection, as otherwise LF signal currents upset the hf's driver signal's purity, and even vice versa.

### 24.1.4.10  Other Networks

Whether the crossover is passive and high level, passive and low level, or active, other components may be associated with each drive unit:

1. DC protection capacitors are connected in series with drive-units to block steady current flow should a DC voltage appear across the box's, the speaker's, or the driver's input terminals. They are not required for hf (and mf) drivers with passive crossovers, where the high and "bandpass" filters already include the required series capacitor as part of the crossover.

In some designs, more complex, active crowbar circuitry is used, in order to obviate the need for a series capacitor.

2. **Zobel** and other "conjugate matching" networks. Comprising networks of capacitors and resistors, and less often inductors, these act to smooth out the impedance variations of the drive-units, either singly or altogether, and as seen by the preceding crossover and also amplifier.

3. **Music overdrive protection**—in some designs, a light bulb, usually a rugged 12-V type, is connected in series with hf drive units, which in practice require protection most of all. At worst, the light bulb will be quicker, easier and less expensive to change than the hf driver or diaphragm. The effect the light bulb has on sonic quality can be small and sonically benign if the lamp is not visibly glowing during normal loud passages. "Auto resetting" "thermal trip devices," alias Ptc (positive temperature coefficient) thermistors, are also used. These are usually in the form of a cement-coated disc. At room temperature, they exhibit a low resistance. When *eventually* tripped by excess current, the hot resistance increases rapidly to about 100-fold, and the protected driver's power dissipation drops 10,000-fold ox pro-rata. The effects on sonic quality of series ptc thermistors are as yet questionable.

### 24.1.4.11  Other Protection

Loudspeakers have also been protected by add-on boxes, containing historic power-reading circuitry, for example, which crudely opens a relay in line with the speaker or line level signal if the drive-unit is seen heading toward a burnout.

## 24.2 The Interrelation of Components

### 24.2.1 What Loudspeakers Look Like to the Amplifier

There is a tacit presumption that most amplifiers can comfortably drive any "reasonable" loudspeaker. Beyond whatever is "reasonable," discomfort may occur to all parties. To the power amplifier, which is nothing but a loudspeaker driver, the most salient information about any loudspeaker it is expected to drive, and the stress that may engender, is that loudspeaker's *impedance*.

#### 24.2.1.1 Impedances

A speaker's nominal impedance is commonly (and oversimplistically) described by a single round figure, usually 15, 8, or 4 ohms for the majority of moving coil drive-units. With ribbon drive-units, or whenever several drive units are paralleled to increase handling or coverage, lower impedances of 3, 2, or 1 ohms or even less are the norm. With electrostatic and piezo (hf) drive-unit types, the load impedance can be higher, but are also more or predominantly *capacitative* (like a capacitor) across the audio range. This can be far more taxing to the amplifier.

#### 24.2.1.2 Low versus High Impedances

At this juncture it is helpful for those unfamiliar with electronics jargon to grasp a counterintuitive fact: that the *lower* impedance, the *heavier* the (current) loading on the amplifier. To remember this and that 4 ohms is *harder* to drive than 16 ohms, think of hill slopes: a 1-in-4 hill is far harder to drive or climb up than a 1-in-16, that is, an impedance in ohms is the reciprocal of the relative loading. Remember also:

A low impedance demands more current, and less signal voltage is needed for a given current.

A high impedance requires more signal voltage to be driven with a given current.

#### 24.2.1.3 Variation versus Frequency

Loudspeakers' impedances nearly always vary over the frequency range of use. Figure 24.8 shows how a nominal 8-ohm, 15" bass drive-unit typically varies from 5.5 ohms at 450 Hz, peaking up to about 40 ohms or so, at the mechanical resonant frequency, which typically lies between 20 to 120 Hz for a bass driver. Here it is 31 Hz. Together, the drive-unit and the speaker enclosure largely determine this. Impedance also rises to a maximum at (and beyond) the highest usable frequency.

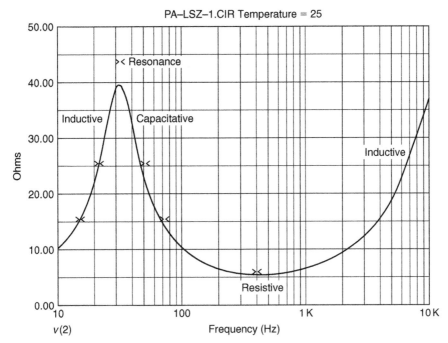

**Figure 24.8: The impedance of a 15″ drive unit mounted on a nominal baffle. In some cabinet designs, there could be two or more resonant peaks. Note the labeling of the resistive, capacitative, and inductive impedance zones.**

At and about the resonant frequency, the impedance variation at the loudspeaker's terminals is due to the reflection of mechanical energy storage, and damping, back to the electrical domain. Figure 24.9 shows that the loading is capacitive on the right side of the resonant peak, where impedance is falling with increasing frequency, while the impedance that slopes upward with increasing frequency, on the left side of the resonant peak, is inductive. Figure 24.9 shows this in another domain. When the phase (lower graph, left scale) is positive, the impedance is capacitive; when negative, it is inductive. When toward the center, it is resistive. Dead center is pure resistance.

The resonant frequency area(s) of any bass speaker system is are commonly highly stressful to amplifiers, and with many BJT amplifiers, when contact is prolonged by a low enough frequency and perhaps an insistently enough pounded note, it has frequently been fatal. Other amplifiers have been known to simply burst into uncontrollable oscillation.

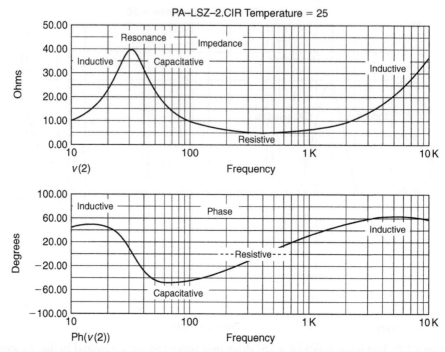

**Figure 24.9: The impedance of Figure 24.8 (upper graph), shown alongside the phase map (lower graph), clearly shows the relationship between pure resistance and inductive and capacitative phase—at least in terms of voltage. In some instances, a plot of current phase might be more appropriate.**

At or above the highest usable frequency, the impedance rise is again inductive. It represents the effects of the voice coil inductance. Eddy currents, as such, or manifest as *skin effect and proximity effect*, may also contribute to the inductivity. Inductive effects are the cause of *back EMFs* ("kick back" voltages) that, unless damped, can upset sound quality and can even destroy an unsound amplifier design.

### 24.2.1.4 Passive Crossover Effects

Most high-performance speakers for domestic and small studio use contain passive (unpowered) crossovers. Such enclosures are driven from a single amplifier. Passive crossover networks are "in line with" the drive units' impedances. The combination is complex; it may increase or decrease peak current demand, hence loading. As an example of the latter, extra parts may be added to create a *conjugate* crossover, which makes the overall loading look resistive, but also absorbs power.

### 24.2.1.5 Static versus Transient

Conventionally, impedance values are taken after applying a steady and repetitive test signal (e.g., continuous sine wave) and allowing a few moments for the recovered signal amplitude that represents impedance to settle. In the short term, impedance can be considerably lower. *At worst, it is possible for an ordinary dynamic (moving coil) type of loudspeaker to demand current as is it had 1/6th of its nominal impedance.* In other words, an 8-ohm speaker can sometimes look like 1.4 ohms. This will not happen all the time or even very often, nor for very long at a time; however, for high-quality sound reproduction, and not forgetting that music involves repetition, the possibility *must be allowed for.*

### 24.2.1.6 Acoustic Contribution

The impedance (load) characteristics of drive units can be affected by cabinet air leaks, and also by reflections in the room, hence the positioning of the enclosure. Horn-loaded drive-units are usually the most sensitive to this.

The upshot is that most loudspeaker loads are a wide variable, not just between different models and types, but depending on program dynamics and excitation frequencies.

## 24.2.2 What Speakers Are Looking For

The fact that most loudspeakers do not employ conjugate impedance compensation, and so they have impedance curves that vary "all over" with frequency, means that for high performance, the amplifier kind the speaker *needs* to see is a "voltage source."

### 24.2.2.1 Why Voltage?

The signal *voltage* must be almost unaffected (ideally far below say 1% change) whether the speaker is connected or not, regardless of whether it's drawing 50 milliamps or 50 amperes. That means a "stiff power source," alias a *low impedance* or "high-current-capable" source.

If the source impedance isn't low (enough), then as the speaker's impedance varies with frequency, the change will be superimposed on its own frequency response as a tonal aberration. A source impedance that is almost as high as speakers' own minimum impedances is a major failing with power amplifiers having low or nil global negative feedback, and the outcome is a tonal anomaly as gross as 5 dB. This may not be all bad, but it will certainly be arbitrary.

### 24.2.2.2 Energy Control

In part, a *voltage source* is required to drive speakers, because loudspeakers store, as well as convert, energy. The fundamental resonance is the place (in the frequency domain) where this is most true. Some of the stored energy "kicks back" and needs to be damped quickly (dissipated) to avoid transient distortion or "smearing." The same reactive effects may also demand surprisingly high peak currents from the amplifier at other times, when driven by music signals.

In both cases, the answer is a high current sourcing and sinking capability. Both of these features are implied but neither are guaranteed by a low source impedance. The overall requirement is a "current-capable-enough voltage source." So far, most power amplifiers throughout history have aimed to be this, but some have come closer than others.

### 24.2.2.3 Damping Factor?

The majority of high-performance amplifiers are solid state and employ global (overall) negative feedback, not least for the unit-to-unit consistency it offers over the wild (e.g., $+/-$ 50%) tolerances of semiconductor parts. One effect of high global NFB (in conventional topologies) is to make the output source impedance ($Z_0$) very low, potentially 100 times lower than the speaker impedance at the amplifier's output terminals. For example, if the amplifier's output impedance is 40 Wohms, then the nominal damping factor with an 8-ohm speaker will be 200, that is, 40 *milli*ohms (0.04) is 1/200th of 8 Q. This "damping factor" is essential for the accurate control of most speakers.

Yet describing an amplifier's ability to damp a loudspeaker with a single number (called "damping factor") is doubtful. This is true even in active systems where there is no passive crossover with their own energy storage effects, complicating especially dynamic behavior.

Figure 24.10 again takes a sine-swept impedance of an 8-ohm, 15″ driver in a nominal box to show how "static" speaker damping varies. Impedance is 70 ohms at resonance but 5.6 ohms at 450 Hz. The bottom part of Figure 24.10 plots the output impedance of a power amplifier that has high negative feedback, and thus the source impedance looking up (or into) it is very low (6 milliohms at 100 Hz), although increasing monotonically above

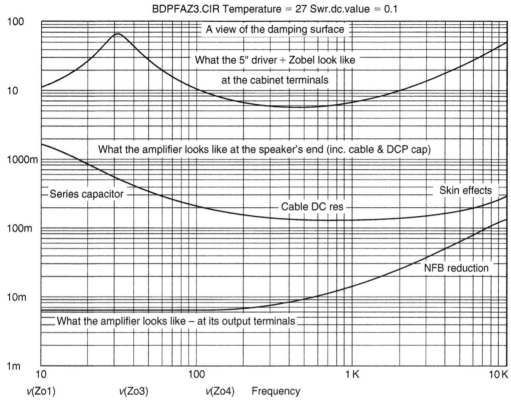

BDPFAZ3.CIR Temperature = 27 Swr.dc.value = 0.1

A view of the damping surface

What the 5" driver + Zobel look like

at the cabinet terminals

What the amplifier looks like at the speaker's end (inc. cable & DCP cap)

Series capacitor                                                    Skin effects

Cable DC res

NFB reduction

What the amplifier looks like – at its output terminals

$v$(Zo1)          $v$(Zo3)          $v$(Zo4)     Frequency

**Figure 24.10: Views of the damping surface in 2D. The lower plot shows the very low steady-state output source impedance of a typical transistor amplifier with high NFB. The middle plot shows how this degrades after passing down a few meters of reasonably rated cable and a series capacitor (which might be the simplest crossover, or for fault protection). The upper plot repeats the impedance versus frequency behavior of the 15" bass driver. The effective damping factor is the smaller and highly variable difference between the upper and the middle plots, not the difference between the highest impedance on the upper plot and the lowest on the lower plot used by amplifier makers!**

1 kHz. The traditional, simplistic "damping factor" takes this ideal impedance at a nominal point (say 100 Hz) and then describes attenuation against an 8-ohm resistor. This gives a damping factor of about three orders, that is, 1000, but up to 10,000 at 30 Hz. Now look at the middle curve of Figure 24.10. This is what the amplifier's damping ability is degraded

to, after it has traversed a given speaker cable and passed through an ideal 10,000-μF series capacitor, as fitted commonly in many professional cabinets for belt' n' braces DC fault protection. The rise at 1 kHz is due to cable resistance, while cable inductance and series capacitance cause the high- and low-end rises, respectively, above 100 milliohms.

We can easily read off static damping against frequency: at 30 Hz, it's about ×100. At midfrequencies, it's about ×50, and again about 100 at 10 kHz. However, instantaneous "dynamic" impedance may dip four times lower, while the DC resistance portion of the speaker impedance increases after hard drive, recovering over tens to thousands of milliseconds, depending on whether the drive-unit is a tweeter or a 24″ shaker.

Even with high NFB, an amplifier's output impedance will be higher with fewer output transistors, less global feedback, junction heating (if the transistors doing the muscle work are MOS-FETs), and more resistive or inductive (longer/thinner) cabling. Reducing the series DC protection capacitor value so it becomes a passive crossover filter will considerably increase source impedance, even in the pass band. The ESR (losses) of any series capacitors and inductors will also increase source impedance, with small, but complex nested variations with drive, temperature, use patterns, and aging. The outcome is that the three curves—and the difference between the upper two that is the map of damping factor—writhe unpredictably.

Full reality is still more complex, as all loudspeakers comprise a number of complex energy storage/release/exchange sections, some interacting with the room space, and each with the others. The conclusion is that damping factor has more dimensions than one number can convey.

### 24.2.2.4 Design Interaction

While high-performance loudspeakers are being designed and optimized, and certainly before they are finalized for production, in-depth listening is a prerequisite. This means that amplifiers are required to drive them while they are being tested and optimized in the design process. Many drive-unit and speaker manufacturers are limited (or limit themselves) to using just one amplifier make to test their designs and production. The situation is rarely publicized.

It follows that many loudspeakers are inevitably looking for that one kind of amplifier that was used when they were "voiced" and "tweaked" by their designer(s). With no less potential for habitual patterns, listeners are looking for the amp that interacts with a loudspeaker in a particular way. The combined behavior or "chemistry" is complex and can be unpredictable and frustratingly unrelated to the type or class of amplifier.

Here is just one reason why the ideal high-performance power amplifier/speaker combination can be determined *only by trying them together* and why quite disparate amplifier designs and topologies may shine equally through a particular speaker.

Knowledgeable "high-end" loudspeaker designers and manufacturing companies employ a well-chosen group of different power amplifiers. One or two will be the best sounding "references"; some will be widely used models and not particularly good performers; others will be niche models, discovered accidentally over the years, that expose loudspeaker problems.

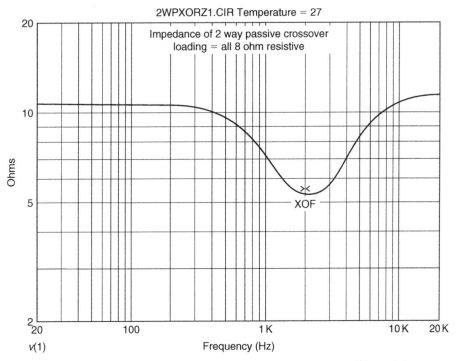

**Figure 24.11: If speakers were simply resistors, the load on the amplifier might appear as here, with the apparent 11-ohm load simply halving to 5.5 ohms at the one point where the two drivers are both drawing substantial current, the crossover point. Note that the crossover dip may as well be a resonance and, as such, adds to the amplifier's load stress.**

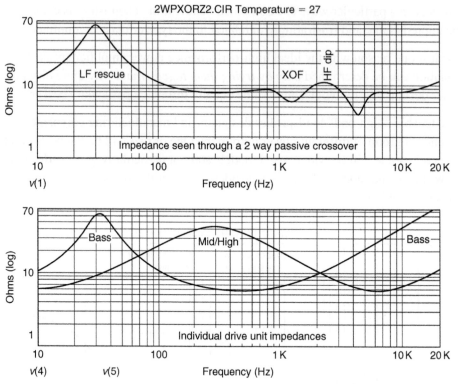

Figure 24.12: Here, the resistors are replaced by drive-units having the impedance characteristics shown in the lower graph. The upper graph shows how the impedance seen by the amplifier has changed—notably two dips where there was one.

### 24.2.3 What Passive Crossovers Look Like to Amplifiers

For conventional full-range loudspeakers with passive crossovers, the crossover components stand between the amplifier and the speaker. Unless the speaker's drive unit(s) is/are blown or have been disconnected or removed, then the crossover won't usually be "seen electrically" (by the amplifier) on its own. Figure 24.11 shows the impedance *that would be seen* by the power amplifier *if* the loudspeakers were 8-ohm resistors and how it drops from about 10.5 ohms to just over 5 ohms at the crossover point. As both drivers are being driven at this point, it's what you might expect. Figure 24.12 shows how much the picture changes when the resistors are replaced by real speakers: now there are two dips.

In the crossover's pass band where there should be no (attenuative) action on the signal voltage at all, the crossover should be "transparent." In practice, one pass band is another driver's stop band. Even a steady-state test signal will experience the added reactive loading at most frequencies. In turn, the crossover is liable to add to the peak current demanded by the drive-units. With music signals, a passive crossover stores energy and can "kick back" like a speaker, potentially adding to the speaker's demands. Overall, for the amplifier's own good, passive crossovers benefit no less than the drive unit from being coupled to an amplifier with very low source impedance, with ample current sourcing *and sinking* capability.

# CHAPTER 25

# Headphones

Ian Sinclair

Not all sound reaches us through loudspeakers, and the advent of the Sony Walkman has led many listeners back to headphone stereo. In this chapter, Ian Sinclair shows the very different type of problems and their solutions as applied to this alternative form of listening.

## 25.1 A Brief History

The history of modern headphones can be traced back to the distant first days of the telephone and telegraph. Then, headphone transducers for both telephone and radio worked on the same moving-iron principle—a crude technique compared to today's sophistication, but one which served so well that they are still to be found in modern telephone receivers.

Moving-iron headphones suffered from a severely restricted frequency response by today's standards, but they were ideal. These sensitive headphones could run directly from a "cat's whisker" of wire judiciously placed on a chunk of galena crystal. Radio could not have become such a success so early in its development without them.

As "cat's whiskers" bristled around the world, these sensitive headphones crackled to the sound of the early radio stations and headphones became as much a part of affluent living as the gramophone. But the invention of sensitive moving-iron loudspeakers was the thin end of the wedge for headphones. Although little more than a telephone earpiece with a horn or radiating cone tagged on (with developments like the balanced armature to increase sensitivity and reduce distortion), they freed individuals or even whole families from the inconvenience of having to sit immobile.

Later, in the 1930s, with the invention of the then revolutionary but insensitive moving-coil loudspeaker and the development of more powerful amplifiers to drive them, headphones were largely relegated to the world of communications, where they stayed until the mid-1950s, when they underwent something of a revival.

Why the change? In a nutshell, stereo. Although stereo was developed in the thirties by Alan Blumlein, it was not to see commercial introduction, via the stereo microgroove record, until the 1950s. Hi-fi, as we now know it, had been developing even before the introduction of the LP, with the 78 rpm record as its source, but the introduction of the microgroove record and then stereo were definitely two large shots in the arm for this emergent science of more accurate sound reproduction. Stereo was also a major stimulus to headphones, and although it may seem an obvious thing now, it took an American Henry Koss to think of the idea of selling stereo headphones as "stereophones" and creating a whole new market for quality stereo headphone listening. Needless to say, Koss has not looked back since and neither has Sony since the introduction of its first Walkman personal cassette player. Ironic, perhaps, because stereo was originally developed by Alan Blumlein for loudspeaker reproduction!

## 25.2  Pros and Cons of Headphone Listening

Good headphones, having no box resonances, can produce a less colored sound than loudspeakers, even today. Headphones have the further acoustic advantage of not exciting room resonances and thus giving the listener a more accurate sense of the recorded acoustics.

However, headphones do not seem capable of producing the sheer impact available from loudspeakers, and unfortunately, stereo images are formed unrealistically inside the head due to the way stereo is recorded for loudspeaker reproduction.

None of these disadvantages matter much if your prime requirement is privacy or, in the case of Walkmans, portability. You can cheerfully blast away at your eardrums without inflicting your musical tastes on others. The closed back type of headphone in particular is very good at containing the sound away from others and insulating the listener from outside sounds.

### 25.2.1  Dummy Heads

One way of overcoming the sound-in-the-head phenomenon that occurs when listening to normal speaker-oriented stereo through headphones is to record the signal completely differently using microphones embedded in artificial ears in a dummy human head.

This "dummy head"-recording technique effectively ensures that each listener's ear receives a replica of the sound at the dummy head's ear positions. The result is much greater realism, with sounds perceivable behind, to the side, and in front of the listener, but not inside the head.

The exact location of sound images depends to some extent on how similar the listeners' ears are to the dummy's, but even using a plastic disc for a head-like baffle between two microphones gives very passable results.

Why then has this not been taken up more widely? In a word, the answer is loudspeakers. Dummy head recordings do not sound at all convincing through loudspeakers due to the blending of sounds between the microphones that occurs around the dummy head during recording. The only way to listen to dummy-head recordings with loudspeakers is to sit immediately between them and listen as if they were headphones, which is not exactly practical.

Nevertheless, listening to good dummy-head recordings through fine headphones is an experience that is not easily forgotten. The sound can be unnervingly real.

### 25.2.2 Cross-Blending

Other ideas used to try to reduce or eliminate the unwanted images in the head with headphone stereo have centered around cross-blending between the two stereo channels to emulate the influence the listener's head would normally have. In other words, to reproduce and imprint electronically the effect of the listener's head on the stereo signals.

The head has two main effects. First, it acts as a baffle, curtailing the high frequencies heard by the ear furthest from the sound. Second, it introduces a time delay (or rotating phase shift with increasing frequency) to the sound heard by the furthest ear from the source.

Naturally such a simulation can only be approximate. Basic circuits simply cross-blend filtered signals between channels to emulate the absorptive factor. More complex circuits have been devised to introduce a time delay, in addition to high-frequency absorption, into the cross-blended signals. This is reported to be more effective, lending a more natural out-of-the-head sound to headphone listening with normal stereo recordings.

### 25.2.3 Biphonics

There were attempts in the 1970s to emulate dummy-head listening when using loudspeakers. The technique was known as biphonics and the idea was to cross-blend

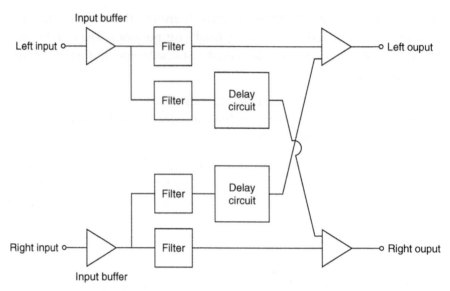

**Figure 25.1: Block diagram of a headphone cross-blend circuit with delay.**

signals feeding two front-positioned loudspeakers to give an all-round effect, with sounds appearing to come from behind the listener as well as in front (Figure 25.1). Surprisingly, the system worked, but only when the listener's head was held within a very limited space in front of the speakers. Predictably, the idea did not catch on.

## 25.3 Headphone Types

### 25.3.1 Moving Iron

Early headphones relied on many turns of very fine wire wound on to a magnetic yoke held close to a stiff disc made of a "soft" magnetic alloy such as Stalloy (Figure 25.2). A permanent magnet pulled the thin disc toward the yoke with a constant force and audio signals fed to the coil caused this force to vary in sympathy with the input. They were very sensitive, needing hardly any power to drive them, and were very poor in sound quality due to the high mass and stiffness of the diaphragm, which caused major resonances and colorations—not to mention distortions due to magnetic nonlinearities.

Currently used in telephone receivers, they are not found today in any application requiring high-quality sound. That is reserved for other more advanced techniques.

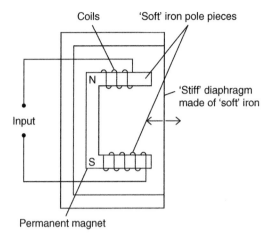

Coils    'Soft' iron pole pieces

N

'Stiff' diaphragm
made of 'soft' iron

Input

S

Permanent magnet

**Figure 25.2: Moving-iron headphone. The current flowing in the coil either strengthens or weakens the force on the soft iron diaphragm. AC audio signals thus vibrate the diaphragm in sympathy.**

### 25.3.2 Moving Coil

Moving-coil headphones (Figure 25.3) work in exactly the same way as moving-coil loudspeakers. A coil of wire, suspended in a radial magnetic field in an annular magnetic gap, is connected to a small radiating cone.

When an alternating audio signal is applied to the coil, the coil vibrates axially in sympathy with the signal, recreating an analogue of the original wave shape. The cone converts this into corresponding fluctuations in air pressure, which are perceived as sound by the listener's nearby ears. Figure 25.4 shows a cut-away view of a Sennheiser unit.

The major difference, of course, between moving-coil headphones and loudspeakers is that the former are much smaller, with lighter and more responsive diaphragms. They can consequently sound much more open and detailed than loudspeakers using the moving-coil principle. They are usually also much more sensitive, which can mean that, in addition to reproducing detail in the signal that is inaudible through loudspeakers, they can also reproduce any background noise more clearly, particularly power amplifier hiss, which is not reduced when the volume is turned down. This is not peculiar to moving-coil headphones and can occur with any sensitive headphone.

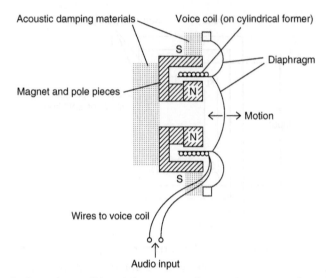

Acoustic damping materials

Voice coil (on cylindrical former)

S

Magnet and pole pieces

Diaphragm

N

N

← ├→ Motion

S

Wires to voice coil

Audio input

**Figure 25.3: Typical moving-coil headphone transducer. The current through the voice coil creates a force on the diaphragm, which vibrates it in sympathy with the audio input.**

**Figure 25.4: Cut-away view of a Sennheiser headphone earpiece showing the diaphragm, magnets, and acoustic damping materials.**

Moving-coil headphones are essentially medium- to low-impedance devices with impedances between eight and a few hundred ohms. They are usually operated via the normal amplifier headphone socket, which simply takes the loudspeaker signal and diverts it through a resistor network to reduce it to a level more suitable for the high sensitivity of the headphones. Alternatively, some high-quality amplifiers provide a separate amplifier to drive the headphones directly.

Many cassette decks also provide a headphone outlet, and some headphones of lower than average sensitivity sometimes do not work very well in this situation due to the limited output available.

### 25.3.3 Electrodynamics Orthodynamic

This type of headphone is essentially in the same family as the moving-coil type, except that the coil has, in effect, been unwound and fixed to a thin, light, plastics diaphragm.

The annular magnetic gap has been replaced by opposing bar magnets, which cause the magnetic field to be squashed more or less parallel to the diaphragm. The "coil" is in fact now a thin conductor zig-zagging or spiraling its way across the surface of the diaphragm, oriented at right angles to the magnetic field so that sending a constant direct current through the conductor results in a more or less equal unidirectional force, which displaces the diaphragm from its rest position. An alternating music signal therefore causes the diaphragm to vibrate in sympathy with it, creating a sound-wave analogue of the music.

The great advantage of the electrodynamic, or flat diaphragm type of headphone, is that the conductor moves the air almost directly, without requiring a cone to carry the vibrations from a coil at one point, to the air at another, with the very real risk of introducing colorations, break-up, distortion, and uneven frequency response.

Unlike a cone, the film diaphragm does not have to be very stiff, although it is sometimes pleated to give it a little more rigidity because the force on it is not entirely uniform. As a result, it can be very thin and light, which results in a lack of stored energy and a very great reduction of the other problems inherent in cones as outlined earlier.

Consequently, good headphones of this type tend to sound more like electrostatics, with an openness and naturalness that eludes even the best moving-coil types.

Electrodynamic headphones tend not to be quite so sensitive as their moving-coil counterparts and are often best used at the output of amplifiers, rather than with cassette decks. Impedance is usually medium to low and is almost entirely resistive.

### 25.3.4 Electrostatic

The electrostatic headphone, like the electrodynamic, uses a thin plastics diaphragm, but instead of a copper track it requires only to be treated to make it very slightly conductive so that the surface can hold an electrostatic charge. It can consequently be very light.

The diaphragm (Figure 25.5) is stretched under low mechanical tension between two perforated conductive plates to which the audio signals are fed via a step-up transformer.

The central diaphragm is kept charged to a very high voltage with respect to the outer plates using a special type of power supply, capable of delivering only a nonlethal, low current, high voltage from the house mains, or, alternatively, by an energizer, which uses some of the audio signal to charge the diaphragm to a similarly high but safe voltage.

The diaphragm experiences electrostatic attraction toward both outer plates. The spacing between the plates and diaphragm, the voltage between them, and the tension on the

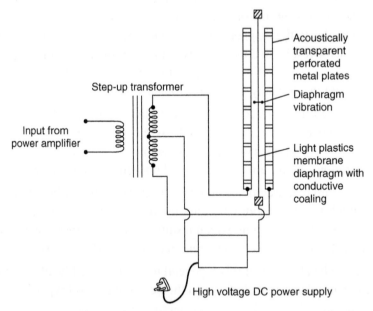

**Figure 25.5: Electrostatic headphone. The transformer steps up the audio signal for feeding to the outer metal plates. The central diaphragm is given a high DC charge with a power supply. An audio signal causes the diaphragm to be attracted alternately to the outer plates.**

diaphragm are all chosen carefully so that the film does not collapse on to either plate. Instead it stays in a stable position between the outer plates, attracted to each one equally during no-signal conditions. When an audio signal is fed to the transformer, it is stepped up at the secondary from a few volts to around a thousand volts. This unbalances the forces on the diaphragm in sympathy with the audio signal, causing it to be attracted alternately to each plate and of course reproducing an analogue of the original sound.

The push–pull action of the transformer and plates effectively produces a linear force on the diaphragm regardless of its position between the plates—unlike normal single-ended electrostatic attraction, which follows an inverse square law and would create large amounts of distortion in a transducer.

The electrostatic headphone is therefore the most linear of all the types available and with its super-light diaphragm, weighing less than several millimeters of adjacent air, it is not surprising that good headphones of this type can offer superb quality sound—the best available. However, this essentially simple technique is the most complex to execute. Not surprisingly, electrostatic headphones are the most costly to manufacture and buy. Figure 25.6 shows details of the Jecklin Float PS-2 type.

They are generally less sensitive than moving-coil types and are usually operated directly from the amplifier's loudspeaker terminals. Due to the capacitive nature of the electrostatic element and the complex inductive/capacitive nature of the transformer, they tend to have more reactive impedance than other types, but this does not usually pose any problem for good amplifiers. Air ionization between the diaphragm and the plates limits the maximum signal and polarizing voltages. Likewise, the signal voltage on the plates must not exceed the polarizing voltage. These factors impose limitations on the maximum sound pressure level that can be achieved.

### 25.3.5 Electrets

Basically, the electret headphone is an electrostatic type but using a material that permanently retains electrostatic charge—the electrostatic equivalent of a permanent magnet. The electret has the advantages of the conventional electrostatic, but does not require an additional external power supply. It is similarly restricted in maximum sound pressure level, although both types produce perfectly adequate sound pressure levels with conventional power amplifiers.

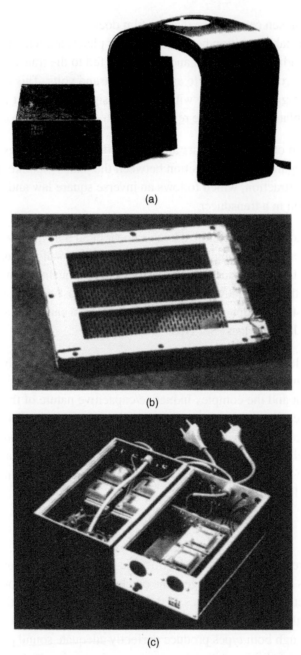

(a)

(b)

(c)

Figure 25.6: (a) Jecklin Float PS-2 electrostatic headphones and energizer unit, (b) Jecklin
Float electrostatic headphone module, and (c) Jecklin Float energizer units.

### 25.3.6 High Polymer

High polymer is basically a generic name to cover piezoelectric plastics films such as polyvinylidene fluoride film.

Piezo-electric materials have been known for many years. They change their dimensions when subjected to an electric field or, conversely generate a voltage when subjected to mechanical strain.

The ceramic barium titanate and crystals such as Rochelle salt and quartz are two materials that have been used for many years in devices such as crystal phono cartridges, ultrasonic transducers, and quartz oscillators but their stiffness is too high and mechanical loss too low for wide-frequency-range audio applications.

High polymer films, however, are very thin, some 8 to 300 μm, and have very low mechanical stiffness, which makes them ideal for transducer diaphragms. The basic film is made piezo electric by stretching it to up to four times its original length, depositing aluminum on each side for electrodes, and polarizing with a high DC electric field at 80–100°C for about an hour.

When voltage is later applied across the film, it vibrates in a transverse direction, becoming alternately longer and shorter. If the material is shaped into an arc, this lengthening and shortening are translated into a pulsating movement, which will generate sound waves in sympathy with the electrical input signal.

It is a relatively simple matter to stretch the high polymer diaphragm across a piece of polyurethane foam pressing against a suspension board to create an arc-shaped diaphragm and make a very simple form of headphone.

High polymer transducers were first developed by Pioneer. Advantages are claimed to be a very low moving mass, similar to electrostatics but without the complexity, and, of course, no power supply. The high polymer headphone is also claimed to be much more sensitive than the electrostatic type and unaffected by humidity, which can reduce the sensitivity of electrostatics. Harmonic distortion is also said to be very low at under 1%.

## 25.4 Basic Headphone Types

Apart from the many different operating principles described earlier, there are two basic categories into which headphones will fall, although some designs will include features of both and will therefore not function purely as one type or the other.

### 25.4.1 Velocity

The open or free-air headphone, known as the velocity type, sits just away from the ear flap, often resting on a pad of light, acoustically transparent reticulated foam. This type of headphone cannot exclude outside sounds, which can intrude on the reproduced music, but the best models can produce a very light, open, airy sound. This type of headphone has been very successful in recent years, particularly since the growth in personal stereos—"Walkman" clones—where their light weight and compact dimensions have made them ideal.

A theoretically perfectly-stiff diaphragm operating in free air, like a velocity headphone, has a frequency response that falls away at low frequencies at 6 dB per octave and so you would expect the system not to work particularly well. However, by juggling with the parameters of mechanical diaphragm stiffness, mass, and acoustic damping, it is possible to get a perfectly adequate low-frequency performance from velocity headphones, which is not dependent on the exact position of the headphones on the ears. The rear face of the diaphragm in velocity headphones is essentially open to the air, which helps impart a more open airy sound, but does not exclude outside noises.

### 25.4.2 Pressure

The other category is the closed or circumaural headphone, known as the pressure type, which totally encloses the ear flap and seals around it with a soft ear cup. The principal advantages of this type of headphone are that sealing around the ear makes it possible to pressure couple the diaphragm to the ear drum from about 700 Hz down to very low frequencies, with a linear response down to 20 Hz easily achievable, as long as the seal is effective (Figure 25.7). The frequency response of the pressure type, unlike the velocity type, is essentially flat down to a low frequency, which is dependent only on the degree of sealing. A poor seal due to inadequate headband pressure, or the wearing of spectacles, for instance, can cause a marked deep bass loss.

The principal disadvantages are that closed headsets tend to be heavier, require greater headband pressure, and can make the ears hot and uncomfortable. Pressure-type headphones can be either closed backed or open backed. Closed-backed headphones offer the exclusion of outside sounds as a distinct advantage in situations that require this, such as recording studios, for instance. However, there is a body of opinion that judges the sound of closed-back headphones to be closed-in compared to open-backed types.

**Figure 25.7: Supra-aural headphones (RE 2390) from Ross Electronics.**

### 25.4.3 Intra-Aural

Another category that has been recently commercialized is the intra-aural type, which actually fits into the ear to provide very lightweight personal listening (Figure 25.8). Obviously the transducer size limits the bass performance, although this is helped by channeling the sound almost directly into the ear canal.

## 25.5 Measuring Headphones

Both types of headphone work into an entirely artificial environment in which the close proximity of the diaphragm and, in the case of the closed headphone, a trapped volume of

**Figure 25.8: Sanyo "turbo" intra-aural earphones with Sanyo's GP600D personal stereo.**

air, modifies the frequency response as perceived by the ear. This situation is additionally confused by the fact that each person's ears are different in shape and at high frequencies will introduce their own pattern of reflections, causing reinforcements and cancellations at different frequencies. Measuring headphones with artificial ears reveals some pretty horrifying curves, but the results on the heads of real listeners are equally alarming.

The fact of the matter is that even if a headphone were made to measure "flat" on a person's head as measured inside the cavity at the entrance to the ear, the headphone would not sound right. This is quite simply because, even when listening without headphones, the sound pressure response at this point, in the ear canal, at the eardrum, or wherever you care to measure it, is simply not flat. We are *not* dealing with an amplifier or an electromechanical transducer but the human ear. The ear and brain have of course figured out that the normally complex pattern of reflections and cancellations with which it has to continually deal are perfectly natural and they sound that way.

Unfortunately, each individual's ears make their own different imprint, but there are some rough trends that can be observed. For instance, there is generally a 2-kHz to

5-kHz boost, a dip around 8 kHz, and all sorts of peaks and dips above 8 kHz. For a natural sound balance with headphones, they should produce measurements that follow this rough trend when tested on real ears. However, measurements carried out using headphones on real ears or artificial ears (designed to mimic real ears for the purposes of testing) reveal differing results that can only be interpreted by the experienced observer. It is clear, though, that published response graphs of headphones should be taken with a very large pinch of salt.

## 25.6 The Future

Guessing on the future is always impossible—one can only be guided by current trends, which are unavoidably based on the past! It is very difficult to see how headphones will develop. New principles of operation are unlikely to spring out of nowhere. Just about all the likely candidates of moving air have been exploited. However, as with the high polymer headphone, the invention of new, better materials can often turn a previously impossible type of headphone into a reality. It is consequently unlikely that tomorrow's headphones will be anything other than developments of today's (unless it becomes possible to inject the audio signals, suitably coded by digital techniques, directly into the auditory nerve or brain).

That may be highly unlikely, not to say impractical and unnecessary, but digital technology could play a part in equalizing the response of headphones. For instance, it is possible to equalize the headphone signal so that the response at the ears when using headphones more closely mimics the headphone-less characteristic. It would be possible to undertake this equalization digitally to exactly compensate for each individual and give improved sound quality. At present this would be fairly expensive, but with processing power falling it should not be long before it is viable commercially. The technology already exists.

Likewise it is perfectly feasible to build digital filters that compensate for each individual's head-baffle effect, thus converting stereo into dummy-head binaural sound for headphone listening. This could all be done at the same time as the earphone-response correction outlined earlier. The only question marks are the very limited market for such equipment and the high cost of developing it. The two conflicting sides (development costs versus economies of manufacturing scale) may not add up to a very balanced equation now, but who knows about the future?

Finally, the diaphragm itself may be driven not by analogue but directly using digital signals. This is not nearly so far-fetched as it sounds. With compact disc as the source, and given the transducer technology, it would be perfectly feasible to retain the signals in digital form from microphone to headphone, with any filtering or correction (as outlined earlier) carried out without the signal having to leave the digital domain.

Guess work may be way off beam, but there is one thing that is certain. There is a great future ahead for headphones.

# Sound Reproduction Systems

# CHAPTER 26

# Tape Recording

Richard Brice

## 26.1 Introduction

Magnetic recording underpins the business of music technology. For all its "glitz" and glamour, the music business, at its most basic, is concerned with one simple function: the recording of music signals onto tape or on disc for subsequent duplication and sale. Before the widespread advent of computer hardware, this technology was pretty well unique to the music industry. Not that this limitation did anything to thwart its proliferation—the cassette player was the second most commonplace piece of technology after the light bulb! Nowadays, with the massive expansion in data-recording products, audio—in the form of digital audio—is just another form of data to be recorded in formats and distributed via highways originally intended for other media. The long-term advantage for music recording applications is the reduction in price brought about by utilizing mass-produced products in high-performance applications that previously demanded a precision, bespoke technology.

A sound recording is made onto magnetic tape by drawing the tape past a recording head at a constant speed. The recording head (which is essentially an electromagnet) is energized by the recording amplifier of the tape recorder. The electromagnet, which forms the head itself, has a small gap so that the magnetic flux created by the action of the current in the electromagnet's coil is concentrated at this gap. The tape is arranged so that it touches the gap in the record head and effectively "closes" the magnetic circuit, as Figure 26.1 illustrates. Because the tape moves and the energizing signal changes with time, a "record" of the flux at any given time is stored on the tape. Replaying a magnetic tape involves dragging the tape back across a similar (or sometimes identical) electromagnet called the playback head. The changing flux detected at the minute gap

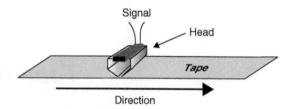

**Figure 26.1: Magnetic tape and the head gap.**

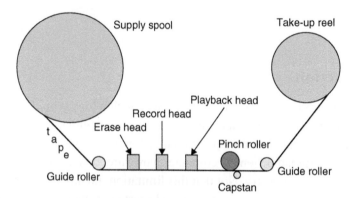

**Figure 26.2: Tape path.**

in the playback head causes a current to flow in the head's coil. This is applied to an amplifier to recover the information left on the tape.

In an analogue tape recorder, the pattern stored on the tape is essentially a stored analogue of the original audio waveform. In a digital recorder the magnetic pattern recorded on the tape is a coded signal that must be decoded by the ensuing operation of the playback electronics. However, at a physical level, analogue and digital recordings using magnetic tape (or discs) are identical.

## 26.2 Magnetic Theory

Figure 26.2 illustrates the path of a magnetic tape through the head assembly of a modern analogue tape recorder. The recording tape is fed from the supply reel across an initial erase head by means of the capstan and pinch roller. The purpose of the erase head is to remove any unwanted previous magnetization on the tape. Next the tape passes the record head, where the audio signal is imprinted upon it, and the playback head, in which the magnetic patterns on the tape are converted back to an audio signal suitable for subsequent amplification and application to a loudspeaker. Finally, the tape is wound onto the take-up reel. When in playback mode, the erase head and the record head are not energized. Correspondingly, in record mode, the playback head may be used to monitor the signal off-tape to ensure that recording levels and so on are correct. Less expensive cassette tape recorders combine the record and playback heads in a composite assembly, in which case off-tape monitoring while recording is not possible.

## 26.3 The Physics of Magnetic Recording

In a tape recording, sound signals are recorded as a magnetic pattern along the length of the tape. The tape itself consists of a polyester-type plastic backing layer, on which is applied a thin coating with magnetic properties. This coating usually contains tiny particles of ferric iron oxide (so-called ferric tapes), although more expensive tapes may use chromium dioxide particles or metal alloy particles, which have superior magnetic properties (so-called chrome or metal tapes, respectively).

The properties of magnetic materials take place as a result of microscopic magnetic domains—each a tiny bar magnet—within the material. In an unmagnetized state, these domains are effectively aligned randomly so that any overall, macroscopic magnetic external field is canceled out. Only when the ferrous material is exposed to an external magnetic field do these domains start to align their axis along the axis of the applied field, the fraction of the total number of domains so aligned being dependent on the strength of the externally applied field. Most significantly, after the external field has been removed, the microscopic domains do not altogether return to their preordered state and the bulk material exhibits external magnetic poles.

The relation between the magnetizing field ($H$) and the resultant induction ($B$) in an iron sample (assumed, initially, to be in a completely demagnetized condition) may be plotted as shown in Figure 26.3. Tracing the path from the origin, note that the first section of the looped curve rises slowly at first (between O and B1), then more rapidly (between B1 and B2), and finally more and more gradually as it approaches a point where only a very few magnetic domains remain left to be aligned. At this point (B3) the ferrous material is said to be saturated. Significantly, when the magnetizing force ($H$) is reduced, the magnetic induction ($B$) does not retrace its path along the curve B3–B2–B1–O, instead it falls along a different path, B3–B4, at which point the magnetizing force is zero again, but the

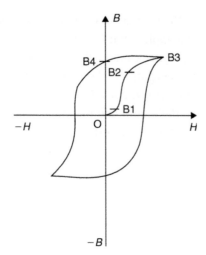

**Figure 26.3: BH curve.**

ferrous material remains magnetized with the residual induction B4. This remaining force is referred to as remnance. For this remnance to be neutralized, an opposite magnetic force must be applied, which accounts for the rest of the looped curve in Figure 26.3. The magnitude of the applied magnetic force required to reduce the remnance to zero is termed coercivity (the ideal magnetic tape exhibiting both high remnance and high coercivity).

## 26.4  Bias

If a sound recording and reproduction system is to perform without adding discernible distortion, a high degree of linearity is required. In the case of tape recording this implies the necessity for a direct relationship between the applied magnetic force and the resultant induction on the tape. Looking again at Figure 26.3, it is apparent that the only linear region over which this relationship holds is between B1 and B2, with the relationship being particularly nonlinear about the origin. The situation may be compared to a transistor amplifier, which exhibits a high degree of nonlinearity in the saturation and cut-off region and a linear portion in between. The essence of good design, in the case of the transistor stage, is appropriately to bias the amplifier in its linear region by means of a steady DC potential. And so it is with magnetic recording. In principle, a steady magnetic force may be applied, in conjunction with the varying force dependent

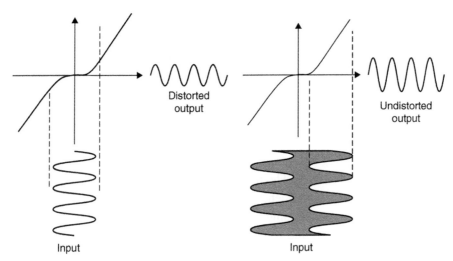

**Figure 26.4: Linearizing effect of AC bias.**

on the audio signal, thereby biasing the audio signal portion of the overall magnetic effect into the initial linear region of the *BH* loop. In practice, such a scheme has a number of practical disadvantages. Instead a system of ultrasonic AC bias is employed, which mixes the audio signal with a high-frequency signal current. This bias signal, as it is known, does not get recorded because the wavelength of the signal is so small that the magnetic domains resulting from it neutralize themselves naturally. It acts solely to ensure that the audio modulation component of the overall magnetic force influences the tape in its linear region. Figure 26.4 illustrates the mechanism.

It is hardly surprising that the amplitude of the superimposed high-frequency bias signal is important in obtaining the best performance from an analogue tape machine and a given tape. Too high an amplitude and high-frequency response suffers; too low a value and distortion rises dramatically. Different tape formulations differ in their ideal biasing requirements, although international standardization work [by the International Electrotechnical Commission (IEC)] has provided recommendations for the formulation of "standard" tape types.

## 26.5 Equalization

For a number of reasons, both the signal that is imprinted upon the tape by the action of the record current in the record head and the signal arising as a consequence of the induced current in the playback head are heavily distorted with respect to frequency and must both be equalized. This is an area where standardization between different manufacturers is particularly important because, without it, tapes recorded on one machine would not be reproducible on another.

In itself, this would not be such a problem were it not for the fact that, due to differences in head geometry and construction, the electrical equalization differs markedly from manufacturer to manufacturer. The IEC provided an ingenious solution to widespread standardization by providing a series of standard prerecorded tapes on which are recorded frequency sweeps and spot levels. The intention was that these must be reproduced (played back) with a level flat-frequency response characteristic, with the individual manufacturer responsible for choosing the appropriate electrical equalization to affect this situation. This appears to leave the situation concerning record equalization undefined, but this is not the case because it is intended that the manufacturer chooses record equalization curves so that tapes recorded on any particular machine must result in a flat-frequency response when replayed using the manufacturer's own IEC standard replay equalization characteristic.

The issue of "portability" should not be overlooked, and any serious studio that still relies on analogue recording must ensure that its analogue tape equipment is aligned (and continues to remain aligned—usually the duty of the maintenance engineer) to the relevant IEC standards. This, unfortunately, necessarily involves the purchase of the relevant standard alignment tapes.

## 26.6  Tape Speed

Clearly another (in fact, the earliest) candidate for standardization was the choice of linear speed of the tape through the tape path. Without this the signals recorded on one machine replay at a different pitch when replayed on another. While this effect offers important artistic possibilities (see later in this chapter), it is clearly undesirable in most operational circumstances. Table 26.1 lists the standard tape speeds in metric (centimeters per second, cm/s) and imperial measures (inches per second, ips) and their likely applications.

## 26.7  Speed Stability

Once standardized, the speed of the tape must remain consistent over both long and short terms. Failure to establish this results in audible effects known, respectively, as wow and

**Table 26.1 Standard tape speeds in metric and imperial measures**

| Tape speed (ips) | cm/s | Application |
|:---:|:---:|:---|
| 30 | 76 | Top professional quality |
| 15 | 38 | Top professional quality |
| 7.5 | 19 | Professional quality (with noise reduction) |
| 3.75 | 9.5 | Semiprofessional quality (with noise reduction) |
| 1.875 | 4.75 | Domestic quality (with noise reduction) |

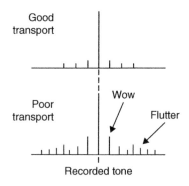

**Figure 26.5: FM sidebands as a result of speed instability.**

flutter. However, these onomatopoeic terms relate to comparatively coarse effects. What is often less appreciated is the action of speed stability upon the purity of audio signals— a fact that is easier to appreciate if speed instability is regarded as a frequency modulation effect. We know that FM modulation results in an infinite set of sidebands around the frequency-modulated carrier. The effect of speed instability in an analogue tape recorder may be appreciated in these terms by looking at the output of a pure sine tone recorded and played back analyzed on a spectrum analyzer, as shown in Figure 26.5. Note that the tone is surrounded by a "shoulder" of sidebands around the original tone.

Happily, the widespread adoption of digital recording has rendered much of the aforementioned obsolete, especially in relation to two-track masters. Where analogue tape machines are still ubiquitous (e.g., in the case of multitrack recorders), engineering excellence is a necessary byword, as is the inevitable high cost that this implies. In addition, alignment and calibration to recognized standards must be performed regularly (as well as regular cleaning) in order to ensure that multitrack tapes can be recorded and mixed in different studios.

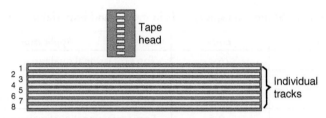

**Figure 26.6: Multiple tape tracks across width of the tape.**

## 26.8  Recording Formats—Analogue Machines

Early tape recorders recorded a single channel of audio across the whole tape width. Pressure to decrease expensive tape usage led to the development of the concept of using "both sides" of a tape by recording the signal across half the tape width and subsequently flipping over the tape to record the remaining unrecorded half in the opposite direction. The advent of stereo increased the total number of audio tracks to four: two in one direction, two in the other. This format is standard in the familiar analogue cassette. From stereo it is a small conceptual step (albeit a very large practical one) to 4, 8, 16, or more tracks being recorded across the width of a single tape. Such a development demanded various technological innovations, the first was the development of composite multiple head assemblies. Figure 26.6 illustrates the general principle. Given the dimensions, the construction of high-quality head assemblies was no mean achievement. The second was the combination of record and replay heads. Without this development, the signal "coming off" tape would be later than the signal recorded onto the tape, a limitation that would make multitrack recording impossible. In early machines, the record head was often made to do temporary duty as playback head during the recording stages of a multitrack session, with its less than perfect response characteristic being adequate as a cue track. The optimized playback head was reserved for mix down only.

Despite this, the number of tracks that it is practical to record across a given width of tape is not governed by head construction limitations only, but by considerations of the signal-to-noise ratio. As shown earlier, the signal recorded onto tape is left as a physical arrangement of magnetic domains. Without an audio signal, these domains remain unmagnetized and persist in a random state. These cause noise when the tape is replayed. Similarly, at some point, when a strong signal is recorded, all the domains are "used up" and the tape saturates. A simple rule applies in audio applications: the more domains onto which the signal is imprinted, the better, up to the point just below saturation. This may be achieved in various ways: by running the tape faster and by using a greater tape width for a given number of tracks. Figure 26.7 illustrates this by depicting the saturation levels of a commercial tape at

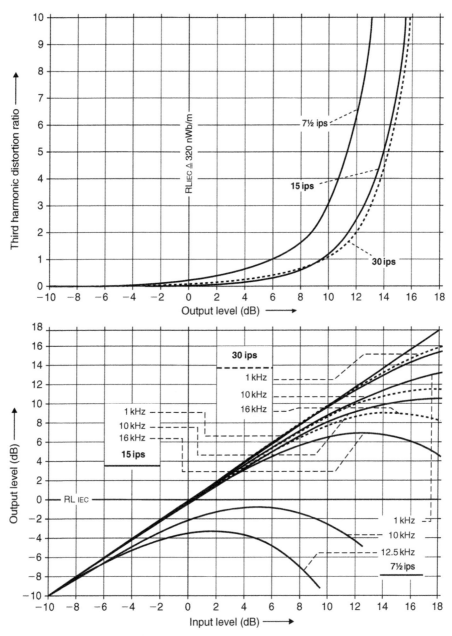

**Figure 26.7: The effects of tape speed on saturation and distortion.**

Table 26.2 Major analogue recording formats

| Tracks | Format | Medium/speed | Application |
|--------|--------|--------------|-------------|
| 2 | ½" stereo | ½" 7.5–30 ips | High-quality mastering |
| 2 | ¼" stereo | ¼" 7.5–30 ips | High-quality mastering |
| 2 | Cassette | ⅛" 15/8" ips | Medium quality replay |
| 4 | ½" four track | ½" 7.5–30 ips | High-quality mastering |
| 4 | Cassette | ⅛" 3.75 ips | Personal multitrack |
| 8 | ¼" multitrack | 7.5–15 ips | Semipro multitrack |
| 16 | ½" multitrack | ½" 15–30 ips | Professional multitrack |
| 16 | 1" multitrack | 1" 30 ips | High-quality multitrack |
| 16–24 | 1" multitrack | 1" 30 ips | High-quality multitrack |
| 24 | 2" multitrack | 2" 30 ips | High-quality multitrack |

**Figure 26.8: Analogue mastering recorder.**

various speeds. This simple principle accounts for the many different tape formats that exist. Each is an attempt to redefine the balance among complexity, sound quality, and tape cost appropriate to a certain market sector. Table 26.2 lists some of the major analogue recording formats. Note that the format of a tape relates to its width, specified in inches.

### 26.8.1 Analogue Mastering

Analogue mastering is now very rare, this office having been made the exclusive domain of digital audio tape. A typical high-quality two-track mastering recorder is illustrated in Figure 26.8.

**Figure 26.9: Cassette-based "notebook" multitrack.**

### 26.8.2 Analogue Multitrack Tape Machines

As mentioned earlier, analogue multitrack machines betray their quality roughly in proportion to the width of the tape utilized for a given number of tracks. A 2-inch tape, 24 track, which utilizes a 2-inch width tape drawn across 24 parallel heads, is therefore better than a 1-inch, 24 track, but not necessarily better than a ½-inch, two track! Not only does a greater head-to-tape contact guarantee a higher tape signal-to-noise ratio (i.e., more domains are usefully magnetized) but it also secures less tape dropout. Dropout is an effect where the contact between tape and head is broken microscopically for a small period during which the signal level falls drastically. Sometimes dropout is due to irregularities in the tape or to the ingress of a tiny particle of dust; whichever, the more tape passing by an individual recording or replay head, the better chance there is of dropouts occurring infrequently. Analogue tape machines are gradually becoming obsolete in multitrack sound recording; however, the huge installed base of these machines means they will be a part of sound recording for many years to come.

### 26.8.3 Cassette-Based Multitracks

Figure 26.9 illustrates a typical analogue cassette-based portable multitrack recorder and mixer combined. This type of low-end "recording studio in a box" is often termed a Portastudio and these units are widespread as personal recording "notebooks." Typically four tracks are available and are recorded across the entire width of the cassette tape (which is intended to be recorded in one direction only). The cassette tape usually runs at twice normal speed, 3.75 ips. Individual products vary but the mixer of the unit illustrated in Figure 26.9 allows for two (unbalanced) microphone inputs and a further four inputs at line level, of which only two are routed to the tape tracks. Each of the first four inputs may be switched between INPUT, OFF, and TAPE (return). Selecting INPUT will (when

**Figure 26.10: Roland digital multitrack.**

the record button is engaged on the tape transport buttons) switch the track to record. The mixer also incorporates two send–return loops and the extra line level inputs mentioned earlier. In addition, an extra monitor mixer is provided, the output of this being selectable via the monitor output. It is thus a tiny split multitrack console. Despite the inevitable compromises inherent in such a piece of equipment, many portable multitrack units are capable of remarkably high quality and many have been used to record and mix release-quality material. Indeed, so popular has this format proved to be that digital versions have begun to appear, products that offer musical notebook convenience with exemplary sound quality. One such is illustrated in Figure 26.10.

# CHAPTER 27

# Recording Consoles

Richard Brice

## 27.1 Introduction

This chapter is about recording consoles, the very heart of a recording studio. Like our own heart, whose action is felt everywhere in our own bodies, consideration of a recording console involves wide-ranging considerations of other elements within the studio system. These, too, are covered in this chapter.

In pop and rock music, as well as in most jazz recordings, each instrument is almost always recorded onto one track of multitrack tape and the result of the "mix" of all the instruments combined together electrically inside the audio mixer and recorded onto a two-track (stereo) master tape for production and archiving purposes. Similarly, in the case of sound reinforcement for rock and pop music and jazz concerts, each individual musical contributor is miked separately and the ensemble sound mixed electrically. It is the job of the recording or balance engineer to control this process. This involves many aesthetic judgements in the process of recording the individual tracks (tracking) and mixing down the final result. However, relatively few parameters exist under her/his control. Over and above the office of correctly setting the input gain control so as to ensure best signal to noise ratio and control of channel equalization, her/his main duty is to judge and adjust each channel gain fader and therefore each contributor's level within the mix. A further duty, when performing a stereo mix, is the construction of a stereo picture or image by controlling the relative contribution each input channel makes to the two, stereo mix amplifiers. In cases of both multitrack mixing and multimicrophone mixing, the apparent position of each instrumentalist within the stereo picture (image) is controlled by a special stereophonic panoramic potentiometer, or pan pot for short.

## 27.2 Standard Levels and Level Meters

Suppose I asked you to put together a device comprising component parts I had previously organized from different sources. And suppose I had paid very little attention to whether each of the component parts would fit together (perhaps one part might be imperial and another metric). You would become frustrated pretty quickly because the task would be impossible. So it would be too for the audio mixer, if the signals it received were not, to some degree at least, standardized. The rationale behind these standards and the tools used in achieving this degree of standardization are the subjects of the first few sections of this chapter.

The adoption of standardized studio levels (and of their associated lineup tones) ensures the interconnectability of different equipment from different manufacturers and ensures that tapes made in one studio are suitable for replay and/or rework in another. Unfortunately, these "standards" have evolved over many years and some organizations have made different decisions, which, in turn, have reflected upon their choice of operating level. National and industrial frontiers exist too, so that the subject of maximum and alignment signal levels is fraught with complication.

Fundamentally, only two absolute levels exist in any electronic system, maximum level, and noise floor. These are both illustrated in Figure 27.1. Any signal that is lower than the noise floor will disappear as it is swamped by noise and signal, which is larger than maximum level will be distorted. All well-recorded signals have to sit comfortably between the "devil" of distortion and the "deep blue sea" of noise. Actually, that's the fundamental job of any recording engineer!

In principle, maximum level would make a good line-up level. Unfortunately, it would also reproduce over loudspeakers as a very loud noise indeed and would therefore, likely

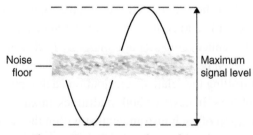

**Figure 27.1: System dynamic range.**

as not, "fray the nerves" of those people working day after day in recording studios! Instead a lower level is used for line-up, which actually has no physical justification at all. Instead it is cleverly designed to relate maximum signal level to the perceptual mechanism of human hearing and to human sight as we shall see. Why sight? Because it really isn't practical to monitor the loudness of an audio signal by sound alone. Apart from anything else, human beings are very bad at making this type of subjective judgement. Instead, from the very earliest days of sound engineering, visual indicators have been used to indicate audio level, thereby relieving the operator from making subjective auditory decisions. There exist two important and distinct reasons to monitor audio level.

The first is to optimize the drive, the gain or sensitivity of a particular audio circuit, so that the signal passing through it is at a level whereby it enjoys the full dynamic range available from the circuit. If a signal travels through a circuit at too low a level, it unnecessarily picks up noise in the process. If it is too high, it may be distorted or "clipped" as the stage is unable to provide the necessary voltage swing, as shown in Figure 27.1.

The second role for audio metering exists in, for instance, a radio or television continuity studio where various audio sources are brought together for mixing and switching. Listeners are justifiably entitled to expect a reasonably consistent level when listening to a radio (or television) station and do not expect one program to be unbearably loud (or soft) in relation to the last. In this case, audio metering is used to judge the apparent loudness of a signal and thereby make the appropriate judgements as to whether the next contribution should be reduced (or increased) in level compared with the present signal.

The two operational requirements described earlier demand different criteria of the meter itself. This pressure has led to the evolution of two types of signal monitoring meter, the volume unit (VU) meter, and the peak program meter (PPM).

### 27.2.1 The VU Meter

A standard VU meter is illustrated in Figure 27.2(a). The VU is a unit intended to express the level of a complex wave in terms of decibels above or below a reference volume, it implies a complex wave—a program waveform with high peaks. A 0 VU reference level therefore refers to a complex-wave power reading on a standard VU meter. A circuit for driving a moving coil VU meter is given in Figure 27.2(b). Note that the rectifiers

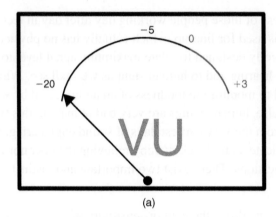

(a)

**Figure 27.2(a): VU meter.**

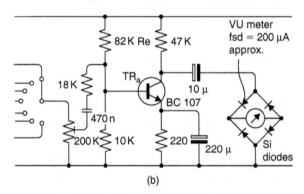

(b)

**Figure 27.2(b): VU meter circuit.**

and meter are fed from the collector of $TR_a$, which is a current source in parallel with Re. Because Re is a high value in comparison with the emitter load of $TR_a$ the voltage gain during the part of the input cycle when the rectifier diodes are not in conduction is very large. This alleviates most of the problem of the Si diodes' offset voltage. From the circuit it is clear that a VU meter is an indicator of the average power of a waveform; it therefore accurately represents the apparent loudness of a signal because the ear too mathematically integrates audio waveforms with respect to time. However, because of this, the VU is not a peak-reading instrument. A failure to appreciate this, and on a practical level this means allowing the meter needle to swing into the red section on transients, means that the mixer is operating with inadequate system headroom. This characteristic has led the VU to be regarded with suspicion in some organizations.

To some extent, this is unjustified because the VU may be used to monitor peak levels, provided the action of the device is properly understood. The usual convention is to assume that the peaks of the complex wave will be 10 to 14 dB higher than the peak value of a sine wave adjusted to give the same reference reading on the VU meter. In other words, if a music or speech signal is adjusted to give a reading of 0 VU on a VU meter, the system must have at least 14 dB headroom—over the level of a sine wave adjusted to give the same reading—if the system is not to clip the program audio signal. In operation, the meter needles should swing only very occasionally above the 0 VU reference level on complex program.

### 27.2.2 The PPM Meter

Whereas the VU meter reflects the perceptual mechanism of the human hearing system, and thereby indicates the loudness of a signal, the PPM is designed to indicate the value of peaks of an audio waveform. It has its own powerful champions, notably the BBC and other European broadcasting institutions. The PPM is suited to applications in which the balance engineer is setting levels to optimize a signal level to suit the dynamic range available from a transmission (or recording) channel. Hence its adoption by broadcasters who are under statutory regulation to control the depth of their modulation and therefore fastidiously to control their maximum signal peaks. In this type of application, the balance engineer does not need to know the "loudness" of the signal, but rather needs to know the maximum excursion (the peak value) of the signal.

It is actually not difficult to achieve a peak reading instrument. The normal approach is a meter driven by a buffered version of a voltage stored on a capacitor, itself supplied by a rectified version of the signal to be measured [see Figure 27.3(a)]. In fact, the main limitation of this approach lies with the ballistics of the meter itself, which, unless standardized, leads to different readings. The PPM standard demands a defined and consistent physical response time of the meter movement. Unfortunately, the simple arrangement is actually unsuitable as a volume monitor due to the highly variable nature of the peak to average ratio of real-world audio waveforms, a ratio known as crest factor. This enormous ratio causes the meter needle to flail about to such an extent that it is difficult to interpret anything meaningful at all! For this reason, to the simple arrangement illustrated in Figure 27.3(a), a logarithmic amplifier is appended as shown at Figure 27.3(b). This effectively compresses the dynamic range of the signal prior to its display; a modification that (together with a controlled decay time constant) enhances the PPM's readability greatly—albeit at the expense of considerable complexity.

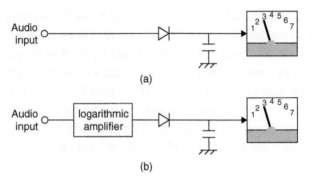

(a)

(b)

Figure 27.3: Peak reading meters.

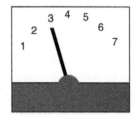

Figure 27.4: BBC style PPM.

The peak program meter of the type used by the BBC is illustrated in Figure 27.4. Note the scale marked 1 to 7, with each increment representing 4 dB (except between 1 and 2, which represents 6 dB). This constant deflection per decade is realized by the logarithmic amplifier. The line-up tone is set to PPM4 and signals are balanced so that peaks reach PPM6, which is 8 dB above the reference level. (The BBC practice is that the peak studio level is 8 dB above the alignment level.) BBC research has shown that the true peaks are actually about 3 dB higher than those indicated on a BBC PPM and that operator errors cause the signal to swing occasionally 3 dB above the indicated permitted maximum, that is, a total of 14 dB above alignment level.

### 27.2.3 PPM Dynamic Performance

### 27.2.3.1 PPM

In BS55428 Part 9, the PPM is stated as requiring, "an integration time of 12 ms and a decay time of 2.8 s for a decay from 7 to 1 on the scale." This isn't entirely straightforward to understand. However, an earlier standard (British Standard

BS4297:1968) defined the rise time of a BBC style PPM in terms of reading relative to 5-kHz tone-burst durations such that, for a steady tone adjusted to read scale 6, bursts of various values should be within the limits given here:

**Table 27.1 Burst duration and the respective meter reading**

| Burst duration | Meter reading (relative to 6) |
|---|---|
| Continuous | 0 dB |
| 100 ms | $0 \pm 0.5$ dB |
| 10 ms | $-2.5 \pm 0.5$ dB |
| 5 ms | $-4.0 \pm 0.5$ dB |
| 1.5 ms | $-9.0 \pm 1.0$ dB |

This definition has the merit of being testable.

### 27.2.3.2 VU Meter

The VU meter is essentially a milliammeter with a 200-mA FSD fed from a full-wave rectifier installed within the case with a series resistor chosen such that the application of a sine wave of 1.228 V RMS (i.e., 4 dB above that required to give 1 mW in 600 R) causes a deflection of 0 VU. Technically, this makes a VU an rms reading volt meter. Of course, for a sine wave the relationship between peak and rms value is known (3 dB or $1/\sqrt{2}$), but no simple relationship exists between rms and pk for real-world audio signals.

In frequency response terms, the response of the VU is essentially flat (0.2 dB limits) between 35 Hz and 10 kHz. The dynamic characteristics are such that when a sudden sine wave type signal is applied, sufficient to give a deflection at the 0 VU point, the pointer shall reach the required value within 0.3 s and shall not overshoot by more than 1.5% (0.15 dB).

### 27.2.4 Opto-Electronic Level Indication

Electronic level indicators range from professional bargraph displays, which are designed to mimic VU or PPM alignments and ballistics, through the various peak-reading displays common on consumer and prosumer goods (often bewilderingly calibrated), to simple peak-indicating LEDs. The latter, can actually work surprisingly well—and actually facilitate a degree of precision alignment, which belies their extreme simplicity.

In fact, the difference between monitoring using VUs and PPMs is not as clear cut as stated. Really, both meters reflect a difference in emphasis: the VU meter indicates loudness—leaving the operator to allow for peaks based on the known, probabilistic nature of real audio signals. The PPM, however, indicates peak, leaving it to the operator to base decisions of apparent level on the known stochastic nature of audio waveforms. However, the latter presents a complication because, although the PPM may be used to judge level, it does take experience. This is because the crest factor of some types of program material differs markedly from others, especially when different levels of compression are used between different contributions. To allow for this, institutions that employ the PPM apply ad hoc rules to ensure continuity of level between contributions and/or program segments. For instance, it is BBC practice to balance different program material to peak at different levels on a standard PPM.

Despite its powerful European opponents, a standard VU meter combined with a peak-sensing LED is very hard to beat as a monitoring device because it both indicates volume and, by default, average crest factor. Any waveforms that have unusually high peak-to-average ratio are indicated by the illumination of the peak LED. Unfortunately, PPMs do not indicate loudness, and their widespread adoption in broadcast accounts for the many uncomfortable level mismatches between different contributions, especially between programs and adverts.

### 27.2.5 Polar CRT Displays

One very fast indicator of changing electrical signals is a cathode ray tube (CRT). With this in mind, there has, in recent years, been a movement to use CRTs as a form of fast audio monitoring, not as in an oscilloscope, with an internal timebase, but as a polar, or XY display. The two-dimensional polar display has a particular advantage over a classic, one-dimensional device like a VU or PPM in that it can be made to display left and right signals at the same time. This is particularly useful because, in so doing, it permits the engineer to view the degree to which the left and right signals are correlated; which is to say the degree to which a stereo signal contains in-phase, mono components and the degree to which it contains out-of-phase or stereo components.

In the polar display, the Y plates inside the oscilloscope are driven with a signal that is the sum of the left and right input signal (suitably amplified). The X plates are driven with a signal derived from the stereo difference signal $(R - L)$, as shown in Figure 27.5. Note that the left signal will create a single moving line along the diagonal L axis as shown. The right signal clearly does the same thing along the R axis. A mono $(L = R)$

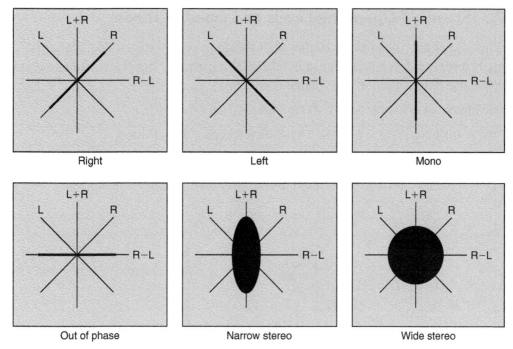

**Figure 27.5: Audio polar displays.**

signal will create a single vertical line, and an out-of-phase mono signal will produce a horizontal line. A stereo signal produces a woolly ball centered on the origin; its vertical extent is governed by the degree of L/R correlation and its horizontal extent is governed by L/R decorrelation. And herein lies the polar display's particular power, that it can be used to assess the character of a stereo signal, alerting the engineer to possible transmission or recording problems, as illustrated in Figure 27.5.

One disadvantage of the polar display methodology is that, in the absence of modulation, the cathode ray will remain undeviated and a bright spot will appear at the center of the display, gradually burning a hole on the phosphor! To avoid this, commercial polar displays incorporate cathode modulation (k mod) so that, if the signal goes below a certain value, the cathode is biased until the anode current cuts off, extinguishing the beam.

## 27.3 Standard Operating Levels and Line-Up Tones

Irrespective of the type of meter employed, it should be pretty obvious that a meter is entirely useless unless it is calibrated in relation to a particular signal level (think about it if rulers had different centimeters marked on them!).

Three important line-up levels exist (see Figure 27.6):

PPM4 = 0 dBu = 0.775 V rms, used by United Kingdom broadcasters.

0VU = +4 dBu = 1.23 V rms, used in commercial music sector.

0VU = −10 dBV = 316 mV rms, used in consumer and "prosumer" equipment.

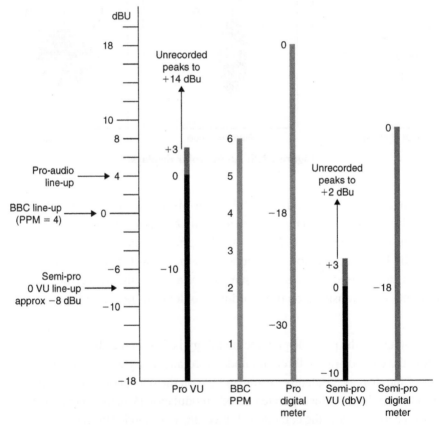

**Figure 27.6: Standard levels compared.**

## 27.4 Digital Line-Up

The question of how to relate 0 VU and PPM 4 to a digital maximum level of 0 dBFS (for 0 dB relative to full scale) has been the topic of hot debate. Fortunately, the situation has crystallized over the last few years to the extent that it is now possible to describe the situation on the basis of widespread implementation in the United States and with European broadcasters. Essentially,

$$0 \text{ VU} = +4 \text{ dBu} = -20 \text{ dBFS (SMPTE RP155)}$$

$$\text{PPM4} = 0 \text{ dBu} = -18 \text{ dBFS (EBU R64-1992)}$$

Sadly, these recommendations are not consistent. And while the EBU recommendation seems a little pessimistic in allowing extra 4-dB headroom above their own worst-case scenario, the SMPTE suggestion looks positively gloomy in allowing 20 dB above the alignment level. Although this probably reflects the widespread, although technically incorrect, methodology, when monitoring with VUs, of setting levels so that peaks often drive the meter well into the red section.

## 27.5 Sound Mixer Architecture and Circuit Blocks

The largest, most expensive piece of capital electronic equipment in any professional studio is the main studio mixer. So much so that in publicity shots of a recording studio this component is always seen to dominate the proceedings! Perhaps that's why there seem to be so many terms for it: mixer, mixing desk, or just desk, audio console, console, and simply "the board," to name but a few. To many people outside of the recording industry, the audio mixer represents the very essence of sound recording. This is partially correct, for it is at the console that many of the important artistic decisions that go toward mixing a live band or a record are made. However, in engineering terms, this impression is misleading. The audio console is a complicated piece of equipment but, in its electronic essence, its duties are relatively simple. In other words, the designer of an audio mixer is more concerned with the optimization of relatively simple circuits, which may then be repeated many, many times, than he/she is with the design of clever or imaginative signal processing. But before we investigate the individual circuit elements of an audio mixer, it is important to understand the way in which these blocks fit together. This is usually termed system architecture.

### 27.5.1 System Architecture

There is no simple description of audio console system architecture. That's because there exist different types of consoles for different duties and because every manufacturer (and there are very many of them) each has their own ideas about how best to configure the necessary features in a manner that is operationally versatile, ergonomic, and maintains the "cleanest" signal path from input to output. However, just as houses all come in different shapes and sizes and yet all are built relying upon the same underlying assumptions and principles, most audio mixers share certain system topologies.

### 27.5.2 Input Strip

The most conspicuous "building block" in an audio console, and the most obviously striking at first glance, is the channel input strip. Each mixer channel has one of these and they tend to account for the majority of the panel space in a large console. A typical input strip for a small recording console is illustrated in Figure 27.9. When you consider that a large console may have 24, 32, or perhaps 48 channels—each a copy of the strip illustrated in Figure 27.9—it is not surprising that large commercial studio consoles look so imposing. But always remember, however "frightening" a console looks, it is usually only the sheer bulk that gives this impression. Most of the panel is repetition and once one channel strip is understood, so are all the others!

Much harder to fathom, when faced with an unfamiliar console, are the bus and routing arrangements that feed the group modules, the monitor, and master modules. These "hidden" features relate to the manner in which each input module may be assigned to the various summing amplifiers within the console. And herein lies the most important thing to realize about a professional audio console; that it is many mixers within one console. First let's look at the groups.

### 27.5.3 Groups

Consider mixing a live rock band. Assume that it is a quintet: a singing bass player, one guitarist, a keyboard player, a saxophonist, and a drummer. The inputs to the mixer might look something like this:

**Table 27.2 Inputs to the mixer**

| | |
|---|---|
| Channel 1 | Vocal mic |
| Channel 2 | Bass amp mic |
| Channel 3 | Lead guitar amp mic |
| Channel 4 | Backing mic (for keyboardist) |
| Channel 5 | Backing mic (for second guitarist) |
| Channel 6 | Sax mic |
| Channel 7 | Hi-hat mic |
| Channel 8 | Snare mic |
| Channel 9 | Bass drum mic |
| Channels 10 and 11 | Drum overheads |
| Channels 12 and 13 | Stereo line piano input |
| Channels 14 and 15 | Sound module line input |

Clearly inputs 7 to 11 are all concerned with the drums. Once these channels have been set so as to give a good balance between each drum, it is obviously convenient to group these faders together so that the drums can be adjusted relative to the rest of the instruments. This is the role of the separate summing amplifiers in a live console, to group various instruments together. That's why these smaller "mixers within mixers" are called groups. These group signals are themselves fed to the main stereo mixing amplifier, the master section. Mixer architecture signal flow is therefore channeled to groups and groups to master output mixer, as shown in Figure 27.7. A block schematic of a simplified live-music mixer is given in Figure 27.8 in which this topology is evident.

### 27.5.4 Pan Control

Stereophonic reproduction from loudspeakers requires that stereo information is carried by interchannel intensity differences alone, as there is no requirement for interchannel delay differences. The pan control progressively attenuates one channel while progressively strengthening the other as the knob is rotated, with the input being shared equally between both channels when the knob is in its center (12 o'clock) position. In the sound mixer shown in Figure 27.8, note that the channel pan control operates in a rather more flexible manner, as a control that may be used to "steer" the input signal between either of the pairs of buses selected by the routing switches. The flexibility of this approach becomes evident when a console is used in a multitrack recording session.

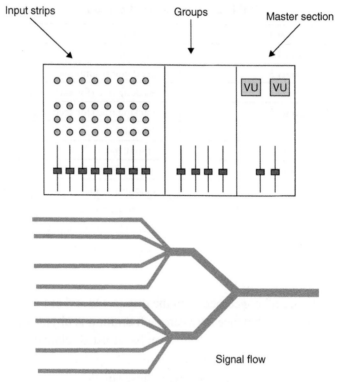

Input strips        Groups        Master section

Signal flow

**Figure 27.7: Signal flow.**

### 27.5.5 Effect Sends and Returns

Not all the functions required by the balance engineer can be incorporated within the audio mixer. To facilitate the interconnection with outboard equipment, most audio mixers have dedicated mix amplifiers and signal injection points called effect sends and returns, which make incorporation of other equipment within the signal flow as straightforward as possible.

### 27.5.6 The Groups Revisited

In a recording situation, the mixing groups may well ultimately be used in the same manner as described in relation to the live console, to group sections of the arrangement so as to make the task of mixing more manageable. But the groups are used in a totally different way during recording or tracking. During this phase, the groups are used to

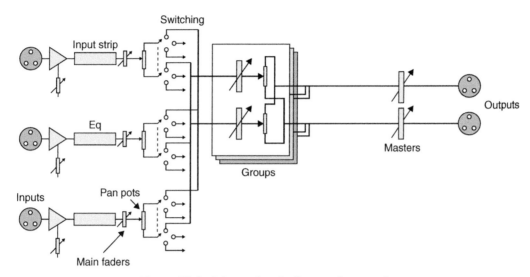

**Figure 27.8: Schematic of a live-music console.**

route signals to the tracks of the multitrack tape machine. From an electronic point of view, the essential difference here is that, in a recording situation, the group outputs are utilized directly as signals and a recording mixer must provide access to these signals. Usually a multitrack machine is wired so that each group output feeds a separate track of the multitrack tape recorder. Often there are not enough groups to do this, in which case, each group feeds a number of tape machine inputs, usually either adjacent tracks or in "groups of groups" so that, for instance, groups 1 to 8 will feed inputs 1 to 8 and 9 to 16 and so on.

### 27.5.7 The Recording Console

So far, this is relatively straightforward. But a major complication arises during the tracking of multitrack recordings because not only must signals be routed to the tape recorder via the groups, tape returns must be routed back to the mixer to guide the musicians as to what to play next. And this must happen at the same time! In fact, it is just possible for a good sound engineer, using "crafty" routing, to cope with this using a straightforward live mixing desk. But it is very difficult. What is really required is a separate mixer to deal with the gradually increasing numbers of tape replay returns, thereby keeping the main mixer free for recording duties. Various mixer designers have solved this problem in different ways. Older consoles (particularly of English origin) have

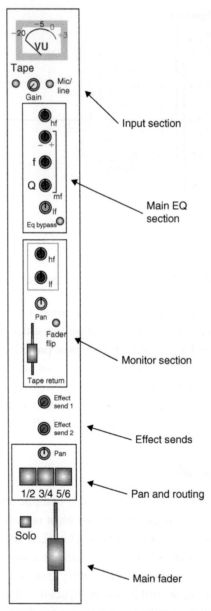

**Figure 27.9: Input strip of an in-line console.**

tended to provide an entirely separate mixer (usually to the right of the group and main faders) devoted to dealing with the return signals. Such a mixer architecture is known as a split console. The alternative approach, which is now very widespread, is known as the in-line console, so named because the tape-return controls are embedded within each channel strip, in line with the main fader. This is the type of console considered in detail later. From an electronic point of view, very little difference exists between these approaches; the difference is more one of operational philosophy and ergonomics.

Both the split and in-line console are yet another example of a "mixer within a mixer." In effect, in the in-line console, the tape returns feed an entirely separate mixer so that each tape return signal travels via a separate fader (sometimes linear, sometimes rotary) and pan control before being summed on an ancillary stereo mix bus known as the monitor bus. The channel input strip of an in-line console is illustrated in Figure 27.9. The monitor bus is exceptionally important in a recording console because it is the output of this stereo mix amplifier that supplies the signal that is fed to the control room power amplifier during the tracking phase of the recording process. (Usually, control room outputs are explicitly provided on the rear panels of a recording console for this purpose.) The architecture is illustrated in Figure 27.10. During mixdown, the engineer will want to operate using the main faders and pan controls, because these are the most operationally convenient controls, being closest to the mixer edge nearest the operator. To this end, the in-line console includes the ability to switch the tape returns back through the main input strip signal path, an operation known as "flipping" the faders. The circuitry for this is illustrated in Figure 27.11.

### 27.5.8 Talkback

Talkback exists so that people in the control room are able to communicate with performers in the studio. So as to avoid sound "spill" from the loudspeaker into open microphones (as well as to minimize the risk of howl-round), performers inside the studio invariably wear headphones and therefore need a devoted signal that may be amplified and fed to special headphone amplifiers. In the majority of instances this signal is identical to the signal required in the control room during recording (i.e., the monitor bus). In addition, a microphone amplifier is usually provided within the desk, which is summed with the monitor bus signal and fed to the studio headphone amplifiers. This microphone amplifier is usually energized by a momentary switch to allow the producer or engineer to communicate with the singer or instrumentalist, but which cannot, thereby, be left open, thus distracting the singer or allowing them to hear a comment in the control room that may do nothing for their ego!

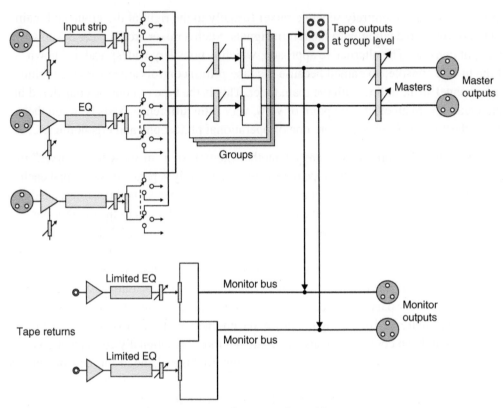

**Figure 27.10: In-line console architecture.**

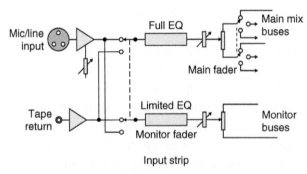

Input strip

**Figure 27.11: Fader "flip" mechanism.**

### 27.5.9 Equalizers

For a variety of reasons, signals arriving at the recording console may require spectral modification. Sometimes this is due to the effect of inappropriate microphone choice or of incorrect microphone position. Sometimes it is due to an unfortunate instrumental tone (perhaps an unpleasant resonance). Most often, the equalizer (or simply EQ) is used in a creative fashion to enhance or subdue a band (or bands) of frequencies so as to blend an instrument into the overall mix or boost a particular element so that its contribution is more incisive.

It is this creative element in the employment of equalization that has created the situation that exists today, that the quality of the EQ is often a determining factor in a recording engineer's choice of one console over another. The engineering challenges of flexible routing, low interchannel cross talk, low noise, and good headroom having been solved by most good manufacturers, the unique quality of each sound desk often resides in the equalizer design. Unfortunately, this state of affairs introduces a subjective (even individualistic) element into the subject of equalization, which renders it very difficult to cover comprehensively. Sometimes it seems that every circuit designer, sound engineer, and producer each has his, or her, idea as to what comprises an acceptable, an average, and an excellent equalizer! A simple equalizer section—and each control's effect—is illustrated in Figure 27.12.

## 27.6 Audio Mixer Circuitry

Now that we understand the basic architecture of the mixer, it is time to look at each part again and understand the function of the electrical circuits in each stage in detail.

The input strip is illustrated in Figure 27.9. Note that below the VU meter, the topmost control is the channel gain trim. This is usually switchable between two regimes: a high gain configuration for microphones and a lower gain line level configuration. This control is set with reference to its associated VU meter. Below the channel gain are the equalization controls; the operation of these controls is described in detail later.

### 27.6.1 Microphone Preamplifiers

Despite the voltage amplification provided by a transformer or preamplifier, the signal leaving a microphone is still, at most, only a few millivolts. This is much too low a level to be suitable for combining with the outputs of other microphones inside an audio mixer.

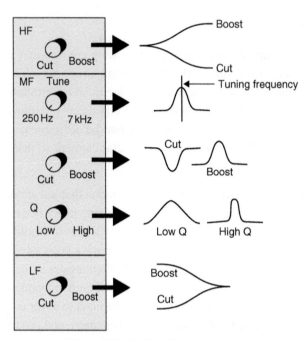

**Figure 27.12: Equalizer controls.**

So, the first stage of any audio mixer is a so-called microphone preamplifier, which boosts the signal entering the mixer from the microphone to a suitable operating level. The example given here is taken from a design (Brice, 1990), which was for a portable, battery-driven mixer—hence the decision to use discrete transistors rather than current-thirsty op-amps. Each of the input stage microphone amplifiers is formed of a transistor "ring of three." The design requirement is for good headroom and a very low noise figure. The final design is illustrated in Figure 27.13.[1] The current consumption for each microphone preamplifier is less than 1 mA.

---

[1]The amplifier shown has an input noise density of 3 nV per root hertz and a calculated input noise current density of 0.3 pA per root hertz ignoring flicker noise. The frequency response and the phase response remain much the same regardless of gain setting. This seems to go against the intractable laws of gain-bandwidth product: as we increase the gain we must expect the frequency response to decrease and vice versa. In fact, the ring-of-three circuit is an early form of "current-mode-feedback" amplifier, which is currently very popular in video applications. The explanation for this lies in the variable gain-setting resistor $R_a$. This not only determines the closed-loop gain by controlling the proportion of the output voltage fed back to the

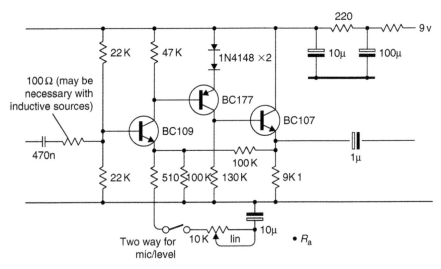

**Figure 27.13: Microphone preamplifier.**

For more demanding studio applications, a more complex microphone preamplifier is demanded. First, the stage illustrated in Figure 27.13 is only suitable for unbalanced microphones, and the majority of high-quality microphones utilize balanced connection, as they are especially susceptible to hum pick-up if this type of circuit is not employed. Second, high-quality microphones are nearly always capacitor types and therefore require a polarizing voltage to be supplied for the plates and for powering the internal electronics. This supply (known as phantom power as mentioned earlier) is supplied via the microphone's own signal leads and must therefore be isolated from the microphone preamplifier. This is one area where audio transformers (Figure 27.14) are still used, and an example of a microphone input stage using a transformer is illustrated, although plenty of practical examples exist where transformerless stages are used for reasons of cost.

Note that the phantom power is provided via resistors, which supply both phases on the primary side of the transformer, with the current return being via the cable screen. The

inverting port but it also forms the dominating part of the emitter load of the first transistor and consequently the gain of the first stage. As the value of $R_a$ decreases, so the feedback diminishes and the closed-loop gain rises. At the same time, the open-loop gain of the circuit rises because TR1's emitter load falls in value. Consequently, the performance of the circuit in respect of phase and frequency response, and consequently stability, remains consistent regardless of gain setting.

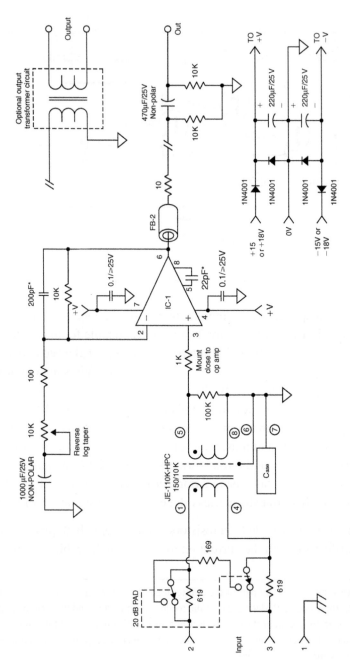

Schematic diagram of typical microphone preamplifier utilizing JE-110K-HPC

Notes:

1. IC-1 = integrated circuit opamp such as MA-332 or NE-5534.
2. Gain range: +24dB → +58dB.
3. Keep traces short between transformer and opamp.
4. All resistors = 1%, metal film.
5. 200pF cap in feedback = 2 μsec compensation.
6. FB-2 = ferrite bead available from Jensen.
7. Capacitors marked* = polystyrene or polypropylene.

## Figure 27.14: Microphone transformer.

transformer itself provides some voltage gain at the expense of presenting a much higher output impedance to the following amplifier. However, the amplifier has an extremely high input impedance (especially so when negative feedback is applied) so this is really of no consequence. In Figure 27.14, amplifier gain is made variable so as to permit the use of a wide range of microphone sensitivities and applications. An ancillary circuit is also provided that enables the signal entering the circuit to be attenuated by 20 dB, which allows for very high output microphones without overloading the electronics.

# Data Sheet

## jensen transformers
### INCORPORATED

# JE-110K-HPC
## MICROPHONE INPUT TRANSFORMER

The JE-110K-HPC is a printed circuit type 150/10K with a winding similar to the JE-115K-E. The multiple interleaved layer winding exhibits very low leakage inductance requiring no series RC network across the 100K ohm secondary load resistor when used with an amplifier incorporating $2\mu$S phase lead compensation in the feedback circuit. Since the PC bobbin contains a smaller stack of laminations than the wire lead JE-115K-E, the JE-110K-HPC uses more total turns of smaller wire. The result is higher maximum level capability at low frequencies and the distortion is the lowest of all types in this size (0.11% @ 20Hz), but the higher series losses increase the noise by 0.9dB, compared to the wire lead version JE-115K-E.

The pin pattern is compatible with the JE-6110K-APC.

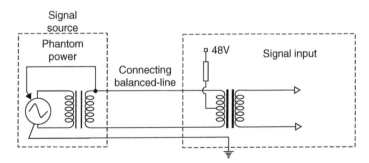

**Figure 27.15: Transformer-based input stage.**

### 27.6.2 Insert Points

After the channel input amplifier, insert points are usually included. These allow external (outboard) equipment to be patched into the signal path. Note that this provision is usually via jack connectors, which are normalized when no plug is inserted (Figure 27.16).

### 27.6.3 Equalizers and Tone Controls

At the very least, a modern recording console provides a basic tone-control function on each channel input, like that shown in Figure 27.17. This circuit is a version of the classic circuit due to Baxandall. However, this type of circuit only provides a fairly crude

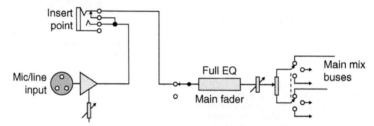

**Figure 27.16: Insert points.**

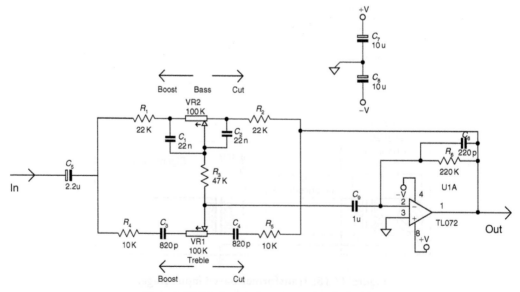

**Figure 27.17: A simple tone control circuit.**

adjustment of bass and treble ranges, as illustrated in Figure 27.12. This type of response (for fairly obvious reasons) is often termed a "shelving" equalizer. So, the Baxandall shelving EQ is invariably appended with a midfrequency equalizer, which is tunable over a range of frequencies, thereby enabling the sound engineer to adjust the middle band of frequencies in relation to the whole spectrum (see also Figure 27.12). A signal manipulation of this sort requires a tuned circuit, which is combined within an amplifier stage, which may be adjusted over a very wide range of attenuation and gain (perhaps as much as $\pm 20\,\text{dB}$). How such a circuit is derived is illustrated in Figure 27.18.

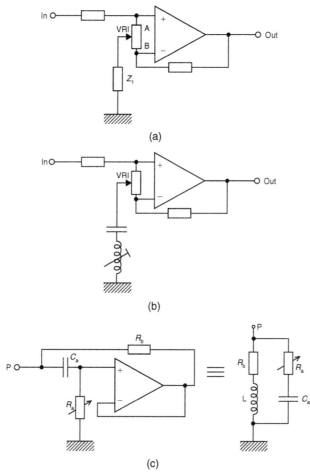

Figure 27.18(a–c): Derivation of midband parametric EQ circuit; see text.

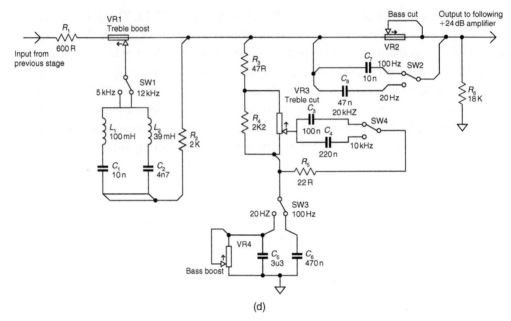

**Figure 27.18(d): Older, passive equalizer circuit.**

Imagine that Z1 is a small resistance. When the slider of VR1 is at position A, the input is heavily attenuated and the lower feedback limb of the op-amp is at its greatest value (i.e., the gain of the stage is low). Now imagine moving the slider of VR1 position B. The situation is reversed; the input is much less attenuated and the gain of the amplifier stage is high. This circuit therefore acts as a gain control because point A is arranged to be at the extreme anticlockwise position of the control. As a tone control this is obviously useless. However, in the second part of Figure 27.18(b), Z1 is replaced with a tunable circuit, formed by a variable inductor and a capacitor. This tuned circuit has high impedance—and therefore little effect—except at its resonant frequency, whereupon it acquires very low dynamic impedance. With the appropriate choice of inductor and capacitor values, the midfrequency equalizer can be made to operate over the central frequency range. The action of VR1 is to introduce a bell-shaped EQ response (as illustrated in Figure 27.12), which may be used to attenuate or enhance a particular range of frequencies, as determined by the setting of the variable inductor.

### 27.6.4 Inductor–Gyrators

As shown, the frequency adaptive component is designated as a variable inductor. Unfortunately, these components do not exist readily at audio frequencies and to construct components of this type expressly for audio frequency equalization would be very expensive. For this reason, the variable inductors in most commercial equalizers are formed by gyrators—circuits that emulate the behavior of inductive components by means of active circuits, which comprise resistors, capacitors, and op-amps. An inductor–gyrator circuit is illustrated in Figure 27.18(c). This is known as the "bootstrap" gyrator and its equivalent circuit is also included within the figure. Note that this type of gyrator circuit (and indeed most others) presents a reasonable approximation to an inductor that is grounded at one end. Floating inductor–gyrator circuits do exist but are rarely seen.

Operation of the bootstrap gyrator circuit can be difficult to visualize, but think about the two frequency extremes. At low frequencies, $C_a$ will not pass any signal to the input of the op-amp. The impedance presented at point P will therefore be the output impedance of the op-amp (very low) in series with $R_b$, which is usually designed to be in the region of a few hundred ohms. Just like an inductor, the reactance is low at low frequencies. Now consider the high-frequency case. At HF, $C_a$ will pass signal so that the input to the op-amp will be substantially that presented at point P. Because the op-amp is a follower, the output will be a low-impedance copy of its input. By this means, resistor $R_b$ will thereby have little or no potential across it because the signal at both its ends is the same. Consequently, no signal current will pass through it. In engineering slang, the low-value resistor $R_b$ is said to have been "bootstrapped" by the action of the op-amp and therefore appears to have a much higher resistance than it actually has. Once again, just like a real inductor, the value of reactance at high frequencies at point P is high. The inductor–gyrator circuit is made variable by the action of $R_a$, which is formed by a variable resistor component. This alters the breakpoint of the RC circuit $C_a/R_a$ and thereby the value of the "virtual" inductor.

In recent years, the fascination in "retro" equipment has brought about a resurgence of interest in fixed inductor–capacitor type equalizers. Mostly outboard units, these are often passive EQ circuits [often of great complexity, as illustrated in Figure 27.18(d)], followed by a valve line amplifier to make up the signal gain lost in the EQ. In a classic LC-type equalizer, the variable-frequency selection is achieved with switched inductors and capacitors.

### 27.6.5 'Q'

Often it is also useful to control the Q of the resonant circuit so that a very broad, or a very narrow, range of frequencies can be affected as appropriate. Hence the inclusion of the Q control as shown in Figure 27.12. This effect is often achieved by means of a series variable resistor in series with the inductor–capacitor (or gyrator–capacitor) frequency-determining circuit.

### 27.6.6 Effect Send and Return

The effect send control feeds a separate mix amplifier. The output of this submix circuit is made available to the user via an effect send output on the back of the mixer. An effect return path is also usually provided so that the submix (suitably "effected"—usually with digital reverberation) can be reinjected into the amplifier at line level directly into the main mix amplifiers.

### 27.6.7 Faders and Pan Controls

Beneath the effect send control is the pan control and the main fader. Each channel fader is used to control the contribution of each channel to the overall mix as described earlier. A design for a fader and pan pot is illustrated in Figure 27.19. The only disadvantage of the circuit is that it introduces loss. However, its use is confined to a part of the circuit where the signal is at a high level and the Johnson noise generated in the network is very low since all the resistances are low. The circuit takes as its starting point that a semilog audio-taper fader can be obtained using a linear potentiometer when its slider tap and "earthy" end is shunted with a resistor one-tenth of the value of the total potentiometer

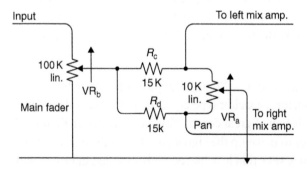

**Figure 27.19: Fader and pan pot circuit.**

resistance. Resistors $R_c$ and $R_d$ and $VR_a$, the pan pot itself, form this one-tenth value network. Because the slider of the pan pot is grounded and the output signal is obtained from either end of the pan pot, it is clear that in either extreme position of $VR_a$, all the signal will appear on one channel and none on the other. It is then only a matter of deciding whether the control law of the pan pot is of the correct type. The calculation of the control law obtained from the circuit is complicated because the signal level fed to left and right channels is not proportional to the resistive value of each part of $VR_a$. This is because the total value of the network $R_c$, $R_d$, and $VR_a$, although reasonably constant, is not the same irrespective of the setting of $VR_a$ and so the resistive attenuator comprising the top part of $VR_b$ and its lower part, shunted by the pan pot network, is not constant as $VR_a$ is adjusted. Furthermore, as $VR_a$ is varied, so the output resistance of the network changes and, since this network feeds a virtual-earth summing amplifier, this effect also has an influence on the signal fed to the output because the voltage-to-current converting resistor feeding the virtual-earth node changes value. The control law of the final circuit is nonlinear: the sum of left and right, when the source is positioned centrally, adding to more than the signal appearing in either channel when the source is positioned at an extreme pan position. This control law is very usable with a source seeming to retain equal prominence as it is "swept across" the stereo stage.

### 27.6.8 Mix Amplifiers

The mix amplifier is the core of the audio mixer. It is here that the various audio signals are combined together with as little interaction as possible. The adoption of the virtual-earth mixing amplifier is universal. An example of a practical stereo mix amplifier is shown in Figure 27.20. Here, the summing op-amp is based on a conventional transistor pair circuit. The only difficult decision in this area is the choice of the value for $R_b$. It is this value, combined with the input resistors, that determines the total contribution each input may make to the final output.

### 27.6.9 Line-Level Stages

Line-level audio stages are relatively straightforward. Signals are at a high level, so noise issues are rarely encountered. The significant design parameters are linearity, headroom, and stability. Another issue of some importance is signal balance, at least in professional line-level stages, which are always balanced. Of these, output-stage stability is the one most often ignored by novice designers.

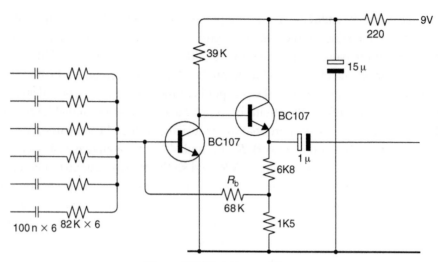

**Figure 27.20: Mix amplifier.**

A high degree of linearity is achieved in modern line-level audio stages by utilizing high open-loop gain op-amps and very large feedback factors. The only issue remaining, once the choice of op-amp has been made, is available headroom. This is almost always determined by choice of power supply rails. Taking a typical audio op-amp (TL072, for example), maximum output swing is usually limited to within a few volts of either rail. So, furnished with 12-V positive and negative supplies, an op-amp could be expected to swing 18 Vpk-pk. This is equivalent to 6.3 V rms, or 16 dBV, easily adequate then for any circuit intended to operate at 0 VU = −10 dBV. In a professional line-output circuit, like that shown in Figure 27.21, this voltage swing is effectively doubled because the signal appears "across" the two opposite phases. The total swing is therefore 12.6 V rms, which is equivalent to 24 dBu. For equipment intended to operate at 0 VU = +4 dBu, such a circuit offers 20 dB headroom, which is adequate. However, ±12-V supplies really represent the lowest choice of rail volts for professional equipment and some designers prefer to use 15-V supplies for this very reason.

Looking again at the output stage circuit illustrated in Figure 27.21, note the inclusion of the small value resistors in the output circuit of the op-amps. These effectively add some real part to the impedance "seen" by the op-amp when it is required to drive a long run of audio cables. At audio frequencies, the equipment interconnection cable looks to the output stage as a straightforward—but relatively large—capacitance. Also, a large

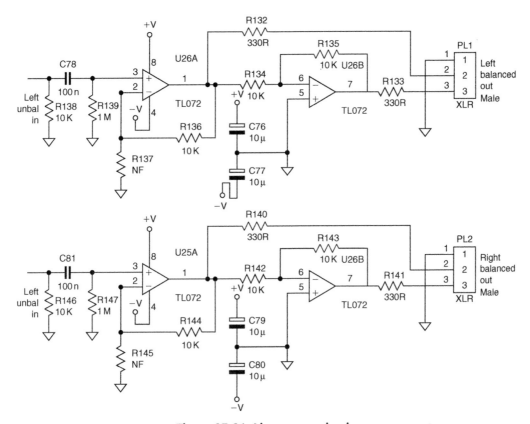

**Figure 27.21: Line output circuit.**

negative reactance is, almost always, an excellent way to destabilize output circuits. Output "padding" resistors, such as R132 and R133, help a great deal in securing a stable performance into real loads.

Line-level input stages present different problems. Note that the function performed by the circuit in Figure 27.21, over and above its duty to drive the output cables, is to derive an equal and opposite signal so as to provide a balanced audio output from a single-ended, unbalanced input signal. This is a common feature of professional audio equipment because, although balanced signals are the norm outside the equipment, internally most signals are treated as single ended. The reasons for this are obvious; without this simplification all the circuitry within, for example, a console would be twice as complex and twice as expensive.

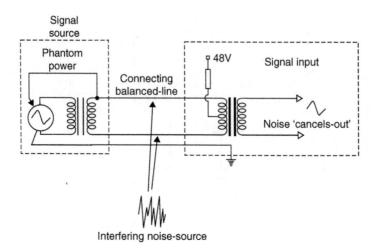

**Figure 27.22: Balanced input circuit, and CM rejection.**

The line-level input stage on professional equipment therefore has to perform a complementary function to the output stage to derive a single-ended signal from the balanced signal presented to the equipment. Conceptually, the simplest circuit is a transformer, like that shown in Figure 27.22. In many ways this is an excellent solution for the following reasons: it provides electrical isolation, it has low noise and distortion, and it provides good headroom, provided the core doesn't saturate. But, most important of all, it possesses excellent common-mode rejection (CMR). That means that any signal that is common (i.e., in phase) on both signal phases is rejected and does not get passed on to following equipment. By contriving the two signal conductors within the signal cable to occupy—as nearly as possible—the same place, by twisting them together, any possible interference signal is induced equally in both phases. Such a signal thereafter cancels in the transformer stage because a common signal cannot cause a current to flow in the primary circuit and cannot, therefore, cause one to flow in the secondary circuit. This is illustrated in Figure 27.22 as well.

Another advantage of a balanced signal interface is that the signal circuit does not include ground. It thereby confers immunity to ground-sourced noise signals. On a practical level it also means that different equipment chassis can be earthed, for safety reasons, without incurring the penalty of multiple signal return paths and the inevitable "hum loops" this creates. However, transformers are not suitable in many applications for a number of reasons. First, they are very expensive. Second, they are heavy, bulky, and tend to be microphonic (i.e., they have a propensity to transduce mechanical vibration into electrical energy!) so that electronically balanced input stages are widely employed instead. These aim to confer all the advantages of a transformer cheaply, quietly, and on a small scale. To some degree, an electronic stage can never offer the same degree of CMR, as well as the complete galvanic isolation, offered by a transformer.

## 27.7 Mixer Automation

Mixer automation consists (at its most basic level) of computer control over the individual channel faders during a mixdown. Even the most dexterous and clear thinking balance engineer obviously has problems when controlling perhaps as many as 24 or even 48 channel faders at once. For mixer automation to work, several things must happen. First, the controlling computer must know precisely which point in the song or piece has been reached in order that it can implement the appropriate fader movements. Second, the controlling computer must have, at its behest, hardware that is able to control the audio level on each mixer channel, swiftly and noiselessly. This last requirement is fulfilled a number of ways, but most often a voltage-controlled amplifier (VCA) is used.

A third requirement of a fader automation system is that the faders must be "readable" by the controlling computer so that the required fader movements can be implemented by the human operator and memorized by the computer for subsequent recall.

A complete fader automation system is shown in schematic form in Figure 27.23. Note that the fader does not pass the audio signal at all. Instead the fader simply acts as a potentiometer driven by a stabilized supply. The slider potential now acts as a control voltage, which could, in theory, be fed directly to the voltage-controlled amplifier, VCA1. But this would miss the point. By digitizing the control voltage, and making this value available to the microprocessor bus, the fader "position" can be stored for later recall. When this happens, the voltage (at the potentiometer slider) is recreated by means of a DAC and this is applied to the VCA, thereby reproducing the operator's original intentions.

One disadvantage of this type of system is the lack of operator feedback once the fader operation is overridden by the action of the VCA; importantly, when in recall mode, the faders fail, by virtue of their physical position, to tell the operator (at a glance) the condition of any of the channels and their relative levels. Some automation systems attempt to emulate this important visual feedback by creating an iconic representation of the mixer on the computer screen. Some even allow these virtual faders to be moved,

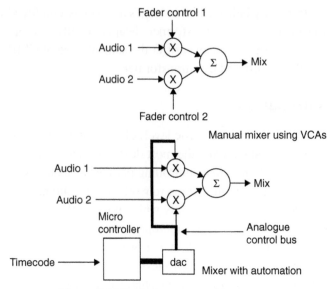

**Figure 27.23: Fader automation system.**

on screen, by dragging them with a mouse. Another more drastic solution, and one that has many adherents on sound quality grounds alone, is to use motorized faders as part of the control system. In this case the faders act electrically as they do in a nonautomated mixer, carrying the audio signal itself. The control system loop is restricted to reading and "recreating" operator fader physical movements. Apart from providing unrivalled operator feedback (and the quite thrilling spectacle of banks of faders moving as if under the aegis of ghostly hands!), the advantage of this type of automation system is the lack of VCAs in the signal path. VCA circuits are necessarily complicated and their operation is beset with various deficiencies, mostly in the areas of insufficient dynamic range and control signal breakthrough. These considerations have kept motorized faders as favorites among the best mixer manufacturers, despite their obvious complexity and cost.

### 27.7.1  Time Code

Time code is the means by which an automation system is kept in step with the music recorded onto tape. Normally, a track of the multitrack tape is set aside from audio use and is devoted to recording a pseudo audio signal composed of a serial digital code.

## 27.8 Digital Consoles

### 27.8.1 Introduction to Digital Signal Processing (DSP)

DSP involves the manipulation of real-world signals (for instance, audio signals, video signals, medical or geophysical data signals) within a digital computer. Why might we want to do this? Because these signals, once converted into digital form (by means of an analogue to digital converter), may be manipulated using mathematical techniques to enhance, change, or display data in a particular way. For instance, the computer might use height or depth data from a geophysical survey to produce a colored contour map or the computer might use a series of two-dimensional medical images to build up a three-dimensional virtual visualization of diseased tissue or bone. Another application, this time an audio one, might be used to remove noise from a music signal by carefully measuring the spectrum of the interfering noise signal during a moment of silence (for instance, during the run-in groove of a record) and then subtracting this spectrum from the entire signal, thereby removing only the noise—and not the music—from a noisy record.

DSP systems have been in existence for many years, but, in these older systems, the computer might take many times longer than the duration of the signal acquisition time to process the information. For instance, in the case of the noise reduction example, it might take many hours to process a short musical track. This leads to an important distinction that must be made in the design, specification, and understanding of DSP systems; that of nonreal time (where the processing time exceeds the acquisition or presentation time) and real-time systems, which complete all the required mathematical operations so fast that the observer is unaware of any delay in the process. When we talk about DSP in digital audio it is always important to distinguish between real-time and nonreal-time DSP. Audio outboard equipment that utilizes DSP techniques is, invariably, real time and has dedicated DSP chips designed to complete data manipulation fast. Nonreal-time DSP is found in audio processing on a PC or Apple Mac where some complex audio tasks may take many times the length of the music sample to complete.

### 27.8.2 Digital Manipulation

So, what kind of digital manipulations might we expect? Let's think of the functions that we might expect to perform within a digital sound mixer. First, there is addition. Clearly, at a fundamental level, that is what a mixer is—an "adder" of signals. Second,

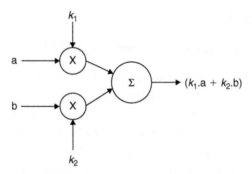

**Figure 27.24: Simple digital audio mixer.**

we know that we want to be able to control the gain of each signal before it is mixed. So multiplication must be needed too. So far, the performance of the digital signal processing "block" is analogous with its analogue counterpart. The simplest form of digital audio mixer is illustrated in Figure 27.24. In this case, two digital audio signals are each multiplied by coefficients ($k_1$ and $k_2$) derived from the position of a pair of fader controls; one fader assigned to either signal. Signals issuing from these multiplication stages are subsequently added together in a summing stage. All audio mixers possess this essential architecture, although it may be supplemented many times over.

But, in fact, the two functions of addition and multiplication, plus the ability to delay signals easily within digital systems, allow us to perform all the functions required within a digital sound mixer, even the equalization functions. That's because equalization is a form of signal filtering on successive audio samples, which is simply another form of mathematical manipulation, even though it is not usually regarded as such in analogue circuitry.

### 27.8.3  Digital Filtering

The simplest form of analogue low-pass filter is shown in Figure 27.25. Its effect on a fast rise-time signal wave front (an "edge") is also illustrated. Note that the resulting signal has its "edges" slowed down in relation to the incoming signal. Its frequency response is also illustrated, with its turnover frequency. Unfortunately, in digital circuits there are no such things as capacitors or inductors, which may be used to change the frequency response of a circuit. However, if you remember, we've come across situations before in sections as diverse as microphones to flanging, phasing, and chorus wherein a frequency response was altered by the interaction of signals delayed with respect to one another. This principle is

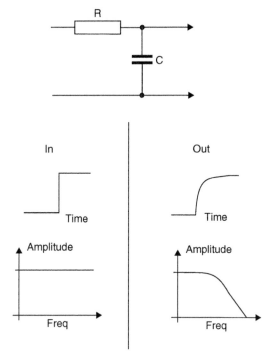

**Figure 27.25: RC low-pass filter.**

the basis behind all digital filtering and may be extended to include several stages of delay as shown in Figure 27.26. By utilizing a combination of adder and variable multiplication factors (between the addition function and the signal taps) it is possible to achieve a very flexible method of signal filtering in which the shape of the filter curve may be varied over a very wide range of shapes and characteristics. While such a technique is possible in analogue circuitry, note that the "circuit" (shown in Figure 27.26) is actually not a real circuit at all, but a notional block diagram. It is in the realm of digital signal processing that such a filtering technique really comes into its own: the DSP programmer has only to translate these processes into microprocessor type code to be run on a microcontroller IC, which is specifically designed for audio applications—a so-called DSP IC. Herein lies the greatest benefit of digital signal processing—that by simply reprogramming the coefficients in the multiplier stages, a completely different filter may be obtained. Not only that, but if this is done in real time too, the filter can be made adaptive, adjusting to demands of the particular moment in a manner that might be useful for signal compression or noise reduction.

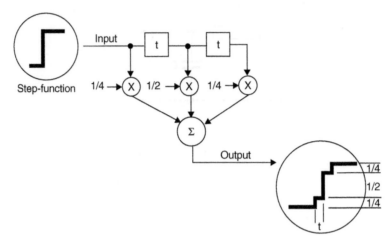

**Figure 27.26: Principle of a digital filter.**

### 27.8.4 Digital Mixer Architecture

Because of the incredible flexibility and "programmability" of digital signal processing-based mixers, architecture is much harder (less necessary!) to define. This, alone, is a great advantage. For instance, digital processing too has blurred the traditional distinction between split and in-line consoles because, with the aid of configurable signal paths and motorized faders, the same small group of faders can be used to "flip" between the role of the recording mixer and that of the playback mixer.

**Box 27.1    Fact Sheet #12: Digital signal processing**

- Architecture of DSP devices
- Convolution
- Impulse response
- FIR and IIR digital filters
- Design of digital filters
- Frequency response
- Derivation of band-pass and high-pass filters
- Digital frequency domain analysis—the z-transform
- Problems with digital signal processing

## Architecture of DSP Devices

The first computers, including those developed at Harvard University, had separate memory space for program and data; this topology being known as Harvard architecture. In fact, the realization, by John von Neumann—the Hungarian-born mathematician—that program instructions and data were only numbers and could share the same "address space" was a great breakthrough at the time and was sufficiently radical that this architecture is often named after its inventor. The advantage of the von Neumann approach was great simplification but at the expense of speed because the computer can only access either an instruction or data in any one processing clock cycle. The fact that virtually all computers follow this latter approach illustrates that this limitation is of little consequence in the world of general computing.

However, the speed limitation "bottleneck," inevitable in the von Neumann machine, can prove to be a limitation in specialist computing applications such as digital audio signal processing. As we have seen in the case of digital filters, digital signal processing contains many, many multiply and add type instructions of the form

$$A = B \cdot C + D$$

Unfortunately, a von Neumann machine is really pretty inefficient at this type of calculation so the Harvard architecture lives on in many DSP chips, meaning that a multiply and add operation can be performed in one clock cycle; this composite operation is termed a Multiply ACcumulate (MAC) function. A further distinction pertains to the incorporation within the DSP IC of special registers that facilitate the managing of circular buffers for the implementation of reverb, phasing, chorus, and flanging effects.

The remaining differences between a DSP device and a general purpose digital microcomputer chip relate to the provision of convenient interfaces, thereby allowing direct connection of ADCs, DACs, and digital transmitter and receiver ICs.

## Convolution

In the simple three-stage digital filter, we imagined the step function being multiplied by a quarter, then by a half, and finally by a quarter again; at each stage, the result was added up to give the final output. This actually rather simple process is given a frightening name in digital signal processing theory, where it is called convolution.

Discrete convolution is a process that provides a single output sequence from two input sequences. In the example given earlier, a time-domain sequence—the step function—was convolved with the filter response yielding a filtered output sequence. In textbooks, convolution is often denoted by the character $^{(*)}$. So if we call the input sequence $h(k)$ and the input sequence $x(k)$, the filtered output would be defined as

$$y(n) = h(k) * x(k)$$

### Impulse Response

A very special result is obtained if a unique input sequence is convolved with the filter coefficients. This special result is known as the filter's impulse response, and the derivation and design of different impulse responses are central to digital filter theory. The special input sequence used to discover a filter's impulse response is known as the "impulse input." (The filter's impulse response being its response to this impulse input.) This input sequence is defined to be always zero, except for one single sample, which takes the value 1 (i.e., the full-scale value). We might define, for practical purposes, a series of samples like this

0, 0, 0, 0, 0, 1, 0, 0, 0, 0

Now imagine these samples being latched through the three-stage digital filter shown earlier. The output sequence will be

0, 0, 0, 0, 0, 1/4, 1/2, 1/4, 0, 0, 0, 0

Obviously the zeros don't really matter, what's important is the central section: 1/4, 1/2, 1/4. This pattern is the filter's impulse response.

### FIR and IIR Digital Filters

Note that the impulse response of the aforementioned filter is finite: in fact, it only has three terms. So important is the impulse response in filter theory that this type of filter is actually defined by this characteristic of its behavior and is named a finite impulse response (FIR) filter. Importantly, note that the impulse response of an FIR filter is identical to its coefficients.

Now look at the digital filter in Figure F27.1. This derives its result from both the incoming sequence and from a sequence that is fed back from the output. Now if we perform a similar thought experiment to the convolution example given earlier and imagine the resulting impulse-response from a filter of this type, it results in

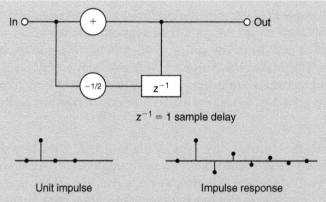

$z^{-1}$ = 1 sample delay

Unit impulse          Impulse response

**Figure F27.1: Infinite impulse response filter.**

an output sequence like that illustrated in the figure: that's to say, an infinitely decaying series of values. Once again, so primordial is this characteristic that this category of filter is termed an infinite impulse response (IIR) filter.

IIR filters have both disadvantages and advantages over the FIR type. First, they are very much more complicated to design because their impulse response is not simply reflected by the tap coefficients, as in the FIR. Second, it is in the nature of any feedback system (like an analogue amplifier) that some conditions may cause the filter to become unstable if it is has not been thoroughly designed, simulated, and tested. Furthermore, the inherent infinite response may cause distortion and/or rounding problems as calculations on smaller and smaller values of data are performed. Indeed, it's possible to draw a parallel between IIR filters and analogue filter circuits: they share the disadvantages of complexity of design and possible instability and distortion, but they also share the great benefit that they are efficient. An IIR configuration can be made to implement complex filter functions with only a few stages, whereas the equivalent FIR filter would require many hundreds of taps with all the drawbacks of cost and signal delay that this implies. (Sometimes FIR and IIR filters are referred to as "recursive" and "nonrecursive," respectively; these terms directly reflect the filter architecture.)

### Design of Digital Filters

Digital filters are nearly always designed from knowledge of the required impulse response. IIR and FIR filters are both designed in this way, although the design of IIR filters is complicated because the coefficients do not represent the impulse response directly. Instead, IIR design involves various mathematical methods, which are used to analyze and derive the appropriate impulse response from the limited number of taps. This makes the design of IIR filters from first principles rather complicated and math heavy! Fortunately, FIRs are easier to understand, and a brief description gives a good deal of insight into the design principles of all digital filters.

We already noted that the response type of the 1/4, 1/2, 1/4 filter was a low-pass; remember it "slowed down" the fast rising edge of the step waveform. If we look at the general form of this impulse response, we will see that this is a very rough approximation to the behavior of an ideal low-pass filter in relation to reconstruction filters. There we saw that the (sin x)/x function defines the behavior of an ideal, low-pass filter and the derivation of this function is given in Figure F27.2. Sometimes termed a sinc function, it has the characteristic that it is infinite, gradually decaying with ever smaller oscillations about zero. This illustrates that the perfect low-pass FIR filter would require an infinite response, an infinite number of taps and the signal would take an infinitely long time to pass through it! Fortunately for us, we do not need such perfection.

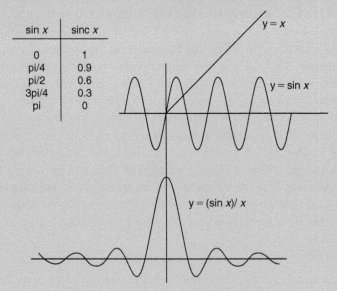

| sin x | sinc x |
|-------|--------|
| 0 | 1 |
| pi/4 | 0.9 |
| pi/2 | 0.6 |
| 3pi/4 | 0.3 |
| pi | 0 |

**Figure F27.2: Derivation of sin *x/x* function.**

However, the 1/4, 1/2, 1/4 filter is really a very rough approximation indeed. So let's now imagine a better estimate to the true sinc function and design a relatively simple filter using a 7-tap FIR circuit. I have derived the values for this filter in Figure F27.2. This suggests a circuit with the following tap values:

0.3, 0.6, 0.9, 1, 0.9, 0.6, 0.3

The only problem these values present is that they total to a value greater than 1. If the input was the unity step-function input, the output would take on a final amplitude of 4.6. This might overload the digital system, so we normalize the values so that the filter's response at DC (zero frequency) is unity. This leads to the following, scaled values:

0.07, 0.12, 0.2, 0.22, 0.2, 0.12, 0.07

The time-domain response of such an FIR filter to a step and impulse response is illustrated in Figure F27.3. The improvement over the three-tap filter is already obvious.

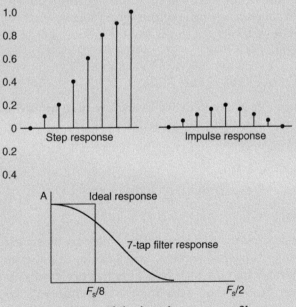

**Figure F27.3: Finite impulse response filter.**

### Frequency Response

But how does this response relate to the frequency domain response? For it is usually with a desired frequency response requirement that filter design begins. This important question is really asking, how can we express something in the time domain in terms of the frequency domain? It should be no surprise by now that such a manipulation involves the Fourier transform. Normal textbook design methods involve defining a desired frequency response and computing (via the Fourier transform) the required impulse response, thereby defining the tap coefficients of the FIR filter.

However, this is a little labor intensive and is not at all intuitive, so here's a little rule of thumb that helps when you're thinking about digital filters. If you count the number of sample periods in the main lobe of the sinc curve and give this the value, $n$, then, very roughly, the cutoff frequency of the low-pass filter will be the sampling frequency divided by $n$,

$$F_c = F_s/n$$

So, for our 7-term, FIR filter above, $n = 8$ and $F_c$ is roughly $F_s/8$. In audio terms, if the sample rate is 48 kHz, the filter will show a shallow roll-off with the turnover at about 6 kHz. The frequency response of this filter (and an ideal response) is shown in Figure F27.3. In order to approach the ideal response, a filter of more than 30 taps would be required.

### Derivation of Band-Pass and High-Pass Filters

All digital filters start life as low-pass filters and are then transformed into band-pass or high-pass types. A high-pass is derived by multiplying each term in the filter by alternating values of $+1$ and $-1$. So, our low-pass filter,

0.07, 0.12, 0.2, 0.22, 0.2, 0.12, 0.07

is transformed into a high-pass like this,

$+0.07, -0.12, +0.2, -0.22, +0.2, -0.12, +0.07$

The impulse response and the frequency of this filter are illustrated in Figure F27.4. If you add up these high-pass filter terms, you'll notice that they come nearly to zero. This demonstrates that the high-pass filter has practically no overall gain at DC, as you'd expect. Note too how the impulse response looks "right," in other words, as you'd anticipate from an analogue type.

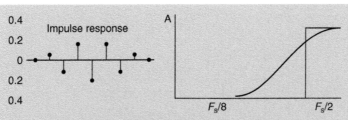

**Figure F27.4: Digital high-pass filter.**

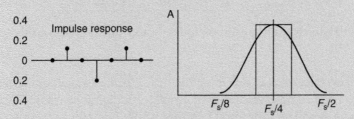

**Figure F27.5 :Digital band-pass filter.**

A band-pass filter is derived by multiplying the low-pass coefficients with samples of a sine wave at the center frequency of the band pass. Let's take our band pass to be centered on the frequency of $F_s/4$. Samples of a sine wave at this frequency will be at the 0 degree point, the 90 degree point, the 180 degree point, the 270 degree point, and so on. In other words,

0, 1, 0, −1, 0, 1, 0, −1

If we multiply the low-pass coefficients by this sequence we get the following,

0, 0.12, 0, −0.22, 0, +0.12, 0

The impulse response of this circuit is illustrated in Figure F27.5. This looks intuitively right too, because the output can be seen to "ring" at $F_s/4$, which is what you'd expect from a resonant filter. The derived frequency response is also shown in the diagram.

### Digital Frequency Domain Analysis—The Z-Transform

The z-transform of a digital signal is identical to the Fourier transform except for a change in the lower summation limit. In fact, you can think of 'z' as a frequency

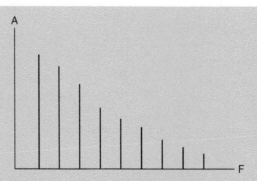

Figure F27.6: Generation of harmonics due to nonlinearity

variable that can take on real and imaginary (i.e., complex) values. When the z-transform is used to describe a digital signal, or a digital process (like a digital filter), the result is always a rational function of the frequency variable z. That's to say, the z-transform can always be written in the form:

$$X(z) = \frac{N(z)}{D(z)} = \frac{K(z - z_1)}{(z - p_1)},$$

where the z's are known as "zeros" and the 'p's are known as "poles."

A very useful representation of the z-transform is obtained by plotting these poles and zeros on an Argand diagram; the resulting two-space representation is termed the "z-plane." When the poles and zeros are plotted in this way, they give us a very quick way of visualizing the characteristics of a signal or digital signal process.

### Problems With Digital Signal Processing

Sampled systems exhibit aliasing effects if frequencies above the Nyquist limit are included within the input signal. This effect is usually no problem because the input signal can be filtered so as to remove any offending frequencies before sampling takes place. However, consider the situation in which a band-limited signal is subjected to a nonlinear process once in the digital domain. This process might be as simple as a "fuzz"-type overload effect, created with a plug-in processor. This entirely digital process generates a new large range of harmonic frequencies (just like its analogue counterpart), as shown in Figure F27.6. The problem arises that many of these new harmonic

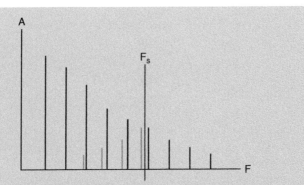

**Figure F27.7: Aliasing of harmonic structure in digital, nonlinear processing.**

frequencies are actually above the half-sampling frequency limit and get folded back into the pass-band, creating a rough quality to the sound and a sonic signature quite unlike the analogue "equivalent" (see Figure F27.7). This effect may account for the imperfect quality of many digital "copies" of classic analogue equipment.

## Reference

1. Brice, R., 'Audio mixer design', *Electronics and Wireless World*, July, 1990.

# CHAPTER 28

# Video Synchronization

Richard Brice

## 28.1 Introduction

Audio, for all its artistic power and technological intricacy, is just one element in today's multimedia entertainment phalanx. Today's recording engineers are more likely to find themselves working in a MIDI studio locked to SMPTE time code than they are to be rigging microphones. Today's composer has a better chance of making his/her living by proactively seeking work for television and film (for which he/she will require an understanding of the medium) than to wait for a commission for a symphony from a rich patron! This chapter contains a description of the principles and concepts that form the technical foundations of an understanding of analogue and digital television.

## 28.2 Persistence of Vision

The human eye exhibits an important property that has great relevance to the film and video industries. This property is known as the persistence of vision. When an image is impressed upon the eye, an instantaneous cessation of the stimulus does not result in a similarly instantaneous cessation of signals within the optic nerve and visual processing centers. Instead, an exponential "lag" takes place with a relatively long time required for total decay. The cinema has exploited this effect for over 100 years. Due to the persistence of vision, if the eye is presented with a succession of still images at a sufficiently rapid rate, each frame differing only in the positions moving within a fixed frame of reference, the impression is gained of a moving image. In a film projector each still frame of film is drawn into position in front of an intense light source while the source of light is shut off by means of a rotating shutter. Once the film frame has stabilized, the light is allowed through—by opening the shutter—and the image on the frame is projected upon a screen by way of an arrangement of lenses. Experiments soon established that a presentation rate of about 12 still frames per second was sufficiently rapid to give a good impression of continuously flowing movement but interrupting the light source at this rate caused unbearable flicker. This flicker phenomenon was also discovered to be related to the level of illumination; the brighter the light being repetitively interrupted, the worse the flicker. Abetted by the low light output from early projectors, this led to the first film frame-rate standard of 16 frames per second (fps). A standard well above that required simply to give the impression of movement and sufficiently rapid to ensure flicker was reduced to a tolerable level when used with early projection lamps. As these lamps improved, flicker became more of a problem until an ingenious alteration to the projector fixed the problem. The solution involved a modification to the rotating shutter so that, once the film frame was drawn into position, the shutter opened, then closed, and then opened again before closing a second time for the next film frame to be drawn into position. In other words, the light interruption frequency was raised to twice that of the frame rate. When the film frame rate was eventually raised to the 24-fps standard, which is still in force to this day, the light interruption frequency was raised to 48 times per second, a rate that enables high levels of illumination to be employed without causing flicker.

## 28.3 Cathode Ray Tube and Raster Scanning

To every engineer, the cathode ray tube (CRT) will be familiar enough from the oscilloscope. The evacuated glass envelope contains an electrode assembly and its terminations at its base whose purpose is to shoot a beam of electrons at the luminescent screen at the other end of the tube. This luminescent screen fluoresces to produce light whenever electrons hit it. In an oscilloscope the deflection of this beam is affected by means of electric fields—a so-called electrostatic tube. In television the electron beam (or beams in the case of color) is deflected by means of magnetic fields caused by currents flowing in deflection coils wound around the neck of the tube where the base section meets the flare. Such a tube is known as an electromagnetic type.

Just like an oscilloscope, without any scanning currents, the television tube produces a small spot of light in the middle of the screen. This spot of light can be made to move anywhere on the screen very quickly with the application of the appropriate current in the deflection coils. The brightness of the spot can be controlled with equal rapidity by altering the rate at which electrons are emitted from the cathode of the electron gun assembly.

This is usually affected by controlling the potential between the grid and the cathode electrodes of the gun. Just as in an electron tube or valve, as the grid electrode is made more negative in relation to the cathode, the flow of electrons to the anode is decreased. In the case of the CRT the anode is formed by a metal coating on the inside of the tube flare. A decrease in grid voltage—and thus anode current—results in a darkening of the spot of light. Correspondingly, an increase in grid voltage results in a brightening of the scanning spot.

In television, the bright spot is set up to move steadily across the screen from left to right (as seen from the front of the tube). When it has completed this journey it flies back very quickly to trace another path across the screen just below the previous trajectory. (The analogy with the movement of the eyes as they "scan" text during reading can't have escaped you!) If this process is made to happen sufficiently quickly, the eye's persistence of vision, combined with an afterglow effect in the tube phosphor, conspires to fool the eye so that it does not perceive the moving spot but instead sees a set of parallel lines drawn on the screen. If the number of lines is increased, the eye ceases to see these as separate too—at least from a distance—and instead perceives an illuminated rectangle of light on the tube face. This is known as a raster. In the broadcast television system employed in Europe, this raster is scanned twice in $\frac{1}{25}$ of a second. One set of 312.5 lines is scanned in the first $\frac{1}{50}$ of a second and a second interlaced set—which is not superimposed but is staggered in the gaps in the preceding trace—is scanned in the second $\frac{1}{50}$. The total number of lines is thus 625. In North America, a total of 525 lines (in two interlaced passes of 262.5) are scanned in $\frac{1}{30}$ of a second.

This may seem like a complicated way of doing things and the adoption of interlace has caused television engineers many problems over the years. Interlace was adopted in order to accomplish a 2 to 1 reduction in the bandwidth required for television pictures with very little noticeable loss of quality. It is thus a form of perceptual coding—what we would call today a data compression technique. Where bandwidth is not so important—as in computer displays—noninterlaced scanning is employed. Note also that interlace is, in some respects, the corollary of the double exposure system used in the cinema to raise the flicker frequency to double the frame rate.

## 28.4 Television Signal

The television signal must do two things, the first is obvious, the second less so. First, it must control the instantaneous brightness of the spot on the face of the CRT in order that the brightness changes that constitute the information of the picture may be conveyed. Second, it must control the raster scanning so that the beam travels across the tube face in synchronism with the tube within the transmitting camera. Otherwise information from the top left-hand side of the televised scene will not appear in the top left-hand side of the screen and so on! In the analogue television signal this distinction between picture information and scan synchronizing information (known in the trade as sync-pulse information) is divided by a voltage level known as black level. All information above black level relates to picture information, whereas all information below relates to sync information. By this clever means, all synchronizing information is "below" black level. The electron beam therefore remains cut off—and the screen remains dark—during the sync information. In digital television the distinction between data relating to picture modulation and sync is established by a unique codeword preamble, which identifies the following byte as a sync byte.

### 28.4.1 Horizontal and Vertical Sync

The analogy between the eye's movement across the page during reading and the movement of the scan spot in scanning a tube face has already been made. Of course the scan spot doesn't move onto another page like the eyes do once they have reached the bottom of the page, but it does have to fly back to start all over again once it has completed one whole set of lines from the top to the bottom of the raster. The spot thus flies back in two possible ways: a horizontal retrace, between lines, and a vertical retrace, once it has completed one whole set of lines and is required to start all over again on another set. Obviously to stay in synchronism with the transmitting camera the television receiver must be instructed to perform both horizontal retrace and vertical retrace at the appropriate times—and furthermore not to confuse one instruction for the other!

It is for this reason that there exist two types of sync information known reasonably enough as horizontal and vertical. Inside the television monitor these are treated separately and respectively initiate and terminate the horizontal and vertical scan generator circuits. These circuits are similar—at least in principle—to the ramp or sawtooth generator circuits. As the current gradually increases in both horizontal and vertical scan coils, the spot is made to move from left to right and top to bottom, the current in the top to bottom circuit growing 312.5 times more slowly than in the horizontal deflection coils so that 312.5 lines are drawn in the time it takes the vertical deflection circuit to draw the beam across the vertical extent of the tube face.

The complete television signal is illustrated in Figures 28.1 and 28.2, which display the signal using two different time bases. Note the amplitude level, which distinguishes the watershed between picture information and sync information. Known as black level, this voltage is set to a standard 0 V. Peak white information is defined not to go beyond a level of 0.7 V above this reference level. Sync information, the line or horizontal sync, 4.7-μs pulse is visible in the figure and should extend 0.3 V below the black reference level. Note also that the picture information falls to the black level before and after the sync pulse. This interval is necessary because the electron beam cannot instantaneously retrace to the left-hand side of the screen to

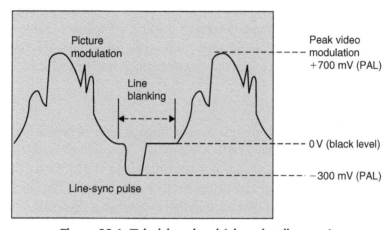

**Figure 28.1: Television signal (viewed at line rate).**

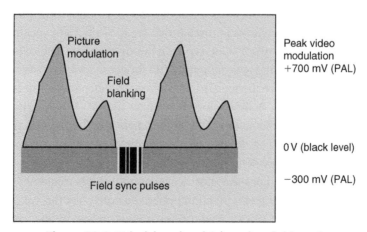

**Figure 28.2: Television signal (viewed at field rate).**

restart another trace. It takes a little time—about 12μs. This period, which includes duration of the 4.7-μs line-sync pulse during which time the beam current is controlled "blacker than black," is known as the line-blanking period. A similar, much longer, period exists to allow the scan spot to return to the top of the screen once a whole vertical scan has been accomplished; this interval is known as the field blanking or vertical interval.

Looking now at Figure 28.2, a whole 625 lines are shown, in two fields of 312.5 lines. Note the wider sync pulses that appear between each field. In order that a monitor may distinguish between horizontal and vertical sync, the duration of line-sync pulses is extended during the vertical interval (the gap in the picture information allowing for the field retrace), and a charge-pump circuit combined with a comparator is able to detect these longer pulses as different from the shorter line-sync pulses. This information is sent to the vertical scan generator to control the synchronism of the vertical scan.

## 28.5 Color Perception

Sir Isaac Newton discovered that sunlight passing through a prism breaks into the band of multicolored light, which we now call a spectrum. We perceive seven distinct bands in the spectrum: red, orange, yellow, green, blue, indigo, and violet. We see these bands distinctly because each represents a particular band of wavelengths. The objects we perceive as colored are perceived thus because they too reflect a particular range of wavelengths. For instance, a daffodil looks yellow because it reflects predominantly wavelengths in the region 570 nm. We can experience wavelengths of different color because the cone cells, in the retina at the back of the eye, contain three photosensitive chemicals, each of which is sensitive in three broad areas of the light spectrum. It is easiest to think of this in terms of three separate but overlapping photochemical processes: a low-frequency (long-wavelength) RED process, a medium-frequency GREEN process, and a high-frequency BLUE process. (Electronic engineers might prefer to think of this as three, shallow-slope band-pass filters!) When light of a particular frequency falls on the retina, the action of the light reacts selectively with this frequency-discriminating mechanism. When we perceive a red object we are experiencing a high level of activity in our long wavelength (low-frequency) process and low levels in our other two. A blue object stimulates the short wavelength or high-frequency process and so on. When we perceive an object with an intermediate color, say the yellow of the egg yoke, we experience a mixture of two chemical processes caused by the overlapping nature of each of the frequency-selective mechanisms. In this case, the yellow light from the egg causes stimulation in both the long wavelength RED process and the medium-wavelength GREEN process (Figure 28.3). Because human beings possess three separate color vision processes we are classified as trichromats. People afflicted with color blindness usually lack one of the three chemical responses in the normal eye; they are known as dichromats, although a few rare individuals are true monochromats. What has not yet been discovered, among people or other animals, is a more-than-three color perception system. This is fortunate for the engineers who developed color television!

The fact that our cone cells only contain three chemicals is why we may be fooled into experiencing the whole gamut of colors with the combination of only three so-called primary colors. The television primaries of red, green, and blue were chosen because each stimulates only one of the photosensitive chemicals found in the cone cells. The great television swindle is that we can, for instance, be duped into believing we are seeing yellow by activating both the red and green tube elements simultaneously—just as would a pure yellow source. Similarly we may be hoodwinked into seeing light blue cyan with the simultaneous activation of green and blue. We can also be made to experience paradoxical colors such as magenta by combining red and blue, a feat that no pure light source could ever do! This last fact demonstrates that our color perception system effectively "wraps around," mapping the linear spectrum of electromagnetic frequencies into a color circle, or a color space. And it is in this way that we usually view the science of color perception: we can regard all visual sense as taking place within a color three space. A television studio vectorscope allows us to view color three space end on, so it looks like a hexagon [Figure 28.4(a)]. Note that each color appears at a different angle, like the numbers on a

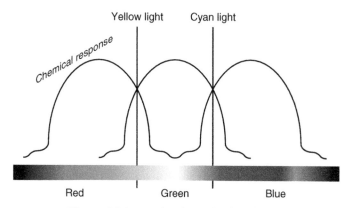

**Figure 28.3: Mechanism of color vision.**

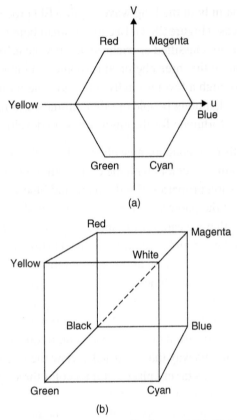

**Figure 28.4: (a) Studio vectorscope display and (b) Color three space.**

clock face. Hue is the term used in image processing and television to describe a color's precise location on this locus. Saturation is the term used to describe the amount a pure color is "diluted" by white light. The dotted axis shown in Figure 28.4(b) is the axis of pure luminance. The more a particular shade moves towards this axis from a position on the boundary of the cube, the more a color is said to be desaturated.

## 28.6 Color Television

From the discussions of the trichromatic response of the eye and of the persistence of vision, it should be apparent that a color scene may be rendered by the quick successive presentation of the red, green, and blue components of a color picture. Provided that these images are displayed frequently enough, the impression of a full color scene is indeed gained. Identical reasoning led to the development of the first color television demonstrations by Baird in 1928 and the first public color television transmissions in America by CBS in 1951. Known as a field-sequential system, in essence the apparatus consisted of a high field-rate monochrome television system with optical red, green, and blue filters presented in front of the camera lens and the receiver screen, which, when synchronized together, produced a color picture. Such an electromechanical system was not only unreliable and cumbersome but also required three times the bandwidth of a monochrome system (because three fields had to be reproduced in the period previously taken by one). In fact, even with the high field rate adopted by CBS, the system suffered from color flicker on saturated colors and was soon abandoned after transmissions started. Undeterred, the engineers took the next most obvious logical step for producing colored images. They argued that instead of presenting sequential fields of primary colors, they would present sequential dots of each primary. Such a (dot sequential) system using the secondary primaries of yellow, magenta, cyan, and black forms the basis of color printing. In a television system, individual phosphor dots of red, green, and blue, provided they are displayed with sufficient spatial frequency, provide the impression of a color image when viewed from a suitable distance.

Consider the video signal designed to excite such a dot-sequential tube face. When a monochrome scene is being displayed, the television signal does not differ from its black and white counterpart. Each pixel (of red, green, and blue) is equally excited, depending on the overall luminosity (or luminance) of a region of the screen. Only when a color is reproduced does the signal start to manifest a high-frequency component, related to the spatial frequency of the phosphor it is designed successively to stimulate. The exact phase of the high-frequency component depends, of course, on which phosphors are to be stimulated. The more saturated the color (i.e., the more it departs from gray), the more high-frequency "colorizing" signal is added. This signal is mathematically identical to a black and white television signal whereupon a high-frequency color-information carrier-signal (now known as a color subcarrier) is superimposed—a single frequency carrier whose instantaneous value of amplitude and phase, respectively, determines the saturation and hue of any particular region of the picture. This is the essence of the NTSC[1] color

---

[1]NTSC stands for National Television Standards Committee, the government body charged with choosing the American color system.

television system launched in the United States in 1953, although, for practical reasons, the engineers eventually resorted to an electronic dot-sequential signal rather than achieving this in the action of the tube. This technique is considered next.

### 28.6.1 NTSC and PAL Color Systems

If you've ever had to match the color of a cotton thread or wool, you'll know you have to wind a length of it around a piece of card before you are in a position to judge the color. That's because the eye is relatively insensitive to colored detail. This is obviously a phenomenon of great relevance to any application of color picture reproduction and coding; that color information may be relatively coarse in comparison with luminance information. Artists have known this for thousands of years. From cave paintings to modern animation studios it is possible to see examples of skilled, detailed monochrome drawings being colored in later by a less skilled hand.

The first step in the electronic coding of an NTSC color picture is color-space conversion into a form where brightness information (luminance) is separate from color information (chrominance) so that the latter can be used to control the high-frequency color subcarrier. This axis transformation is usually referred to as RGB to YUV conversion and it is achieved by mathematical manipulation of the form:

$$Y = 0.3R + 0.59G + 0.11B$$

$$U = m(B - Y)$$

$$V = n(R - Y)$$

The Y (traditional symbol for luminance) signal is generated in this way so that it as nearly as possible matches the monochrome signal from a black and white camera scanning the same scene. (The color green is a more luminous color than either red or blue and red is more luminous than blue.) Of the other two signals, U is generated by subtracting Y from B: for a black and white signal this evidently remains zero for any shade of gray. The same is true of $R - Y$. These signals therefore denote the amount the color signal differs from its black and white counterpart. They are therefore dubbed color difference signals. (Each color difference signal is scaled by a constant.) These signals may be a much lower bandwidth than the luminance signal because they carry color information only, to which the eye is relatively insensitive. Once derived, they are low-pass filtered to a bandwidth of 0.5 MHz.[2] These two signals are used to control the amplitude and phase of a high-frequency subcarrier superimposed onto the luminance signal. This chrominance modulation process is implemented with two balanced modulators in an amplitude-modulation-suppressed-carrier configuration—a process that can be thought of as multiplication. A clever technique is employed so that U modulates one carrier signal and V modulates another carrier of identical frequency but phase shifted with respect to the other by 90°. These two carriers are then combined and result in a subcarrier signal that varies its phase and amplitude dependent on the instantaneous value of U and V. Note the similarity between this and the form of color information noted in connection with the dot-sequential system: amplitude of high-frequency carrier dependent on the depth—or saturation—of the color and phase dependent on the hue of the color. (The difference is that in NTSC, the color subcarrier signal is coded and decoded using electronic multiplexing and demultiplexing of YUV signals rather than the spatial multiplexing of RGB components attempted in dot-sequential systems.) Figure 28.5 illustrates the chrominance coding process.

While this simple coding technique works well, it suffers from a number of important drawbacks. One serious implication is that if the high-frequency color subcarrier is attenuated (for instance, due to the low pass action of a long coaxial cable), there is a

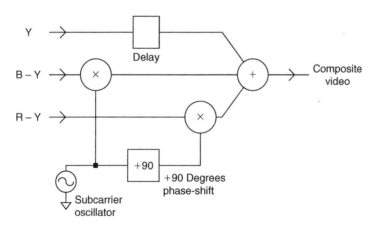

**Figure 28.5: NTSC color-coding process.**

---

[2]In NTSC systems or 1 MHz in PAL systems.

resulting loss of color saturation. More serious still, if the phase of the signal suffers from progressive phase disturbance, the color in the reproduced color is likely to change. This remains a problem with NTSC where no means are taken to ameliorate the effects of such a disturbance. The phase alternation line (PAL) system takes steps to prevent phase distortion having such a disastrous effect by switching the phase of the V subcarrier on alternate lines. This really involves very little extra circuitry within the coder but has design ramifications, which means the design of PAL decoding is a very complicated subject indeed. The idea behind this modification to the NTSC system (for that is all PAL is) is that, should the picture—for argument's sake—take on a red tinge on one line, it is cancelled out on the next when it takes on a complementary blue tinge. The viewer, seeing this from a distance, just continues to see an undisturbed color picture. In fact, things aren't quite that simple in practice but the concept was important enough to be worth naming the entire system after this one notion: phase alternation line. Another disadvantage of the coding process illustrated in Figure 28.5 is because of the contamination of luminance information with chrominance and vice versa. Although this can be limited to some degree by complementary band-pass and band-stop filtering, a complete separation is not possible, which results in the swathes of moving colored bands (cross-color), which appear across high-frequency picture detail on television—herringbone jackets proving especially potent in eliciting this system pathology.

In the color receiver, synchronous demodulation is used to decode the color subcarrier. One local oscillator is used and the output is phase shifted to produce the two orthogonal carrier signals for the synchronous demodulators (multipliers). Figure 28.6 illustrates the block schematic of an NTSC color decoder. A PAL decoder is much more complicated.

Mathematically, we can consider the PAL and NTSC coding process, thus

$$\text{NTSC colour signal} = Y + 0.49(B - Y)\sin \omega t + 0.88(R - Y)\cos \omega t$$
$$\text{PAL colour signal} = Y + 0.49(B - Y)\sin \omega t + 0.88(R - Y)\cos \omega t$$

Note that following the demodulators, the U and V signals are low-pass filtered to remove the twice frequency component and that the Y signal is delayed to match the processing delay of the demodulation process before being combined with the U and V signals in a reverse color space conversion. In demodulating the color subcarrier, the regenerated carriers must not only remain spot-on frequency, but also maintain a precise phase

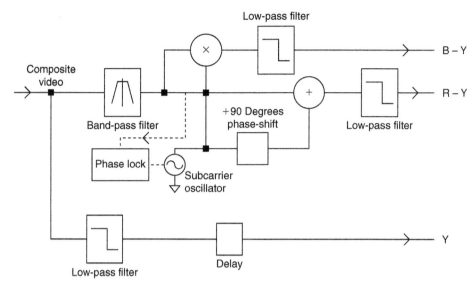

**Figure 28.6: NTSC color decoder.**

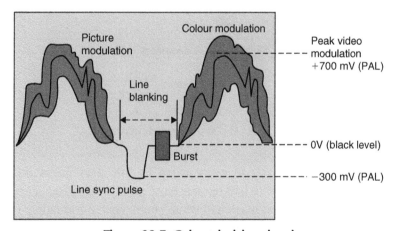

**Figure 28.7: Color television signal.**

relationship with the incoming signal. For these reasons the local oscillator must be phase locked and for this to happen the oscillator must obviously be fed a reference signal on a regular and frequent basis. This requirement is fulfilled by the color burst waveform, which is shown in the composite color television signal displayed in Figure 28.7.

**Table 28.1**

|                                  | NTSC      | PAL        |
|----------------------------------|-----------|------------|
| Field frequency                  | 59.94 Hz  | 50 Hz      |
| Total lines                      | 525       | 625        |
| Active lines                     | 480       | 575        |
| Horizontal resolution[a]         | 440       | 572        |
| Line frequency                   | 15.75 kHz | 15.625 kHz |

[a]Horizontal resolutions are calculated for an NTSC bandwidth of
4.2 MHz and 52-µs line period; PAL, 5.5 MHz bandwidth and
52-µs period.

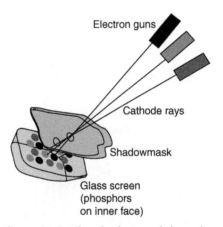

Electron guns

Cathode rays

Shadowmask

Glass screen
(phosphors
on inner face)

**Figure 28.8: The shadowmask in action.**

The reference color burst is included on every active TV line at a point in the original black and white signal given over to line retrace. Note also the high-frequency color information superimposed on the "black and white" luminance information. Once the demodulated signals have been through a reverse color space conversion and become RGB signals once more, they are applied to the guns of the color tube (Table 28.1).

As you watch television, three colors are being scanned simultaneously by three parallel electron beams, emitted by three cathodes at the base of the tube and all scanned by a common magnetic deflection system. But how to ensure that each electron gun only

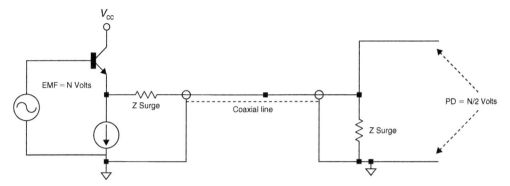

**Figure 28.9: Video interconnection.**

excites its appropriate phosphor? The answer is the shadowmask—a perforated, sheet-steel barrier that masks the phosphors from the action of an inappropriate electron gun. The arrangement is illustrated in Figure 28.8. For a color tube to produce an acceptable picture at reasonable viewing distance, there are about half a million phosphor red, green, and blue triads on the inner surface of the screen. The electron guns are set at a small angle to each other and aimed so that they converge at the shadowmask. The beams then pass through one hole and diverge a little between the shadowmask and the screen so that each strikes only its corresponding phosphor. Waste of power is one of the very real drawbacks of the shadowmask color tube. Only about a quarter of the energy in each electron beam reaches the phosphors. Up to 75% of the electrons do nothing but heat up the steel!

## 28.7 Analogue Video Interfaces

Due to their wide bandwidth, analogue television signals are always distributed via coaxial cables. The technique known as matched termination is universally applied. In this scheme, both the sender impedance and the load impedance are set to match the surge impedance of the line itself. This minimizes reflections. Standard impedance in television is 75 $\Omega$. A typical interconnection is shown in Figure 28.9. Note that matched termination has the one disadvantage that the voltage signal arriving across the receiver's termination is half that of the signal EMF provided by the sender. Standard voltage levels (referred to earlier) always relate to voltages measured across the termination impedance.

## 28.8  Digital Video

In order to see the forces that have led to the rapid adoption of digital video processing and interfacing throughout the television industry in the 1990s, it is necessary to look at some of the technical innovations in television during the late 1970s and early 1980s. The NTSC and PAL television systems described earlier were developed primarily as transmission standards, not as television production standards. As we have seen, because of the nature of the NTSC and PAL signal, high-frequency luminance detail can easily translate to erroneous color information. In fact, this cross-color effect is an almost constant feature of the broadcast standard television pictures and results in a general "business" to the picture at all times. That said, these composite TV standards (so named because the color and luminance information travel in a composite form) became the primary production standard mainly due to the inordinate cost of "three-level" signal processing equipment (i.e., routing switchers, mixers), which operated on the red, green, and blue or luminance and color-difference signals separately. A further consideration, beyond cost, was that it remained difficult to keep the gain, DC offsets, and frequency response (and therefore delay) of such systems constant, or at least consistent, over relatively long periods of time. Systems that did treat the R, G, and B components separately suffered particularly from color shifts throughout the duration of a program. Nevertheless, as analogue technology improved, with the use of integrated circuits as opposed to discrete semiconductor circuits, manufacturers started to produce three-channel, component television equipment that processed the luminance, $R-Y$ and $B-Y$ signals separately. Pressure for this extra quality came particularly from graphics areas, which found that working with the composite standards resulted in poor quality images that were tiring to work on, and where they wished to use both fine detail textures, which created cross-color, and heavily saturated colors, which do not produce well on a composite system (especially NTSC).

So-called analogue component television equipment had a relatively short stay in the world of high-end production largely because the problems of intercomponent levels, drift, and frequency response were never ultimately solved. A digital system, of course, has no such problems. Noise, amplitude response with respect to frequency, and time are immutable parameters "designed into" the equipment—not parameters that shift as currents change by fractions of milliamps in a base-emitter junction somewhere! From the start, digital television offered the only real alternative to analogue composite processing and, as production houses were becoming dissatisfied with the production value obtainable with composite equipment, the death knell was dealt to analogue processing in television.

### 28.8.1 The 4:2:2 Protocol Description—General

Just as with audio, so with video, as more and more television equipment began to process the signals internally in digital form, so the number of conversions could be kept to a minimum if manufacturers provided a digital interface standard allowing various pieces of digital video hardware to pass digital video information directly without recourse to standard analogue connections. This section is a basic outline of the 4:2:2 protocol (otherwise known as CCIR 601), which has been accepted as the industry standard for digitized component TV signals. The data signals are carried in the form of binary information coded in 8- or 10-bit words. These signals comprise the video signals themselves and timing reference signals. Also included in the protocol are ancillary data and identification signals. The video signals are derived by coding of the analogue video signal components. These components are luminance (Y) and color difference (Cr and Cb) signals generated from primary signals (R, G, B). The coding parameters are specified in CCIR Recommendation 601 and the main details are reproduced in Table 28.2.

#### 28.8.1.1 Timing Relationships

The digital active line begins at 264 words from the leading edge of the analogue line synchronization pulse; this time is specified between half amplitude points. This relationship is shown in Figure 28.10. The start of the first digital field is fixed by the

**Table 28.2: Encoding Parameter Values for the 4:2:2 Digital Video Interface**

| Parameters | 525-line, 60 field/s systems | 625-line, 50 field/s systems |
| --- | --- | --- |
| 1 Coded signals: Y, Cb, Cr. These signals are obtained from gamma precorrected RGB signals. | | |
| 2 Number of samples per total line: | | |
| • Luminance signal (Y) | 858 | 864 |
| • Each color-difference signal (Cb, Cr) | 429 | 432 |
| 3 Sampling structure | Orthogonal line, field and picture repetitive Cr and Cb samples cosited with odd (1st, 3rd, 5th, etc.) Y samples in each line. | |

(Continued)

**Table 28.2: Continued**

| Parameters | 525-line, 60 field/s systems | 625-line, 50 field/s systems |
|---|---|---|
| 4 Sampling frequency: | | |
| • Luminance signal | 13.5 MHz | |
| • Each color-difference signal | 6.75 MHz | |
| The tolerance for the sampling frequencies should coincide with the tolerance for the line frequency of the relevant color television standard. | | |
| 5 Form of coding | Uniformly quantized PCM, 8 bits per sample, for the luminance signal and each color-difference signal. | |
| 6 Number of samples per digital active line: | | |
| • Luminance signal | 720 | |
| • Each color-difference signal | 360 | |
| 7 Analogue to digital horizontal timing relationship: | | |
| • From end of digital active line to 0 H | 16 luminance clock periods (NTSC) | 12 luminance clock periods (PAL) |
| 8 Correspondence between video signal levels and quantization levels: | | |
| • Scale | 0 to 255 | |
| • Luminance signal | 220 quantization levels with the black level corresponding to level 16 and the peak white level corresponding to level 235. The signal level may occasionally excurse beyond level 235. | |
| • Color-difference signal | 225 quantization levels in the center part of the quantization scale with zero signal corresponding to level 128. | |
| 9 Codeword usage | Codewords corresponding to quantization levels 0 and 255 are used exclusively for synchronization. Levels 1 to 254 are available for video. | |

Note that the sampling frequencies of 13.5 MHz (luminance) and 6.75 MHz (color difference) are integer multiples of 2.25 MHz, the lowest common multiple of the line frequencies in 525/60 and 625/50 systems, resulting in a static orthogonal sampling pattern for both. The luminance and the color-difference signals are thus sampled to 8- (or 10-) bit depth with the luminance signal sampled twice as often as each chrominance signal (74 ns as against 148 ns). These values are multiplexed together with the structure as follows:

Cb, Y, Cr, Y, Cb, Y, Cr ... etc.

where the three words (Cb, Y, Cr) refer to cosited luminance and color-difference samples and the following word Y corresponds to a neighboring luminance only sample. The first video data word of each active line is Cb.

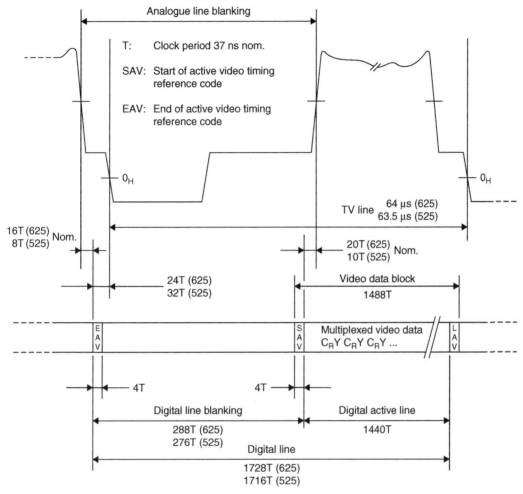

**Figure 28.10: Relationships between timing in digital and analogue TV.**

position specified for the start of the digital active line: the first digital field starts at 24 words before the start of the analogue line No. 1. The second digital field starts 24 words before the start of analogue line No. 313.

### 28.8.1.2 Video Timing Reference Signals (TRS)

Two video TRS are multiplexed into the data stream on every line, as shown in Figure 28.10, and retain the same format throughout the field blanking interval. Each timing reference signal consists of a four-word sequence, with the first three words being a fixed preamble and the fourth containing the information defining:

First and second field blanking;

State of the field blanking;

Beginning and end of the line blanking.

This sequence of four words can be represented, using hexadecimal notation, in the following manner:

FF 00 00 XY

in which XY represents a variable word. In binary form, this can be represented in the following form:

| Data bit number | First word (FF) | Second word (00) | Third word (00) | Fourth word (XY) |
|---|---|---|---|---|
| 7 | 1 | 0 | 0 | 1 |
| 6 | 1 | 0 | 0 | F |
| 5 | 1 | 0 | 0 | V |
| 4 | 1 | 0 | 0 | H |
| 3 | 1 | 0 | 0 | P3 |
| 2 | 1 | 0 | 0 | P2 |
| 1 | 1 | 0 | 0 | P1 |
| 0 | 1 | 0 | 0 | P0 |

The binary values of F, V, and H characterize the three items of information listed earlier:

F = 0 for the first field;

V = 1 during the field-blanking interval;

H = 1 at the start of the line-blanking interval.

The binary values P0, P1, P2, and P3 depend on the states of F, V, and H in accordance with the following table and are used for error detection/correction of timing data:

| F | V | H | P3 | P2 | F1 | P0 |
|---|---|---|----|----|----|----|
| 0 | 0 | 0 | 0 | 0 | 0 | 0 |
| 0 | 0 | 1 | 1 | 1 | 0 | 1 |
| 0 | 1 | 0 | 1 | 0 | 1 | 1 |
| 0 | 1 | 1 | 0 | 1 | 1 | 0 |
| 1 | 0 | 0 | 0 | 1 | 1 | 1 |
| 1 | 0 | 1 | 1 | 0 | 1 | 0 |
| 1 | 1 | 0 | 1 | 1 | 0 | 0 |
| 1 | 1 | 1 | 0 | 0 | 0 | 1 |

### 28.8.1.3  Clock Signal

The clock signal is at 27 MHz, there being 1728 clock intervals during each horizontal line period (PAL).

### 28.8.1.4  Filter Templates

The remainder of CCIR Recommendation 601 is concerned with the definition of the frequency response plots for presampling and reconstruction filter. The filters required by Recommendation 601 are practically difficult to achieve and equipment required to meet this specification has to contain expensive filters in order to obtain the required performance.

### 28.8.2  Parallel Digital Interface

The first digital video interface standards were parallel in format. They consisted of 8 or 10 bits of differential data at ECL data levels and a differential clock signal again as an ECL signal. Carried via a multicore cable, the signals terminated at either end in a standard D25 plug and socket. In many ways this was an excellent arrangement and is well suited to connecting two local digital videotape machines together over a short distance. The protocol for the digital video interface is shown in Table 28.3. Clock transitions are specified to take place in the center of each data-bit cell.

Problems arose with the parallel digital video interface over medium/long distances, resulting in misclocking of input data and visual "sparkles" or "zits" on the picture.

Table 28.3: Parallel Digital Video Interface

| Pin No. | Function |
|---------|----------|
| 1 | Clock+ |
| 2 | System ground |
| 3 | Data 7 (MSB)+ |
| 4 | Data 6+ |
| 5 | Data 5+ |
| 6 | Data 4+ |
| 7 | Data 3+ |
| 8 | Data 2+ |
| 9 | Data 1+ |
| 10 | Data 0+ |
| 11 | Data − 1 + 10-bit systems only |
| 12 | Data − 2 + 10-bit systems only |
| 13 | Cable shield |
| 14 | Clock − |
| 15 | System ground |
| 16 | Data 7 (MSB) − |
| 17 | Data 6 − |
| 18 | Data 5 − |
| 19 | Data 4 − |
| 20 | Data 3 − |
| 21 | Data 2 − |
| 22 | Data 1 − |
| 23 | Data 0 − |
| 24 | Data − 1–10-bit systems only |
| 25 | Data − 2–10-bit systems only |

Furthermore, the parallel interface required expensive and nonstandard multicore cable (although over very short distances it could run over standard ribbon cable) and the D25 plug and socket are very bulky. Today, the parallel interface standard has been largely superseded by the serial digital video standard, which is designed to be transmitted over relatively long distances using the same coaxial cable as used for analogue video signals. This makes its adoption and implementation as simple as possible for existing television facilities converting from analogue to digital video standards.

### 28.8.3 Serial Digital Video Interface

SMPTE 259M specifies the parameters of the serial digital standard. This document specifies that parallel data in the format given in the previous section be serialized and transmitted at a rate 10 times the parallel clock frequency. For component signals, this is

$$27 \text{ Mbits/s} \times 10 = 270 \text{ Mbits/s}.$$

Serialized data must have a peak-to-peak amplitude of $800\,\text{mV}$ ($\pm 10\%$) across $75\,\Omega$, have a nominal rise time of $1\,\text{ns}$, and have a jitter performance of $\pm 250\,\text{ps}$. At the receiving end, signals must be converted back to parallel in order to present original parallel data to the internal video processing. (Note that no equipment processes video in its serial form, although digital routing switchers and DAs, where there is no necessity to alter the signal, only buffer it or route it, do not decode the serial bit stream.)

Serialization is achieved by means of a system illustrated in Figure 28.11. Parallel data and a parallel clock are fed into input latches and then to a parallel to serial conversion circuit. The parallel clock is also fed to a phase-locked loop, which performs parallel clock multiplication (by 10 times). A sync detector looks for TRS information and ensures that this is encoded correctly irrespective of 8- or 10-bit resolution. Serial data are fed out of the serializer and into the scrambler and NRZ to NRZI circuit. The scrambler circuit uses a linear feedback shift register, which is used to pseudo-randomize incoming serial data. This has the effect of minimizing the DC component of the output serial data stream, the NRZ to NRZI circuit converts long series of ones to a series of transitions. The resulting signal contains enough information at clock rate and is sufficiently DC free that it may be sent down existing video cables. It may be then be reclocked, decoded, and converted back to parallel data at the receiving equipment. Because of its very high data rate, serial video must be carried by ECL circuits. An illustration of a typical ECL gate is given in Figure 28.12. Note that standard video levels are commensurate with data levels in ECL logic. Clearly the implementation of such a high-speed interface is a highly specialized task. Fortunately, practical engineers have all the requirements for interface encoders and decoders designed for them by third-party integrated circuit manufacturers (Figure 28.13).

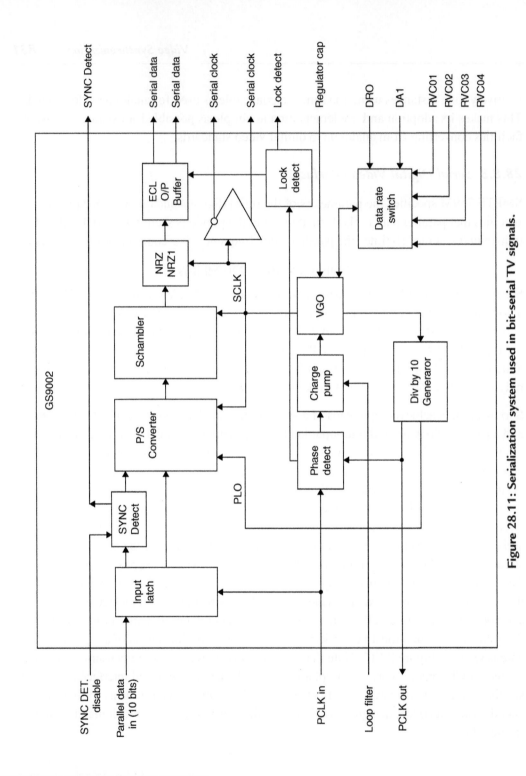

**Figure 28.11: Serialization system used in bit-serial TV signals.**

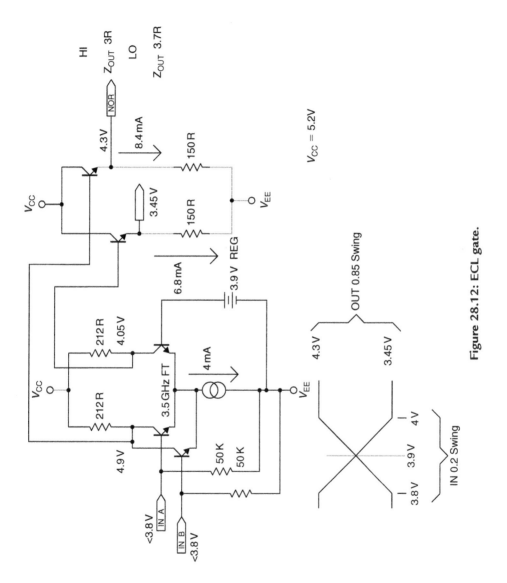

**Figure 28.12: ECL gate.**

**Figure 28.13: Gennum Corp. serializer chip.
(Photos courtesy of Gennum Corporation.)**

## 28.9 Embedded Digital Audio in the Digital Video Interface

So far, we have considered the interfacing of digital audio and video separately. Manifestly, there exist many good operational reasons to combine a television picture with its accompanying sound "down the same wire." The standard that specifies the embedding of digital audio data, auxiliary data, and associated control information into the ancillary data space of the serial digital interconnect conforming to SMPTE 259M in this manner is the proposed SMPTE 272M standard.

The video standard has adequate "space" for the mapping of a minimum of 1 stereo digital audio signal (or two mono channels) to a maximum of 8 pairs of stereo digital audio signals (or 16 mono channels). The 16 channels are divided into 4 audio signals in 4 "groups." The standard provides for 10 levels of operation (suffixed A to J), which allow for various different and extended operations over and above the default synchronous 48-kHz/20-bit standard. The audio may appear in any and/or all the line blanking periods and should be distributed evenly throughout the field. Consider the case of one 48-kHz audio signal multiplexed into a 625/50 digital video signal. The number of samples to be transmitted every line is:

$$(48000)/(15625),$$

which is equivalent to 3.072 samples per line. The sensible approach is taken within the standard of transmitting 3 samples per line most of the time and transmitting 4 samples per line occasionally in order to create this noninteger average data rate. In the case of 625/50, this leads to 1920 samples per complete frame. (Obviously a comparable calculation can be made for other sampling and frame rates.) All that is required to achieve this "packeting" of audio within each video line is a small amount of buffering

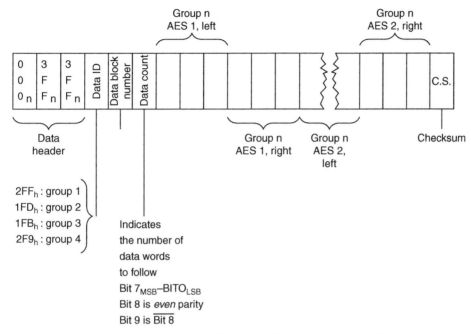

**Figure 28.14: Data format for digital audio packets in SDV bit stream.**

either end and a small data overhead to "tell" the receiver whether it should expect 3 or 4 samples on any given line.

Figure 28.14 illustrates the structure of each digital audio packet as it appears on preferably all, or nearly all, the lines of the field. The packet starts immediately after the TRS word for EAV (end of active line) with the ancillary data header 000,3FF,3FF. This is followed by a unique ancillary data ID, which defines which audio group is being transmitted. This is followed with a data-block number byte. This is a free-running counter counting from 1 to 255 on the lowest 8 bits. If this is set to zero, a deembedder is to assume that this option is not active. The 9th bit is even parity for b7 to b0 and the 10th is the inverse of the 9th. It is by means of this data-block number word that a vertical interval switch could be discovered and concealed. The next word is a data count, which indicates to a receiver the number of audio data words to follow. Audio subframes then follow as adjacent sets of three contiguous words. The format in which each AES subframe is encoded is illustrated in Figure 28.15. Each audio data packet terminates in a checksum word.

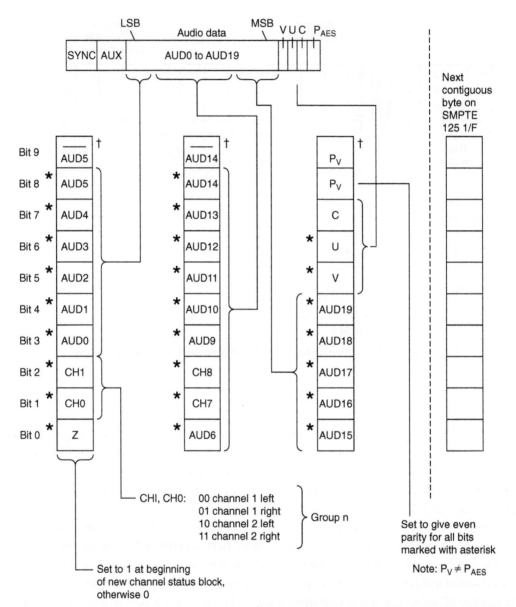

Figure 28.15: Each EAS subframe is encoded as three contiguous data packets.

The standard also specifies an optional audio control packet. If the control packet is not transmitted, a receiver defaults to 48 kHz, synchronous operation. For other levels, the control byte must be transmitted in field interval.

## 28.10 Time Code

### 28.10.1 Longitudinal Time Code (LTC)

As we have seen, television (like movie film) gives the impression of continuous motion pictures by the successive, swift presentation of still images, thereby fooling the eye into believing it is perceiving motion. It is probably therefore no surprise that time code (deriving as it does from television technology) operates by "tagging" each video frame with a unique identifying number called a time code address. The address contains information concerning hours, minutes, seconds, and frames. This information is formed into a serial digital code, which is recorded as a data signal onto one of the audio tracks of a videotape recorder. (Some videotape recorders have a dedicated track for this purpose.)

Each frame's worth of data is known as a word of time code and this digital word is formed of 80 bits spaced evenly throughout the frame. Taking the European Broadcasting Union (EBU) time code[3] as an example, the final data rate therefore turns out to be 80 bits × 25 frames per second = 2000 bits per second, which is equivalent to a fundamental frequency of 1 kHz; easily low enough, therefore, to be treated as a straightforward audio signal. The time code word data format is illustrated (along with its temporal relationship to a video field) in Figure 28.16. The precise form of the electrical code for time code is known as Manchester biphase modulation. When used in a video environment, time code must be accurately phased to the video signal. As defined in the specification, the leading edge of bit '0' must begin at the start of line 5 of field 1 ($\pm 1$ line). Time address data are encoded within the 80 bits as 8, 4-bit BCD (binary coded decimal) words (i.e., 1, 4-bit number for tens and 1 for units). Like the clock itself, time address data are only permitted to go from 00 hours, 00 minutes, 00 seconds, 00 frames to 23 hours, 59 minutes, 59 seconds, 24 frames.

However, a 4-bit BCD number can represent any number from 0 to 9, so in principle, time code could be used to represent 99 hours, 99 minutes, and so on. But, as there are no

---

[3]The EBU time code is based on a field frequency of 25 frames per second.

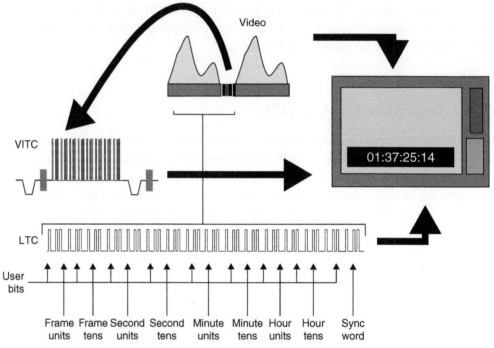

**Figure 28.16: Format of LTC and VITC time code.**

hours above 23, no minutes or seconds above 59, and no frames above 24 (in PAL), the time code possesses potential redundancy. In fact, some of these extra codes are exploited in other ways. The basic time address data, and these extra bits are assigned their position in the full 80-bit time code word, are like this:

| 0–3 | Frame units |
|------|-------------|
| 4–7 | First binary group |
| 8–9 | Frame tens |
| 10 | Drop frame flag |
| 11 | Color frame flag |
| 12–15 | Second binary group |
| 16–19 | Seconds units |
| 20–23 | Third binary group |
| 24–26 | Seconds tens |

| 27 | Unassigned |
|---|---|
| 28–31 | Fourth binary group |
| 32–35 | Minutes units |
| 36–39 | Fifth binary group |
| 40–42 | Minutes tens |
| 43 | Unassigned |
| 44–47 | Sixth binary group |
| 48–51 | Hours units |
| 52–55 | Seventh binary group |
| 56–57 | Hours tens |
| 58–59 | Unassigned |
| 60–63 | Eighth binary group |
| 64–79 | Synchronizing sequence |

### 28.10.2 Vertical Interval Time Code (VITC)

LTC is a quasi-audio signal recorded on an audio track (or hidden audio track dedicated to time code). VITC, however, encodes the same information within the vertical interval portion of the video signal in a manner similar to a Teletext signal. Each has advantages and disadvantages; LTC is unable to be read while the player/ recorder is in pause, while VITC cannot be read while the machine is in fast forward or rewind modes. It is advantageous that a videotape has both forms of time code recorded. VITC is also illustrated in Figure 28.16. Note how time code is displayed "burned in" on the monitor.

### 28.10.3 PAL and NTSC

Naturally, time code varies according to the television system used, and for NTSC (SMPTE) there are two versions of time code in use to accommodate the slight difference between the nominal frame rate of 30 frames per second and the actual frame rate of NTSC of 29.97 frames per second. While every frame is numbered and no frames are ever actually dropped, the two versions are referred to as "drop"- and "nondrop"-frame time code. Nondrop-frame time code will have every number for every second present, but will drift out of relationship with clock time by 3.6 seconds every hour. Drop-frame time code drops numbers from the numbering system in a predetermined sequence, so that the time code-time and clock time remain in synchronization. Drop frame is important in broadcast work, where actual program time is important.

### 28.10.4  User Bits

Within the time code word there is provision for the hours, minutes, seconds, frames, and field ID that we normally see and "user bits," which can be set by the user for additional identification. Use of user bits varies with some organizations using them to identify shoot dates or locations and others ignoring them completely.

# CHAPTER 29

# Room Acoustics

Ian Sinclair

The reproduction of natural instrumental sound begins with microphones, but the behavior of microphones cannot be separated from the acoustics of the studio. Ian Sinclair shows here the principles and practices of studio acoustics as used today.

## 29.1 Introduction

Over the past few years the performance, sophistication, and quality of the recording medium and ancillary hardware and technology have advanced at a considerable rate. The 50- to 60-dB dynamic range capability of conventional recording and disc technology suddenly has become 90 to 100 dB with the introduction of digital recording and the domestic digital compact disc (CD). In the foreseeable future a dynamic range of 110 dB could well become commonplace for digital mastering.

The increased dynamic range, coupled with other advances in loudspeaker and amplifier technology, now means that audio, broadcast, and hi-fi systems can offer a degree of resolution and transparency to the domestic market today that was unachievable only a few years ago, even with the best professional equipment.

The acoustic environments of the studios in which the majority of the recordings or broadcasts originate from have become correspondingly more critical and important. Recordings can no longer be made in substandard environments. Control rooms and studios exhibiting an uneven or colored acoustic response or too high a level of ambient noise, which previously could be lost or masked by traditional analogue recording process, can no longer be tolerated. The transparency of the digital or FM broadcast medium immediately highlights such deficiencies.

To many, the subject of studio acoustics is considered a black art, often surrounded by considerable hype and incomprehensible terminology. However, this is no longer the case. Today, there certainly is an element of art in achieving a desirable and a predictable acoustic environment, but it is very much based on well-established scientific principles and a comprehensive understanding of the underlying physics of room acoustics and noise control.

Studios and control rooms create a number of acoustic problems that need to be overcome. Essentially, they can be divided into two basic categories: noise control and room acoustics, with the latter including the interfacing of the monitor loudspeakers to the control room.

## 29.2  Noise Control

The increased dynamic range of the hi-fi medium has led to corresponding decreases in the levels of permissible background noise in studios and control rooms.

A single figure rating (e.g., 25 dBA) is not generally used to describe the background noise requirement as it is too loose a criterion. Instead a set of curves that take account of the spectral (frequency) content of the noise are used. The curves are based on an octave or 1/3 octave band analysis of the noise and also take account of the ear's reduced sensitivity to lower frequency sounds.

The curves used most frequently are the NC (noise criterion) and NR (noise rating) criteria. Figures 29.1 and 29.2 graphically present the two sets of criteria in terms of the octave band noise level and frequency. (Although as Figures 29.1 and 29.2 show, criteria are not exactly the same; the corresponding target criteria, e.g., NC20 or NR20, are frequently interchanged.)

The NC system is intended primarily for rating air conditioning noise, while the NR system is more commonplace in Europe and can be used to rate noises other than air conditioning. (An approximate idea of the equivalent dBA value can be obtained by adding 5 to 7 dB to the NC/NR level).

Table 29.1 presents typical design targets for various studio and recording formats. Many organizations, for example, the BBC, demand even more stringent criteria, particularly at low frequencies. From Table 29.1 and curves shown in Figures 29.1 and 29.2, it can be seen that the background noise level requirements are pretty stringent, typically being

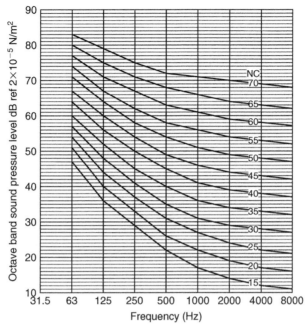

**Figure 29.1: Noise criterion (NC) curves.**

25 dBA or less, which subjectively is very quiet. (The majority of domestic living rooms, considered by their occupants to be quiet, generally measure 30–40 dBA.)

The background noise level in a studio is made up from a number of different sources and components and for control purposes these may be split up into four basic categories:

- External airborne noise,

- External structure borne noise,

- Internally and locally generated noise,

- Internal noise transfer,

Each source of noise requires a slightly different approach in terms of its containment or control.

Achieving low noise levels in modern studios is a complex and generally costly problem but any short cuts or short circuits of the noise control measures will destroy the integrity of the low noise countermeasures and allow noise into the studio.

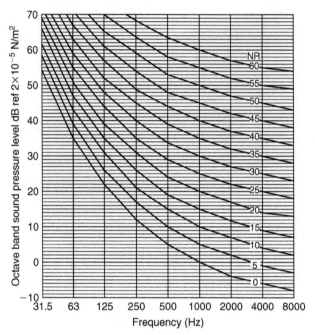

**Figure 29.2: NC curves.**

**Table 29.1**

| Studio/recording environment | NC/NR requirement |
|---|---|
| Studio with close mic technique (300 mm) (bass cut may be required) | 30 |
| TV studio control rooms | 25 |
| TV studios and sound control rooms | 20 |
| Nonbroadcast–preproduction | 25 |
| Broadcast studio and listening rooms | 15–20 |
| Music sound studio | 15–20 |
| Concert halls | 15–20 |
| Drama studios (broadcast) | 15 |
| Sound effects and postproduction (preferred) | 5 |
| (max.) | 15 |
| Commercial recording and pop studios (depending on type and budget) | 15–25 |

### 29.2.1  External Airborne Noise

Here we are typically concerned with noise from local road traffic, aircraft, and railways. Any of these sources can generate levels of over 100 dB at low frequencies at the external facade of a building. Controlling such high levels of low-frequency noise is extremely difficult and costly. Therefore, if possible, when planning a sensitive acoustic area such as a studio, it is obviously preferable to avoid locations or buildings exposed to high external levels of noise. An extensive noise survey of the proposed site is therefore essential.

Wherever possible the walls to a studio or control room should not form part of the external envelope of the building. Instead a studio should be built within the main body of the building itself so a buffer area between the studio and external noise source can be created, for example, by corridors or offices or other noncritical areas. Similarly, it is not good practice to locate a studio on the top floor of a building. However, if it is not possible to create suitable buffer areas, a 1-m minimum separation should be created between external and studio walls or roof and ceiling etc.

### 29.2.2  Internally and Locally Generated Noise

Apart from being transmitted by sound pressure waves within the air, noise can also be transmitted by the building structure itself. Structural noise can be induced either directly by the vibration of the structure, for example, by locating the building over or adjacent to a railway or underground train line, or sound pressure waves (particularly at low frequencies) can cause the structure itself to vibrate. Again, the best method of control is to separate the sensitive areas from such potential noise sources as structural isolation is required. One solution is to spend the available budget on the equipment of rubber pads, or literally steel springs if really effective low frequency isolation is required. However, the best solution is to choose the site carefully in the first place and spend the available budget on the equipment and acoustic finishes rather than on expensive structural isolation.

If the studio is to be built in a building with other occupants/activities, again a thorough survey should be conducted to establish any likely sources of noise, which will require more than usual treatment (e.g., machinery coupled into the structure on floors above or below the studio. This includes lifts and other normal "building services" machinery such as air conditioning plant, etc.).

### 29.2.3  Internal Noise Transfer

The biggest noise problem in most studios is noise generated or transmitted by the air conditioning equipment. Studios must provide a pleasant atmosphere to work in, which, apart from providing suitable lighting and acoustics, also means that the air temperature and humidity need to be carefully controlled. Sessions in pop studios may easily last 12 to 15 h. The studio/control room must therefore have an adequate number of air changes and sufficient fresh air supply in order for the atmosphere not to become "stuffy" or filled with cigarette smoke and so on.

It should be remembered that the acoustic treatments also tend to provide additional thermal insulation and in studios with high lighting capacity (e.g., TV or film) or where a large number of people may be working, the heat cannot easily escape, but is contained within the studio, requiring additional cooling capacity over that normally expected.

Large air conditioning ducts with low air flow velocities are therefore generally employed in studios with special care being taken to ensure that the air flow through the room diffuser grilles is low enough not to cause air turbulence noise (e.g., duct velocities should be kept below 500 ft$^3$/min, and terminal velocities below 250 ft$^3$/min). Studios should be fed from their own air conditioning plant, which is not shared with other parts of the building.

Ducts carrying air to and from the studios are fitted with attenuators or silencers to control the noise of the intake or extract fans. Cross talk attenuators are also fitted in ducts that serve more than one area, for example, studio and control room, to prevent the air conditioning ducts acting as large speaking tubes or easy noise transmission paths.

Studio air conditioning units should not be mounted adjacent to the studio but at some distance away to ensure that the air conditioning plant itself does not become a potential noise source. Duct and pipe work within the studio complex will need to be mounted with resilient mountings or hangers to ensure that noise does not enter the structure or is not picked up by the ductwork and transmitted to the studio.

Water pipes and other mechanical services ducts or pipes should not be attached to studio or control room walls, as again noise can then be transmitted readily into the studio.

Impact noise from footfall or other movement within the building must also be dealt with if it is not be transmitted through the structure and into the studio. A measure as simple as carpeting adjacent corridors or office floors can generally overcome the majority of such problems. Wherever possible, corridors adjacent to studios should not be used at critical times or their use should be strictly controlled. Main thoroughfares should not pass through studio areas.

### 29.2.4  Sound Insulation

Having created a generally suitable environment for a studio, it is then up to the studio walls themselves to provide the final degree of insulation from the surrounding internal areas.

The degree of insulation required will very much depend on the following

1. Uses of the studio, for example, music, speech, and drama.

2. The internal noise level target, for example, NC20.

3. Noise levels of the surrounding areas.

4. The protection required to the surrounding area (e.g., in a pop music studio, the insulation is just as much there to keep the noise in so that it will not affect adjoining areas or studios as it is to stop noise getting into the studio itself).

Appropriate insulation or separation between the control room and studio is also another very important aspect that must be considered here. The minimum value of sound insulation required between control room and studio is probably about 55–60 dB. Monitoring levels are generally considerably higher in studio control rooms than in the home. For example, 85 dBA is quite typical for TV or radio sound monitoring, whereas pop studio control rooms will operate in the high 90s and above.

Sound insulation can only be achieved with mass or with mass and separation. Sound-absorbing materials are not sound insulators, although there is much folklore that would have you believe otherwise. Table 29.2 sets out the figures, showing that only very high values of sound absorption produce usable sound reductions.

To put the figures used in Table 29.2 in perspective, a single 4.5-in. (112-mm) brick wall has a typical sound insulation value of 45 dB, and a typical domestic brick cavity wall 50–53 dB—and you know how much sound is transmitted between adjoining houses! Furthermore, typical sound-absorbing materials have absorption coefficients in the region of 0.7 to 0.9, which is equivalent to a noise reduction of only 5–10 dB.

Table 29.2: Sound Absorption

| Absorption coefficient | Percentage absorbed (%) | Insulation produced (dB) | Incident sound energy transmitted (%) |
|---|---|---|---|
| 0.1 | 10 | 0.5 | 90 |
| 0.5 | 50 | 3 | 50 |
| 0.7 | 70 | 5 | 30 |
| 0.9 | 90 | 10 | 10 |
| 0.99 | 99 | 20 | 1 |
| 0.999 | 99.9 | 30 | 0.1 |
| 0.9999 | 99.99 | 40 | 0.01 |

Figure 29.3 shows the so-called mass law of sound insulation, which shows that sound insulation increases by approximately 5 dB for every doubling of the mass. A number of typical constructions and building elements have been drawn on for comparison purposes.

It should be noted that the sound-insulating properties of a material are also affected by the frequency of the sound in question. In fact the insulation generally increases by 5 dB for every doubling of frequency. This is good news as far as the insulation of high-frequency sound is concerned, but bad news for the lower frequencies.

Generally, a single value of sound insulation is often quoted—this normally refers to the average sound insulation achieved over the range 100 Hz to 3.15 KHz and is referred to as the sound reduction index (SRI). Sound transmission class (STC) in the United States is based on the average between 125 Hz and 4 KHz.)

An approximate guide to the performance of a material can be obtained from the 500-Hz value of insulation as this is often equivalent to the equated SRI. For example, our 112-mm brick wall will tend to have an SRI of 45 dB and have a sound insulation of 45 dB at 500 Hz, 40 dB at 250 Hz, and only 35 dB at 125 Hz, which begins to explain why it is always the bass sound or beat that seems to be transmitted. In practice, therefore, the studio designer or acoustic consultant first considers controlling break-in (or break-out) of the low-frequency sound components, as overcoming this problem will invariably automatically sort out any high-frequency problems.

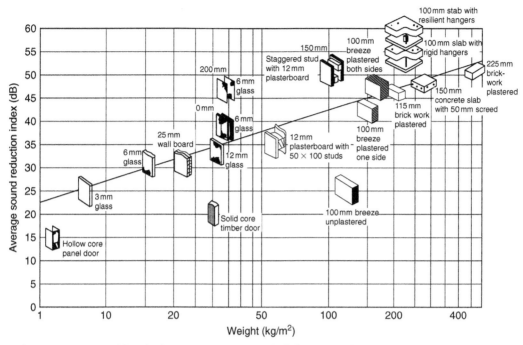

**Figure 29.3: Sound insulation performance of building materials compared with mass law.**

From the graph shown in Figure 29.3, it can be seen that some pretty heavy constructions are required in order to achieve good sound separation/insulation. However, it is possible to also achieve good sound insulation using lightweight materials by using multilayered construction techniques.

Combinations of layers of plasterboard and softboard, separated by an airspace, are used frequently. Figure 29.4 shows a typical construction. The softboard is used to damp out the natural resonances of the plasterboard, which would reduce its sound insulation. The airspace may also be fitted with loose acoustic quilting (e.g., fiber-glass or mineral wool, to damp out any cavity resonances).

Note that an airtight seal has to be created for optimum sound insulation efficiency. Also note that different sized airspaces are used.

A double-leaf construction as shown in Figure 29.4 can achieve a very good level of sound insulation, particularly at mid and high frequencies.

**Table 29.3: Typical Sound Absorption Coefficients**

| Material | Hz | | | | | |
| --- | --- | --- | --- | --- | --- | --- |
| | 125 | 250 | 500 | IK | 2K | 4K |
| Drapes | 0.07 | 0.37 | 0.49 | 0.31 | 0.65 | 0.54 |
| • Light velour | 0.07 | 0.31 | 0.49 | 0.75 | 0.70 | 0.60 |
| • Medium velour | 0.07 | 0.35 | 0.55 | 0.72 | 0.70 | 0.65 |
| • Heavy velour | 0.14 | 0.35 | 0.55 | 0.72 | 0.70 | 0.65 |
| Carpet | | | | | | |
| • Heavy on concrete | 0.02 | 0.06 | 0.14 | 0.37 | 0.60 | 0.65 |
| • Heavy on underlay or felt | 0.08 | 0.24 | 0.57 | 0.70 | 0.70 | 0.73 |
| • Thin | 0.01 | 0.05 | 0.10 | 0.20 | 0.45 | 0.65 |
| Cork floor tiles and linoleum, plastic flooring | 0.02 | 0.04 | 0.05 | 0.05 | 0.10 | 0.05 |
| Glass (windows) | | | | | | |
| • 4 mm | 0.30 | 0.20 | 0.10 | 0.07 | 0.05 | 0.02 |
| • 6 mm | 0.10 | 0.08 | 0.04 | 0.03 | 0.02 | 0.02 |
| Glass, tile, and marble | 0.01 | 0.01 | 0.01 | 0.01 | 0.02 | 0.02 |
| Glass fiber mat | | | | | | |
| • 80 kg/m$^3$–25 mm | 0.10 | 0.30 | 0.55 | 0.65 | 0.75 | 0.80 |
| • 80 kg/m$^3$–50 mm | 0.20 | 0.45 | 0.70 | 0.80 | 0.80 | 0.80 |
| Wood fiber ceiling tile | 0.15 | 0.40 | 0.70 | 0.80 | 0.80 | 0.80 |
| Plasterboard over deep airspace | 0.20 | 0.15 | 0.10 | 0.05 | 0.05 | 0.05 |

Typically, such a partition might achieve the following insulation values:

| Frequency | (Hz) | 63 | 125 | 250 | 500 | 1K | 2K | 4K |
| --- | --- | --- | --- | --- | --- | --- | --- | --- |
| Insulation | (dB) | 40 | 50 | 60 | 65 | 70 | 75 | 80 |

However, the final value will very much depend on the quality of the construction and also on the presence of flanking transmission paths, that is, other sound paths, such as ducting or trunking runs, that short circuit or bypass the main barrier.

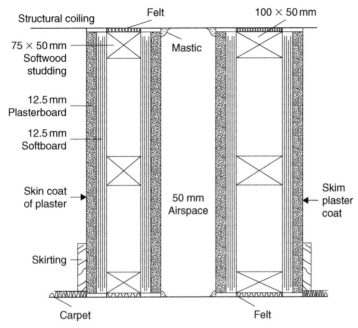

**Figure 29.4: Typical construction of a double-skin lightweight compartment.**

Triple layer constructions and multiple layer heavy masonry constructions can be employed to achieve improved levels of insulation, particularly at the lower frequencies.

Studios can also be built from modular construction assemblies based on steel panels/ cavities lined with sound-absorptive material.

The following table compares a double leaf steel panel construction with an equivalent thickness (300 mm) of masonry wall construction.

| Frequency (Hz) | 63 | 125 | 250 | 500 | IK | 2K | 4K |
|---|---|---|---|---|---|---|---|
| Masonry | 28 | 34 | 34 | 40 | 56 | 73 | 76 |
| Steel paneling | 40 | 50 | 62 | 71 | 80 | 83 | 88 |

Again it must be remembered that when it comes to sound insulation, large differences frequently occur between the theoretical value (i.e., what can be achieved) and the actual on-site value (i.e., what is actually achieved). Great care has to be taken to the sealing of all joints and to avoid short-circuiting panels by the use of wall ties or debris within the cavities. Furthermore, where doors or observation windows penetrate a wall, particular care must be taken to ensure that these do not degrade the overall performance. For example, the door or window should have the same overall sound insulation capabilities as the basic studio construction.

Studio doors should therefore be solid core types, and are frequently fitted with additional lead or steel sheet linings. Proprietary seals and door closers must be fitted to form an airtight seal round the edge of the door. Soft rubber compression or magnetic sealing strips are commonly employed while doors are rebated to help form a better joint. (Wherever possible a sound lock should be employed, i.e., two doors with an acoustically treated space between, such that one door is closed before the other opened.) Other methods of improving door insulation include the use of two separate doors on a single frame, each fitted with appropriate seals.

Control room windows may be either double or triple glazed depending on the performance required. Large airspaces, for example, 200–300 mm plus, are necessary in order to achieve appropriate insulation. Tilting one of the panes not only helps to cut down visual reflections, but also breaks up single frequency cavity resonances, which reduce the potential insulation value significantly. The reveals to the windows should also be lined with sound-absorbing material to help damp out cavity resonances.

Window glass needs to be much thicker/heavier than typical domestic glazing. For example, a typical 0.25-in., 6-mm domestic window would have a sound insulation performance of around 25–27 dB (21–23 dB at 125 Hz), whereas a 200-mm void fitted with 12- and 8-mm panes could produce an overall SRI of 52 dB (31 dB at 125 Hz).

Special laminated windows and panes can also be used to improve performance.

Studios are frequently built on the box within a box principle, whereby the second skin is completely isolated from the outer. This is achieved by building a "floating" construction. For example, the floor of the studio is isolated from the main structure of the building by constructing it on resilient pads or on a continuous resilient mat. The inner walls are then built up off the isolated floor with minimal connection to the outer leaf. The ceiling is supported from the inner leaves alone and so does not bridge onto the outer walls, which would short-circuit the isolation and negate the whole effect. With such constructions, particular care has to be taken with the connection of the various services into the studio. Again attention to detail is of tantamount importance if optimum performance is to be achieved. Note that the sealing of any small gaps or cracks is extremely important, as shown by the plastered and unplastered walls in Table 29.4.

**Table 29.4: Summary of Some Typical Building Materials**

| Materials | SRI (dB) |
|---|---|
| *Walls* | |
| Lightweight block work (not plastered) | 35 |
| (plastered) | 40 |
| 200-mm lightweight concrete slabs | 40 |
| 100-mm solid brickwork (unplastered) | 42 |
| (plastered 12 mm) | 45 |
| 110-mm dense concrete (sealed) | 45 |
| 150-mm dense concrete (sealed) | 47 |
| 230-mm solid brick (unplastered) | 48 |
| (plastered) | 49 |
| 200-mm dense concrete (blocks well sealed/plastered) | 50 |
| 250-mm cavity brick wall (i.e., 2 × 110) | 50 |
| 340-mm brickwork (plastered both sides 12 mm) | 53 |
| 450-mm brick/stone plastered | 55 |
| 112-mm brick + 50-mm partition of one layer plasterboard and softboard per side | 56 |
| 225 brick—50 mm cavity—225 brick | 65 |
| 325 brick—230 mm cavity—225 brick (floated) | 75 |
| 12-mm plasterboard on each side of 50-mm stud frame | 33 |
| 12-mm plasterboard on each side of 50-mm stud frame with quilt in cavity | 36 |
| Two × 12-mm plasterboard on 50-mm studs with quilt in cavity | 41 |
| As above with 75-mm cavity | 45 |
| Three layers of 12-mm plasterboard on separate timber frame with 225-mm air gap with quilt | 49 |

(Continued)

**Table 29.4: Continued**

| Materials | SRI (dB) |
|---|---|
| *Floors* | |
| 21-mm +1 & g boards or 19-mm chipboard | 35 |
| 110-mm concrete and screed | 42 |
| 21-mm +1 & g boards or 19-mm chipboard with plasterboard below and 50-mm sand pugging | 45 |
| 125-mm reinforced concrete and 150–mm screed | 45 |
| 200-mm reinforced concrete and 50–mm screed | 47 |
| 125-mm reinforced concrete and 50–mm screed on glass fiber or mineral wool quilt | 50 |
| 21-mm +1 & g boards or 19-mm chipboard in form of raft on mineral fiber quilt with plasterboard and 50–mm sand pugging | 50 |
| 150-mm concrete on specialist floating raft | 55–60 |
| *Windows* | |
| 4-mm glass well sealed | |
| 23–25 | |
| 6-mm glass well sealed | 27 |
| 6-mm glass—12-mm gap—6-mm glass | 28 |
| 12-mm glass well sealed | 31 |
| 12-mm glass laminated | 35 |
| 10-mm glass—80-mm gap—6-mm glass | 37 |
| 4-mm glass—200-mm gap—4-mm glass | 39 |
| 10-mm glass—200-mm gap—6-mm glass | 44 |
| 10-mm glass—150-mm gap—8-mm glass | 44 |
| 10-mm glass—200-mm gap—8-mm glass | 52 |

## 29.3 Studio and Control Room Acoustics

Today the acoustic response of the control room is recognized as being as equally important, if not more important, than that of the studio itself. This is partly due to improvements in monitor loudspeakers, recent advances in the understanding of the underlying psychoacoustics, and the recent trend of using the control room as a recording space, for example, for synthesizers, whereby better contact can be maintained among musician, producer, and recording engineer.

For orchestral or classical music recording, the studios need to be relatively large and lively, with reverberation times lying somewhere between 1.2 and 2 s depending on their size and use. Pop studios tend to be much more damped than this, perhaps typically varying from 0.4 to 0.7 s.

Broadcast studios would typically have reverberation times as follows:

- Radio talks        0.25 s        ($60 \, m^3$)

- Radio drama        0.4 s        ($400 \, m^3$)

- Radio music
    - Small        0.9 s        ($600 \, m^3$)
    - Large        2.0 s        ($1000 \, m^3$)

- Television        0.7 s        ($7000 \, m^3$)

- Control rooms        0.25 s

While it has been recognized for some time now that reverberation time alone is not a good descriptor of studio acoustics, it is difficult to determine any other easily predictable or measurable parameters that are. Reverberation time therefore continues to be used.

Reverberation time is effectively a measure of the sound absorption within a room or studio, etc. Its importance therefore really lies within the way it alters with frequency rather than on its absolute value. It is therefore generally recognized that reverberation time should be maintained essentially constant within tight limits (e.g., 10%) across the audio band of interest with a slight rise at the lower bass frequencies generally being permissible. By maintaining the reverberation time constant, the room/studio is essentially affecting all frequencies equally—at least on a statistical basis. However, in practice, just because a studio or control room has a perfectly flat reverberation time characteristic does not necessarily guarantee that it will sound all right, but it is a good baseline from which to begin.

Studios are generally designed so that they can either be divided up, for example, by use of portable screens or by having their acoustic characteristics altered/adjusted using hinged or fold over panels, for example, with one side being reflective and the other absorptive or by the use of adjustable drapes and so on. Isolation booths or voice-over booths are also frequently incorporated in commercial studios to increase separation. Drama studios frequently have a live end and a dead end to cater for different dramatic requirements and scenes. Recent advances in digital reverberation, room simulation, and effects have, however, added a new dimension to such productions, enabling a wide variety of environments to be electronically simulated and added in at the mixing stage. The same is the case with musical performances/recordings.

It is important not to make studios and control rooms too dead, as this can be particularly fatiguing and even oppressive.

TV studios have to cater for a particularly wide range of possible acoustic environments, ranging from the totally reverberation-free outdoors to the high reverberant characteristics of sets depicting caves, cellars, or tunnels, with light entertainment and orchestral music and singing somewhere in between the two.

As the floor of the TV studio has to be ruler flat and smooth for camera operations and the roof is a veritable forest of lighting, the side walls are the only acoustically useful areas left. In general, these are covered with wideband acoustic absorption, for example, 50-mm mineral wool/glass fiber over a 150-mm partitioned airspace and covered in wire mesh to protect it. Modular absorbers are also used and allow a greater degree of final tuning to be carried out should this be desired.

### 29.3.1 Absorbers

Proprietary modular absorbers are frequently used in commercial broadcast/radio studios. These typically consist of 600-mm square boxes of various depths in which cavities are created by metal or cardboard dividers over which a layer of mineral wool is fixed together with a specially perforated faceboard.

By altering the percentage perforations, thickness/density of the mineral wool, and depth of the cavities, the absorption characteristics can be adjusted and a range of absorbers created, varying from wideband mid- and high-frequency units to specifically tuned low-frequency modules that can be selected to deal with any particularly strong or difficult room resonances or colorations, particularly those associated with small rooms where the modal spacing is low or where coincident modes occur.

### 29.3.2 Resonances

Room modes or Eigentones are a series of room resonances formed by groups of reflections that travel back and forth in phase within a space, for example, between the walls or floor and ceiling.

At certain specific frequencies, the reflections are perfectly in phase with each other and cause a resonance peak in the room frequency response. The resonant frequencies are directly related to the dimensions of the room, the first resonance occurring at a frequency corresponding to the half wavelength of sound equal to the spacing between the walls or floor and so on (see Figure 29.5). Resonances also occur at the harmonics of the first (fundamental) frequencies. At frequencies related to the quarter wavelength dimension, the incident and reflected waves are out of phase with each other and, by destructive wave interference, attempt to cancel each other out, resulting in a series of nulls or notches in the room frequency response. Room modes therefore give rise to a harmonic series of peaks and dips within the room frequency response.

At different positions throughout the room, completely different frequency responses can occur depending on whether a peak or null occurs at a given location for a given

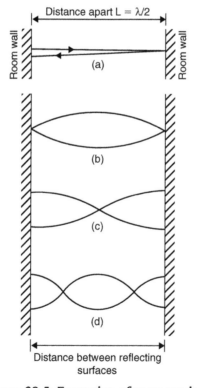

**Figure 29.5: Formation of room modes.**

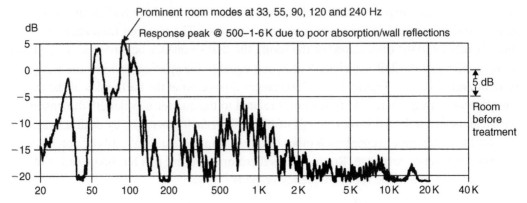

**Figure 29.6: Frequency response trace of a monitor loudspeaker in a poorly treated control room.**

frequency. Variations of 10–15 dB within the room response are quite common unless these modal resonances are brought under control. Figure 29.6 presents a frequency response trace of a monitor loudspeaker in a poorly treated control room.

There are three types of room mode: the axial, oblique, and tangential, but in practice it is the axial modes that generally cause the most severe problems. Axial modes are caused by in and out of phase reflections occurring directly between the major axes of the room, that is, floor to ceiling, side wall to side wall, and end wall to end wall.

Figure 29.7 shows the variation in sound pressure level (loudness) that occurred within a poorly treated control room at the fundamental and first harmonic modal frequencies. Subjectively, the room was criticized as suffering from a gross bass imbalance, having no bass at the center but excess at the boundaries, which is exactly what Figure 29.7 shows, there being a 22-dB variation in the level of the musical pitch corresponding to the fundamental modal frequency and 14 dB at the first harmonic.

When designing a control room or studio, it is vitally important to ensure that the room dimensions are chosen such as not to be similar or multiples of each other, as this will cause the modal frequencies in each dimension to coincide, resulting in very strong resonance patterns and an extremely uneven frequency response anywhere within the room.

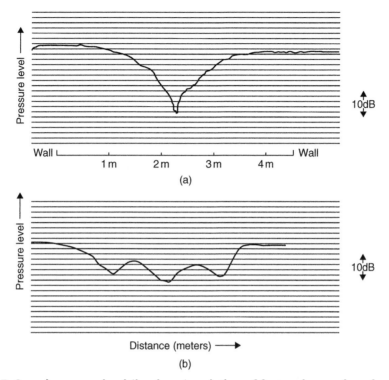

**Figure 29.7: Sound pressure level (loudness) variation of first and second modes measured in a poorly treated control room.**

The room (studio) dimensions should therefore be selected to give as even as possible distribution of the Eigentones. A number of ratios have been formulated that give a good spread. These are sometimes referred to as golden ratios. Typical values include:

$$\begin{array}{lll} H & W & L \\ 1: & 1.14: & 1.39 \\ 1: & 1.28: & 1.54 \\ 1: & 1.60: & 2.33 \end{array}$$

Figure 29.8 presents a graphical check method for ensuring an optimal spread of room resonances.

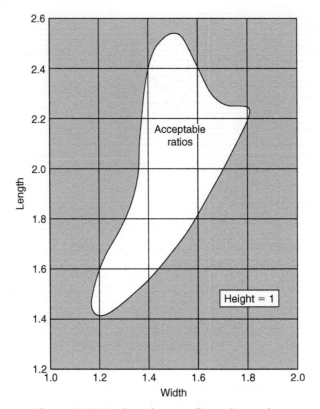

**Figure 29.8: Preferred room dimension ratios.**

The frequencies at which main axial room resonances occur can be simply found using the following formula:

$$F_{res} = \frac{172(n)}{L},$$

where $L$ is the room dimension (meters),

$$n = 1,2,3,4 \text{ etc.}$$

The graph presented in Figure 29.9 provides a quick method of estimating room mode frequencies.

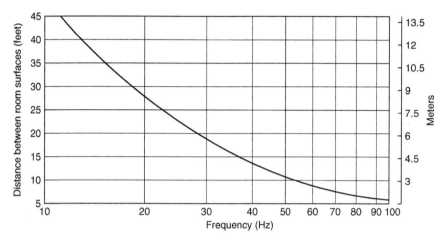

**Figure 29.9: Axial mode frequency versus room dimension.**

The complete formula for calculating the complete series of modes is as follows:

$$F = 172\left[\left(\frac{nx}{Lx}\right)^2 + \left(\frac{ny}{Ly}\right)^2 + \left(\frac{nz}{Lz}\right)^2\right]^{\frac{1}{2}}.$$

At low frequencies, the density of room modes is low, causing each to stand out and consequently be more audible. However, at higher frequencies, the density becomes very much greater, forming a continuous spectrum, that is, at any given frequency, a number of modes will occur, which counteract each other. Room modes therefore generally cause problems at low or lower midfrequencies (typically up to 250–500 Hz). While this is primarily due to the low modal density at these frequencies, it is also exacerbated by the general lack of sound absorption that occurs at low frequencies in most rooms.

After determining that the room/studio dimensions are appropriate to ensure low coincidence of modal frequencies, control of room modes is brought about by providing appropriate absorption to damp down the resonances.

### 29.3.3 Absorber Performance

Sound absorbers effectively fall into four categories:

1. High and mediumfrequency dissipative porous absorbers.
2. Low-frequency panel or membrane absorbers.

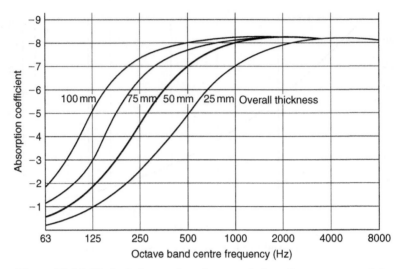

**Figure 29.10: Typical absorption characteristics of porous material.**

3. Helmholtz-tuned frequency resonator/absorber.

4. Quadratic residue/phase interference absorbers/diffusers.

Porous absorbers include many common materials, including such items as drapes, fiberglass/mineral wool, foam, carpet, and acoustic tile.

Figure 29.10 illustrates the frequency range over which porous absorbers typically operate. Figure 29.10 clearly shows the performance of these types of absorber to fall off at medium to low frequencies—unless the material is very thick (comparable in fact to 1/4 wavelength of the sound frequency to be absorbed).

Panel or membrane absorbers (Figure 29.11) therefore tend to be used for low-frequency absorption. The frequency of absorption can be tuned by adjusting the mass of the panel and the airspace behind it. By introducing some dissipative absorption into the cavity to act as a damping element, the absorption curve is broadened. Membrane absorption occurs naturally in many studios and control rooms where relatively lightweight structures are used to form the basic shell, for example, plasterboard and stud partitions, plasterboard or acoustic tile ceilings with airspace above, and so on.

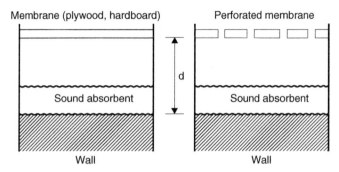

**Figure 29.11: Typical panel or membrane absorber construction.**

The frequency at which maximum absorption occurs can be calculated from the following formula:

$$f = \frac{60}{\sqrt{md}},$$

where $m$ is mass of panel in kg/m$^2$ and $d$ is airspace in meters.

The action of porous and membrane type absorbers is often combined to form a single wide frequency band absorber or to extend the low-frequency characteristics of porous materials.

Bass traps are generally formed by creating an absorbing cavity or layer of porous absorption equivalent to a quarter wavelength of the sound frequency or bass note in question. Table 29.5 presents the dimensions required for the range 30–120 Hz.

Figure 29.12 shows basic principles and typical absorption characteristics of a Helmholtz, or cavity, absorber.

The frequency of maximum absorption occurs at the natural resonant frequency of the cavity, which is given by the following formula:

$$F_{res} = 55\sqrt{\frac{S}{lV}}\,\text{Hz},$$

where $S$ is the cross-sectional area of the neck (m$^2$), $l$ is the length of the neck (m), and $V$ is the volume of air in the main cavity (m$^3$).

**Table 29.5: Frequency 1/4 Wavelength Dimension**

| Hz | m | ft |
|----|------|-------|
| 30 | 2.86 | 11.38 |
| 40 | 2.14 | 7.04 |
| 50 | 1.72 | 5.64 |
| 60 | 1.43 | 4.69 |
| 70 | 1.23 | 4.03 |
| 80 | 1.07 | 3.52 |
| 90 | 0.96 | 3.13 |
| 100 | 0.86 | 2.82 |
| 120 | 0.72 | 2.35 |

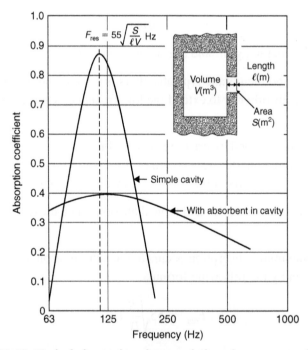

**Figure 29.12: Typical absorption characteristics of a resonant absorber.**

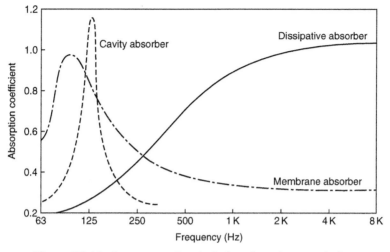

**Figure 29.13: Summary of typical absorber characteristics.**

Often membrane absorbers are combined with cavity absorbers to extend their range. Some commercial modular absorbers also make use of such techniques and, by using a range of materials/membranes/cavities, can provide a wide frequency range of operation within a standard size format. Figure 29.13 illustrates general characteristics of membrane, cavity, and dissipative/porous absorbers.

A newer type of "absorber" is the quadratic residue diffuser. This device uniformly scatters or diffuses sound striking it so that, although individual reflections are attenuated and controlled, the incident sound energy is essentially returned to the room. This process can therefore be used to provide controlled low-level reflections or reverberation enabling studios or control rooms to be designed without excessive absorption or subjective oppressiveness that often occurs when trying to control room reflections and resonances. When designing a studio or control room, the various absorptive mechanisms described earlier are taken into account and combined to produce a uniform absorption/frequency characteristic, that is, reverberation time.

### 29.3.4 Reverberation and Reflection

Reverberation times can be calculated using the following simple Sabine formula:

$$RT = \frac{0.16lV}{A},$$

where $A$ is the total sound absorption in m$^2$, which is computed by multiplying each room surface by its sound absorption coefficient and summing these to give the total absorption present, or

$$RT = \frac{0.16lV}{S\bar{a}},$$

where $\bar{a}$ is the average absorption coefficient.

In acoustically dead rooms, the following Eyring formula should be used:

$$RT = \frac{0.16lV}{-S\log_e(1 - \bar{a}) + MV},$$

where $V$ is the volume of the room in m$^3$ and $M$ is an air absorption constant.

Achieving a uniform room reverberation time (absorption) characteristic does not necessarily ensure good acoustics; the effect and control of specific reflections must also be fully taken into account. For example, strong reflections can strongly interfere with the recorded or perceived live sound, causing both tonal colorations and large frequency response irregularities. Such reflections can be caused by poorly treated room surfaces, by large areas of glazing, by off doors, or by large pieces of studio equipment including the mixing console itself.

Figure 29.14 illustrates the effect well, being a frequency response plot of a monitor loudspeaker (with a normally very flat response characteristic) mounted near to a reflective side wall.

Apart from causing gross frequency response irregularities and colorations, side wall reflections also severely interfere with stereo imaging precision and clarity. Modern studio designs go to considerable lengths to avoid such problems by either building the monitor loudspeakers into, but decoupled from, the structure and treating adjacent areas with highly absorbing material or by locating the monitors well away from the room walls and again treating any local surfaces. Near field monitoring overcomes many of these problems, but reflections from the mixing console itself can produce comb filtering irregularities. However, hoods or careful design of console-speaker geometry/layout can be used to help overcome this.

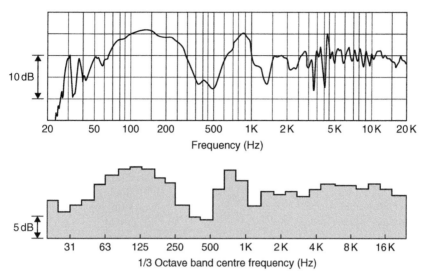

**Figure 29.14: Frequency response plot of a monitor loudspeaker mounted adjacent to wall.**

The so-called live end dead end approach to control room design uses the aforementioned techniques to create a reflection-free zone, enabling sound to be heard directly from the monitors with a distinct time gap before any room reflections arrive at the mixing or prime listening position. Furthermore, such reflections that do occur are carefully controlled to ensure that they do not cause sound colorations nor affect the stereo image, but just add an acoustic liveliness or low-level reverberant sound field, which acts to reduce listening fatigue and provide better envelopment within the sound field.

# PART 7

# Audio Test and Measurement

# CHAPTER 30

# Fundamentals and Instruments

John Linsley Hood

Audio equipment is, by definition, ultimately intended to provide sounds that will be heard by a listener. Despite rather more than a century of experience in the electrical transmission and reception of audible signals, the relationships between the electrical waveforms into which sound patterns are transformed and the sound actually heard by the listener are still not fully understood.

This situation is complicated by the observable fact that there is a considerable variation from person to person in sensitivity to, and preferences in respect of, sound characteristics, particularly where these relate to modifications or distortions of the sound. Also, since the need to describe or indeed the possibility of producing such modified or distorted sounds is a relatively new situation, we have not yet evolved a suitable and agreed vocabulary by which we can define our sensations.

There is, however, some agreement, in general, about the types of defect in electrical signals that, in the interests of good sound quality, the design engineer should seek to avoid or minimize. Of these, the major ones are those associated with waveform distortion, under either steady-state or transient conditions; the intrusion of unwanted signals; relative time delays in certain parts of the received signal in relation to others; or changes in the pitch of the signal, known as "wow" or "flutter," depending on its frequency. However, the last kind of defect is only likely to occur in electromechanical equipment, such as turntables or tape drive mechanisms, used in the recording or replay of signals.

It is sometimes claimed that, at least so far as the design of purely electronic equipment is concerned, the performance can be calculated sufficiently precisely that it is unnecessary to make measurements on a completed design for any other reason than simply to confirm that the target specification is met. Similarly, it is argued that it is absurd to attempt to endorse or reject any standard of performance by carrying out listening trials. This is so since, even if the results of calculations were not adequate to define the performance, instrumental measurements are so much more sensitive and reproducible than any purely "subjective" assessments that no significant error could escape instrumental detection.

Unfortunately, all these assertions remain a matter of some dispute. With regard to the first of these points—the need for instrumental measurements—the behavior patterns of many of the components, both "passive" and "active," used in electronic circuit design are complex, particularly under transient conditions, and it may be difficult to calculate precisely what the final performance of any piece of audio equipment will be over a comprehensive range of temperatures or of signal and load conditions. However, appropriate instrumental measurements can usually allow a rapid exploration of the system behavior over the whole range of interest.

On the second point, the usefulness of subjective testing, the problem is to define just how important any particular measurable defect in the signal process is likely to prove in the ear of any given listener. So where there is any doubt, recourse must be had to carefully staged and statistically valid comparative listening trials to try to determine some degree of consensus. These trials are expensive to stage, difficult to set up, and hard to purge of any inadvertent bias in the way they are carded out. They are therefore seldom done, and even when they are, the results are disputed by those whose beliefs are not upheld.

## 30.1  Instrument Types

An enormous range of instruments is available for use in the test laboratory, among which, in real-life conditions, the actual choice of equipment is mainly limited by considerations of cost and of value for money in respect to the usefulness of the information that it can provide.

Although there is a wide choice of test equipment, much of the necessary data about the performance of audio gear can be obtained from a relatively restricted range of instruments, such as an accurately calibrated signal generator, with sinewave and square-wave outputs, a high input impedance, a wide bandwidth AC voltmeter, and some instrument for measuring waveform distortion—all of which would be used in conjunction with a high-speed double trace cathode-ray oscilloscope. I have tried, in the following pages, to show how these instruments are used in audio testing, how the results are interpreted, and how they are made. Since some of the circuits that can be used are fairly simple, I have given details of the layouts needed so that they could be built if required by the interested user.

## 30.2 Signal Generators

### 30.2.1 Sinewave Oscillators

Variable frequency sinusoidal input test waveforms are used for determining the voltage gain, the system bandwidth, the internal phase shift or group delay, the maximum output signal swing, and the amount of waveform distortion introduced by the system under test. For audio purposes, a frequency range of 20 Hz to 20 kHz will normally be adequate, although practical instruments will usually cover a somewhat wider bandwidth than this. Except for harmonic distortion measurements, a high degree of waveform purity is probably unnecessary, and stability of output as a function "of time and frequency" is probably the most important characteristic for such equipment.

It is desirable to be able to measure the output signal swing and voltage gain of the equipment under specified load conditions. In, for example, an audio power amplifier, this would be done to determine the input drive requirements and output power that can be delivered by the amplifier. For precise measurements, a properly specified load system, a known frequency source, and an accurately calibrated RMS reading AC voltmeter would be necessary, together with an oscilloscope, to monitor the output waveform to ensure that the output waveform is not distorted by overloading.

Some knowledge of the phase errors (the relative time delay introduced at any one frequency in relation to another) can be essential for certain uses—for example, in long-distance cable transmission systems—but in normal audio usage such relative phase errors are not noticeable unless they are very large. This is because the ear is generally able to accept without difficulty the relative delays in the arrival times of sound pressure waves due to differing path lengths caused by reflections in the route from the speaker to the ear.

Oscillators designed for use with audio equipment will typically cover the frequency range 10 Hz to 100 kHz, with a maximum output voltage of, perhaps, 10 V rms. For general purpose use, harmonic distortion levels in the range of 0.5–0.05% will probably be adequate, although equipment intended for performance assessments on high-quality audio amplifiers will usually demand waveform purity (harmonic distortion) levels at 1 kHz in the range from 0.02% down to 0.005 %, or lower. In practice, with simpler instruments, the distortion levels will deteriorate somewhat at the high- and low-frequency ends of the output frequency band.

A variety of electronic circuit layouts have been proposed for use as sinewave signal generators, of which by far the most popular is the "Wien bridge" circuit shown in Figure 30.1. It is a requirement for continuous oscillation in any system that the feedback from output to input shall have zero (or some multiple of 360°) phase shift at a frequency where the feedback loop gain is very slightly greater than unity. Although to avoid waveform distortion, it is necessary that the gain should fall to unity at some value of output voltage within its linear voltage range.

In the Wien bridge, if $R_1 = R_2 = R$ and $C_1 = C_2 = C$, the condition for zero phase shift in the network is met when the output frequency $f_0 = 1/(2\pi RC)$. At this frequency the attenuation of the $RC$ network, from $Y$ to $X$, in Figure 30.1, is 1/3. The circuit shown will therefore oscillate at $f_0$ if the gain of the amplifier, $A_m$, is initially slightly greater than three times. The required gain level can be obtained automatically by the use of a thermistor ($TH_1$) in the negative feedback path and the correct choice of the value of $R_3$.

Since $C_1R_1$ and $C_2R_2$ are the frequency-determining elements, the output frequency of the oscillator can be made variable by using a twin-gang variable resistor as $R_1/R_2$ or a twin gang capacitor as $C_1/C_2$. If a modern, very low distortion, operational amplifier, such as the LM833, the NE5534, or the OP27, is used as the amplifier gain block ($A_1$) in this circuit, the principal source of distortion will be that caused by the action of the thermistor ($TH_1$) used to stabilize the output signal voltage, where, at lower frequencies, the waveform peaks will tend to be flattened by its gain–reduction action. With an RS Components "RA53" type thermistor as $TH_1$, the output voltage will be held at approximately 1 V rms, and the THD at 1 kHz will be typically of the order of 0.008%.

The output of almost any sinewave oscillator can be converted into square-wave form by the addition of an amplifier that is driven into clipping. This could be an opamp. A string of CMOS inverters, or, preferably, a fast voltage comparator IC, such as that also shown in Figure 30.1, where $RV_1$ is used to set an equal mark to space ratio in the output waveform. An alternative approach used in some commercial instruments is simply to

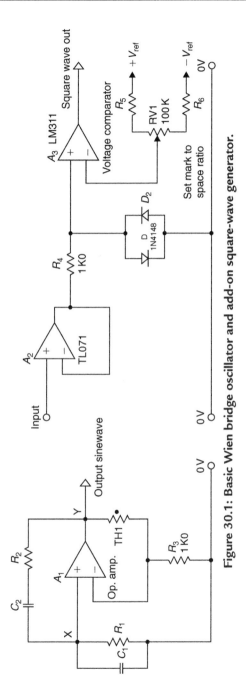

**Figure 30.1: Basic Wien bridge oscillator and add-on square-wave generator.**

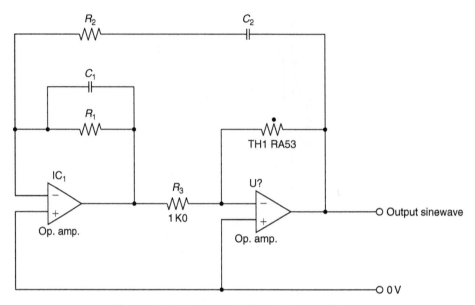

**Figure 30.2: Improved Wien bridge oscillator.**

use a high-speed analogue switch, operated by a control signal derived from a frequency stable oscillator, to feed one or an other of a pair of preset voltages, alternately, to a suitably fast output buffer stage.

An improved Wien bridge oscillator circuit layout of my own, shown in Figure 30.2 (*Wireless World*, May 1981, pp. 51–53) in which the gain blocks $A_1$ and $A_2$ are connected as inverting amplifiers, thereby avoiding "common mode" distortion, is capable of a THD below 0.003% at 1 kHz with the thermistor-controlled amplitude stabilization layout shown in Figure 30.2, and about 0.001% when using the improved stabilization layout, using an LED and a photo-conductive cell, described in the article.

As a general rule the time required (and, since this relates to a number of waveform cycles, it will be frequency dependent) for an amplitude-stabilized oscillator of this kind to "settle" to a constant output voltage, following some disturbance (such as switching on, or alteration to its output frequency setting), will increase as the harmonic distortion level of the circuit is reduced. This characteristic is a nuisance for general purpose use where the THD level is relatively unimportant. In this case, an alternative output voltage

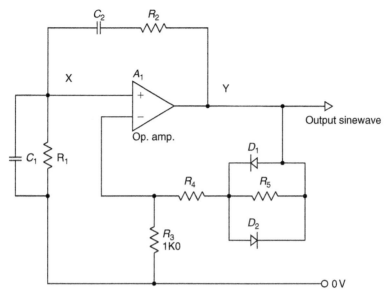

**Figure 30.3: Diode-stabilized oscillator.**

stabilizing circuit, such as the simple back-to-back connected silicon diode peak limiter circuit shown in Figure 30.3, would be preferable, despite its relatively modest (0.5% at 1 kHz) performance in respect of waveform distortion.

Rather greater control of the output signal amplitude can be obtained by more elaborate systems, such as the circuit shown in Figure 30.4. In this circuit, the output sinewave is fed to a high-input impedance rectifier system ($A_2/D_1/D_2$), and the DC voltage generated by this is applied to the gate of an FET used as a voltage-controlled resistor. The values chosen for $R_6/R_7$ and $C_3/C_4/C_5$ determine the stabilization time constant and the output signal amplitude is controlled by the ratio of $R_8:R_9$. In operation, the values of $R_4$ and $R_3$ are chosen so that the circuit will oscillate continuously with the FET ($Q_1$) in zero-bias conducting mode. Then, as the −ve bias on the $Q_1$ gate increases as a result of the rectifier action of $Q_1/Q_2$, the amplitude of oscillation will decrease until an equilibrium output voltage level is reached.

In commercial instruments, a high-quality small-power amplifier would normally be interposed between the output of the oscillator circuit and the output take-off point to

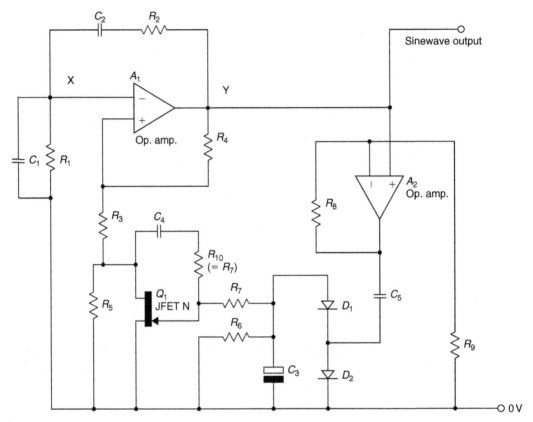

**Figure 30.4: Amplitude control by FET.**

isolate the oscillator circuit from the load and to increase the output voltage level to, say, 10 V rms. An output attenuator of the kind shown in Figure 30.5 would then be added to allow a choice of maximum output voltage over the range 1 mV to 10 V rms, at a 600-$\Omega$ output impedance.

A somewhat improved performance in respect of THD is given by the "parallel T" oscillator arrangement shown in Figure 30.6, and a widely used and well-respected low-distortion oscillator was based on this type of frequency-determining arrangement. This differs in its method of operation from the typical Wien bridge system in that the network gives zero transmission from input to output at a frequency determined by the values of the resistors $R_x$, $R_y$, and $R_z$ and the capacitors $C_x$, $C_y$, and $C_z$. If $C_x = C_y = C_z/2 = C$ and

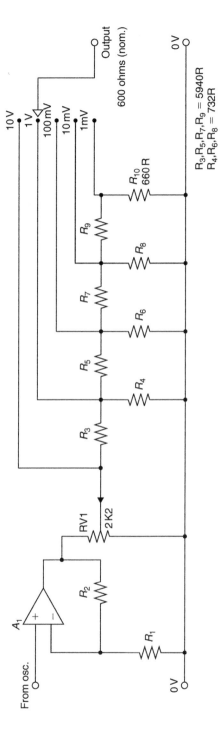

**Figure 30.5: Output level control.**

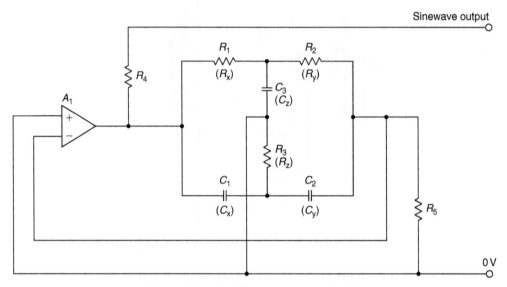

**Figure 30.6: Oscillator using parallel T.**

$R_x = R_y = 2R_z = R$, the frequency of oscillation will be $1/2\pi CR$, as in the Wien bridge oscillator.

If the parallel T network is connected in the negative feedback path of a high gain amplifier $(A_1)$, oscillation will occur because there is an abrupt shift in the phase of the signal passing through the 'T' network at frequencies close to the null, and this, and the inevitable phase shift in the amplifier $(A_1)$, converts the nominally negative feedback signal derived from the output of the 'T' network into a positive feedback, oscillation-sustaining one.

A problem inherent in the parallel T design is that in order to alter the operating frequency it is necessary to make simultaneous adjustments to either three separate capacitors or three separate resistors. If fixed capacitor values are used, then one of these simultaneously variable resistors is required to have half the value of the other two. Alternatively, if fixed value resistors are used, then one of the three variable capacitors must have a value that is, over its whole adjustment range, twice that of the other two. This could be done by connecting two of the "gangs" in a four-gang capacitor in parallel, although, for normally available values of capacitance for each gang, the resistance values needed for the 'T' network will be in the megohm range. Also, it is necessary that the drive shafts of $C_x$ and $C_y$ be isolated from that of $C_z$.

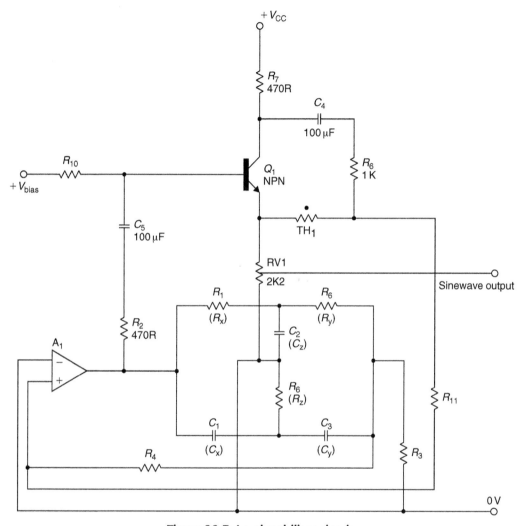

**Figure 30.7: Level stabilizer circuit.**

These difficulties are lessened if the oscillator is only required to operate at a range of fixed "spot" frequencies, and a further circuit of my own of this kind (*Wireless World*, July 1979, pp. 64–66) is shown in simplified form in Figure 30.7. The output voltage stabilization used in this circuit is based on a thermistor/resistor bridge connected across a transistor, $Q_1$. The phase of the feedback signal derived from this, and fed to $A_1$, changes

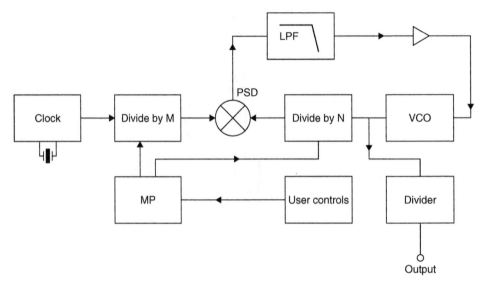

**Figure 30.8: Phase-locked loop oscillator.**

from +ve to −ve as the output voltage exceeds some predetermined output voltage level. The THD given by this oscillator approaches 0.0001% at 1 kHz, worsening to about 0.0003% at the extremes of its 100-Hz to 10-kHz operating frequency range.

It is expected in modem wide-range low-distortion test bench oscillators that they will offer a high degree of both frequency and amplitude stability. This is difficult to obtain using designs based on resistor/capacitor or inductor/capacitor frequency control systems, which has encouraged the development of designs based on digital waveform synthesis, and other forms of digital signal processing.

### 30.2.2  Digital Waveform Generation

Because of the need for a precise, stable, and reproducible output signal frequency in a test oscillator, a number of circuit arrangements have been designed in which use is made of the frequency drift-free output obtainable from a quartz crystal oscillator. Since this will normally provide only a single spot-frequency output, some arrangement is needed to derive a variable frequency signal from this fixed frequency reference source.

One common technique makes use of the "phase locked loop" (PLL) layout shown in Figure 30.8. In this, the outputs from a highly stable quartz crystal "clock" oscillator and from a variable-frequency "voltage-controlled oscillator" (VCO) are taken to a "phase-sensitive detector" (PSD)—a device whose output consists of the "sum" and "difference" frequencies of the two input signals. If the sum frequency is removed by filtration, and if the two input signals should happen to be at the same frequency, the difference frequency will be zero, and the PSD output voltage will be a DC potential whose sign is determined by the relative phase angle between the two input signals.

If this output voltage is amplified (having been filtered to remove the unwanted "sum" frequencies) and then fed as a DC control voltage to the VCO (a device whose output frequency is determined by the voltage applied to it), then, providing that the initial operating frequencies of the clock and the VCO are within the frequency "capture" range determined by the loop low-pass filter, the action of the circuit will be to force the VCO into frequency synchronism (but phase quadrature) with the clock signal: a condition usually called "lock." Now if, as in Figure 30.8, the clock and VCO signals are passed through frequency divider stages, having values of ÷M and ÷N, respectively, when the loop is in lock the output frequency of the VCO will be $F_{out} = F_{ck}$ (N/M). If the clock frequency is sufficiently high, appropriate values of M and N can be found to allow the generation of virtually any desired VCO frequency. In an audio band oscillator, since the VCO will probably be a "varicap"-controlled LC oscillator, operating in the MHz range, the output signal will normally be obtained from a further variable ratio frequency divider, as shown in Figure 30.8. For the convenience of the user, once the required output frequency is keyed in, the actual division ratios required to generate the chosen output frequency will be determined by a microprocessor from ROM-based look-up tables, and the output signal frequency will be displayed as a numerical readout.

Given the availability of a stable, controllable frequency input signal, the generation of a low-distortion sinewave can, again, be done in many ways. For example, the circuit arrangement shown in Figure 30.9 is quoted by Horowitz and Hill (*The Art of Electronics*, 2nd ed. 667). In this, a logic voltage level step is clocked through a parallel output shift register connected to a group of resistors whose outputs are summed by an amplifier ($A_1$). The output is a continuous waveform, of staircase type character, at a frequency of $F_{ck}/16$.

If the values of the resistors $R_1$–$R_7$ are chosen correctly, the output will approximate to a sinewave, the lowest of whose harmonic distortion components is the 15th, at −24 dB. This distortion can be further reduced by low-pass filtering the output waveform. A more precise

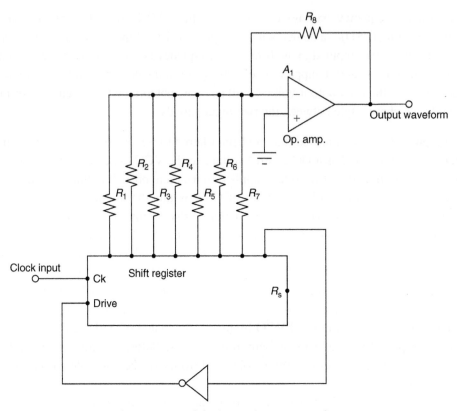

**Figure 30.9: Digital waveform generation.**

waveform, having smaller, higher frequency staircase steps, could be obtained by connecting two or more such shift registers in series, with appropriate values of loading resistors.

Like all digitally synthesized systems, this circuit will have an output frequency stability that is as good as that of the clock oscillator, which will be crystal controlled. Also the output frequency can be displayed numerically and there will be no amplitude "bounce" on switch on, or on changing frequency.

A more elegant digitally synthesized sinewave generator is shown in Figure 30.10. In this the quantized values of a digitally encoded sine waveform are drawn from a data source, which could be a numerical algorithm, of the kind used, for example, in a "scientific" calculator, but, more conveniently, would be a ROM-based "look-up" table. These are

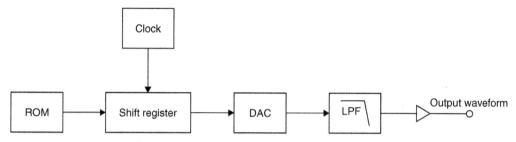

**Figure 30.10: ROM-based waveform synthesis.**

then clocked through a shift register into a 16-bit digital-to-analogue converter (DAC). If data chosen are those for a 16-bit encoded sinusoidal waveform, the typical intrinsic purity of the output signal will be of the order of 0.0007%, improved by the use of low-pass, sample-and-hold filtering.

Moreover, if digital filtering is used, prior to the DAC, this can be made to track the frequency of the output signal. As in the previous design (Figure 30.9) the output signal frequency is related to, and controlled by, the clock frequency.

## 30.3 Alternative Waveform Types

A range of waveforms, including square- and rectangular-wave shapes, as well as triangular and "ramp" type outputs, are typically provided by a "function generator." The outputs from this kind of instrument will usually also include a sinusoidal waveform output having a wide frequency range but only a modest degree of linearity.

An IC that allows the provision of all these output waveform types is the ICL '8038' and its homologues, for which the recommended circuit layout is shown in Figure 30.11. The output from this is free from amplitude "bounce" on frequency switching or adjustment and can be set to give a 1-kHz distortion figure of about 0.5% by adjustment of the twin-gang potentiometers RV4 and RV5. As shown, the frequency coverage, by a single control (RV2), is from 20 Hz to 20 kHz.

Since a square-wave signal contains a very wide range of odd-order harmonics of its fundamental frequency, a good quality signal of this kind, with fast leading edge (rise) and trailing edge (fall) times, and negligible overshoot or "ripple," allows the audio systems engineer to make a rapid assessment both of the load stability of an amplifier and of the frequency response of a complete audio system.

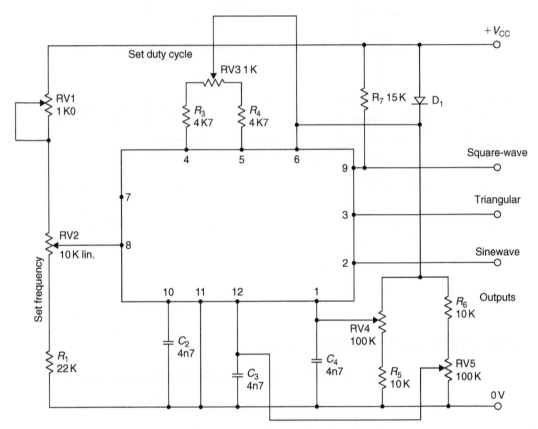

**Figure 30.11: ICL 8038 application circuit.**

With an input such as that shown in Figure 30.12(a), an output waveform of the type shown in Figure 30.12(b) would indicate a relatively poor overall stability by comparison with that shown in Figure 30.12(c), which would merely indicate some loss of high frequency. The type of waveform shown in Figure 30.12(d) would imply that the system was only conditionally stable and unlikely to be satisfactory in use.

The type of oscilloscope waveform shown in Figure 30.12(e) would indicate a fall off in gain at low frequencies, whereas that of Figure 30.12(f) would show an increase in gain at LF. Similarly, the response shown in Figure 30.12(g) would be due to an excessive HF gain. With experience, the engineer is likely to recognize the types of output waveform

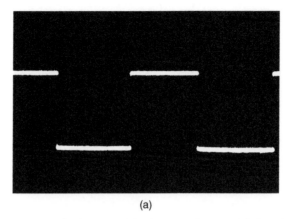

(a)

**Figure 30.12(a): Square-wave input waveform (wide bandwidth).**

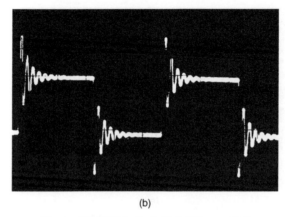

(b)

**Figure 30.12(b): Poor amplifier stability.**

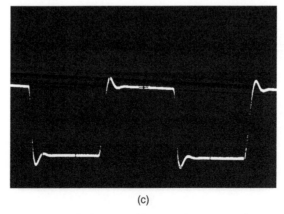

(c)

**Figure 30.12(c): Relatively rapid (−12 dB/octave) loss of HF gain.**

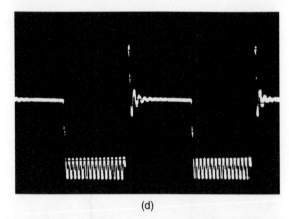

(d)

**Figure 30.12(d): Output showing conditional stability.**

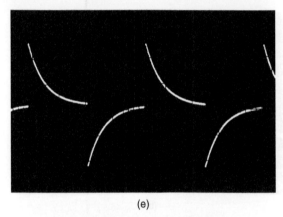

(e)

**Figure 30.12(e): Poor LF gain.**

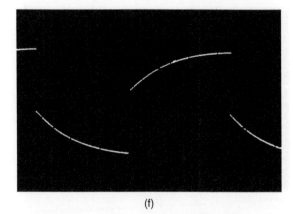

(f)

**Figure 30.12(f): Increased gain at LF.**

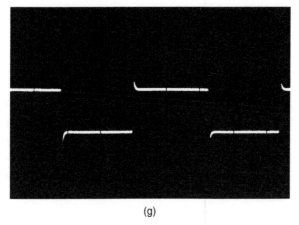

(g)

**Figure 30.12(g): Excessive HF gain.**

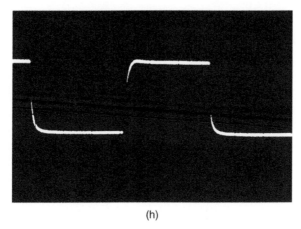

(h)

**Figure 30.12(h): Loss of HF gain.**

associated with many of the common gain/frequency characteristics or design or performance problems.

The other types of waveform that can be provided by a function generator have specific applications. The sinewave output, although too poor in linearity to allow amplifier THD measurements to be made with any accuracy, is usually quite free from amplitude variation, and the "single knob" wide-range frequency control of such an instrument allows fast checking of the performance of such circuits as low- or high-pass filters or tone controls.

## 30.4 Distortion Measurement

One of the more important characteristics of any audio system is the extent to which the signal waveform is distorted during its passage through the system. Where the input signal (fin) is of a single frequency, and this distortion is due to some nonlinearity in the transfer characteristics of the signal handling stages, this nonlinearity will generate a series of further spurious signals occurring at frequencies that are multiples of the input frequency (or frequencies).

# Index

Printed and bound by CPI Group (UK) Ltd, Croydon, CR0 4YY

03/10/2024

01040343-0016